# Lecture Notes in Computer Science　　12680

More information about this subseries at http://www.springer.com/series/7409

Christian S. Jensen · Ee-Peng Lim ·
De-Nian Yang · Chia-Hui Chang ·
Jianliang Xu · Wen-Chih Peng ·
Jen-Wei Huang · Chih-Ya Shen (Eds.)

# Database Systems for Advanced Applications

## DASFAA 2021 International Workshops

BDQM, GDMA, MLDLDSA, MobiSocial, and MUST
Taipei, Taiwan, April 11–14, 2021
Proceedings

 Springer

*Editors*
Christian S. Jensen (ID)
Aalborg University
Aalborg, Denmark

De-Nian Yang
Academia Sinica
Taipei, Taiwan

Jianliang Xu
Hong Kong Baptist University
Kowloon Tong, Hong Kong

Jen-Wei Huang (ID)
National Cheng Kung University
Tainan City, Taiwan

Ee-Peng Lim (ID)
Singapore Management University
Singapore, Singapore

Chia-Hui Chang
National Central University
Taoyuan City, Taiwan

Wen-Chih Peng (ID)
National Chiao Tung University
Hsinchu, Taiwan

Chih-Ya Shen
National Tsing Hua University
Hsinchu, Taiwan

ISSN 0302-9743          ISSN 1611-3349   (electronic)
Lecture Notes in Computer Science
ISBN 978-3-030-73215-8          ISBN 978-3-030-73216-5   (eBook)
https://doi.org/10.1007/978-3-030-73216-5

LNCS Sublibrary: SL3 – Information Systems and Applications, incl. Internet/Web, and HCI

This Springer imprint is published by the registered company Springer Nature Switzerland AG
The registered company address is: Gewerbestrasse 11, 6330 Cham, Switzerland

# Preface

Along with the main conference, the Database Systems for Advanced Applications (DASFAA) workshops provide international forums for researchers and practitioners to introduce and discuss research results and open problems, aiming at more focused problem domains in the database areas. This year, five workshops were held in conjunction with DASFAA 2021:

- The 6th International Workshop on Big Data Quality Management (BDQM 2021)
- The 5th International Workshop on Graph Data Management and Analysis (GDMA 2021)
- The 1st International Workshop on Machine Learning and Deep Learning for Data Security Applications (MLDLDSA 2021)
- The 6th International Workshop on Mobile Data Management, Mining, and Computing on Social Networks (MobiSocial 2021)
- The 3rd International Workshop on Mobile Ubiquitous Systems and Technologies (MUST 2021)

All the workshops were selected through a public Call-for-Proposals process, and each of them focused on a specific area that contributed to the main themes of DASFAA 2021. After the proposals were accepted, each workshop proceeded with its own call for papers and a review of the submissions. In total, 29 papers were accepted, including 5 papers for BDQM 2021 and 6 papers for each of the other workshops.

We would like to thank all of the members of the Workshop Organizing Committees, along with their Program Committee members, for their tremendous efforts in making the DASFAA 2021 workshops a success. In addition, we are grateful to the main conference organizers for their generous support and help.

February 2021

Chia-Hui Chang
Jianliang Xu
Wen-Chih Peng

# Organization

## BDQM 2021 Organization

### Chair and Co-chair

Dongjing Miao      University of Illinois at Chicago, China
Weitian Tong      Eastern Michigan University, USA

### Program Committee

Wenjie Zhang      University of New South Wales, Australia
Yajun Yang      Tianjin University, China
Jiannan Wang      Simon Fraser University, Canada
Zhaonian Zou      Harbin Institute of Technology, China
Kedong Yan      Nanjing University of Science and Technology, China
Quan Chen      Guangdong University of Technology, China
Xu Zheng      University of Electronic Science and Technology
     of China, China
Xueli Liu      Tianjin University, China
Guilin Li      Xiamen University, China
Shengfei Shi      Harbin Institute of Technology, China

## GDMA 2021 Organization

### General Chair

Lei Zou      Peking University, China

### Program Chairs

Xiaowang Zhang      Tianjin University, China
Weiguo Zheng      Fudan University, China

### Program Committee

Liang Hong      Wuhan University, China
Peng Peng      Hunan University, China
Meng Wang      Southeast University, China
Zhe Wang      Griffith University, Australia
Guohui Xiao      Free University of Bozen-Bolzano, Italy
Xiaowang Zhang      Tianjin University, China
Weiguo Zheng      Fudan University, China
Chengzhi Piao      Chinese University of Hong Kong, China

# MLDLDSA 2021 Organization

## Steering Committee

Wenzhong Guo    Fuzhou University, China
Chin-Chen Chang   Feng-Chia University, Taiwan

## General Chairs

Chi-Hua Chen    Fuzhou University, China
Brij B. Gupta    National Institute of Technology Kurukshetra, India

## Session Chairs

Feng-Jang Hwang   University of Technology Sydney, Australia
Fuquan Zhang    Minjiang University, China
Yu-Chih Wei    National Taipei University of Technology, Taiwan
K. Shankar     Alagappa University, India
Cheng Shi     Xi'an University of Technology, China

## Technical Program Committee

Haishuai Wang    Fairfield University & Harvard University, USA
Eyhab Al-Masri    University of Washington Tacoma, USA
Xianbiao Hu     Missouri University of Science and Technology, USA
Victor Hugo C. de   University of Fortaleza, Brazil
 Albuquerque
Xiao-Guang Yue   European University Cyprus, Cyprus
Hanhua Chen    Huazhong University of Science and Technology,
         China
Ching-Chun Chang   Tsinghua University, China
Chunjia Han     University of Greenwich, UK
Doris Xin      Newcastle University, UK
Lingjuan Lyu    National University of Singapore, Singapore
Ting Bi       Maynooth University, Ireland
Fang-Jing Wu    Technische Universität Dortmund, Germany
Paula Fraga-Lamas   Universidade da Coruña, Spain
Usman Tariq     Prince Sattam bin Abdulaziz University, Saudi Arabia
Hsu-Yang Kung    National Pingtung University of Science
         and Technology, Taiwan
Chin-Ling Chen    Chaoyang University of Technology, Taiwan
Hao-Chun Lu    Chang Gung University, Taiwan
Yao-Huei Huang   Fu-Jen Catholic University, Taiwan
Hao-Hsiang Ku    National Taiwan Ocean University, Taiwan
Hsiao-Ting Tseng   National Central University, Taiwan
Chia-Yu Lin     Yuan Ze University, Taiwan
Bon-Yeh Lin     Chunghwa Telecom Co. Ltd., Taiwan
Jianbin Qin     Shenzhen University, China

| | |
|---|---|
| Liang-Hung Wang | Fuzhou University, China |
| Fangying Song | Fuzhou University, China |
| Genggeng Liu | Fuzhou University, China |
| Chan-Liang Chung | Fuzhou University, China |
| Lianrong Pu | Fuzhou University, China |
| Ling Wu | Fuzhou University, China |
| Xiaoyan Li | Fuzhou University, China |
| Mingyang Pan | Dalian Maritime University, China |
| Chih-Min Yu | Yango University, China |
| Lei Xiong | Guangzhou Academy of Fine Arts, China |
| Bo-Wei Zhu | Macau University of Science and Technology, Macau |
| Insaf Ullah | Hamdard University, Pakistan |

## MobiSocial 2021 Organization

### Organizers

| | |
|---|---|
| Wang-Chien Lee | Pennsylvania State University, USA |
| De-Nian Yang | Academia Sinica, Taiwan |
| Hong-Han Shuai | National Chiao Tung University, Taiwan |
| Chih-Ya Shen | National Tsing Hua University, Taiwan |

### Program Chairs

| | |
|---|---|
| Chih-Hua Tai | National Taipei University, Taiwan |
| Yi-Ling Chen | National Taiwan University of Science and Technology, Taiwan |
| Yixiang Fang | University of New South Wales, Australia |

### Program Committee

| | |
|---|---|
| Cheng-Te Li | National Cheng Kung University, Taiwan |
| Lifang He | Weill Cornell Medical College, USA |
| Jiawei Zhang | Florida State University, USA |
| Chun-Ta Lu | Google, USA |
| Keng-Pei Lin | National Sun Yat-sen University, Taiwan |
| Arwen Teng | Academia Sinica, Taiwan |
| Bay-Yuan Hsu | University of California, Santa Barbara, USA |
| Chia-Yu Lin | National Chiao Tung University, Taiwan |
| Lo-Yao Yeh | National Center of High-performance Computing, Taiwan |
| Jay Chen | Avalanche Computing, Taiwan |

# MUST 2021 Organization

## General Co-chairs

Junping Du                Beijing University of Posts and Telecommunications,
                          China
Xiaochun Yang             Northeast University, China

## Program Co-chairs

Yongxin Tong              Beihang University, China
Yafei Li                  Zhengzhou University, China

## Program Committee

Qian Chen                 Google, USA
Lei Chen                  Huawei Noah's Ark Lab, China
Peipei Yi                 Lenovo Machine Intelligence Center, China
Cheng Xu                  Simon Fraser University, Canada
Shangwei Guo              Nanyang Technology University, Singapore
Chengcheng Dai            Grab Holdings Inc., Malaysia
Qidong Liu                Nanyang Technology University, Singapore
Xiaoyi Fu                 Hong Kong Baptist University, China
Ji Wan                    Beihang University, China
Hao Su                    Beihang University, China
Guanglei Zhu              Zhengzhou University, China

# Contents

# The 6th International Workshop on Big Data Quality Management

# ASQT: An Efficient Index for Queries on Compressed Trajectories

Binghao Wang[✉], Hongbo Yin, Kaiqi Zhang, Dailiang Jin, and Hong Gao[✉]

Harbin Institute of Technology, Harbin 150001, Heilongjiang, China
{wangbinghao,hongboyin,zhangkaiqi,jdl,honggao}@hit.edu.cn

**Abstract.** Nowadays, the amount of GPS-equipped devices is increasing dramatically and they generate raw trajectory data constantly. Many location-based services that use trajectory data are becoming increasingly popular in many fields. However, the amount of raw trajectory data is usually too large. Such a large amount of data is expensive to store, and the cost of transmitting and processing is quite high. To address these problems, the common method is to use compression algorithms to compress trajectories. This paper proposes a high efficient spatial index named ASQT, which is a quadtree index with adaptability. And based on ASQT, we propose a range query processing algorithm and a top-$k$ similarity query processing algorithm. ASQT can effectively speed up both the trajectory range query processing and similarity query processing on compressed trajectories. Extensive experiments are done on a real dataset and results show the superiority of our methods.

**Keywords:** Compressed trajectory · ASQT index · Trajectory range query · Top-$k$ trajectory similarity query

## 1 Introduction

Nowadays, with the development of GPS-equipped devices, location-based services are becoming increasingly popular in many fields. The amount of trajectory data collected is usually very large. For example, Didi Chuxing, the largest ride-sharing platform in China, receives hundreds of millions trajectories and processes more than 3 million travel requests every day [1]. The data quality management is very important [2,3]. The trajectory data is transmitted to the cloud to support various services, such as route planning, travel time prediction, and traffic management, etc.

However, such a large amount of trajectory data brings some serious problems. It is very expensive to be stored, and the cost of transmitting and processing is quite high. If each GPS device records its latest position every 10 s,

The National Key Research and Development Program of China (Grant No. 2019YFB2101902), Joint Funds of the National Natural Science Foundation of China under Grant No. U19A2059.

C. S. Jensen et al. (Eds.): DASFAA 2021 Workshops, LNCS 12680, pp. 3–15, 2021.
https://doi.org/10.1007/978-3-030-73216-5_1

1 GB storage space is needed for 4 thousand GPS-equipped devices to store the trajectory data every day [4]. Most existing trajectory query processing and data mining algorithms are executed in the main memory, which makes trajectories inconvenience to be handled when the amount is too large.

To solve the above problems, the common method is to apply some trajectory compression algorithm [4–13] to compress the trajectories. The storage space, communication bandwidth as well as processing cost of the queries can be reduced obviously by using compressed trajectory data since the trajectories are compressed into a much smaller size.

In most location-based services, there are two types of basic but essential trajectory queries, i.e. trajectory range queries and top-$k$ trajectory similarity queries. The trajectory range query is to find all trajectories that intersect a given area, and the top-$k$ trajectory similarity query returns the $k$ most similar trajectories to the query trajectory. Range queries can be used in the lost and found service to search all passing taxis in the area. Similarity queries can be used in the taxi detour detection to determine whether its trajectory is an outlier.

We need to verify all trajectories, if no acceleration strategy is used in the trajectory range query processing. Similarly, it is also necessary to calculate the similarity between the query trajectory and all trajectories in the top-$k$ similarity query processing. Hence, the computing cost is quite high. To solve the problem, we propose ASQT, an efficient index, to speed up the query processing. The main contributions of this paper are summarized as follows:

- We propose an Adaptive Spatial Quadtree index for compressed Trajectories (ASQT for short). ASQT is an adaptive index, which can speed up the processing of trajectory queries efficiently.
- Based on this high efficient index, we propose a trajectory range query processing algorithm and a top-$k$ trajectory similarity query processing algorithm.
- We conducted extensive experiments on a real-life trajectory dataset to verify the performance of our methods. The results show that all our methods are of great efficiency.

## 2   Related Work

### 2.1   Trajectory Compression Algorithm

Trajectory compression algorithms fall into two categories, i.e. lossless compression algorithms and lossy compression algorithms. Although more raw trajectory information can be retained by using the lossless compression algorithms, only a limited compression ratio can be obtained. Therefore, more research is devoted to the lossy compression algorithms. A much higher compression ratio can be achieved by using the lossy while retaining the raw trajectory information within the error tolerance. The lossy compression methods can be also divided into two categories, i.e. the batch mode [6–8] and the online mode [9–13]. Algorithms in the batch mode require that all the points of a trajectory must be loaded in the main memory before compression, which means that the main memory must be large enough to hold the whole trajectory. Therefore, the space complexities of these algorithms are all at least $O(N)$, which limits their applications in resource-constrained environments. By contrast, algorithms in the online mode only need a small size of local buffer to compress trajectories in an online processing manner. Therefore, the online mode can be used in more application scenarios.

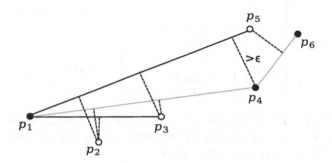

**Fig. 1.** An example of OPW algorithm

In recent years, many lossy compression algorithms were proposed. Since this paper focuses on the queries on compressed trajectories, we choose Open Window algorithm (OPW), one of the most accurate compression algorithms, as our compression method. OPW processes each trajectory point sequentially. Assume $p_i$ as the starting point of the current trajectory line segment, and $\epsilon$ as the threshold. When processing point $p_j$, connect $p_i, p_j$ and judge whether the vertical distances from $p_{i+1}, ..., p_{j-1}$ to line segment $p_i p_j$ are all less than $\epsilon$. If satisfied, then $p_i, ...p_j$ can be approximately represented by line segment $p_i p_j$; otherwise, we can use the line segment $p_i p_{j-1}$ to approximately represent points $p_i, ..., p_{j-1}$, and $p_{j-1} p_j$ is the new current line segment with $p_{j-1}$ as the new starting point. The next point $p_{j+1}$ is processed iteratively in the same way until there are no unprocessed trajectory points. For example in Fig. 1, these points are compressed into line segments $p_1 p_4$ and $p_4 p_6$.

## 2.2 Existing Similarity Measures

When processing trajectory similarity queries, measuring the similarity between two trajectories is the most fundamental operation. Most similarity measures evaluate the similarity by calculating the distance between the matched point pairs of two trajectories. We use $r_i$ and $s_j$ to represent the $i$-th and $j$-th trajectory point of trajectory $R$ and $S$ respectively.

A classic similarity measure is Euclidean distance [14], which requires the two trajectories to have the same number of trajectory points. The calculation formula is as follows:

$$ED(R, S) = \sqrt{\sum_{i=1}^{n} (r_{i,x} - s_{i,x})^2 + (r_{i,y} - s_{i,y})^2} \qquad (1)$$

Some more complex similarity measures, such as DTW [15], LCSS [16], EDR [17] and Swale [18], adopt the Dynamic Programming (DP) method. DTW can be recursively calculated as:

$$DTW(R, S) = \begin{cases} 0 & if\ |R| = |S| = 0 \\ \infty & if\ |R| = 0\ or\ |S| = 0 \\ min \begin{cases} dist(r_1, s_1) + \\ DTW(Rest(R), \\ Rest(S)) \\ DTW(Rest(R), S) \\ DTW(R, Rest(S)) \end{cases} & otherwise \end{cases} \qquad (2)$$

where $Rest(R)$ and $Rest(S)$ respectively represent the remaining part of $R$ and $S$ after removing $r_1$ and $s_1$. $dist(r_1, s_1)$ is the Euclidean distance between $r_1$ and $s_1$. LCSS can be recursively calculated as:

$$LCSS(R, S) = \begin{cases} 0 & if\ |R| = 0\ or\ |S| = 0 \\ 1 + LCSS(Rest(R), Rest(S)) & if\ dist(r_1, s_1) \leq \epsilon \\ max \begin{cases} LCSS(Rest(R), S) \\ LCSS(R, Rest(S)) \end{cases} & otherwise \end{cases} \qquad (3)$$

It assigns 0 to non-matched points, otherwise 1. In other words, LCSS ignores dissimilar point pairs, and only focuses on the number of similar point pairs. The larger the $LCSS(R, S)$, the more similar $R$ and $S$ are.

## 3   The ASQT Index

### 3.1   Index Construction

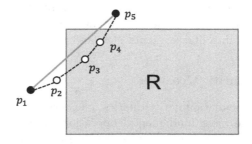

**Fig. 2.** Trajectory segment overlapped with $R$

We aim to process queries on compressed trajectories. In previous work, each trajectory is thought to be a set of discrete points. A raw trajectory is overlapped with a rectangle region $R$ iff at least one point in this trajectory falls in $R$. However, a line segment of a compressed trajectory may represent a large number of points in the raw trajectory [19]. That is to say, although neither of the two endpoints of a line segment is not in $R$, as shown in Fig. 2, the discarded points between them may fall in $R$. To address this, each compressed trajectory is regarded as a set of continuous line segments. If a line segment of a compressed trajectory intersects the rectangular area, this trajectory is considered to be overlapped with $R$.

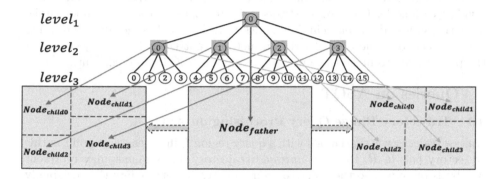

**Fig. 3.** An overview of the ASQT index

In order to speed up query processing on compressed trajectories, we propose a high efficient index, Adaptive Spatial Quadtree index for compressed Trajectories (ASQT). For each trajectory, we only store its ID in the index to reduce space overhead. And the Minimum Bounding Rectangle (MBR for short) is stored for each trajectory. MBR is the smallest rectangle that can completely contain the trajectory. At the beginning, there is only one root node which contains all compressed trajectories in the corresponding region. The median of center points' x coordinates (y coordinates) of all the MBRs is calculated, which is expressed as $median(x)$ $(median(y))$. Then, the root node is divided into two parts by using $median(x)$ $(median(y))$. Furthermore, the medians of center points' y coordinates (x coordinates) of all the MBRs in each part are calculated, which are expressed as $median(y_1)$ and $median(y_2)$ $(median(x_1)$ and $median(x_2))$. As shown in Fig. 3, the root is divided into 4 disjoint child nodes with $madian(x)$, $median(y_1)$ and $median(y_2)$ $(madian(y), median(x_1), median(x_2))$. Therefore, there are two methods to divide the root node. Meanwhile, a trajectory is distributed to a child node if the region of this node completely contains the MBR of the trajectory, otherwise it is left in the original tree node. In these two division methods, the method that makes the root node retain fewer trajectories is adopted. Then, each child node is divided recursively in the same way, and until the number of trajectories within the child node is less than the threshold $\xi$, the partition process of this node stops.

## 3.2   The Properties of ASQT

**Approximate Balance.** The partition method using the medians can ensure that the number of trajectories fallen into each child node is roughly equal to each other. This makes ASQT approximately balanced.

**Adaptability.** During the partition process of a tree node, the trajectories completely contained by a child node are distributed to the corresponding node. However, some trajectories which are not contained by any child are remained in the parent node. As shown in Fig. 3, there are two partition methods when dividing each node. The fewer the trajectories remained in the parent node is, the more trajectories distributed in child nodes are, which makes the effect of pruning better. Therefore, choose the partition method that keeps fewer trajectories in the parent node to make the query processing on ASQT more efficient.

# 4   Queries on ASQT

## 4.1   Trajectory Range Query Processing on ASQT

An raw trajectory is overlapped with a query region $R$ iff at least one point in this trajectory falls in $R$. But a compressed trajectory line segment may represent hundreds of points, so there may be some raw trajectory points that originally fall in R are lost after compression. Hence, such a trajectory is missing in the result set possibly. Inspired by this, we define the range query processing on compressed trajectories as follows:

**Definition 1 (Range Query).** *Given the compressed trajectory dataset* $\mathbb{T}$ *and the query region $R$, the range query result $Q_r(R, \mathbb{T})$ contains all such compressed trajectories in* $\mathbb{T}$*, at least one of whose trajectory line segments overlaps $R$, i.e.,*

$$Q_r(R, \mathbb{T}) = \{T \in \mathbb{T} \mid \exists p_i p_j \in T, \ s.t. \ p_i p_j \ overlaps \ R\} \tag{4}$$

For simplicity, we consider the query regions as two-dimensional rectangles.

The brute force method to verify whether a compressed trajectory $T$ is in the query result $Q_r(R, \mathbb{T})$ is to determine whether there exists a line segment of $T$ overlapped with $R$. If so, $T$ is added to the result $Q_r(R, \mathbb{T})$. But with the help of ASQT, most trajectories can be filtered out by traversing the index. We adopt Depth First Search (DFS) to traverse ASQT, and apply pruning strategies to speed up the query:

(1) If the query region $R$ doesn't overlap the corresponding region of an ASQT node, then all the trajectories stored on the subtree rooted at this node must not overlap $R$. Therefore, we skip the search for this subtree.
(2) If the query region $R$ contains the corresponding region of an ASQT node, then all the trajectories stored on the subtree rooted at this node fall into $R$ absolutely. Hence, all these trajectories stored on this subtree should be put into $Q_r(R, \mathbb{T})$ directly.
(3) Otherwise, verify whether each trajectory in the node overlaps $R$ sequentially. If this node is a non-leaf node, its childs are handled recursively in the same way.

## 4.2 Trajectory Similarity Query on ASQT

Given the query trajectory $T_q$ and a compressed trajectory dataset $\mathbb{T}$, the goal of top-$k$ similarity query processing is to find the $k$ compressed trajectories most similar to $T_q$ in $\mathbb{T}$. We formally define the top-$k$ similarity query on compressed trajectories as follows:

**Definition 2 (Top-$k$ Similarity Query).** *Given the query trajectory $T_q$ and the compressed trajectory dataset $\mathbb{T}$, the top-k similarity query processing returns a set $Q_k(T_q, \mathbb{T})$ consisting of k trajectories, which is defined as:*

$$Q_k(T_q, \mathbb{T}) = \{T \in \mathbb{T} \mid \forall T' \in Q_k(T_q, \mathbb{T}), \ T'' \in (\mathbb{T} - Q_k(T_q, \mathbb{T})), \\ s.t. \ Dist(T_q, T') \leq Dist(T_q, T'')\} \quad (5)$$

*where $Dist(T_q, T')$ represents the distance between $T_q$ and $T'$ under a certain similarity measure.*

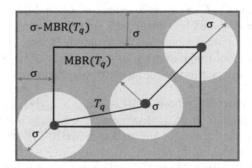

**Fig. 4.** The region of similarity candidate set

If no acceleration strategy is adopted, the similarity between $T_q$ and each compressed trajectory in $\mathbb{T}$ must be calculated to answer the top-$k$ similarity query. The time cost of this process is quite high. If the distance between two points is too far, the matching of the two points has no positive effect on the similarity. So given a threshold $\sigma$, the similarity of two trajectories is greater than 0 iff there are at least a pair of points with a distacne less than $\sigma$ in these two trajectories. Based on this idea, the trajectories in $\mathbb{T}$ are filtered to generate a candidate set, which is represented by $C'(T_q, \mathbb{T}, \sigma)$. As shown in Fig. 4, $C'(T_q, \mathbb{T}, \sigma)$ consists of all trajectories overlapping the gray regions. Only the similarity between each trajectory in $C'(T_q, \mathbb{T}, \sigma)$ and $T_q$ needs to be calculated.

However, the number of the grey regions is the same as the number of trajectories points, which makes it expensive to calculate $C'(T_q, \mathbb{T}, \sigma)$. To address the problem, we extand the candidate set to a set of trajectories which overlap the MBR of all the grey regions. Before introducing the new candidate set, we define the $\sigma$-MBR of $T_q$ as follows:

**Definition 3 ($\sigma$-MBR($T_q$)).** *Given the distance threshold $\sigma$ and the MBR of a compressed trajectory $T_q$ which is represented by $[x_{min}, x_{max}] * [y_{min}, y_{max}]$, $\sigma$-MBR($T_q$) is defined as $[x_{min} - \sigma, x_{max} + \sigma] * [y_{min} - \sigma, y_{max} + \sigma]$,*

The $\sigma$-MBR($T_q$) is the whole orange region shown in Fig. 4. $\sigma$-MBR($T_q$) is the smallest MBR that happens to completely contain all the grey regions. We use $C(T_q, \mathbb{T}, \sigma)$ to represent the new candidate set. And for a query trajectory $T_q$, $C(T_q, \mathbb{T}, \sigma)$ consists of all trajectories overlapping $\sigma$-MBR($T_q$). We formally define the candidate set of the top-$k$ similarity query processing algorithm as follows:

**Definition 4 (Similarity Candidate Set).** *Given the distance threshold $\sigma$, the query trajectory $T_q$ and the compressed trajectory dataset $\mathbb{T}$, the candidate set is defined as:*

$$C(T_q, \mathbb{T}, \sigma) = \{T \in \mathbb{T} \mid \exists p_i p_j \in T, \ s.t. \ p_i p_j \ overlaps \ \sigma\text{-MBR}(T_q)\} \quad (6)$$

That is, $C(T_q, \mathbb{T}, \sigma)$ is equal to $Q_r(\sigma\text{-MBR}(T_q), \mathbb{T})$. Because $\sigma$-MBR($T_q$) contains the grey region, $C'(T_q, \mathbb{T}, \sigma)$ is a subset of $Q_r(\sigma\text{-MBR}(T_q), \mathbb{T})$. The new candidate set can be obtained by executing a trajectory range query and a lot of unnecessary similarity calculations can be avoided. Hence, ASQT can improve the execution efficiency of range queries, as well as that of the similarity queries. The top-$k$ similarity query processing algorithm is summarized as follows:

(1) For a query trajectory $T_q$, whose MBR is $[x_{min}, x_{max}] * [y_{min}, y_{max}]$, calculate its $\sigma$-MBR($T_q$): $[x_{min} - \sigma, x_{max} + \sigma] * [y_{min} - \sigma, y_{max} + \sigma]$.
(2) By executing a range query for the region $\sigma$-MBR($T_q$), we can get the similarity candidate set $C(T_q, \mathbb{T}, \sigma)$.
(3) Calculate the similarity between $T_q$ and each trajectory in $C(T_q, \mathbb{T}, \sigma)$, and select the $k$ most similar trajectories as the similarity query result $Q_k(T_q, \mathbb{T})$.

## 5  Experiments

### 5.1  Experiment Settings

**Experiment Environment.** Experiments were all conducted on a machine with a 64-bit, 8-core, 3.6GHz Intel(R) Core (TM) i9-9900K CPU and 32GB memory. Our codes were all implemented in C++ on Ubuntu 18.04 operating system.

**Experiment Dataset.** The experiments were conducted on Planet [20] dataset. It contains more than 8 million raw trajectories and 2.67 billion trajectory points. The average sampling rate of these trajectories is 4.273 s, and the average distance between two continuous points is 8.589 m. We select all 265303 trajectories in the geographic coordinate range $[7, 14] * [46, 53]$ as our test data.

## 5.2  Experiment Results

For a range query, let $Q_{or}$ and $Q_r$ denote the query result on the raw trajectories and the compressed trajectories respectively. The precision, recall and $F_1$ score are defined respectively as follows:

$$Precision = \frac{|Q_{or} \cap Q_r|}{|Q_r|} \tag{7}$$

$$Recall = \frac{|Q_{or} \cap Q_r|}{|Q_{or}|} \tag{8}$$

$$F_1 = \frac{2 * Precision * Recall}{Precision + Recall} \tag{9}$$

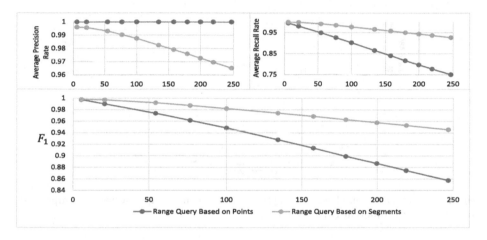

**Fig. 5.** Comparison of *Precision*, *Recall*, and $F_1$ score of trajectory range query processing algorithms based on points and line segments on compressed trajectories

In all the experiments below, the region size of each trajectory range query was fixed as $16\,\text{km}^2$, and the region of each trajectory range query was randomly generated. The trajectory range query processing methods which consider a trajectory as discrete points and continuous line segments were used to execute 10,000 trajectory range queries on same regions respectively. We evaluated the average *Precision*, *Recall* and $F_1$ score on the compressed dataset w.r.t. varying the compression ratio, and the result is shown in Fig. 5. When each compressed trajectory is seen as discrete points, the *Precision* is always 1, since the points in each compressed trajectory must be a subset of the corresponding raw trajectory points. The result shows that the algorithm which considers a trajectory as continuous line segments is much more accurate than that which considers a trajectory as discrete points.

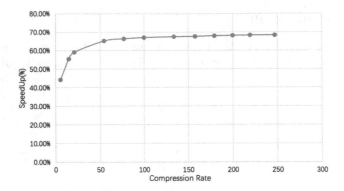

**Fig. 6.** The acceleration effect of executing range queries on compressed trajectories with different compression ratios

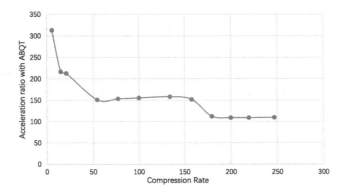

**Fig. 7.** The acceleration effect of ASQT when executing range queries on compressed trajectories with different compression ratios

In order to measure the acceleration effect of executing range queries on compressed trajectories, we executed point-based and line-segment-based trajectory range query processing algorithms on raw trajectories and compressed trajectories with different compression ratios respectively. Each range query processing algorithm is executed for the same 10000 randomly generated query regions, and the average execution time is calculated respectively. In this experiment, ASQT was not used for the acceleration of queries. The ratio of the average execution time is shown in Fig. 6. As we can see, the efficiency of the range query processing can be significantly improved by executing queries on compressed trajectories. And as the compression ratio increases, the acceleration effect tends to stabilize.

To study the impact of ASQT, we executed 10,000 trajectory range queries on compressed trajectories with different compression ratios in both cases of using and not using ASQT to accelerate. Then the average execution time was calculated respectively. Figure 7 shows the acceleration effect of ASQT. It can be seen that the filtering effect of ASQT on irrelevant trajectories is very obvious when executing trajectory range queries on compressed trajectories.

**Table 1.** The effect of the threshold $\xi$ on the average height of ASQT and the query time of trajectory range queries

| $\xi$ | Average height | Average range query time (ms) |
|---|---|---|
| 200 | 9.141037 | 0.047437 |
| 400 | 8.081274 | 0.057316 |
| 800 | 8.004260 | 0.058540 |
| 1600 | 7.006801 | 0.111759 |
| 3200 | 7 | 0.112808 |
| 6400 | 6 | 0.213536 |

Table 1 shows the influence of the threshold $\xi$ on the average tree height of ASQT and the average processing time of range queries. In this experiment, the compression ratio of the queried trajectories was fixed as 10. From the result, we can see that ASQT has approximate balance. When $\xi$ is high, by examining the structure of ASQT, it is found to be a full quadtree. As the height of the tree increases, the processing time of range queries decreases. The reason is that as the depth of the leaf node increases, the filtering effect of ASQT on range query processing becomes better.

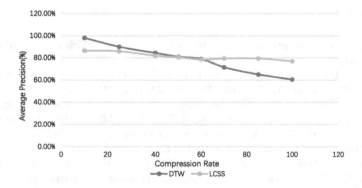

**Fig. 8.** The Precision of top-$k$ similarity queries under different similarity measures on compressed trajectories with different compression ratios

Then we used DTW and LCSS to execute top-$k$ similarity queries on compressed trajectories with different compression ratios respectively. Use the query results on raw trajectories as the standard, and compare the *Precision* of the algorithm under different compression ratios. Since the size of the result set is

$k$, the *Precision*, *Recall* and $F_1$ score are the same to each other, so only one of them is sufficient. The parameter $k$ is fixed as 100. As shown in Fig. 8, the *Precision* of trajectory similarity queries performed on compressed trajectories using different similarity measures is quite high, especially when the compression ratio of the trajectories is less than 25.

## 6 Conclusion

It is meaningful that our paper regards the compressed trajectory as continuous trajectory line segments instead of discrete points. In this paper, we propose an efficient index named ASQT, which is adaptive. Experiment results show that ASQT is approximately balanced. Based on ASQT, we propose a range query processing algorithm and a top-$k$ similarity query processing algorithm. The range query processing algorithm considers a trajectory as continuous line segments, rather than discrete points. Experiments shows that when querying on compressed trajectories, the performance of the range query processing algorithm proposed in this paper is significantly better than that of the algorithm which regards a trajectory as discrete points. The range query processing algorithm can accelerate the similarity query processing. Hence, ASQT can improve the execution efficiency of range queries, as well as that of the top-$k$ similarity queries. Experiments show that ASQT has a significant acceleration effect when querying on compressed trajectories. This paper proposes a series of query techniques on compressed trajectories including index structure and query processing algorithms, and has achieved satisfactory results.

## References

1. Kosoff, M.: Uber is getting dwarfed by its Chinese rival, which gives 3 million rides a day (2015)
2. Miao, D., Cai, Z., Li, J., Gao, X., Liu, X.: The computation of optimal subset repairs. Proc. VLDB Endow. **13**(11), 2061–2074 (2020)
3. Miao, D., Cai, Z., Li, J.: On the complexity of bounded view propagation for conjunctive queries. IEEE Trans. Knowl. Data Eng. **30**(1), 115–127 (2018)
4. Meratnia, N., de By, R.A.: Spatiotemporal compression techniques for moving point objects. In: Bertino, E., et al. (eds.) EDBT 2004. LNCS, vol. 2992, pp. 765–782. Springer, Heidelberg (2004). https://doi.org/10.1007/978-3-540-24741-8_44
5. Keogh, E., Chu, S., Hart, D., Pazzani, M.: An online algorithm for segmenting time series. In Proceedings 2001 IEEE International Conference on Data Mining, pp. 289–296 (2001)
6. Chen, M., Xu, M., Franti, P.: A fast O(N) multiresolution polygonal approximation algorithm for GPS trajectory simplification. IEEE Trans. Image Process. **21**(5), 2770–2785 (2012)
7. Long, C., Wong, R.C.W., Jagadish, H.V.: Direction-preserving trajectory simplification. Proc. VLDB Endow. **6**(10), 949–960 (2013)
8. Long, C., Wong, R.C.W., Jagadish, H.V.: Trajectory simplification: on minimizing the direction-based error. Proc. VLDB Endow. **8**(1), 49–60 (2014)

9. Liu, J., Zhao, K., Sommer, P., Shang, S., Kusy, B., Jurdak, R.: Bounded quadrant system: error-bounded trajectory compression on the Go. In: 2015 IEEE 31st International Conference on Data Engineering, pp. 987–998 (2015)
10. Muckell, J., Olsen, P.W., Hwang, J.H., Lawson, C.T., Ravi, S.: Compression of trajectory data: a comprehensive evaluation and new approach. GeoInformatica 18(3), 435–460 (2014). https://doi.org/10.1007/s10707-013-0184-0
11. Katsikouli, P., Sarkar, R., Gao, J.: Persistence based online signal and trajectory simplification for mobile devices. In: Proceedings of the 22nd ACM SIGSPATIAL International Conference on Advances in Geographic Information Systems, pp. 371–380 (2014)
12. Ke, B., Shao, J., Zhang, D.: An efficient online approach for direction-preserving trajectory simplification with interval bounds. In: 2017 18th IEEE International Conference on Mobile Data Management (MDM), pp. 50–55 (2017)
13. Ke, B., Shao, J., Zhang, Y., Zhang, D., Yang, Y.: An online approach for direction-based trajectory compression with error bound guarantee. In: Li, F., Shim, K., Zheng, K., Liu, G. (eds.) APWeb 2016. LNCS, vol. 9931, pp. 79–91. Springer, Cham (2016). https://doi.org/10.1007/978-3-319-45814-4_7
14. Faloutsos, C., Ranganathan, M., Manolopoulos, Y.: Fast subsequence matching in time-series databases. ACM Sigmod Rec. 23(2), 419–429 (1994)
15. Berndt, D.J., Clifford, J.: Using dynamic time warping to find patterns in time series. In: KDD Workshop, vol. 10, no. 16, pp. 359–370 (1994)
16. Vlachos, M., Kollios, G., Gunopulos, D.: Discovering similar multidimensional trajectories. In: Proceedings 18th International Conference on Data Engineering, pp. 673–684 (2002)
17. Chen, L., Ozsu, M.T., Oria, V.: Robust and fast similarity search for moving object trajectories. In: Proceedings of the 2005 ACM SIGMOD International Conference on Management of Data, pp. 491–502 (2005)
18. Morse, M.D., Patel, J.M.: An efficient and accurate method for evaluating time series similarity. In: Proceedings of the 2007 ACM SIGMOD International Conference on Management of Data, pp. 569–580 (2007)
19. Lee, J.G., Han, J., Whang, K.Y.: Trajectory clustering: a partition-and-group framework. In: Proceedings of the 2007 ACM SIGMOD International Conference on Management of data, pp. 593–604 (2007)
20. https://wiki.openstreetmap.org/wiki/Planet.gpx

# ROPW: An Online Trajectory Compression Algorithm

Sirui Li[1]([⊠]), Kaiqi Zhang[1], Hongbo Yin[1], Dan Yin[2], Hongquan Zu[1], and Hong Gao[1]([⊠])

[1] Harbin Institute of Technology, Harbin 150001, Heilongjiang, China
{kuwylsr,zhangkaiqi,hongboyin,zuhq,honggao}@hit.edu.cn
[2] Harbin Engineering University, Harbin 150001, Heilongjiang, China
yindan@hrbeu.edu.cn

**Abstract.** In smart phones, vehicles and wearable devices, GPS sensors are ubiquitous, which can collect a large amount of valuable trajectory data by tracking moving objects. Analysis of this valuable trajectory data can benefit many practical applications, such as route planning and transportation optimization. However, unprecedented large-scale GPS data poses a challenge to the effective storage of trajectories. Therefore, the necessity of trajectory compression (also called trajectory sampling) is reflected. However, the latest compression methods usually perform unsatisfactorily in terms of space-time complexity or compression rate, which leads to rapid exhaustion of memory, computing, storage, and energy. In response to this problem, this paper proposes an online trajectory compression algorithm (ROPW algorithm) with error bounded that traverses the sliding window backwards. This algorithm has significantly improved the trajectory compression rate, and its average time complexity and space complexity is $O(NlogN)$ and $O(1)$ respectively. Finally, we conducted experiments on three real data sets to verify that the ROPW algorithm performed very well in terms of compression rate and time efficiency.

**Keywords:** Online algorithm · Trajectory compression · Sliding window · Reverse traversal

## 1 Introduction

The universal application of global positioning system sensors (GPS) in smart phones, vehicles and wearable devices enables the collection of a large amount of trajectory data by tracking moving objects, and these data can be applied to many fields, such as traffic information services, navigation services, bus arrival time services, etc.

According to the statistics, assuming that GPS collects information every 5s, the data generated by a vehicle in a day requires about 70 MB of storage space,

National Natural Science Foundation of China under Grant 61702132, The National Key Research and Development Program of China (Grant No. 2019YFB2101902), Joint Funds of the National Natural Science Foundation of China under Grant No. U19A2059.

C. S. Jensen et al. (Eds.): DASFAA 2021 Workshops, LNCS 12680, pp. 16–28, 2021.
https://doi.org/10.1007/978-3-030-73216-5_2

and the storage space required by all vehicles in a city will be very large. It can be seen that GPS will generate unprecedented GPS data with time stamps, and it is urgently needed for trajectory database for effective storage.

Secondly, for long-term trajectory data with scattered data points, such as wild animal data, bicycle data and other data sets, they provide high-resolution trajectory data to achieve better management and services. In addition, for foxes, pigeons and other animals, the size and weight of trajectory equipment will be limited, which will bring great challenges to the acquisition of trajectory information details. Therefore, the combination of long-term operating requirements and limited resources requires an intelligent online processing algorithm that can process the input position information in time, that is, it can process in a constant time and space to achieve a higher compression rate.

Therefore, designing the trajectory compression algorithm in online mode will become extremely important and meaningful. Fortunately, these problems can be solved or greatly reduced by trajectory compression technology. Among them, the method based on line simplification is widely used due to its unique advantages [1–14,16]. The characteristics of this type of algorithm are as follows: (a) the algorithm is simple and easy to implement, (b) it does not require additional knowledge and it is suitable for objects that move freely, (c) it have a better compression rate under the premise of bounded error.

The most famous algorithm based on line segment simplification is the Douglas-Peucker algorithm [16] invented in the 1870s, which is used to reduce the number of points required to represent digitized line segments in the context of computer graphics and image processing. The most primitive Douglas-Peucker algorithm is a batch processing algorithm, and its time complexity is $O(N^2)$, where N is the number of trajectory points in the trajectory to be compressed. After that, several DP-based LS algorithms appeared, for example, by combining DP with sliding/opening windows [10] for online processing. However, these algorithms still have a high time overhead in general, and the compression rate is not very good, which makes them perform poorly in resource-constrained mobile devices.

Recently, an algorithm called BQS [11] was proposed, which uses a new method of distance checking, which uses an open window based on the convex hull to select eight special points and use them to carry out corresponding inspections and calculations. In many cases, it only needs to calculate the distance from a special point to a straight line, rather than all the data points in the window. In the worst case, the time complexity of the BQS algorithm is still $O(N^2)$. However, its simplified version of the FBQS algorithm [12] directly outputs line segments and starts a new window when the eight special points cannot constrain all the points considered so far. Indeed, the FBQS algorithm has linear time complexity, but when it is used in actual data sets for trajectory compression, the time performance is average and its compression rate is low.

In addition, a lot of work [17,18] is being done in terms of data quality.

In summary, our contributions are two-fold:

1. Using the idea of sliding/open window, a Reverse OPening Window algorithm (ROPW algorithm for short) based on the reverse traversal of trajectory points in the window and accelerated by jumping strategy is designed (ROPW

algorithm is introduced in Sect. 4.3). Although the time complexity is still $O(N^2)$ and the space complexity is $O(N)$, in practical applications, we have confirmed through experiments that the upper bound of the time complexity will be reached in rare cases and its average time complexity is $O(NlogN)$. In addition, the algorithm can achieve extremely high compression rate under the premise of given error accuracy.

2. We compared the ROPW algorithm with the OPW algorithm, the OPERB algorithm, etc. on three real trajectory data sets (Animal data set, Indoor data set and Planet data set), and conducted extensive experimental research (Sect. 5). We found that the ROPW algorithm has the highest compression rate under the premise of the same error precision bound, and it also has a good performance in terms of time efficiency and error.

## 2    Related Work

Trajectory compression algorithms can generally be divided into two categories, namely lossless trajectory compression algorithms and lossy trajectory compression algorithms. Lossless compression methods refer to a compression method that can reconstruct the original trajectory data without losing information. For example, the delta algorithm [1] is a lossless trajectory compression algorithm with a time complexity of $O(N)$. The limitation of lossless compression is its relatively poor compression rate. In contrast, lossy compression methods allow errors or deviations compared to lossless trajectory compression algorithms. This kind of algorithm usually tends to select relatively important data points on the trajectory and remove redundant data points from the trajectory. They focus on achieving trajectory compression with a higher compression ratio within an acceptable error range. They can be roughly divided into two categories: Line simplification based methods and Semantics based methods. This article focuses on the trajectory compression method based on line simplification.

The trajectory compression method based on line simplification is currently the more popular trajectory compression method. This method not only has a satisfactory compression ratio and error limit, but the algorithm is also easy to implement, so it is widely used in practice. According to the way they process trajectory data, this method can be divided into: batch mode methods and online mode methods.

The trajectory compression method in batch mode needs to load all the trajectory point data before officially starting to compress the trajectory. The existing batch trajectory compression algorithms include Bellman [7], DP [8], DPhull, TD-TR, MRPA [9] and so on. Since this mode can get all the data of the trajectory history, its goal is to achieve a trade-off between the compression rate and the loss of data information.

The trajectory compression method in online mode does not need to have the entire trajectory data before it officially starts to compress the trajectory, so it is very suitable for compressing the data provided by various sensors. These data are stored in a local buffer, and the online simplification algorithm must determine the trajectory points to be discarded. Because the online algorithm cannot know all the information of the entire trajectory, it is difficult to obtain the ideal result in the problem of minimizing errors and maximizing the compression rate. However, whether it is batch mode or online mode, they have a remarkable feature that we can set an error bound $\epsilon$ to ensure that the error of any discarded point does not exceed $\epsilon$. The existing online trajectory compression algorithm [10–13] usually uses a fixed or open sliding window to compress the sub-trajectories in the window, but its performance in terms of compression rate is relatively general. Therefore, we propose an online trajectory compression method that reversely traverses the trajectory points in the sliding window and uses a jumping strategy to accelerate. Furthermore, through experimental analysis, our method has better time efficiency and higher compression rate in most cases.

## 3 Preliminaries

In this section, we mainly introduce some basic concepts of trajectory compression and some online trajectory compression algorithms related to this research.

### 3.1 Basic Notations

**Definition 1.** *(Points (P)): A data point is defined as a triple $P(x, y, t)$, which represents that a moving object is located at longitude $x$ and latitude $y$ at time $t$.*

**Definition 2.** *(Trajectories (T)): A trajectory $T = \{p_1, p_2, ..., p_N\}$ is a sequence of trajectory points in a monotonically increasing order of their associated time values (i.e., $t_1 < t_2 < ... < t_N$). $T[i] = p_i$ is the ith trajectory point in $T$.*

**Definition 3.** *(Compressed Trajectory (T$'$)): Given a trajectory $T = \{p_1, p_2, ..., p_N\}$ and one set of $T$'s corresponding consecutive trajectory segments, the compressed trajectory $T'$ of $T$ is a set of consecutive line segments of all trajectory segments in $T$ and $T'$ can be denoted as $T' = \{p_{i_1}, p_{i_2}, p_{i_2}, ..., p_{i_{n-1}}, p_{i_n}\}(p_{i_1} = p_1, p_{i_n} = p_N)$*

**Definition 4.** *(Compression Rate (r)): Given a raw trajectory $T = \{p_1, p_2, ..., p_N\}$ with N raw trajectory points and its compressed trajectory $T' = \{p_{i_1}, p_{i_2}, p_{i_2}, ..., p_{i_{n-1}}, p_{i_n}\}(p_{i_1} = p_1, p_{i_n} = p_N)$ with $n - 1$ consecutive line segments, the compression rate is*
$$r = N/n.$$

**Definition 5.** *(Directed line segments ($\mathcal{L}$)): A directed line segment (or line segment for simplicity) $\mathcal{L}$ is defined as $\overrightarrow{P_s P_e}$, which represents the closed line segment that connects the start point $P_s$ and the end point $P_e$.*

**Definition 6.** *(Distances (d)): Given a point $P_i$ and a directed line segment $\mathcal{L} = \overrightarrow{P_s P_e}$, the distance of $P_i$ to $\mathcal{L}$, denoted as $d(P_i, \mathcal{L})$, is the Euclidean distance from $P_i$ to the line $\overrightarrow{P_s P_e}$, commonly adopted by most existing LS methods.*

## 3.2   Line Simplification Algorithms

The line simplification algorithm is an important and widely used trajectory compression method. Next, we briefly introduce an online trajectory compression algorithm OPW algorithm [10] based on line simplification.

Given a trajectory $T = \{p_1, p_2, ..., p_N\}$ and an error bound $\epsilon$, algorithm OPW [10] maintains a window $W[P_s, ..., P_k]$, where $P_s$ and $P_k$ are the start and end points, respectively. Initially, $P_s = P_0$ and $P_k = P_1$, and the window $W$ is gradually expanded by adding new points one by one. OPW tries to compress all points in $W[P_s, ..., P_k]$ to a single line segment $\mathcal{L}(P_s, P_k)$. If the distances $d(P_i, \mathcal{L}) \leq \epsilon$ for all points $P_i (i \in [s, k])$, it simply expands $W$ to $[P_s, ..., P_k, P_{k+1}](k+1 \leq n)$ by adding a new point $P_{k+1}$. Otherwise, it produces a new line segment $\mathcal{L}(P_s, P_{k-1})$, and replaces $W$ with a new window $P_{k-1}, ..., P_{k+1}$. The above process repeats until all points in $T$ have been considered. The worst case of the algorithm occurs when every trajectory point can be compressed, so the time complexity of the algorithm is $O(N^2)$, which also makes the algorithm not suitable for compressing long trajectories. OPW is formally described in Algorithm 1.

---

**Algorithm 1: OPW**

---

**Input:** Raw trajectory $T = \{p_1, p_2, ..., p_N\}$ , error tolerance $\epsilon$

**Output:** $\epsilon$ - Error-bounded trajectory
$$T' = \{p_{i_1}, p_{i_2}, ..., p_{i_n}\}(p_{i_1} = p_1, p_{i_n} = p_N)$$

$e = 0$;
$originalIndex = 0$;
$T' = originalIndex$ ;
**while** $e < T.length() - 1$ **do**
 $CurPoint = originalIndex + 1$;
 $flag = true$;
 **while** $CurPoint > e$ && $flag$ **do**
  **if** $PED(StartPoint, CurPoint, LastPoint) > epsilon$ **then**
   $flag = false$;
  **else**
   $CurPoint + +$;
  **end**
 **end**
 **if** $!flag$ **then**
  $originalIndex = CurPoint$;
  $T'.Append(originalIndex)$;
  $e = originalIndex + 2$;
 **else**
  $e++$;
 **end**
**end**
$T'.Append(T.length() - 1)$;
return $T'$;

---

# 4    Algorithm ROPW

## 4.1    Motivation

Before officially introducing the ROPW algorithm, let's make a simple analysis of the OPW algorithm [10] described in Sect. 3.2. Through the execution process of the OPW algorithm, it is not difficult to find that the OPW algorithm is a forward progressive process. This process of forward progression means that when compressing a trajectory, all the distances between the starting point and the "current" end point to the target straight line are required to be less than the threshold, until a trajectory point that does not meet the threshold requirement appears. This point is the end of the current track segment. Indeed, this processing method can ensure that the perpendicular Euclidean distance from each compressed trajectory point to the compressed trajectory line segment must be less than the set threshold, but we intuitively feel that such processing may reduce the compression rate of the trajectory. The reason is that the OPW algorithm not only guarantees that the perpendicular Euclidean distance from each compressed point $P_i(s < i < k)$ to the trajectory segment $\mathcal{L}(P_s, P_k)$ formed by the start and end points after compression meets $d(P_i, \mathcal{L}) \leq \epsilon$, but it also additionally restricts that when each compressed trajectory point $P_i$ serves as the end point of the current sliding window, the distance from each compressed track point $P_m(s < m < i)$ which is between $P_i$ and the starting point $P_s$ in the current window to the line segment $\mathcal{L}(P_s, P_i)$ meets restriction $d(P_m, \mathcal{L}) \leq \epsilon$.

Let's use an example to verify our conjecture. As shown in Fig. 1, the trajectory points $P_1, P_2, P_3$ are three trajectory points that arrive in chronological order. When we execute the OPW algorithm, it is not difficult to find that when the track point $P_3$ enters the window, the vertical Euclidean distance from the track point $P_2$ to the track line segment $\mathcal{L}(P_1, P_3)$ meets $d(P_2, \mathcal{L}) > \epsilon$, so the track point $P_2$ will not be compressed. However, assuming that we can predict a certain trajectory point $P_N$ that exists later, we can find that when $P_N$ is the last point of the current window, the distances from trajectory point $P_1$ and trajectory point $P_2$ to trajectory line segment $\mathcal{L}(P_1, P_N)$ meet the error requirements, Which means that the track point $P_2$ will be compressed at this time. Based on such a thinking, we designed the ROPW algorithm.

## 4.2    Error-Based Metrics

In the process of trajectory compression, we use a set of continuous line segments to approximate each original trajectory. In order to obtain a good compression algorithm, the deviation between these line segments and the original trajectory should be as small as possible. Therefore, the design of a distance error measurement method is an important criterion to measure the quality of the compression algorithm. This paper uses the more traditional perpendicular Euclidean distance measurement error, which is suitable for most existing trajectory compression algorithms. It is calculated by calculating the Euclidean distance from each compressed trajectory point to the line segment corresponding to the trajectory point.

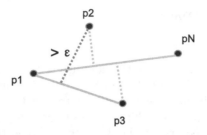

**Fig. 1.** Example of verifying conjecture.

**Definition 7.** *(Perpendicular Euclidean Distance): Given a trajectory segment T, the trajectory segment F is the compressed form of the trajectory T, then for each trajectory point P discarded in the trajectory T, the PED error of the trajectory point p is calculated as follows:*

$$PED(P_m) = \frac{\left\| \overrightarrow{P_sP_m} \times \overrightarrow{P_sP_e} \right\|}{\left\| \overrightarrow{P_sP_e} \right\|}$$

*Where × is the symbol of cross multiplication in vector operations, and ‖ ‖ is the length of a vector.*

### 4.3  ROPW

The proposal of the Reverse OPening Window algorithm greatly avoids the occurrence of the situation described in Sect. 4.1, and therefore maximizes the compression rate of the algorithm. Although its time complexity is still $O(N^2)$, it is difficult to reach the upper bound of its time complexity in practical applications and its average time complexity is $O(NlogN)$. Therefore, its time performance is also better. As the name implies, by reversely traversing the points in the window as the current trajectory end point, the ROPW algorithm avoid the additional conditions generated by the "current end point". In addition, the algorithm uses a jump strategy to find the first trajectory point that does not meet the error limit requirement from back to front, and let it be compressed as the "current trajectory end point", which greatly improves the efficiency of the algorithm. However, because the online trajectory compression algorithm cannot get all the data of the original trajectory, we also use the sliding window method to optimize the implementation of the ROPW algorithm by setting the window size.

Reverse OPening Window algorithm. The input of the algorithm is an original trajectory $T$ and a given error limit $\epsilon$. The output of the algorithm is the compressed trajectory $T'$. The execution process of the ROPW algorithm is given below: (The pseudo code of the algorithm is shown in Algorithm 2).

Assume that the sliding window size is the default size. First, initialize the start and end points of the current compressed trajectory (lines 1–2). Second, put the starting point into the compressed track point set (line 3). Then, check whether the middle track point meets the perpendicular Euclidean distance error requirement from back to front (lines 8–13). If the requirements are not met, the trajectory point that does not meet the requirements is taken as the end point of the current compressed trajectory. Otherwise, set the start point as the current end point, add the end point to the compressed track point set, and set the end point as the last point of the current window (lines 15–21). Line 23 is used to add the last line segment to the final result set.

---

**Algorithm 2:** ROPW

---

**Input**: Raw trajectory $T = \{p_1, p_2, ..., p_N\}$ , error tolerance $\epsilon$

**Output**: $\epsilon$ - Error-bounded trajectory

$$T' = \{p_{i_1}, p_{i_2}, ..., p_{i_n}\}(p_{i_1} = p_1, p_{i_n} = p_N)$$

$LastPoint = T[T.length() - 1];$

$StartPoint = T[0];$

$T' = [StartPoint];$

**while** $StartPoint \neq T.length() - 1$ **do**

    $CurPoint = end - 1;$

    $flag = true;$

    **while** $CurPoint > StartPoint$ $\&\&$ $flag$ **do**

        **if** $PED(StartPoint, CurPoint, LastPoint)$ ¿ $epsilon$ **then**

            $flag = false;$

            $break;$

        **else**

            $CurPoint - -;$

        **end**

    **end**

    **if** $!flag$ **then**

        $LastPoint = CurPoint;$

    **else**

        $StartPoint = LastPoint;$

        $T'.Append(StartPoint);$

        $LastPoint = T[T.length() - 1];$

    **end**

**end**

$T'.Append(T[T.length() - 1]);$

return $T'$;

---

An example of the ROPW algorithm is given below to illustrate the execution process of the algorithm (as shown in Fig. 2). Assume that there are trajectory points $P_1, P_2, ..., P_6$ that arrive in chronological order in the current sliding window. The ROPW algorithm first takes the first point in the window as the starting point and the last point as the "current end point" for compression. Then, the algorithm forms the line segment $\mathcal{L}(P_1, P_6)$ as the current target compressed trajectory, and calculates the perpendicular Euclidean distance $d(P_i, \mathcal{L})$ from the trajectory point $P_5, P_4, P_3$ to the target straight line from back to front. We find that the trajectory point $P_3$ does not meet the error requirements. At this time, the algorithm will continue to compress $P_3$ as the "current end point" (the trajectory point before the trajectory point $P_3$ will not be calculated). Suppose that when $P_1, P_2, P_3$ is compressed into a line segment $\mathcal{L}(P_1, P_3)$ (track point $P_2$ is compressed), at this time, the starting point becomes $P_3$, and the "current end point" becomes $P_6$, and the compression process is continued. Until the starting point becomes the last point in the window, the execution of the trajectory compression algorithm ends.

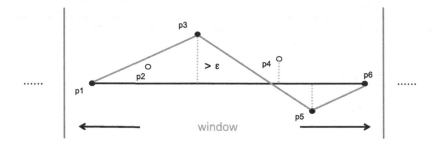

**Fig. 2.** ROPW algorithm execution example.

## 5   Experiments

In this section, we conduct a comprehensive experimental analysis and verification of the ROPW algorithm. We conducted three experiments on three real data and evaluated the performance of each algorithm in terms of compression ratio, compression time, and perpendicular Euclidean distance error under the same error limit.

### 5.1   Experimental Data Set

In this experiment, we selected three real data sets, namely Animal data set, Indoor data set and Planet data set. The trajectories in the three data sets have different characteristics, which can more comprehensively evaluate the quality of the algorithm. The attribute characteristics of the three data sets are shown in Table 1.

**Table 1.** Statistics of data sets.

| | Number of trajectories | Number of points | Average length of trajectories(m) | Average sampling rate(s) | Average distance between two sampling points(m) | Total size |
|---|---|---|---|---|---|---|
| Animal | 327 | 1558407 | 1681350 | 753.009 | 352.872 | 298 MB |
| Indoor | 3578257 | 3634747297 | 43.876 | 0.049 | 0.043 | 224 GB |
| Planet | 8745816 | 2676451044 | 111051.670 | 4.273 | 8.589 | 255 GB |

Animal data set [19], this data set is provided by the Movebank database, which records the migration trajectory data of eight populations of white storks from 2013 to 2014. The trajectories in this data set have the following characteristics: extremely long length, low sampling rate, and long distance between two trajectory points. This data set can reflect the characteristics of sparse trajectory points and fast moving trajectories.

The Indoor data set [20], which contains the trajectories of visitors in Osaka ATC shopping mall. The trajectories in this data set have the following characteristics: short length, high sampling rate and very close distance between two trajectory points. This data set can reflect the characteristics of dense trajectory points and slower moving trajectories.

Planet data set [21], which is GPS data collected through satellites provided by OpenStreetMap. These data are a very large collection, containing the movement data of target objects in several countries. These objects may be walking people, people using vehicles, or some other creatures. The trajectory in this data set has the following characteristics: Long length, the average speed of the object is about $2\,\mathrm{m/s}$, and the speed is relatively moderate. This data set can reflect the characteristics of a more mixed trajectory.

## 5.2  Experiment Settings

In this experiment, we compared the performance of the ROPW algorithm and two currently popular online trajectory compression algorithms on three real data sets. These two algorithms are also error-bounded online trajectory compression algorithms based on perpendicular Euclidean distance measurement error (the two algorithms are BOPW algorithm [10] and OPERB algorithm [13]). In order to conduct a more precise and fair experiment, we used C++ to rewrite all the algorithms to be compared. The implementation of the rewrite is completely based on the original operating logic, and after rigorous testing, it can get exactly the same result.

The data used in the experiment comes from the 3 data sets introduced in Sect. 4.2 (Planet data set, Indoor data set, Animal data set). However, due to the slow running time of some algorithms on some data sets, we randomly sampled the Planet data set and the Indoor data set, and finally got 567 original data tracks on the Planet data set and 1602 on the Indoor data set. Original data track. After we processed the data set, the size of the three data sets was about 57 MB.

This experiment was performed on a 64-bit 4-core 2.1 GHz AMD(R) Ryzen 5 3500u CPU and 8 GB RAM Linux machine. Our algorithm was implemented on Ubuntu 18.04 through C++.

## 5.3  Experiment Results

In this part of the experiment, we mainly evaluate the algorithm in terms of time performance, compression rate, and trajectory error after compression. In the evaluation of time performance, in order to measure the accuracy of the results, we only calculate the time consumed by the algorithm compression process. The compression rate is calculated by dividing the total number of trajectory points in the original trajectory in the data set by the total number of trajectory points after compression. The error of trajectory compression is calculated by the perpendicular Euclidean distance calculation formula.

In the first experiment, we evaluated the compression ratio of the three algorithms under different thresholds. The experimental results are shown in Fig. 3. It is not difficult to find that the compression rate of the ROPW algorithm is significantly higher than other algorithms by given different error precision limits, that is to say, the ROPW algorithm performs more prominently in terms of compression efficiency.

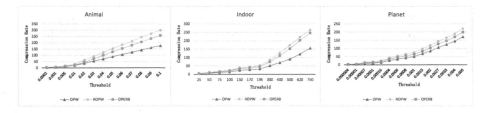

**Fig. 3.** Given different thresholds, the performance of different algorithm compression ratios.

In the second experiment, we evaluate the time performance of the algorithm. By controlling the compression ratio of the algorithm (with an error of $\pm$ 0.5), we observe the changes in the execution time of the algorithm. The experimental results are shown in Fig. 4. By observing the experimental results, there is an intersection between the ROPW algorithm curve and the OPW algorithm curve. In other words, when the threshold is less than the target threshold, the OPW algorithm has higher time performance, and when the threshold is greater than the target threshold, the ROPW algorithm performs better. The reason for this phenomenon is that the worst case of OPW algorithm in time performance occurs when each track point of the input trajectory can be compressed, that is to say, when the threshold is large enough, the OPW algorithm will reach the upper bound of the time complexity. However, the ROPW algorithm is the opposite. The worst case of its time performance occurs when each trajectory point of the input trajectory cannot be compressed, that is to say, the threshold is small enough, the ROPW algorithm will reach the upper bound of the time complexity. In short, the OPW algorithm will decrease the time performance of the algorithm with the increase of the threshold, while the time consumption of ROWP will decrease. The OPERB algorithm is always the fastest. The reason is that the data points are checked only once during the entire process of trajectory simplification. In the following experiments, we can find that the error is relatively large.

In the third experiment, we evaluated the average PED error of each algorithm under the same compression ratio (error within $\pm$ 0.5). The experimental

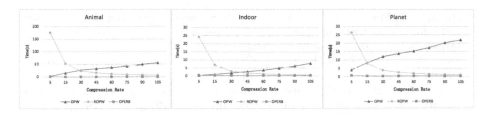

**Fig. 4.** Given different compression ratios, the performance of different algorithm compression ratios.

results are shown in Fig. 5. By observing the experimental results, we can find that in the three real data sets, the average PED error of the ROPW algorithm is optimal, while the average PED error of the OPERB algorithm is relatively large.

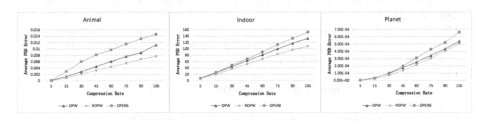

**Fig. 5.** Given different compression ratios, different algorithms PED error performance.

# 6    Conclusion

In this paper, we designed the ROPW algorithm by using the method of traversing the trajectory points in the window in reverse order to avoid the too strict constraints of OPW and other algorithms. The ROPW algorithm is an error-bounded algorithm on the PED error. Due to the reverse traversal feature of the algorithm in the sliding window, it is more suitable for scenes with relatively standardized moving trajectories and dense trajectory sampling. In this scenario, time performance and compression performance can be greatly improved. And the experiment proved that the algorithm has good performance in terms of compression rate, time efficiency and PED error.

# References

1. Nibali, A., He, Z.: Trajic: an effective compression system for trajectory data. IEEE Trans. Knowl. Data Eng. **27**(11), 3138–3151 (2015)
2. Chen, Y., Jiang, K., Zheng, Y., Li, C., Yu, N.: Trajectory simplification method for location-based social networking services. In: SIGSPATIAL GIS Workshop on Location-Based Social Networks, SIGSPATIAL (2009)

3. Hung, C.C., Peng, W.C., Lee, W.C.: Clustering and aggregating clues of trajectories for mining trajectory patterns and routes. VLDB J. **24**(2), 169–192 (2015). https://doi.org/10.1007/s00778-011-0262-6
4. Song, R., Sun, W., Zheng, B., Zheng, Y.: PRESS: a novel framework of trajectory compression in road networks. Proc. VLDB Endow. **7**(9), 661–672 (2014)
5. Richter, K.F., Schmid, F., Laube, P.: Semantic trajectory compression: representing urban movement in a nutshell. J. Spat. Inf. Sci. **4**(2012), 3–30 (2012)
6. Long, C., Wong, R.C.W., Jagadish, H.V.: Direction-preserving trajectory simplification. Proc. VLDB Endow. **6**(10), 949–960 (2013)
7. Bellman, R.: On the approximation of curves by line segments using dynamic programming. Commun. ACM **4**(6), 284 (1961)
8. Practice, E: A Digitized Line or its Caricature. Class, Cartogr (2011)
9. Chen, M., Xu, M., Fränti, P.: A fast O(N) multiresolution polygonal approximation algorithm for GPS trajectory simplification. IEEE Trans. Image Process. **21**(5), 2770–2785 (2012)
10. Keogh, E., Chu, S., Hart, D., Pazzani, M.: An online algorithm for segmenting time series. In: Proceedings of IEEE International Conference Data Mining, ICDM, pp. 289–296 (2001)
11. Liu, J., Zhao, K., Sommer, P., Shang, S., Kusy, B., Jurdak, R.: Bounded quadrant system: error-bounded trajectory compression on the go. In: Proceedings of International Conference on Data Engineering, vol. 2015-May, pp. 987–998 (2015)
12. Liu, J., et al.: A novel framework for online amnesic trajectory compression in resource-constrained environments. IEEE Trans. Knowl. Data Eng. **28**(11), 2827–2841 (2016)
13. Lin, X., Jiang, J., Zuo, Y.: Dual error bounded trajectory simplification. In: Data Compression Conference Proceedings, vol. Part F1277, p. 448 (2017)
14. Cao, W., Li, Y.: DOTS: an online and near-optimal trajectory simplification algorithm. J. Syst. Softw. **126**(M), 34–44 (2017)
15. Zhang, D., Ding, M., Yang, D., Liu, Y., Fan, J., Shen, H.T.: Trajectory simplification: an experimental study and quality analysis. Proc. VLDB Endow. **11**(9), 934–946 (2018)
16. Douglas, D.H., Peucker, T.K.: Algorithms for the reduction of the number of points required to represent a digitized line or its caricature. Can. Cartogr. **10**(2), 112–122 (1973)
17. Cai, Z., Miao, D., Li, Y.: Deletion propagation for multiple key preserving conjunctive queries: approximations and complexity. In: ICDE, pp. 506–517 (2019)
18. Miao, D., Cai, Z., Liu, X., Li, J.: Functional dependency restricted insertion propagation. Theor. Comput. Sci. **819**, 1–8 (2020)
19. https://doi.org/10.5441/001/1.78152p3q
20. https://irc.atr.jp/crest2010_HRI/ATC_dataset/
21. https://wiki.openstreetmap.org/wiki/Planet.gpx

# HTF: An Effective Algorithm for Time Series to Recover Missing Blocks

Haijun Zhang$^{(\boxtimes)}$, Hong Gao$^{(\boxtimes)}$, and Dailiang Jin

Harbin Institute of Technology, Harbin 150001, Heilongjiang, China
1160300327@stu.hit.edu.cn, {honggao,jdl}@hit.edu.cn

**Abstract.** With the popularity of time series analysis, failure during data recording, transmission, and storage makes missing blocks in time series a problem to be solved. Therefore, it is of great significance to study effective methods to recover missing blocks in time series for better analysis and mining. In this paper, we focus on the situation of continuous missing blocks in multivariate time series. Aiming at the blackout missing block pattern, we propose a method called hankelized tensor factorization (HTF), based on singular spectrum analysis (SSA). After the hankelization of the time series, this method decomposes the intermediate result into the product of time-evolving embedding, time delaying embedding, and hidden variables embedding of multivariate variables in the low-dimensional space, to learn the essence of time series. In an experimental benchmark containing 5 data sets, the recovery effect of HTF and other baseline methods in three missing block patterns are compared to evaluate the performance of HTF. Results show that when the missing block pattern is blackout, the HTF method achieves the best recovery effect, and it can also have good results for other missing patterns.

**Keywords:** Multivariate time series · Missing block pattern · Missing value recovery · Tensor factorization

## 1 Introduction

With the rapid development of 5G, big data, and the internet of things, time series data from various sensors, financial markets, meteorological centers, industry monitoring system and the internet is growing at an unprecedented rate. People expect to exploit the huge value that can reveal the development trends in the field of interest behind the data, making the analysis, mining and forecasting of time series becoming a popular topic. Unfortunately, all of this requires time series Completeness as a prerequisite that data in real scenarios often lacks due to network failure, storage equipment malfunction and other situations from time to time. Data missing in a period of time, that is, the existence of missing

The National Key Research and Development Program of China (Grant No. 2019YFB2101902), Joint Funds of the National Natural Science Foundation of China under Grant No. U19A2059.

C. S. Jensen et al. (Eds.): DASFAA 2021 Workshops, LNCS 12680, pp. 29–44, 2021.
https://doi.org/10.1007/978-3-030-73216-5_3

blocks in time series has already been a common and urgent problem waiting to be solved.

Data Quality is a very popular topic recently, and there are a lot of good study result, i.e. [4]. However, when facing the data quality problem with time series, things are different.

In the course of practice, people gradually reach a consensus that traditional statistical method (such as interpolation, multiple imputation [15], etc.) to fill in missing blocks will obscure the hidden pattern of the original data, destroying the dynamic trend of data, followed by the recovery result not conducive to subsequent analysis and prediction [1]. As a result, new methods are gradually emerging.

In view of positions where missing blocks appear, Khayati et al. [9] put forward three kinds of missing patterns: disjoint, overlap and blackout, and then a detailed comparison of the previous work in these three patterns was presented. Their work pointed out that the recovery effect of the same method on the same time series in different missing patterns can have a huge difference, which previous study did not pay attention to. As a result of their research, they suggested the study of missing block recovery in time series data should focus on the performance in different missing patterns, especially in the case of blackout where there is a lack of good methods.

Based on the work of Khayati et al. [9], in particular, there is a gap in the topic of missing block recovery in the blackout missing pattern. An effective algorithm named hankelized tensor factorization (HTF) is proposed, to solve the situation that existing works usually have a bad performance in the case of blackout. The Inspiration of HTF is from the singular spectrum analysis (SSA), a powerful time series analysis technique, decomposes the time data to the weighted sum of a series of independent and explainable components, so as to reconstruct the original sequence well. However, singular spectrum analysis is not suitable for the case of large missing blocks. This paper drawing on the idea of singular spectrum analysis, decomposes the sequence into the product of time-evolving embedding, time delaying embedding and the hidden variables embedding of multi-dimensions in the low-dimensional space. And then a reconstruction of the original sequence is displayed, filling missing values by their reconstruction estimation. In order to obtain the recovery effect of the method, we conducted experiments under a benchmark of 5 datasets, evaluating the performance of HTF and other baseline methods. In conclusion, the contribution of this paper are listed below:

- HTF, an effective algorithm to recover the missing blocks in time series is proposed. With inspiration from SSA, we solve the problem of missing block recovery by an approach named tensor factorization that decomposes high-dimensional data to low-dimensional embeddings to learning high order temporal correlations among the sequential data.
- Based on the work of Khayati et al., the recovery effect of HTF is compared with other baseline methods in three different missing patterns on the

benchmark of 5 datasets inherited from their study, including deep learning methods which was not contained before.

## 2 Related Work

Study of recovery effect under different missing blocks patterns has not been widely paid attention to. The work of Simeng Wu et al. [17] is most similar to this paper, they proposed the HKMF method which focuses on how to fill missing values of blackout in time series, but their work only suits the case that the time series is univariate. One contribution of this paper can be regarded as extending their work to a multivariate case by another technique. More importantly, we find the theoretical foundation that our work may be the best result we can achieve when the time series satisfies linear recursive formula by explaining the fact that HTF and singular spectrum analysis is the same thing in some sense.

Apart from the work mentioned above, Li et al. [10] suggested Dynammo, a method based on the linear dynamic system which makes use of kalman filter to model the time series with the assumption that the observed time sequence is generated by the linear evolution of hidden variables. Kevin et al. [16] recommended a sequence matching method using the most similar subsequence to generate an imputation, by selecting k complete sequence most similar to the sequence just before the missing blocks. Haiang-Fu Yu et al. [18] put forward a matrix factorization method named TRMF to handle the case when the dimension of time series is high. The main contribution of their work is a new kind of regularization that can be explained as a constraint over a temporal graph. Wei Cao et al. [3] proposed a new deep learning model based on the original bidirectional LTSM, modifying the structure to make it more adaptive to the characteristic of time series. A new loss function that minimizes the error both in forward imputation and backward imputation was presented at the same time.

The following sections are arranged as below: Sect. 3 introduces the problem definition and missing block patterns. Section 4 introduces the HTF method and its relation to SSA. At last, in Sect. 5, we evaluate the recovery effect and performance of our method in the benchmark.

## 3 Backgroud

As a type of data, time series can have various auxiliary data associated with it, and timestamp is the most common among them. In this paper, we do not discuss the situation this additional information are contained. Only considering observed multivariate time series sampled evenly can help us simplify the problem and find the essence of this type of data.

### 3.1 Definition

The general problem of missing block recovery for time series is defined as follows:

**Given:** A time series of length $T$, $\boldsymbol{X} = \langle \boldsymbol{x}_1, \boldsymbol{x}_2, ..., \boldsymbol{x}_T \rangle$, where $\boldsymbol{x}_t \in \mathbb{R}^D$ is a D-dimensional column vector for data at time t; an indicating matrix $\boldsymbol{W} = \{0,1\}^{D \times T}$, with $\boldsymbol{W}_{i,t} = 0$ if the i-th dimension of $\boldsymbol{x}_t$ is missing, and $\boldsymbol{W}_{i,t} = 1$ if the data is present.

**Solve:** Estimate the missing values in $\boldsymbol{X}$ indicated by $\boldsymbol{W}$.

We say $\boldsymbol{W}_{i,j}.\boldsymbol{W}_{i,j+1}, ..., \boldsymbol{W}_{j+l-1}$ is a missing block of length $l$, if $\boldsymbol{W}_{i,j}.\boldsymbol{W}_{i,j+1}, ..., \boldsymbol{W}_{j+l-1} = 0$ . If $\boldsymbol{W}_{1,j}, \boldsymbol{W}_{2,j}, ..., \boldsymbol{W}_{D,j} = 0$, we state there is a blackout in the j-th column. The main topic this paper study is when the length $l \geq 20$. The missing pattern of data is defined by the indicating matrix $\boldsymbol{W}$, we notate the max length over all missing blocks as $l_M$, and $h_M$ is the notation of the max height over all the columns of missing blocks that $\boldsymbol{W}_{i,j}, \boldsymbol{W}_{i+1,j}, ..., \boldsymbol{W}_{i+h-1,j} = 0$.

### 3.2   Patterns of Missing Blocks

The pattern of missing blocks is determined by factors like the relative position, the length and the height of missing blocks, and so on. Khayati et al. argued the following three possible pattern types of missing, as shown in Fig. 1, $X_1, X_2, X_3$ are the univariate time series of same length column by column, the missing blocks in every sequence is marked as grey:

1) Disjoint: In this case, missing blocks do not intersect on each other in the same time span. That is to say for each missing block, the sequences from other variables at the same time is complete, equivalent to $h_M = 1$. It is the simplest case in all three types.
2) Overlap: In this case, missing blocks intersect each other partly, and there would be at most $D - 1$ variables missing at the same time, equivalent to $2 \leq h_M < D$. It is a relatively complex case.
3) Blackout: In this case, missing blocks coincide with each other at some time interval, equivalent to $h_M = D$. Generally speaking, this case is very hard to recover when the missing length is large. So in this paper, we constrain $20 \leq l_M \leq 100$. It is the most challenging case.

In the real scenarios, these three patterns of missing seldom appear alone, a problem that all three types mix together usually happens when data transmission in a bad network.

## 4   Hankelized Tensor Factorization

The Hankelized Tensor Factorization algorithm we proposed in this paper, mainly focuses on the Blackout missing pattern. In this case, the indicating matrix $\boldsymbol{W}$ contains some continuous columns all filled with zero, like $\boldsymbol{W}_{:,j}, \boldsymbol{W}_{:,j+1}, ..., \boldsymbol{W}_{:,j+l-1} = 0$.

The details of HTF are listed below. First, apply hankelization on the input time series matrix $\boldsymbol{X} \in \mathbb{R}^{D \times T}$ carrying the missing position. Then, a novel tensor factorization approach inspired by gradient descent will be carried out on the

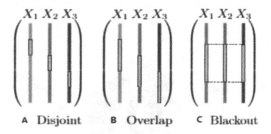

**Fig. 1.** Different types of missing block pattern (the missing marked as grey)

result of the first step, causing the high dimensional data with missing transformed to some low dimensional embeddings. Thirdly, a reconstruction based on the embedding obtained just before will be multiplied to make a reconstruction of the hankelized time series with missing imputed. In the end, we will do an inverse version of hankelization to turn the reconstruction to an estimation matrix, with each missing value in the original data matrix being imputed.

### 4.1 Methodology

In the following subsections, the detail of the procedures listed above will be explained. For the simplicity of mathematical expression, we give the following arithmetic definition:

**Definition 1.** $A, B$ *are matrices of size* $m \times n$,
$$(A \odot B)_{ij} \triangleq A_{ij} \times B_{ij}$$

**Definition 2.** $x, y$ *are vectors of length* $m$ *and* $n$,
$$x \otimes y \triangleq \begin{bmatrix} x_1 y_1 & x_1 y_2 & \cdots & x_1 y_n \\ x_2 y_1 & x_2 y_2 & \cdots & x_2 y_n \\ . & . & \cdots & . \\ x_m y_1 & x_m y_2 & \cdots & x_m y_n \end{bmatrix}$$

#### A. Tensor Hankelization

The first step of HTF is to hankelize the input time series matrix $X$. Compared to the way Hassani et al. [7] proposed that does hankelization for the sequence $X_{i,:}$ of each variable i, then merge to a compound matrix by put all the results in the row order or in the column order, generating a much larger matrix compared with the original $X$. We give a way that makes the matrix to a tensor, just like what Hassani et al. do when it hankelizes the univariate sequence as defined. The difference between these two ways is shown in Fig. 2. We believe the result tensor can be represented by more explainable, effective and independent components in our approach, so the essential can be learned better and the reconstruction result has a better recovery effect. The comparison experiments in Sect. 5.2.C support our point of view.

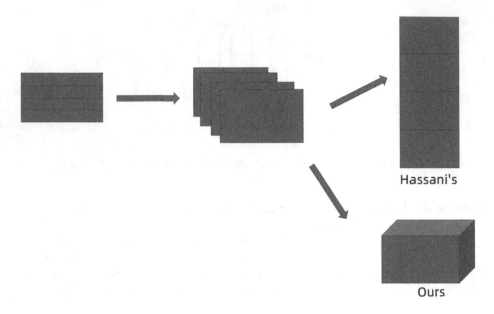

**Fig. 2.** Different ways of hankelization (Hassani's and ours)

Tensor hankelization can be thought of as doing the process of formula (1) for the sequence vector $X_{i,:}$ of each variable in time series matrix $X$, and then put the result in the third dimension *height*, vertical to the dimension of time-evolving (row) and multi-variables (column). We notate $z_i, z_{i+1}, ..., z_{i+L-1}$ as time evolving vector list of length L, and $z_j, z_{j+1}, ..., z_{j+K-1}$ as time delaying vector list of length K. The value of K must satisfy $K \geq l_M + 1$, which $l_M$ has been defined in the Sect. 3.1. For the recovery effect of our method, the value of K should be appropriately large. The result of this step can be seen as a map from all matrix elements to all of the tensor, which is shown in Eq. (2).

$$H(z) = H([z_1, z_2, ..., z_3]) = \begin{bmatrix} z_1 & z_2 & ... & z_L \\ z_2 & z_3 & ... & z_{L+1} \\ . & . & ... & . \\ z_K & z_{K+1} & ... & z_T \end{bmatrix} \quad (1)$$

$$H_K(X)_{i,j,k} = X_{i,j+k-1}, H_K(X) \in \mathbb{R}^{D \times K \times L} \quad (2)$$

**Tensor Factorization**

The next step of HTF is to decompose the tensor coming from the tensor hankelization. By decomposing the result to time-evolving embedding, time delaying embedding and multi-variate embedding, the essential of time series can be represented independently and explainably. Another important advantage of factorization is that the embeddings which map the components of time series to low dimensional space are dense despite the original in the high dimension is sparse, the principle behind similar to that the product $Z = xy^T$ of vector $x$ ,

$y$ of length $m$ and $n$ generates a matrix of $m \times n$ elements, while solving $x, y$ does not need as many as $m \times n$ elements of $Z$. This paper takes the following formula to do tensor factorization,

$$H_K(X) = Y \approx \sum_{r=1}^{R} A_{:,r} \otimes B_{:,r} \otimes C_{:,r} \tag{3}$$

which $A, B$ and $C$ represent the time-evolving trend, time delaying trend and the hidden variables of multi-variables in the low R-dimension space.

In order to solve $A, B, C$, we select the method based on the stochastic gradient descent from machine learning, compared with other approaches to solve tensor factorization problem, such as alternating least squares [8]. The reason is that other methods may be harmful to the result because of their default initialization of zero when dealing with missing values which is no difference between a tensor with no missing values but is filled with zero and the tensor which is missing at the same position and initialized with zero. We propose the following function:

$$\mathcal{L}_{A,B,C(Y)} = \sum_{(i,j,k)\in\Omega} (Y_{i,j,k} - \sum_{r=1}^{R} A_{i,r} B_{j,r} C_{k,r})^2 + \lambda_s R_T(A, B, C) + \lambda_r R(A, B, C) \tag{4}$$

where $\Omega$ is the set of indexes correspond to the observed elements in $Y$, which is generated from $W$. $Y_{i,j,k}$ is the $(i, j, k)$th element of $Y$. $R_T(A, B, C), R(A, B, C)$ are two regularizers defined in Eq. (6) and (7), respectively, with $\lambda_r, \lambda_s$ being the coefficients. Given Eq. (4), the task of learning $A, B, C$ is achieved by solving:

$$\langle A, B, C \rangle = argmin_{A,B,C} \mathcal{L}_{A,B,C}(Y) \tag{5}$$

More specifically, the objective function in Eq. (4) contains four components as follows:

First, $\sum_{(i,j,k)\in\Omega} (Y_{i,j,k} - \sum_{r=1}^{R} A_{i,r} B_{j,r} C_{k,r})^2$ quantifies the error of $Y \approx \sum_{r=1}^{R} A_{:,r} \otimes B_{:,r} \otimes C_{:,r}$.

Second, $R_T(A, B, C)$ is the temporal regularizer defined as

$$R_T(A, B, C) = \sum_{i=1}^{D} \sum_{j=1}^{K} \sum_{k=1}^{L} (\sum_{r=1}^{R} A_{i,r} B_{j,r} (C_{k,r} - C_{k-1,r}))^2 \tag{6}$$

which restricts the adjacent element of the solution of $\hat{Y} \triangleq \sum_{r=1}^{R} A_{:,r} \otimes B_{:,r} \otimes C_{:,r}$ should be close.

Third, $R(A, B, C)$ is a $L_2$ regularizer defined as:

$$R(A, B, C) = ||A||_F^2 + ||B||_F^2 + ||C||_F^2 \tag{7}$$

which solve the overfitting problem of machine learning.

Solving Rule

To find the optimized $A, B, C$ that minimizes the objective function as shown in Eq. (4), we adopt the stochastic gradient descent approach with the following update rules:

$$\begin{aligned}
A_{i,:}^{new} = A_{i,:} + \eta[err_{i,j,k}(B_{j,:} \odot C_{k,:}) - \lambda_r A_{i,:} \\
- \lambda_s(\Gamma \odot (\Delta \odot C_{k,:})) \\
- \lambda_s(\Gamma' \odot (\Delta' \odot C_{k,:}))]
\end{aligned} \tag{8}$$

$$\begin{aligned}
B_{j,:}^{new} = B_{j,:} + \eta[err_{i,j,k}(A_{i,:} \odot C_{k,:}) - \lambda_r B_{j,:} \\
- \lambda_s(\Gamma \odot (A_{i,:} \odot C_{k,:})) \\
- \lambda_s(\Gamma' \odot (A_{i,:} \odot C_{k,:}))]
\end{aligned} \tag{9}$$

$$\begin{aligned}
C_{k,:}^{new} = C_{k,:} + \eta[err_{i,j,k}(A_{i,:} \odot B_{j,:}) - \lambda_r C_{k,:} \\
- \lambda_s(\Gamma \odot (A_{i,:} \odot \Delta)) \\
- \lambda_s(\Gamma' \odot (A_{i,:} \odot \Delta'))]
\end{aligned} \tag{10}$$

which:

$$\Delta = B_{j,:} - B_{j-1,:}$$

$$\Gamma = A_{i,:}(\Delta \odot C_{k,:})$$

$$\Delta' = B_{j+1,:} - B_{j,:}$$

$$\Gamma' = A_{i,:}(\Delta' \odot C_{k,:})$$

$$err_{i,j,k} = Y_{i,j,k} - \sum_{r=1}^{R} A_{i,r} B_{j,r} C_{k,r}$$

## 4.2   Relation to Singular Spectrum Analysis

The HTF algorithm this paper put forward can be thought of as the tensor factorization version of singular spectrum analysis. Hassani et al. [6] pointed out, as a novel and powerful time series analysis technique, which can be used to process the time series from dynamic system, signal processing, economy and many other spheres, singular spectrum analysis decomposes the original data into the sum of a series of independent and explainable components, such as smooth trend, period components, quasi-period components and non-structural noise.

It generally consists of two steps: decomposition and reconstruction. The first step is to hankelize the time series, then do singular value decomposition: $H(X) \approx \lambda_1 U_1 V_1 + \lambda_2 U_2 V_2 + ... + \lambda_n U_n V_n$. The second step is to select eigenvalues that have large impact on the reconstruction by $\sum_{l=1}^{r} \lambda_l / \sum_{l=1}^{n} \lambda_l \geq$ *threshold*, with the purpose of denoising data. So the reconstruction of $H(X)$ is $r < n, \hat{H}(X) = \lambda_1 U_1 V_1 + ... + \lambda_r U_r V_r$. At the last, get the estimation of $X$ by diagonal average approach.

Golyandina et al. [5] suggested singular spectrum analysis is the optimal reconstruction for the time series satisfying the Eq. (11):

$$x_n = c_1 x_{n-1} + c_2 x_{n-2} + \dots + c_k x_{n-k} \qquad (11)$$

This property makes it powerful for lots of time series since many time-evolving processes in real scenarios can be represented or approximately represented by this formula.

Due to the characteristic of singular value decomposition (SVD) that treats the missing value of matrix to decompose as zero or some constant, singular spectrum analysis always takes this default initialization of missing as part of the trend in data, resulting in the lack of capacity of the method to recover missing blocks. Although Mahmoudvand et al. [13] proposed a kind of improvement based on singular spectrum analysis, it has a strict demand on the shape of missing, which can not deal with arbitrary shapes of missing blocks. However, the tensor factorization method proposed by this paper which uses stochastic gradient descent does not take missing into consideration, making it adapt to the problem.

Since apart from the approach to decompose the original matrix in the procedure of decomposition, the rest steps of the SSA and HTF are almost the same, there is no doubt that the HTF has the same capacity as SSA if tensor factorization technique can reach the effect of SVD which is shown in [12].

# 5   Experiments

## 5.1   Experiment Setting

In this section, we conducted experiments using real-world data sets to evaluate the performance of our approach based on the benchmark of Khayati et al., appending the comparison of deep learning approach to get a more exhaustive evaluation.

The environment of the experiments is a 4-core Intel i5-7300HQ CPU, NVIDIA GTX 1050 GPU. The implementation of all methods mentioned in experiments are coded with Pytorch1.5 and Python3.6.

**A. Data Sets and Experiment Methodology**
Five real-world data sets are used for the experiments[1]:

- Air Quality: contains air quality data in a city of Italy from 2004 to 2005. There is a periodic trend and jumping changes in the data. We cut it to 10 variables, 1000 time points.
- Electricity: contains family electricity usage in France per minute from 2006 to 2010. There is a strong time varying property in it. We cut it to 20 variables, 2000 time points.

---

[1] All 5 datasets can be found on https://archive.ics.uci.edu/ml/datasets.php.

- Temperature: contains temperature data sampled in weather stations across China from 1960 to 2012. There is a high correlation between variables. We cut it to 50 variables, 5000 time points.
- Gas: contains gas data collected by a chemical laboratory in the United States in a gas extraction platform from 2007 to 2011. There is a huge difference in correlation between variables. We cut it to 100 variables, 1000 time points.
- Chlorine: contains chlorine data from 166 intersections in a water system, sampled every 5 min during 15 days. There is a very smooth trend in data, with a strong periodicity. We cut it to 50 variates, 1000 time points.

The evaluation metric we used in this paper is:

$$NRMSE(\boldsymbol{X}, \hat{\boldsymbol{X}}) = \sqrt{\frac{1}{|\Omega_{test}|} \sum_{(i,j)\in\Omega_{test}} (\boldsymbol{X}_{i,j} - \hat{\boldsymbol{X}}_{i,j})^2 /}$$

$$(\frac{1}{|\Omega_{test}|} \sum_{(i,j)\in\Omega_{test}} |\boldsymbol{X}_{i,j}|)$$

where $\hat{\boldsymbol{X}}$ is the result that the missing in the original matrix is filled with estimation. $\Omega_{test}$ is the set of all indices that value is missing. The smaller the NRMSE of the algorithm is, the better the recovery effect is.

The reason why use NRMSE is that not only the recovery effect of different methods on the same data set can be compared, but also the effect of one method on different datasets can be evaluated. It has wide use in papers like [11,14,18], becoming a popular metric to evaluate the effect of a method to recover missing values.

To demonstrate the effectiveness of HTF, we compare its performance against the following baseline approaches: 1) TKCM [16]; 2) Dynammo; [10]; 3) TRMF [18]; 4) BRITS [3]. The brief statement of their work has been introduced in Sect. 2.

The original dataset in the benchmark is complete, we simulate the different missing patterns by generating the related indicating matrix $\boldsymbol{W}$, and the missing blocks only appear in the middle of the series to avoid the end case which is similar to forecasting tasks that the methods we experiment usually perform badly. Because of the randomness of result caused by the position of missing generated randomly (There is a large difference between the smooth sequence piece and steep sequence piece in the recovery effect.), we conduct a duplicate experiment way to reduce the randomness.

## B. Empirical Parameters Setting

$K, L = T + 1 - K$ in the step of hankelization, $\lambda_r, \lambda_s, \eta$ in the step of update rule, and the number of iterations $iter\_num$ that applying update rule are all the hyper-parameters that we need to set empirically.

Before we compare our method with others, we carried out several trials to select these hyper-parameters of HTF to avoid bad performance of our method. So in the end, we set $\lambda_s = \lambda_r \in 0.0001, 0.001, 0.01$ for different datasets, $R =$

$\lceil 1.2 * D \rceil$, $K = \lceil 1.5 * l_M \rceil$, and the number of iterations of gradient descent is $iter\_num = 100$. One process that we did experiments to find a good parameter is shown in Sect. 5.2.D. Also, we do the same process for each baseline method if it has selective parameters.

## 5.2   Recovery Effect

### A. Performance on Disjoint and Overlap

In this set of experiments, when generating indicating matrix $W$, the number of involved variables is at least 4, at most about 40% of total. In the pattern of Overlap, $h_M = 2$.

From Table 1 and Table 2, it can be concluded that overlap case is more difficult to recover effectively than disjoint, and almost all methods have the a degree of performance reduction on all data sets. Many methods can achieve good results when complete sequences of variables exists. The reason is that the complex dependence of variables in time series at each timestamp can be learned based on the complete sequence of certain variables as a reference. Among them, the BRITS algorithm based on deep learning achieved the best performance, and HTF did not get a good enough result. The essential of this phenomenon is that HTF focuses on learning the autocorrelation of variables while other methods more focus on the correlations between variables.

**Table 1.** Performance on disjoint pattern

| Dataset | TKCM | Dynammo | TRMF | BRITS | HTF |
|---|---|---|---|---|---|
| Air quality | 0.916 | 0.707 | 0.494 | **0.018** | 0.620 |
| Electricity | 0.838 | 0.867 | 0.811 | **0.683** | 0.704 |
| Chlorine | 0.323 | 0.034 | **0.015** | 0.047 | 0.129 |
| Gas | 1.740 | 0.384 | 0.081 | **0.068** | 0.467 |
| Temperature | 0.426 | 0.133 | 0.111 | **0.099** | 0.281 |

**Table 2.** Performance on overlap pattern

| Dataset | TKCM | Dynammo | TRMF | BRITS | HTF |
|---|---|---|---|---|---|
| Air quality | 1.244 | 0.770 | 0.672 | **0.578** | 0.714 |
| Electricity | 0.863 | 1.347 | 0.873 | **0.489** | 0.674 |
| Chlorine | 0.335 | 0.095 | **0.061** | 0.081 | 0.165 |
| Gas | 1.726 | 0.476 | 0.383 | **0.073** | 0.452 |
| Temperature | 0.442 | 0.481 | 0.430 | **0.140** | 0.288 |

## B. Performance on Blackout

In this set of experiments, generating indicating matrix $W$, the number of involved variables is about 40% of total. the length of each missing blocks is 100.

From Table 3, we can find the HTF algorithm we proposed achieved the best NRMSE in all datasets. All other methods perform really bad, they lost the ability to recover the missing, which can be confirmed both in nrmse result and Fig. 3. In [2], authors discussed the reason why dynammo fails in the blackout pattern. Due to only extracting the most similar subsequence to fill the missing, TKCM can not adapt the local dynamic well, so the shifting phenomenon is very serious. The matrix factorization technique TRMF takes can not handle the situation that there is a column in the matrix that is missing, which caused one column of embedding not having information to update its value. So the imputation may look like white noise with little variance with random initialization. For BRITS, long intermediate process missing, a kind of gradient vanishing is hard to solve for neural networks, which is the inherent defect of deep learning, and there may only be fluctuation values near the end of missing blocks, while the middle of the block is smooth.

Figure 3 shows the picture of what the recovery result looks like compared with the original data in the blackout missing pattern on the Electricity dataset, which is evidence of the analysis we give before. The time points of the missing block are from 325 to 425, and the red one in the figure is the original data, while the blue is the image of recovery. It is not hard to see only our method return the result looks reflecting the characteristic of original time series and there is no much difference between the recovery and truth.

**Table 3.** Performance on blackout pattern

| Dataset | TKCM | Dynammo | TRMF | BRITS | HTF |
|---------|------|---------|------|-------|------|
| Air quality | 1.254 | 1.070 | 1.252 | 1.797 | **0.763** |
| Electricity | 1.235 | 1.165 | 1.273 | 1.604 | **0.652** |
| Chlorine | 0.521 | 0.691 | 0.517 | 0.516 | **0.275** |
| Gas | 1.459 | 1.235 | 1.202 | 1.737 | **0.606** |
| Temperature | 1.108 | 0.542 | 1.123 | 0.593 | **0.219** |

## C. Influence of Different Ways of Hankelization

The two ways of hankelization introduced in 4.1.A is valid in theory, and the effect can hardly be analyzed by mathematical. So we compared these two ways in experiments, we name the way we use tensor hankelization HTF, and the way we use the matrix hankelization HTF-M. Which can be seen from Table 4, HTF outperforms HTF-M in almost all cases, though the advantage is not obvious, this kind of difference is enough in Statistical.

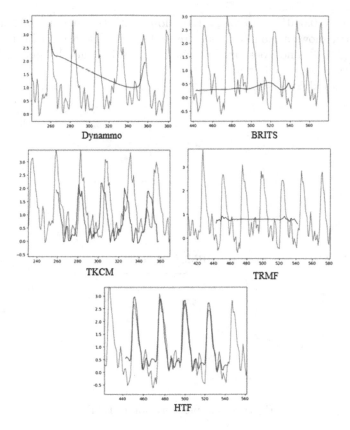

**Fig. 3.** Recovery effect of different methods on Electricity dataset

**Table 4.** Influence of different ways of hankelization

| Dataset | HTF | HTF-M |
|---|---|---|
| Air quality | **0.763** | 0.798 |
| Electricity | **0.652** | 0.658 |
| Chlorine | **0.275** | 0.293 |
| Gas | **0.606** | 0.632 |
| Temperature | **0.219** | 0.239 |

## D. Influence of parameter K

The selection of K may have a big impact in performance , and this set of experiments aims to find a good K. Because of K should satisfy $K \geq l_M + 1$, we set $K = \lceil \xi l_M \rceil (\xi > 1)$. By adjust the value of $\xi$, we can see the impact of different K.

Figure 4 shows the result of the influence of different $\xi$ on NRMSE and time cost every iteration in Air Quality and Electricity data with parameters fixed with $\lambda_s = \lambda_r = 0.0001, iter\_num = 100$. along with the increase of $\xi$, time cost each iteration shows a monotonous increase trend, while the NRMSE decreases when $\xi \in (1, 1.6]$, and goes smooth when $\xi > 1.6$. A good value of k should balance both recovery effect and time cost the program does, so $\xi = 1.5, K = 1.5 * l_M$ is a good selection.

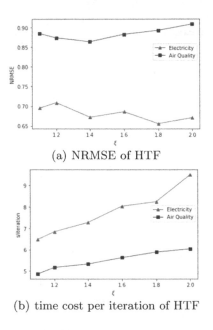

(a) NRMSE of HTF

(b) time cost per iteration of HTF

**Fig. 4.** The recovery effect and time cost of HTF related to $\xi$

### 5.3   Result Conclusion

In conclusion, the HTF algorithm achieved an excellent result in blackout pattern compared to other baseline methods, while also performs well in other missing patterns. It is an effective algorithm for time series to recover missing blocks.

## 6   Conclusion

This paper presents a novel tensor factorization-based approach called HTF to address the challenging problem of estimating the values of missing blocks in time series, especially the blackout missing pattern by decomposing multivariate data sequence into time-evolving embedding, time delaying embedding and multivariate embedding. Following this idea, the method first transforms a time series matrix into a hankelized version. Through the experiments on the benchmark inherited from the previous work, we demonstrate the effectiveness of HTF by comparing its performance against state-of-art baseline approaches.

For future work, we plan: 1) design a deep learning method to handle the Blackout missing pattern Specifically, aiming to deal with the case that time series does not satisfy linear recursive formula. 2) Solving the problem when data is sampled unevenly.

# References

1. Aydilek, I.B., Arslan, A.: A hybrid method for imputation of missing values using optimized fuzzy c-means with support vector regression and a genetic algorithm. Inf. Sci. (Ny) **233**, 25–35 (2013)
2. Cai, Y., Tong, H., Fan, W., Ji, P.: Fast mining of a network of coevolving time series, pp. 298–306 (2015). https://doi.org/10.1137/1.9781611974010.34
3. Cao, W., Wang, D., Li, J., Zhou, H., Li, Y., Li, L.: Brits: bidirectional recurrent imputation for time series. In: Proceedings of the 32nd International Conference on Neural Information Processing Systems, NIPS 2018, pp. 6776–6786. Curran Associates Inc., Red Hook (2018)
4. Miao, D., Cai, Z., Li, J., Gao, X., Liu, X.: The computation of optimal subset repairs. Proc. VLDB Endow. **13**, 2061–2074 (2020)
5. Golyandina, N., Korobeynikov, A.: Basic singular spectrum analysis and forecasting with R. Comput. Stat. Data Anal. **71**, 934–954 (2014). https://doi.org/10.1016/j.csda.2013.04.009
6. Hassani, H.: Singular spectrum analysis: Methodology and comparison. University Library of Munich, Germany, MPRA Paper 5 (2007)
7. Hassani, H., Mahmoudvand, R.: Multivariate singular spectrum analysis: a general view and new vector forecasting approach. Int. J. Energy Stat. **01**, 55–83 (2013). https://doi.org/10.1142/S2335680413500051
8. Hidasi, B., Tikk, D.: Fast ALS-based tensor factorization for context-aware recommendation from implicit feedback. In: Flach, P.A., De Bie, T., Cristianini, N. (eds.) ECML PKDD 2012. LNCS (LNAI), vol. 7524, pp. 67–82. Springer, Heidelberg (2012). https://doi.org/10.1007/978-3-642-33486-3_5
9. Khayati, M., Lerner, A., Tymchenko, Z., Cudre-Mauroux, P.: Mind the gap: an experimental evaluation of imputation of missing values techniques in time series. Proc. VLDB Endow. **13**, 768–782 (2020). https://doi.org/10.14778/3377369.3377383
10. Li, L., Mccann, J., Pollard, N., Faloutsos, C.: DynaMMo : mining and summarization of coevolving sequences with missing values. In: KDD 2009 Proceedings of the 15th ACM SIGKDD International Conference on Knowledge Discovery and Data Mining, pp. 507–516 (2009). https://doi.org/10.1145/1557019.1557078
11. Li, Z., Ye, L., Zhao, Y., Song, X., Teng, J., Jin, J.: Short-term wind power prediction based on extreme learning machine with error correction. Prot. Control Modern Power Syst. **1** (2016). https://doi.org/10.1186/s41601-016-0016-y
12. Maehara, T., Hayashi, K., Kawarabayashi, K.T.: Expected tensor decomposition with stochastic gradient descent. In: Proceedings of the Thirtieth AAAI Conference on Artificial Intelligence, AAAI 2016, pp. 1919–1925. AAAI Press (2016)
13. Mahmoudvand, R., Rodrigues, P.: Missing value imputation in time series using singular spectrum analysis. Int. J. Energy Stat. **04**, 1650005 (2016). https://doi.org/10.1142/S2335680416500058
14. Pennekamp, F., et al.: The intrinsic predictability of ecological time series and its potential to guide forecasting. Ecol. Monogr. **89**, e01359 (2019)

15. Sterne, J.A.C., et al.: Multiple imputation for missing data in epidemiological and clinical research: potential and pitfalls. BMJ **338** (2009). https://doi.org/10.1136/bmj.b2393. https://www.bmj.com/content/338/bmj.b2393
16. Wellenzohn, K., Böhlen, M.H., Dignös, A., Gamper, J., Mitterer, H.: Continuous imputation of missing values in streams of pattern-determining time series. In: EDBT (2017)
17. Wu, S., Wang, L., Wu, T., Tao, X., Lu, J.: Hankel matrix factorization for tagged time series to recover missing values during blackouts. In: 2019 IEEE 35th International Conference on Data Engineering (ICDE), pp. 1654–1657 (2019). https://doi.org/10.1109/ICDE.2019.00165
18. Yu, H.F., Rao, N., Dhillon, I.S.: Temporal regularized matrix factorization for high-dimensional time series prediction. In: Lee, D.D., Sugiyama, M., Luxburg, U.V., Guyon, I., Garnett, R. (eds.) Advances in Neural Information Processing Systems, vol. 29, pp. 847–855. Curran Associates, Inc. (2016). http://papers.nips.cc/paper/6160-temporal-regularized-matrix-factorization-for-high-dimensional-time-series-prediction.pdf

# LAA: Inductive Community Detection Algorithm Based on Label Aggregation

Zhuocheng Ma[1], Dan Yin[1]([✉]), Chanying Huang[2], Qing Yang[1],
and Haiwei Pan[1]

[1] Harbin Engineering University, Harbin, China
{dymzcc,yindan,yangqing,panhaiwei}@hrbeu.edu.cn
[2] Nanjing University of Science and Technology, Nanjing, China
hcy@njust.edu.cn

**Abstract.** The research task of discovering nodes sharing the same attributes and dense connection is community detection, which has been proved to be a useful tool for network analysis. However, the existing approaches are transductive, even for original networks with structures or attributes changed, retraining was required to get the results. The rapid changes and explosive growth of information makes real-world application have great expectations for inductive community detection models that can quickly obtain results. In this paper, we proposed **L**abel **A**ggregation **A**lgorithm (LAA), an inductive community detection algorithm based on label aggregation. Like the traditional label propagation algorithm, LAA uses labels to indicate the community to which the node belongs. The difference is that LAA takes the advantages of network representation learning's ability for information aggregation to generate nodes' final labels by aggregating the labels propagated from local neighbors. The experimental results show that LAA has excellent generalization capabilities to handle overlapping community detection task.

## 1 Introduction

Networks provide a natural and powerful way to represent relational information among data objects from complex real-world systems. Due to the flexibility of using networks to model data, many real-world applications mine information by analyzing networks. One way to analyze networks is to identify the groups of nodes that share highly similar properties or functions, such as proteins with similar functions in biological networks, users with close relationships in social networks, papers in the same scientific fields in citation networks. The research task of discovering such groups of nodes is called the community discovery problem [7].

In recent years, rapid progress in network representation learning has fueled the research of community detection [2]. Since network data are high-dimensional and non-Euclidean, it is not easy to use them directly in real-world applications. However, by representation learning, every node in the network can be embedded

C. S. Jensen et al. (Eds.): DASFAA 2021 Workshops, LNCS 12680, pp. 45–56, 2021.
https://doi.org/10.1007/978-3-030-73216-5_4

into a low-dimensional space, i.e. learning vector representations for each node. The learned node representation vectors can benefit a wide range of network analysis tasks such as node classification, link prediction, community detection and so on.

Several methods have been proposed, which combine community detection with network representation learning did show great promise in designing more accurate and more scalable algorithms [15]. However, community detection in previous models focuses on a single and fixed network. Even for dynamic networks with the same sizes and attributes, retraining is required to get the results. The rapid changing and explosive growth of information make real-world applications have great expectations for inductive community detection models that can quickly obtain results.

To address this problem, we propose LAA, an inductive community detection algorithm based on label aggregation. LAA uses AGM (Affiliation Graph Model) to assign a label for each node, and the label indicates the membership of the node belonging to the community. Each node receives the label propagated from the local neighborhood according to the similarity of the neighboring nodes. Through training a series of aggregation functions, the received label and its own label are aggregated as the final label of the node. The aggregation function explains the influence of the membership information of neighboring nodes on which communities the target node will be divided into, i.e., the relationship between the generation of communities and the network topology. We are based on the assumption that no matter how the structure changes, such a relationship will not change. In addition, in the networks with the same size and attributes, such relationships are supposed to be similar. Therefore, through the trained aggregation function, the labels corresponding to the untrained nodes in the networks can be obtained, and the communities corresponding to the nodes in the network are determined.

## 2    Related Work

**Community Detection.** In recent years, several community detection models have been proposed based on different perspectives. One of the popular directions is to design some indicators to measure the quality of the community [16]. By optimizing these indicators, the community structure can be discovered. The most common method is to optimize the modularity, and obtain the global optimal modularity by making a locally optimal choice. Another popular direction is to use matrix factorization technology to decompose the adjacency matrix of the network or other correlation matrices to obtain the node-community correlation matrix [9,10,14]. However, due to the limitation of the capacity of the bilinear model and the complexity of matrix factorization, these models cannot be extended to lager networks. In addition, some researchers use generative models to discover communities [1,18]. The basic idea is using generative models to describe the process of network generation, and transform community detection problem into reasoning problem. However, due to the complex reasoning, the computational complexity of these methods is also high.

**Label Propagation Algorithm.** Label propagation algorithm (LPA) [12] has proved to be a popular choice for community detection task for its efficiency and ease of implementation. The basic idea of LPA is that nodes propagate labels to neighboring nodes according to the similarity of nodes, and neighboring nodes update their own labels according to the received labels. LPA detects communities by the iterative propagation of node labels until convergence, and nodes with the same label belong to a community. However, LPA still has some shortcomings: it cannot obtain accurate results when dealing with the overlapping community; the randomness of label propagation will lead to unstable results. In order to overcome the shortcoming of LPA, some enhanced algorithms have been proposed, such as SLPA [17], COPRA [3].

**Network Representation Learning.** Network representation learning is one of the most popular research directions in recent years, and many network representation learning models have been proposed. Most of these models can be divided into two categories. One preserves the information of the network by the distance between the node vectors, and the nodes' representation vectors usually do not contain any information, such as Deepwalk, Node2vec, LINE [4,11,13]. Another one aggregates the information of the neighborhood based on the structure of the network so that the representation vectors of nodes not only represent the attributes of the nodes, but also retains the structure information of the network, such as GraphSAGE, GCN [5,8].

**Community Detection Based on Network Representation Learning.** Most of existing community detection models based on network representation learning use the network representation learning algorithm to obtain the representation vector of the nodes, and then performs clustering calculation or matrix decomposition calculation. However, in the process of network representation learning, existing a problem of structural information loss. Therefore, when dealing with dense and overlapping networks, it is often impossible to achieve ideal performance. In order to address this problem, some researchers have carried out network representation learning and community detection tasks at the same time, and combine community structure and node representation for optimization, such as CommunityGAN [6]. However, such algorithms are often more complex, costly in training, and do not have the ability to infer and summarize.

## 3   Problem Definition

Given a network $\mathcal{G} = (\mathcal{V}, \mathcal{E})$, where $\mathcal{V} = (v_1...v_{|\mathcal{V}|})$ represents the nodes in the network, $\mathcal{E} = \{e_{ij}\}$ are their connections. $N(v)$ represents the neighborhood of node $v$, and $N_s(v)$ represents the sampled neighborhood of node v based on the node similarity sampling. We denote $\mathcal{L}_v$ as the label of node $v$ which indicate the community to which the node v belongs. We denote $\mathcal{L}_v^k$ as the label of node $v$ after $k$ labels aggregation, and $\mathcal{L}_v^0$ as the initial label. LAA model takes the initial label $\mathcal{L}^0 = (\mathcal{L}_1^0, ...\mathcal{L}_{|\mathcal{V}|}^0)$ as input, aggregates the labels through a serious of trained aggregation function, and get the final label $\mathcal{L}^k = (\mathcal{L}_1^k, ...\mathcal{L}_{|\mathcal{V}|}^k)$.

# 4  LAA Framework

The main idea of LAA is to use the powerful information aggregation capabilities of the network representation learning algorithm to aggregate label information, and understand how labels propagating in the network, so as to obtain the generalization ability of the model. In this section, we present the LAA model. The main idea of our model is to combine the efficiency and ease of implementation of label propagation and the information aggregation power of the network representation learning.

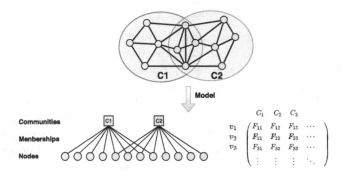

**Fig. 1.** AGM framework

## 4.1  AGM

In traditional label propagation algorithms, the definition and initialization of labels are important, and directly affect the results. The LAA model draws on the related concepts and ideas of AGM when defining and initializing labels. AGM is proposed to solve the problem of community detection in overlapping networks. It is based on the idea that a network is generated according to the community affiliation model. The more common communities that two nodes belong to, the greater the probability that they are connected. The framework of AGM is shown in Fig. 1, which can be expressed in the form of a non-negative membership matrix, or in the form of a bipartite network between nodes and communities. In AGM, each node can belong to 0, 1, or multiple communities, and AGM denotes $F_v = F_{vc_1}...F_{vc_n}$ as the affiliation vector of node $v$, where $n$ is the number of the communities in the network, and $F_{vc_1}$ indicates the membership strength of node $v$ to community $C_1$. The higher $F_{vc_1}$ means the greater the probability that node v belongs to community $C_1$. LAA use the affiliation vector $F_v$ as the label $L_v$ of node $v$.

## 4.2  Label Aggregation

After defining and initializing the label of each node, the next step is to carry out label information propagation and label information aggregation.

### 4.2.1   Accept Labels from the Local Neighborhood

In each layer, each node will propagate labels to neighboring nodes, and each node will also receive labels from neighboring nodes. Since the degree of each node and the number of neighboring nodes are different, the number of labels received is naturally different. In order to make the algorithm expandable, LAA adopts different label receiving strategies for different aggregation functions: Mean-aggregation and Attention-aggregation. Attention aggregation can adaptively handle variable-scale inputs, focusing on the most relevant part of the input in decision-making. Mean aggregation requires probabilistic sampling in the first-order neighborhood and second-order neighborhood of the target node according to the similarity of the nodes to generate a fixed-scale input, as shown in Fig. 2. The node similarity is defined as follows:

$$J(v, u) = \frac{|N(v) \cap N(u)| + |E(v, u)|}{d_v + d_u},$$

where $N(v)$ and $N(u)$ represents the neighborhood nodes set of node $v$ and $u$, $d_v$ and $d_u$ represents the degrees of node $v$ and $u$. $E(v, u)$ is the judgment whether the node v and u are connected, if they are connected, $E(v, u) = 1$, else $E(v, u) = 0$. For all nodes in the network, we calculate the similarity of any two nodes. For each node, we normalize the similarity of its neighboring nodes. And probability sampling is performed according to the normalization result to obtain the neighborhood aggregation $N_s$ of a fixed size. The process is shown in Fig. 3.

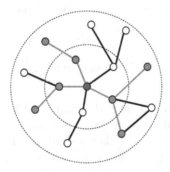

**Fig. 2.** Neighborhood sampling based on the node similarity

### 4.2.2   Label Aggregation

In each layer, after each node receives the label propagated by the node in the sampling neighborhood $N_s$, LAA uses the aggregation function to aggregate the received label and its own label as the label in the next layer. The process is shown in Fig. 3. LAA implements two aggregate functions:

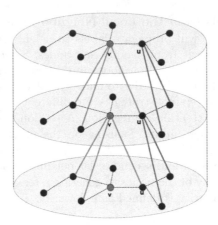

**Fig. 3.** Label aggregation

## 1. Mean-aggregation

Mean aggregation is to take the average of the label information of the neighborhood, and then aggregate it with the label of the target node. As shown in the Algorithm 1, mean aggregation is to take the average of the label information of the neighborhood, and then aggregate it with the label of the target node. In the k layer, we first take the average of the label $\mathcal{L}_u^k$ propagated by the nodes in the sampling neighborhood $N_s(v)$, then splice the result and the label of the target node $\mathcal{L}_v^k$. After multiplying with the weight matrix $W^k$, we perform a non-linear transformation, and use the result as the label of the node $v$ in the $k+1$ layer, i.e. $\mathcal{L}_v^{k+1}$.

$$\mathcal{L}_{N_s(v)}^k = Mean(\mathcal{L}_u^k, \forall u \in N_s(v))$$

$$\mathcal{L}_v^{k+1} = \sigma(W^k \cdot concat(\mathcal{L}_v^k, \mathcal{L}_{N_s(v)}^k))$$

## 2. Attention-aggregation

Attention aggregation is to calculate the similarity between the label information of the neighboring node and the label information of the target node. The label with high similarity has a higher weight during aggregation. As shown in Algorithm 2, attention aggregation is to calculate the similarity between the label information of the neighboring node and the label information of the target node. The label information with high similarity has a higher weight during aggregation. In the $k$ layer, we first calculate the similarity between the label vector propagated by nodes in the neighborhood and the label vector of the target node $v$, then normalize the similarity coefficient with softmax function to obtain the attention coefficient:

$$\alpha_{vu}^k == \frac{a^T(concat(W^k\mathcal{L}_v^k, W\mathcal{L}_u^k)))}{\sum_{w \in N(v)} exp(a^T(concat(W^k\mathcal{L}_v^k, W\mathcal{L}_w^k)))}$$

---

**Algorithm 1.** Mean aggregation for label aggregation algorithm

---

**Input:** Graph $\mathcal{G}(\mathcal{V}, \mathcal{E})$; initial label $\mathcal{L}^0 = (\mathcal{L}_1^0 ... \mathcal{L}_\mathcal{V}^0)$; number of label aggregation times $K$; weight matrices $W^k, \forall k \in \{1, ..., K\}$; non-linearity $\sigma$; sampling neighborhood $N_s(v)$.

**Output:** Aggregated label $\mathcal{L}_v^k$ for all nodes $v \in \mathcal{V}$.

1: **for** $k = 1, ... K$ **do**
2:     **for** $v \in \mathcal{V}$ **do**
3:         $\mathcal{L}_{N_s(v)}^{k-1} = Mean(\mathcal{L}_u^{k-1}, \forall u \in N_s(v))$
4:         $\mathcal{L}_v^k = \sigma(W^k \cdot concat(\mathcal{L}_v^{k-1}, \mathcal{L}_{N_s(v)}^{k-1}))$
5:     **end**
6: **end**
7: $\mathcal{L}_v^k \leftarrow \mathcal{L}_v^k / \|\mathcal{L}_v^k\|$
8: **return** $\mathcal{L}_v^k$

---

where, $\boldsymbol{a}^T$ is a vector of dimension $2n$, and $n$ is the number of the communities, i.e. the dimension of the label vector. $\boldsymbol{a}^T$ can map the spliced vector to a real number, which is the similarity coefficient. After the normalized attention coefficient is obtained, it is linearly combined with the corresponding label, and the result is subjected to a nonlinear transformation, which is used as the label of the node $v$ in the $(k+1)$th layer.

$$\mathcal{L}_v^{k+1} = \sigma(\sum_{u \in N(v)} \alpha_{vu}^k W \mathcal{L}_u^k)$$

---

**Algorithm 2.** Attention aggregation for label aggregation algorithm

---

**Input:** Graph $\mathcal{G}(\mathcal{V}, \mathcal{E})$; initial label $\mathcal{L}^0 = (\mathcal{L}_1^0 ... \mathcal{L}_\mathcal{V}^0)$; number of label aggregation times $K$; weight matrices $W^k, \forall k \in \{1, ..., K\}$; non-linearity $\sigma$; sampling neighborhood $N_s(v)$.

**Output:** Aggregated label $\mathcal{L}_v^k$ for all nodes $v \in \mathcal{V}$.

1: **for** $k = 1, ... K$ **do**
2:     **for** $v \in \mathcal{V}$ **do**
3:         **for** $u \in N(v)$ **do**
4:             $\alpha_{vu}^{k-1} == \dfrac{a^T(concat(W^{k1}\mathcal{L}_v^{k-1}, W\mathcal{L}_u^k)))}{\sum_{w \in N(v)} exp(a^T(concat(W^{k-1}\mathcal{L}_v^{k-1}, W\mathcal{L}_w^{k-1})))}$
5:         **end**
6:         $\mathcal{L}_v^k = \sigma(\sum_{u \in N(v)} \alpha_{vu}^{k-1} W \mathcal{L}_u^{k-1})$
7:     **end**
8: **end**
9: $\mathcal{L}_v^k \leftarrow \mathcal{L}_v^k / \|\mathcal{L}_v^k\|$
10: **return** $\mathcal{L}_v^k$

---

### 4.2.3  Determining Community Membership

After getting the final label vector of each node, we define a threshold to determine the membership relationship between the node and the community. The threshold is defined as follows:

$$\delta = \sqrt{-\log(1 - 2|\mathcal{E}|/|\mathcal{V}|(|\mathcal{V}| - 1))}$$

If the membership vector $F_{v_i c} > \delta$, then the node $v_i$ is considered to belong to the community $c$.

## 4.3  Parameter Training

**1. Supervised parameter training**

In real-world network datasets, the definition of the community often has strong practical significance, and sometimes it may not have a tight topology. In this case, if the community detection task is only based on the network topology, the desired result may not be obtained. Therefore, it is necessary to know part of the information of the network in advance. LAA uses a multi-label classification cross-entropy loss function, through a small number of real samples for parameter training, making the learned aggregation function more realistic.

**2. Unsupervised parameter training**

However, in many application scenarios, we cannot obtain enough information of the network in advance, so it's difficult to obtain a sufficient number of training sets for supervised parameter training. In order to address these problems, traditional network representation algorithms define such loss function for unsupervised learning:

$$Loss = -logS - log(1 - \widetilde{S})$$

Where $S = \sigma(\mathcal{L}_i^T \mathcal{L}_j), j \in N(i)$, and $\widetilde{S} = \sigma(\mathcal{L}_i^T \mathcal{L}_k), k \notin N(i)$. By optimizing the parameters to minimize the loss, the distance between the nodes which are connected tightly is smaller, and the distance between the nodes which are connected not tightly nodes is larger. However, in the experiment process, due to the sparsity of the input label vector, such loss functions cannot achieve a good training effect. The method adopted by LAA is to use the initialized label as the real label, and perform parameter training through the multi-label classification cross-entropy loss function. The aim of the LAA model is to learn the way of label aggregation. When other unsupervised algorithms are used to initialize the labels, the labels have learning value because they have certain accuracy. And aggregation with better performance will be obtained when a better initialization algorithm is used. Therefore, the LAA model under unsupervised training can be seen as an extension for other community detection algorithms.

## 5  Experiments

In this section, we evaluate the performance of the LAA model through a series of datasets with real communities and compare it with other classic algorithms.

## 5.1 Datasets

We evaluate the performance of LAA based on datasets with ground-truth communities, the detailed information is shown in Table 1.

**Table 1.** Datasets statistics. $|V|$: number of nodes; $|E|$: number of edges; $|C|$: number of communities; $|A|$: community numbers for per node.

| Datasets | $|V|$ | $|E|$ | $|C|$ | $|A|$ |
|---|---|---|---|---|
| LFR-1 | 1000 | 1766 | 11 | 1.2 |
| LFR-2 | 1000 | 1747 | 11 | 1.2 |
| LFR-3 | 1000 | 1737 | 11 | 1.2 |
| Amazon | 3225 | 10262 | 100 | 3.03 |

**LFR datasets:** are social networks generated by LFR algorithm. Nodes represent individuals and edges represent the relationship between them. The dataset contains three social networks that are generated by using the same parameters.

**Amazon dataset:** is collected from the Amazon website. Nodes represent products, edges represent the co-purchase relationships, and communities represent the categories of products.

## 5.2 Evaluation

F1-score and NMI (Normalized Mutual Information) are commonly used indicators for evaluating community detection algorithms. Since it is difficult to determine the corresponding relationship between the detected community and the real community, we used the improved F1-score:

$$F1(C^*, \hat{C}) = \frac{1}{2}\left(\frac{1}{C^*} \sum_{c_i^* \in C^*} \max_{\hat{c}_j \in \hat{C}} F1(c_i^*, \hat{c}_j) + \frac{1}{\hat{C}} \sum_{\hat{c}_j \in \hat{C}} \max_{c_i^* \in \hat{C}^*} F1(c_i^*, \hat{c}_j)\right)$$

where $C^*$ is the set of detected communities and $\hat{C}$ is the set of real communities. Similar to the F1-score, we extend the definition of NMI so that it can assess the quality of overlapping communities.

## 5.3 Baseline Methods

We compare the performance of LAA with the following baseline methods:

**Infomap:** is designed to solve this problem: How to use the shortest code to describe the path generated by a random walk. It divides nodes into different communities, then encodes communities and nodes respectively. And a good community division can make the random walk sequence have a shorter code, so by implementing a shorter code, a better community division can be achieved.

**AGM:** is a generative model for networks. It is based on the idea that the more communities two nodes belong to, the greater the probability that they are connected. It divides nodes into different communities and generates the edges according to the common communities. Then it fit the model to networks to detect communities.

**LPA:** is used to assign labels to unlabeled samples. LPA constructs an edge-weighted network according to the similarity of all nodes, and then each node performs label propagation to its neighboring nodes. And nodes with the same label belong to the same community.

### 5.4   Setup

For Mean aggregation and Attention aggregation, we use two layers to aggregate labels, i.e. $K = 2$. And we sample 8 nodes from 1-hop neighborhood and 15 nodes from 2-hop neighborhood for each node to generate the sampled neighborhood. Due to the performance of LAA is deeply influenced by the label initialization algorithm, we use different label initialization algorithms to initialize the label for different datasets in order to get the best initialization effect, Infomap for LFR datasets and AGM for Amazon datasets.

Based on two aggregation methods, mean aggregation and attention aggregation, we have implemented a total of four LAA models under supervised training and unsupervised training respectively: Sup-Mean, Unsup-Mean, Sup-Attention and Unsup-Attention.

In order to verify the generalization ability of the algorithm, we conducted two experiments. In the first experiment, we use LFR(Lancichinetti Fortunato Radicchi) algorithm to generate three datasets with the same parameters so that we can obtain three networks with the same size and attributes. Then, we sample 25% nodes of one network for training, and test the generalization ability of the LAA on the remaining 75% nodes and two other networks (None nodes are trained). In the second experiment, we test the performance of LAA in the real-world datasets. We sample 10% nodes of the dataset for training, and use the remaining nodes for testing.

### 5.5   Experiment Results

For each algorithm and each dataset, we calculate the extended F1-score and NMI. The performance of 7 algorithms on the series of networks is shown in Table 2. As we can see: 1) Performance of traditional LPA is poor which means it is difficult to find high-quality communities of practical significance if only based on the topology information of the networks especially for dealing with overlapping communities. 2) Infomap initialization algorithm for LFR datasets and AGM initialization algorithm for amazon datasets do make a good initial division of the communities. 3) The performance of LAA is excellent especially when dealing with the dense overlapping and large-scale dataset, and compared with the initialization algorithm, it has a significant improvement effect. 4) Performance of LAA on LFR-2 and LFR-3 dataset shows LAA understands the deeper reasons for the formation of the community by learning the method of label aggregation. And we confirmed the generalization ability of LAA.

**Table 2.** F1-score and NMI on community detection

| Model | LFR-1 | | LFR-2 | | LFR-3 | | Amazon | |
|---|---|---|---|---|---|---|---|---|
| | F1 | NMI | F1 | NMI | F1 | NMI | F1 | NMI |
| Sup-Mean | 0.8347 | 0.6571 | 0.7199 | 0.4957 | 0.7985 | 0.5746 | 0.8461 | 0.8305 |
| Unsup-Mean | 0.8048 | 0.5786 | 0.712 | 0.4796 | 0.7814 | 0.5408 | 0.7233 | 0.6474 |
| Sup-Attention | 0.8929 | 0.7286 | 0.7086 | 0.4779 | 0.7986 | 0.5661 | 0.8547 | 0.8472 |
| Unsup-Attention | 0.8169 | 0.593 | 0.7428 | 0.5211 | 0.8043 | 0.5708 | 0.7363 | 0.6699 |
| AGM | – | – | – | – | – | – | 0.7046 | 0.6095 |
| Infomap | 0.8035 | 0.5717 | 0.7779 | 0.5322 | 0.7803 | 0.5341 | – | – |
| LPA | 0.1679 | 0.0243 | 0.154 | 0.0122 | 0.1519 | 0.0165 | 0.3298 | 0.1141 |

## 6   Conclusions

In this paper, we proposed LAA, an inductive community detection algorithm based on label aggregation. We use AGM to generate the membership vectors of nodes as labels. Each node propagates labels to neighboring nodes, and also receives labels from neighboring nodes. We learn how labels are aggregated by training a series of aggregation functions. Excellent performance of community detection is obtained through faster and cheaper initialization methods. Besides, when the network structures change after initialization, we can get results without retraining and reinitialization. We evaluate the performance of LAA through two categories of datasets. Experiments have verified the generalization ability of LAA. In the future, we would like to conduct more in-depth experiments on more densely overlapping and large-scale graphs. In addition, we will try more efficient initialization algorithms.

**Acknowledgment.** This work is partially supported by National Natural Science Foundation of China under Grant 61702132, 61802183, and the China Postdoctoral Science Foundation No. 2018M631913.

## References

1. Cai, Z., Miao, D., Li, Y.: Deletion propagation for multiple key preserving conjunctive queries: approximations and complexity. In: 2019 IEEE 35th International Conference on Data Engineering (ICDE), pp. 506–517. IEEE (2019)
2. Cui, P., Wang, X., Pei, J., Zhu, W.: A survey on network embedding. IEEE Trans. Knowl. Data Eng. **31**(5), 833–852 (2018)

3. Fortunato, S., Hric, D.: Community detection in networks: a user guide. Phys. Rep. **659**, 1–44 (2016)
4. Grover, A., Leskovec, J.: node2vec: scalable feature learning for networks. In: Proceedings of the 22nd ACM SIGKDD International Conference on Knowledge Discovery and Data Mining, pp. 855–864 (2016)
5. Hamilton, W., Ying, Z., Leskovec, J.: Inductive representation learning on large graphs. In: Advances in Neural Information Processing Systems, pp. 1024–1034 (2017)
6. Jia, Y., Zhang, Q., Zhang, W., Wang, X.: CommunityGAN: community detection with generative adversarial nets. In: The World Wide Web Conference, pp. 784–794 (2019)
7. Khan, B.S., Niazi, M.A.: Network community detection: a review and visual survey. arXiv preprint arXiv:1708.00977 (2017)
8. Kipf, T.N., Welling, M.: Semi-supervised classification with graph convolutional networks. arXiv preprint arXiv:1609.02907 (2016)
9. Ma, X., Dong, D., Wang, Q.: Community detection in multi-layer networks using joint nonnegative matrix factorization. IEEE Trans. Knowl. Data Eng. **31**(2), 273–286 (2018)
10. Miao, D., Cai, Z., Li, J., Gao, X., Liu, X.: The computation of optimal subset repairs. Proc. VLDB Endow. **13**(12), 2061–2074 (2020)
11. Perozzi, B., Al-Rfou, R., Skiena, S.: DeepWalk: online learning of social representations. In: Proceedings of the 20th ACM SIGKDD International Conference on Knowledge Discovery and Data Mining, pp. 701–710 (2014)
12. Raghavan, U.N., Albert, R., Kumara, S.: Near linear time algorithm to detect community structures in large-scale networks. Phys. Rev. E **76**(3), 036106 (2007)
13. Tang, J., Qu, M., Wang, M., Zhang, M., Yan, J., Mei, Q.: Line: large-scale information network embedding. In: Proceedings of the 24th International Conference on Worldwide Web, pp. 1067–1077 (2015)
14. Wu, W., Kwong, S., Zhou, Y., Jia, Y., Gao, W.: Nonnegative matrix factorization with mixed hypergraph regularization for community detection. Inf. Sci. **435**, 263–281 (2018)
15. Wu, Z., Pan, S., Chen, F., Long, G., Zhang, C., Philip, S.Y.: A comprehensive survey on graph neural networks. IEEE Trans. Neural Netw. Learn. Syst. (2020)
16. Xiang, J., et al.: Local modularity for community detection in complex networks. Physica A **443**, 451–459 (2016)
17. Xie, J., Szymanski, B.K., Liu, X.: SLPA: uncovering overlapping communities in social networks via a speaker-listener interaction dynamic process. In: 2011 IEEE 11th International Conference on Data Mining Workshops, pp. 344–349. IEEE (2011)
18. Yang, J., Leskovec, J.: Community-affiliation graph model for overlapping network community detection. In: 2012 IEEE 12th International Conference on Data Mining, pp. 1170–1175. IEEE (2012)

# Modeling and Querying Similar Trajectory in Inconsistent Spatial Data

Weijia Feng[1,2]([⊠]), Yuran Geng[1], Ran Li[1], Maoyu Jin[1], and Qiyi Tan[1]

[1] Tianjin Normal University, Binshui Xi Road 393, XiQing District, Tianjin, China
WeijiaFeng@tjnu.edu.cn
[2] Postdoctoral Innovation Practice Base Huafa Industrial Share Co., Ltd.,
Changsheng Road 155, ZhuHai, China

**Abstract.** Querying clean spatial data is well-studied in database area. However, methods for querying clean data often fail to process the queries on inconsistent spatial data. We develop a framework for querying similar trajectories inconsistent spatial data. For any given entity, our method will provide a way to query its similar trajectories in the inconsistent spatial data. We propose a dynamic programming algorithm and a threshold filter for probabilistic mass function. The algorithm with the filter reduces the expensive cost of processing query by directly using the existing similar trajectory query algorithm designed for clean data. The effectiveness and efficiency of our algorithm are verified by experiments.

**Keywords:** Similar trajectory query · Spatial data · Database

## 1 Introduction

In real world, spatial data is often inconsistent, such as data collected from location-based services [1, 2], data integration [3, 12], and objects monitoring [4]. Methods for querying such kind of spatial data are not yet well developed. Query processing in inconsistent data is extremely expensive by a straightforward extension of existing methods for clean data [6, 9]. To develop efficient query processing methods for inconsistent spatial data, we in this paper study frequent nearest neighbor query which is a very time costing problem even in the context of clean data.

### 1.1 Modeling Inconsistent Spatial Data

Inconsistent spatial data overruns everywhere. For example, due to the inaccuracy of measurements, it is hard to obtain the concrete location of an entity [5, 6, 8]. To deal with this, the spatial information of an entity is modeled as a distribution over some local area. Besides, data is often integrated from different sources, thus causing the hardness to obtain the right distribution of an entity. Therefore, in real applications, a spatial entity is usually represented by a finite set of potential specific spatial distributions. For example, a real-life entity $E$ is composed of $N$ inconsistent distributions, that is, $E = \{e_1, \cdots, e_N\}$, where $e_i$ represents the distribution from the $i$-th data source.

C. S. Jensen et al. (Eds.): DASFAA 2021 Workshops, LNCS 12680, pp. 57–67, 2021.
https://doi.org/10.1007/978-3-030-73216-5_5

In this paper, we use this popular model, *i.e.*, probabilistic data, to model the inconsistent data and then study the query algorithm on it. We associate $e_i$ area with the data collected in $t$ time probability $reg^t(e_i)$ model. In this paper, we use a collection of multiple discrete sampling locations where $s_i^t$ is the entity data collected in $t$ time. Every sampling location $S_{i,j} \in s_i^t$ has a probability of its occurrence which can be written as $\mathbf{Pr}(S_{i,j}) \in (0, 1]$. The occurrence probability $\mathbf{Pr}(S_{i,j})$ should satisfy the requirement $\sum_{j=1}^{m} \mathbf{Pr}(S_{i,j}) = 1$. The sets of entity collection data in $E$ constitute the probability database $D^p$, and the data set collected by the entity at a given time $T$ is called the sketch of an entity at time $T$.

In the real application, probability data often associates with regional correlation. For example, in the spatial data, the physical quantities such as temperature and light measured from the nearby locations are probabilistic, and these data are also very similar, which is the regional correlation of the data. Regional association probability entity data contains several non-overlapping association regions. The entity $e_i \in E$ belongs to only one Inconsistent Regional Circle (*IRC*), and the entity set in the same *IRC* as $e_i$ is denoted as the regional association set $IRC(e_i)$, where $IRC(e_i)$ contains $e_i$. We assume that entities in the same *IRC* are independent of each other, and entities in different *IRC* are independent of each other, and therefore, the data acquired from practices are independent of each other.

We use Bayesian network to express the regional correlation of data. Bayesian networks are able to represent the dependencies between entities as acyclic directed graphs, and each entity has a conditional probability table, which represents the conditional probability under the joint distribution with the parent of the entity.

In probabilistic data model, the probabilistic entity $e_i$ uncertainly becomes the nearest neighbor of a query position $q$. The probability $\mathbf{Pr}_{sim\text{-}point}(e_i, q)$ is used to represent the probability that $e_i$ becomes the nearest neighbor of the query location $q$. Then the probability nearest neighbor query *sim-point* on the regional correlation probability data is to find the entity whose $\mathbf{Pr}_{sim\text{-}point}$ exceeds the given threshold constraint. A trivial method is to access the conditional probability table of $e_i$, and using the variable elimination method to obtain the joint probability, and then decide if $e_i$ belongs to the query result.

## 1.2 Nearest Entities in Inconsistent Spatial Data

In the real world, spatial data is usually changing. For example, data obtained from sensor nodes for the environment monitoring keeps changing over time. At different time, the data collected by sensor nodes is different. Results of a *sim-traj* query in sensor network is the entities with high probability frequently appearing in the result of probabilistic nearest neighbor query in multiple sketches represented at different moments.

This problem can be described as follows. Given the positive integer $k > 0$ and the threshold $\delta > 0$, then input $t$ sketches $D_1^p, \cdots, D_t^p$ with the query positions $q_1, \cdots q_t$. The *sim-traj* query outputs a collection of entities such that for each entity in the collection, the probability of being the nearest neighbor to the query location greater than or equal to $k$ is greater than the threshold $\delta$.

Let $N^{e_i}$ be the number of sketches where $e_i$ becomes the nearest neighbor of the query locations. For example, $N^{e_1} = 3$ implies the event "$e_i$ is becomes the nearest neighbor of the query location three times".

We suppose that $\mathbf{Pr}(N^{e_1} \geq k) = \sum_{i=k}^{\infty} \mathbf{Pr}(N^{e_1} = i)$ which is the probability mass function of $N^{e_1}$. The result of a *sim-traj* query is a collection of entities that satisfies $\mathbf{Pr}(N^{e_1} \geq k) \geq \delta$. A *sim-point* query is a query carried out on the sketch $D_t^P$ corresponding to a moment $t$ in the probabilistic data set $D^P$. Therefore, a *sim-point* query can be resolved as follows: let be $\alpha = 0$ be the threshold of the *sim-traj* query, then evaluate the query over each sketch by *find-sim-traj* for entities $e_j$ in any sketch $D_i^P$ to obtain the probability $p_{ji}$ of becoming the nearest neighbor of query position $q_i$. Then, According to the result of $t$ and $2^t$ different combinations of calculate the probability distribution of random variables $N^{e_i}$, and thus decide if entity $e_i$ is in the query result. However, such a trivial method may lead to a large time overhead, reasons can be listed as follows,

1. Let the value of *sim-traj* threshold be $\alpha = 0$, *find-sim-traj* may degenerate to trivial search on indexes due to the failure of the upper bound pruning method.
2. Accessing conditional probability tables results in a lot of time overhead, as we can see from the experiments on the pruning method proposed in this paper, a lot of access is unnecessary.
3. Each entity $e_i$ requires an exponential time overhead to compute the probability distribution of $N^{e_i}$.

As discussed above, the existing work cannot effectively handle *sim-traj* queries. The inadequacy of the existing work implies the significance of our method for the *sim-traj* query processing.

## 1.3  Contributions

In this paper, we study *sim-traj* query problem. We use probabilistic data to model inconsistent spatial data. A kind of frequent probabilistic nearest neighbor query is studied in-depth in this paper. We study the probabilistic data of regional association by using multiple sketches and propose a framework to find frequent probability nearest neighbor. A dynamic programming algorithm is designed to compute the square time of the probabilistic mass function. Based on the DP algorithm for *sim-traj* queries, a basic processing algorithm is proposed; And then we develop an efficient pruning methods which are used to reduce the search space of *sim-traj* queries. Finally, a series of experiments are conducted on both the synthetic data set and the real data set, and experimental results show that the efficiency of our algorithm.

## 2  Problem Definition

**Definition 1** (*similar-point-probability*). Let $E = \{e_i\}_{i=1}^{N}$ be a collection of probabilistic entities, and $D^P = \{s_i\}_{i=1}^{N}$ the sample data of entity set $E$ at some time. Given position $q$, let $r_1$ and $r_2$ be the possible nearest and farthest distance between probabilistic entity $e_i$ and $q$ (depends on the sampling situation of the corresponding region $reg(e_i)$) and

$L_2$-norm $d(\cdot, \cdot)$. Then $\mathbf{Pr}_{sim\text{-}point}\ (q, e_i)$ is the probability that entity $e_i$ becomes the nearest neighbor of query location $q$ as

$$\mathbf{Pr}_{sim\text{-}point}(q, e_i) = \int_{r_2}^{r_1} \mathbf{Pr}(d(q, e_i) = r) \times \mathbf{Pr}\left(\wedge_{\forall u \in IRC(e_i)\setminus\{e_i\}} d(q, u) \ge r|r\right) \times \mathbf{Pr}\left(\wedge_{\forall v \in E\setminus IRC(e_i)} d(q, v) \ge r\right) \mathbf{d}r$$

**Definition 2 (*similar-point*).** Let $E = \{e_i\}_{i=1}^N$ be a collection of probabilistic entities, and $D^P = \{s_i\}_{i=1}^N$ the sample data of entity set $E$ at some time. Given threshold $\alpha$ and query location $q$, the result of a *sim-point* query $(q, \alpha)$ is a set of probabilistic entities such that $\gamma = \{e_i : \mathbf{Pr}_{sim\text{-}point}(q, e_i) > \alpha, i \in [1, N]\}$.

In this paper, the random indicator variable $I_{ij} = 1$ is used to indicate that $e_j$ becomes *sim-point* of $q_i$. At the same time, we denote $\mathbf{Pr}(I_{ij} = 1)$ as $\mathbf{Pr}(I_{ij})$ for short, hence, $\mathbf{Pr}(I_{ij}) = \mathbf{Pr}_{sim\text{-}point}(q, e_i)$. Now, we formally define the frequent probabilistic nearest neighbor query on regional associated probabilistic data.

*Sim-traj* query. Let $E = \{e_i\}_{i=1}^N$ be a collection of probabilistic entities, and $D^P = \{s_i\}_{i=1}^N$ the sample data of entity set $E$ at some time. Given a positive integer $k$, the threshold value $\delta$, and the set containing $t$ query locations $Q = \{q_i\}_{i=1}^t$, the result of a *sim-traj* query $Q$ is $\gamma = \{e_i : \mathbf{Pr}_{sim\text{-}point}(N^{e_i} \ge k) \ge \delta, i \in [1, N]\}$.

Obviously, when the parameters $t = 1$, $k = 1$, $\alpha = \delta$, any *sim-traj* query is equivalent to a *sim-point* query. Therefore, the problem studied in this paper is actually a generalized version of *sim-point* query.

## 3    Frequent Probabilistic Nearest Neighbor Query Processing

We begin with an overview of query processing that deals with frequent probability nearest neighbors,

1. For the given $t$ sketches $D_1^p, \cdots, D_t^p$ to establish corresponding R-star index tree $index_1, \cdots, index_t$. In detail, and in all probability will all *IRC* entities into the corresponding minimum circumscribed rectangle, insert each *IRC*, so as to establish indexes;
2. For the current set of input query location, the *find-sim-point* method proposed in literature [7] was used, by setting $\alpha = 0$. Traverse the indexes and obtaining the upper bound of *sim-point* probability of all entities relative to $t$ query locations. Use the first pruning condition given below in the next subsection to filter out the invalid entities.
3. Access and compute the conditional probability table for the remaining entities, to obtain the *sim-point* probability relative to $t$ query locations
4. At last, for every candidate entity left $e_j$, compute the exact probability $\mathbf{Pr}(N^{e_j} \ge k)$ that it becomes a *sim-traj*, and return the query results.

Next, we detail how to calculate the probability $\mathbf{Pr}(N^{e_j} \ge k)$, in other words, we give a method to calculate the joint nearest neighbor probability efficiently under the condition that the *sim-point* probability $\mathbf{Pr}(I_{ij})$ of $e_j$ is known in advance.

### 3.1 The Computation of the Nearest Neighbor Probability

As mentioned earlier, a trivial algorithm leads to an exponential time cost. In this section, we design the dynamic programming algorithm which can derive the probability in quadratic time while avoiding the computational cost.

First, the cumulative probability $\mathbf{Pr}_{\geq s,l}(e_j)$ is defined for each probabilistic entity $e_j$, its meaning is that there are more than $s$ positions in the given $l$ positions (respectively $q_1 \cdots, q_l$), then, the sim-point probability of $e_j$ can be formulated as

$$\mathbf{Pr}_{\geq s,l}(e_j) = \sum_{Q' \subseteq Q_l, |Q'| \geq s} \left( \prod_{q_i \in Q'} \mathbf{Pr}(I_{ji}) \times \prod_{q_i \in Q_l \setminus Q'} \left(1 - \mathbf{Pr}(I_{ji})\right) \right)$$

Then, let $\mathbf{Pr}_{\geq 0,l}(e_j) = 1, \forall 0 \leq l \leq t$, and $\mathbf{Pr}_{\geq s,l}(e_j) = 1, \forall 0 \leq l \leq s$.

**Time Complexity.** For each candidate probability entity $e_j$, the dynamic programming algorithm costs $O(t^2)$ space cost and $O(t^2)$ time to calculate $\mathbf{Pr}(N^{e_j} \geq k)$.

### 3.2 Frequent Probability Nearest Neighbor Lookup Basic Query Algorithm

The literature [7] proposed *sim-point* query processing algorithm *find-sim-point*. Given the probabilistic entity database $D^P$, R-star index tree, location query $q$ and threshold $\alpha$, it takes the best first search strategy to traverse the index. During the traversal, the algorithm pruned the invalid entities by using the upper and lower bounds of the probability recorded in the index, and finally calculated the exact value of $\mathbf{Pr}_{sim\text{-}point}$ for the remaining candidate entities successively. The aim of this paper is to calculate the exact value of $\mathbf{Pr}_{sim\text{-}point}$ for all entities. For this purpose, let the threshold $\alpha = 0$. This algorithm can get the accurate value $\mathbf{Pr}_{sim\text{-}point}$ for each entity. After all the probability values are obtained by calling the above algorithm, the dynamic programming algorithm introduced in the previous section is used to calculate the $\mathbf{Pr}_{sim\text{-}point}$. We present a basic algorithm **Baseline** to compute a *sim-traj* query as shown in Fig. 1.

**Time Complexity of Baseline Algorithm.** The dynamic programming algorithm in steps 5 to 10 has a time cost of $O(Nt^2)$. The time overhead of steps 1 to 3 comes from the computation of $N_t$ probability values $\mathbf{Pr}(I_{ij})$. This needs to access the conditional probability table in the regional association probability database which is very expensive. Therefore, we next propose the pruning methods.

### 3.3 Probability Nearest Neighbor Advanced Query Algorithm

The main time overhead of the Baseline algorithm is related to the number of entities to be precisely calculated for the probability. Therefore, the next section focuses on pruning in steps 2 and 4 of query processing. In order to reduce substantial time, the number of entities to be precisely calculated is reduced at a lower cost as possible. Next, we explain the pruning conditions in steps 2 and 4, which are called first pruning and second pruning, in order to improve the basic algorithm mentioned above.

> **Input:**    entity set $E = \{e_i\}_{i=1}^N$, sketch set $\{D_i^P\}_{i=1}^t$, R star tree index $\{index_i\}_{i=1}^t$,
>           query location set $\{q_i\}_{i=1}^t$, positive integer $k$, threshold $\delta \in (0,1]$
> **Output:** *sim-traj* query result collection $\gamma$
> 1 **foreach** $i \leq t$ **do**
> 2     Call subprocess *find-sim-point*$\left(index_i, \ q_i, \ 0\right)$ to get all $p\left(I_{ji}\right)$
> 3 **end**
> 4 **foreach** $e_j \in E$ **do**
> 5     Calculate $\mathbf{Pr}_{sim\text{-}traj}\left(e_j\right)$ (dynamic programming algorithm)
> 6     **if** $p_{sim\text{-}traj}\left(e_j\right) \geq \delta$ **then**
> 7         $\gamma = \gamma \cup \{e_j\}$
> 8     **end**
> 9 **end**
> 10 **return** $\gamma$

**Fig. 1.** Baseline algorithm

To calculate the exact value of each $\mathbf{Pr}\left(I_{ij}\right)$. The algorithm in step 2 must consume a lot of access time to access and calculate the relevant conditional probability table. Therefore, it is necessary to propose efficient pruning methods to improve performance.

As long as the pruning condition filters most of the candidate entities, it can significantly reduce the number of entities that need to be accurately calculated for the probability $\mathbf{Pr}\left(I_{ji}\right)$. Every time a candidate entity is pruned, the time saved is at least $t$ times the calculation time of the exact value. We can see from the follow-up experiments in this paper, the pruning effect of this method is very obvious, which greatly improves the query efficiency. Step 2 in the index phase, we can prune a large number of probability entities to calculate the exact value of *sim-point* by using the upper bound of *sim-point* derived below. When the upper bound of probability $\mathbf{Pr}\left(I_{ji}\right)$ can be calculated with less online overhead, the pruning method can be used, hence, the baseline algorithm is improved. An index pruning criterion based on Chernoff bounds can be expressed as follows.

**Theorem 1.** For the probability entity $e_j$, given $k (\leq t)$, threshold $\delta \in (0,1]$ and upper bound set $\left\{\widehat{\mathbf{Pr}}_{ji}\right\}_{i=1}^t$. If $\sum_{i=1}^t \widehat{\mathbf{Pr}}_{ji} < k$ and $\sum_{i=1}^t \widehat{\mathbf{Pr}}_{ji} \geq k\left(\ln\left(\sum_{i=1}^t \widehat{\mathbf{Pr}}_{ji}\right) - \ln k + 1\right) - \ln \delta$, then $e_j \notin \gamma$.

Each pruning of an entity saves the time cost of accurate calculation of $I_{ji}$ and verification of $\mathbf{Pr}(N^{e_j}) \geq k$. Therefore, by the above Theorem 1, the query processing algorithm is improved as follows: (1) For each entity $e_j$ and the corresponding query location $q_i$, the upper bound $\widehat{\mathbf{Pr}}_{ji}$ is quickly calculated when traversing the index. (2) Using the formula given in Theorem 1 to prune the invalid entity.

The final experiment in this paper verifies that the actual performance of pruning Theorem 1 is efficient.

Algorithm 3–2 gives the frequent probability neighbor query processing algorithm *find-sim-traj*.

1. Line 1 to 3 initialize each probability upper bound as 0.
2. Lines 4 to 9 traverse the index of each sketch by calling the *find-sim-point* method. In this way, the upper bound of probability of each entity can be updated in the process of traversing, and the corresponding upper bound of probability of each entity can be obtained after traversing $t$ indexes.
3. Line 10 uses Theorem 1 for the pruning.
4. From line 11 to 13, the accurate probability of the remaining entities is calculated by accessing the sketch data.
5. Lines 14 to 19 run the dynamic programming algorithm, compute the corresponding probability of entities that are not pruned during the second time, and output the query result $\gamma$.

---

**Input:**   entity set $E = \{e_i\}_{i=1}^{N}$, sketch set $\{D_i^P\}_{i=1}^{t}$, R star tree index $\{index_i\}_{i=1}^{t}$, query location set $\{q_i\}_{i=1}^{t}$, positive integer $k$, threshold $\delta \in (0,1]$

**Output:** *sim-traj* query result collection $\gamma$

1 **foreach** $e_j \in E$ **do**
2      Initialization $\widehat{\mathbf{Pr}}_{total}(e_j) \leftarrow 0$
3 **end**
4 **foreach** $i \le t$ **do**
5      $\beta$=*find-sim-point*$\left(index_i,\ q_i,\ 0\right)$
6      **foreach** $e_j \in \beta$ **do**
7          $\hat{\mathbf{Pr}}_{total}\ (e_j) = \hat{\mathbf{Pr}}_{total}\ (e_j) + \hat{\mathbf{Pr}}_{ji}$
8      **end**
9 **end**
10 Perform the first pruning by Theorem 1 and the remaining $\beta$
11 **foreach** $e_j$   $\beta$ **do**
12      Calculate the value of $\mathbf{Pr}(I_{ji})$
13 **end**
14 **foreach** $e_j \in \beta$ **do**
15      **if** $\mathbf{Pr}_{sim\text{-}traj}(e_j) \ge \delta$ **then**
16          $\gamma = \gamma \cup \{e_j\}$
17      **end**
18 **end**
19 **return** $\gamma$

Algo. 3-2 *sim-traj* query processing algorithm

## 3.4   The Calculation of the Probability Upper Bounds

As mentioned before, if the above theorem can be applied in practice, we also need to know how to get the upper bound of *sim-traj* probability for each entity to become a query location by fast online calculation. Therefore, two off-line precomputation methods are presented in the next section. By embedding the structure in the index to save the online time of calculating the upper bound, the above pruning method is used to make the query processing algorithm run in practice.

The upper bound of the exact value of probability $\mathbf{Pr}\left(I_{ij}\right)$ of entity $e_j$ in quick sketch $D_i^P$ is

$$\widehat{\mathbf{Pr}}_{ji} = \sum\nolimits_{\forall s_{jl} \in s_j^i \wedge \lambda = max\left\{\lambda' | \lambda' \leq d(q_i, s_{jl}) - d\left(q_i, piv_{s_{jl}}\right)\right\}} \mathbf{Pr}_J\left(s_{jl}, \lambda\right),$$

where the joint probability is

$$\mathbf{Pr}_J\left(s_{jl}, \lambda\right) = \mathbf{Pr}\left\{\wedge_{\forall u \in IRC(e_j)} d\left(piv_{s_{jl}}, u\right) \geq d\left(q_i, s_{jl}\right) - d\left(q_i, piv_{s_{jl}}\right)\right\}$$

Here, if the query location $q_i$ is not given, then the exact value of the inequality variable cannot be obtained. However, we find that the endpoints of the numerical range can be pre-estimated. That is to say, for each sampling position $s_{jl}$ of each entity $e_j$, first select a value $\lambda$ in the interval $[\lambda_{min}, \lambda_{max}]$, and at the same time, select a pivot $piv_{s_{jl}}$, then $\mathbf{Pr}_J\left(s_{jl}, \lambda\right)$ can be calculated offline in advance. When the query location $q_i$ arrives, the algorithm only calculates the value of $b = d\left(q_i, s_{jl}\right) - d\left(q_i, piv_{s_{jl}}\right)$ temporarily, then it can find the required $\mathbf{Pr}_J\left(s_{jl}, \max_{\lambda}\{\lambda < b\}\right)$. The sum of such upper bound probability values corresponding to each sampling position is the upper probability bound $\hat{p}_{ji}$ of Theorem 1.

In order to select the pivot, the data space is divided into rectangles with side length $\varepsilon \in (0, \varepsilon_{max}]$. Each sampling position $s_{jl}$ selects four rectangular endpoints as alternative pivots for offline prediction. When the query position $q_i$ comes, the algorithm can select the candidate points which are in the same quadrant as $q_i$ relative to $s_{jl}$ as pivots, and calculate the joint probability $\mathbf{Pr}_J\left(s_{jl}, \lambda\right)$, so as to obtain the upper bound value. Specifically, the algorithm usually selects $c$ preset $\varepsilon$ value $\varepsilon_1, \ldots, \varepsilon_c$ as an alternative parameter in a small interval $(0, \varepsilon_{max}]$ uniformly, randomly and without playback in real data. Then, corresponding to each position, the optional pivots are calculated offline under each optional parameter, and the number is $4 c$ for each optional parameter in turn. Once the query position $q_i$ is given, the algorithm first finds the pivots in the same block as the query position in the different alternative parameters $\varepsilon$ and then temporarily computes their corresponding upper bounds, which results in $c$ alternative upper bounds.

## 4   Experiments

In this paper, extensive experiments are conducted on real and synthetic regional association probability datasets to investigate the proposed query processing method.

### 4.1   Experimental Configuration

All experimental configurations are the same as in the previous section. The generation method given in the literature [11] is used here to generate the artificial data set. The distribution of a given number of query positions has a great impact on the results of a *sim-traj* query Querying randomly on *t* randomly *sim-traj* generated sketches is usually not practical. Thus, in order to facilitate us to comprehensively investigate the performance of the algorithm, we experimentally query *sim-traj* on *t* replicas of the dataset $D^p$. Again, we examine the query on *t* queries with a close distribution of positions. Thus, in the experiment, we generate randomly *t* points with a high concentration of positions in a relatively small interval *r*. Let $\delta$ be 0.1/0.2/0.4/0.8, *k* be 5/10/20, *t* be 20, *r* be 0.01, *N* be 10, and *d* be 5.

To generate the sketch set $D^p$, 200, 000 points are randomly selected in the interval $R = [0, 10]^d$ and these points correspond to the reference points of *IRC*. On this basis, we construct the dimension of the entity, randomly select the real number *a* from the interval [1, 4], and set a to the region size of the *IRC*. We further randomly generate *n* probabilistic entities in the region of *IRC* according to two distributions, where the probability distributions we choose are the most common random probability distribution and the Gaussian probability distribution. Finally, we construct an acyclic directed graph of the *IRC* on the *IRC* and generate a joint probability distribution of the entities. We examine the performance of the algorithm on the U.S. traffic data whose total number of NR is 20,000 [11]. The experiment takes the geographic coordinates as the center point of the *IRC* and expands it as above into a set of available probabilistic data. Based on different distribution functions, the generated datasets can be classified into uniform synthetic data (UniS), traffic data (UniT) and Gaussian traffic data (GauT).

## 4.2 Analysis of Experimental Results

The comparison algorithm chosen for the experiments in this paper is the **Baseline** Algorithm and its corresponding improvement algorithm. The experiments focus on the efficiency and speed-up ratio of the algorithm for different environmental parameters. Here, we define the speed-up ratio $\eta = \frac{t_{Baseline}}{t_{sim-traj}}$, where $t_{Baseline}$ represents the runtime of the **Baseline** algorithm and $t_{sim-traj}$ represents the runtime of the *sim-traj* query processing algorithm.

Here, we examine the pruning effect of the pruning strategy on different datasets. Fig. 2 clearly shows that the pruning strategy proposed in this paper can greatly improve the efficiency of the algorithm and reduce the computation time. When pruning a candidate set for the first time using the upper boundary condition, the wrong entities selected by the **Baseline** algorithm can be pruned out, avoiding unnecessary time spent on subsequent calculations of joint probability distributions and boundary probabilities.

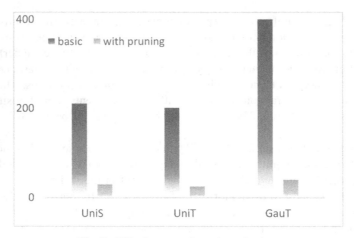

**Fig. 2.** Effectiveness of pruning criterions

## 5  Conclusion

In this paper, a general processing framework is proposed including dynamic programming algorithm and threshold filtering method for probabilistic mass function which solves the problem that it is expensive to deal with the query directly by using the existing nearest neighbor query algorithm based on traditional clean data. The effectiveness and efficiency of the algorithm are verified by experiments . Therefore, the work of this paper overcomes the shortcoming that the existing methods are too strict to query results, as many query results as possible are given with custom quality assurance are be ensured.

**Acknowledgement.** This research is funded by the Natural Science Foundation of China (NSFC): 61602345; National Key Research and Development Plan: 2019YFB2101900; Application Foundation and Advanced Technology Research Project of Tianjin (15JCQNJC01400).

## References

1. Chen, L., Thombre, S., Järvinen, K., et al.: Robustness, security and privacy in location-based services for future IoT: a survey. IEEE Access **5**, 8956–8977 (2017)
2. Zheng, X., Cai, Z., Li, J., et al.: Location-privacy-aware review publication mechanism for local business service systems. In: IEEE INFOCOM 2017-IEEE Conference on Computer Communications, pp. 1–9. IEEE (2017)
3. Stonebraker, M., Ilyas, I.F.: Data integration: the current status and the way forward. IEEE Data Eng. Bull. **41**(2), 3–9 (2018)
4. Cheng, S., Cai, Z., Li, J.: Approximate sensory data collection: a survey. Sensors **17**(3), 564 (2017)
5. Miao, D., Cai, Z., Li, J., et al.: The computation of optimal subset repairs. Proc. VLDB Endowment **13**(12), 2061–2074 (2020)

6. Bertossi, L.: Database repairs and consistent query answering: origins and further developments. In: Proceedings of the 38th ACM SIGMOD-SIGACT-SIGAI Symposium on Principles of Database Systems, pp. 48–58 (2019)

7. Lian, X., Chen, L., Song, S.: Consistent query answers in inconsistent probabilistic databases. In: Proceedings of the 2010 ACM SIGMOD International Conference on Management of Data, pp. 303–314 (2010)

8. Wijsen, J.: Foundations of query answering on inconsistent databases. ACM SIGMOD Rec. **48**(3), 6–16 (2019)

9. Greco, S., Molinaro, C., Trubitsyna, I.: Computing approximate query answers over inconsistent knowledge bases. IJCAI **2018**, 1838–1846 (2018)

10. Mitzenmacher, M., Upfal, E.: Probability and Computing: Randomization and Probabilistic Techniques in Algorithms and Data Analysis. Cambridge university press (2017)

11. Chen, L., Lian, X.: Query processing over uncertain and probabilistic databases. In: Lee, S.-G., Peng, Z., Zhou, X., Moon, Y.-S., Unland, R., Yoo, J. (eds.) DASFAA 2012. LNCS, vol. 7239, pp. 326–327. Springer, Heidelberg (2012). https://doi.org/10.1007/978-3-642-29035-0_32

12. Miao, D., Liu, X., Li, J.: On the complexity of sampling query feedback restricted database repair of functional dependency violations. Theoret. Comput. Sci. **609**, 594–605 (2016)

# The 5th International Workshop on Graph Data Management and Analysis

# ESTI: Efficient $k$-Hop Reachability Querying over Large General Directed Graphs

Yuzheng Cai and Weiguo Zheng$^{(\boxtimes)}$

Fudan University, Shanghai, China
{yzcai17,zhengweiguo}@fudan.edu.cn

**Abstract.** As a fundamental task in graph data mining, answering $k$-hop reachability queries is useful in many applications such as analysis of social networks and biological networks. Most of the existing methods for processing such queries can only deal with directed acyclic graphs (DAGs). However, cycles are ubiquitous in lots of real-world graphs. Furthermore, they may require unacceptable indexing space or expensive online search time when the input graph becomes very large. In order to solve $k$-hop reachability queries for large general directed graphs, we propose a practical and efficient method named *ESTI* (Extended Spanning Tree Index). It constructs an extended spanning tree in the offline phase and speeds up online querying based on three carefully designed pruning rules over the built index. Extensive experiments show that *ESTI* significantly outperforms the state-of-art in online querying, while ensuring a linear index size and stable index construction time.

**Keywords:** $k$-hop reachability queries · General directed graphs · Extended spanning tree

## 1 Introduction

Graph is a flexible data structure representing connections and relations among entities and concepts, which has been widely used in real world, including XML documents, cyber-physical systems, social networks, biological networks and traffic networks [1–3,9,12]. Nowadays, the size of graphs such as knowledge graphs and social networks is growing rapidly, which may contain billions of vertices and edges. $k$-hop reachability query in a directed graph is first discussed by Cheng et al. [1]. It asks whether a vertex $u$ can reach $v$ within $k$ hops, i.e., whether there exists a directed path from $u$ to $v$ in the given directed graph and the path is not longer than $k$. Note that the input general directed graph is not necessary to be connected. Take the graph $G$ in Fig. 1(a) as an example, vertex $a$ can reach vertex $e$ within 2 hops, but $a$ cannot reach vertex $d$ within 1 hop.

Efficiently answering $k$-hop reachability queries is helpful in many analytical tasks such as wireless networks, social networks and cyber-physical systems

© Springer Nature Switzerland AG 2021
C. S. Jensen et al. (Eds.): DASFAA 2021 Workshops, LNCS 12680, pp. 71–89, 2021.
https://doi.org/10.1007/978-3-030-73216-5_6

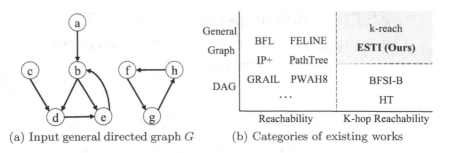

(a) Input general directed graph $G$        (b) Categories of existing works

**Fig. 1.** Illustration of input graph and existing works

[1,2,12]. Several methods for $k$-hop reachability has been proposed, providing different techniques to solve this kind of queries. However, existing methods suffer some shortcomings, which make them not practical or general enough to answer $k$-hop reachability queries efficiently. To the best of our knowledge, $k$-reach [1,2] is the only method aiming at dealing with $k$-hop reachability queries for general directed graph, which builds an index based on vertex cover of the graph. It is infeasible to build such an index for large graphs due to the huge space cost. Thus a partial coverage is employed in [2]. However, partial coverage technique is also not practical enough since most queries may fall into the worst case, which requires online BFS search.

A bunch of methods have been proposed to solve $k$-hop reachability queries in DAGs. *BFSI-B* [12] builds a compound index, containing both FELINE index [10] and breadth-first search index (BFSI). *HT* [3] works on 2-hop cover index, which selects some high-degree nodes in the DAG as hop nodes. Experiments have shown that both of them are practical and efficient to answer $k$-hop reachability queries. However, they are developed only for dealing with DAGs, which are not general enough since most graphs in real applications may have cycles, such as social networks and knowledge graphs.

A simple version of $k$-hop rechability query is reachability query. Given a graph $G$, reachability query can be taken as a specific case of $k$-hop reachability queries, since they are actually equivalent when $k \geq \lambda(G)$, where $\lambda(G)$ represents the length of the longest simple path in graph $G$. Note that for a general directed graph, we can obtain the corresponding DAG by condensing each strongly connected component (SCC) as a supernode, such that the reachability information in original graph can be completely preserved in the constructed DAG. Although lots of methods have been proposed to handle reachability queries [4,6,8,10,11,13], they cannot be directly used for $k$-hop reachability queries since more information such as distance is missing in the transformation above.

We categorize the methods related to $k$-hop reachability queries [1–4,6,8,10–13], as shown in Fig. 1(b). Clearly, right-top corner represents $k$-hop reachability in general directed graphs, which is the most general one. As discussed above, *k-reach*, the only existing method in this research area, is not practical enough to handle very large graphs. Hence, we develop a practical method named *ESTI* to answer $k$-hop reachability queries efficiently.

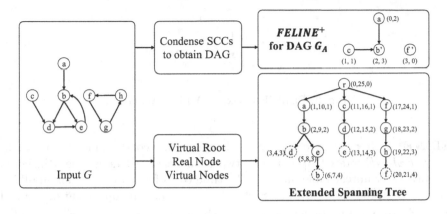

**Fig. 2.** Overview of Extended Spanning Tree Index (ESTI)

Our proposed approach, *ESTI*, follows the offline-and-online paradigm. It builds an index for a given graph in the offline phase, and answers arbitrary $k$-hop reachability queries in the online phase. In offline indexing process, both *FELINE$^+$* index and Extended Spanning Tree Index (ESTI) are constructed. We introduce the concept of *Real Node* and *Virtual Node* to build the extended spanning tree with both BFS and DFS. As for online querying, the offline index helps to answer $k$-hop reachability queries efficiently, and three pruning strategies are devised to further speed up query process.

**Paper Organiztion.** This paper is organized as follows. Section 3 explains the details of *ESTI* offline index, followed by the querying process as discussed in Sect. 4. Section 5 shows the results of experiments comparing *ESTI* with other $k$-hop reachability methods. In Sect. 6, some exciting works related to $k$-hop reachability queries are presented. Finally, Sect. 7 concludes the paper.

## 2 Problem Definition and Overview

### 2.1 Problem Definition

In this paper, the input general directed unweighted graph is represented as $G = (V, E)$, where $V$ denotes the set of vertices and $E$ denotes the set of edges. $|V|$ and $|E|$ denote the number of vertices and edges in $G$, respectively. For any two vertices $u, v \in V$ and $u \neq v$, we say that $u$ can reach $v$ within $k$ hops if there exists a directed path from $u$ to $v$ in $G$ which is not longer than $k$. Let $u \xrightarrow{?k} v$ represent a query asking whether $u$ can reach $v$ within $k$ hops in $G$.

### 2.2 Overview

*ESTI* follows the offline-and-online paradigm, and Fig. 2 presents the overview of our offline index structure. For better understanding, we briefly introduce our basic ideas and techniques for answering arbitrary $k$-hop reachability queries.

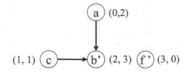

**Fig. 3.** FELINE index $(X, Y)$ in DAG $G_A$

**FELINE$^+$ Index.** Since reachablity is the neccessary condition for $k$-hop reachability, *FELINE* index [10] including two topological orders can be utilized to efficiently filter unreachable queries. The time cost of generating index in offline phase is $O(|V|log|V| + |E|)$. In Sect. 3.1, we present an optimization named *FELINE$^+$* to speed up index generation, which costs $O(|V|log(Deg_m^{(out)}) + |E|)$ time, where $Deg_m^{(out)}$ is the maximum outgoing degree of a vertex.

**Extended Spanning Tree Index.** In order to preserve as much information as possible for answering queries, we introduce *Virtual Root*, *Real Nodes* and *Virtual Nodes* to constuct an extended spanning tree from the input graph $G$ in Sect. 3.2. Also, pre- and postorders and global level are assigned to nodes in the tree, which helps to efficiently answer $k$-hop queries online.

**Online Querying.** Given arbitrary query $u \xrightarrow{?k} v$, the constructed index is utilized to directly return the correct answer or prune search space. In Sect. 4.2, three pruning strategies are developed to further accelerate online querying.

## 3   Offline Indexing

### 3.1   FELINE$^+$ Index

If $u$ cannot reach $v$ in $G$, the answer of query $u \xrightarrow{?k} v$ is apparently *False*. To efficiently filter those unreachable queries in online querying phase, *FELINE* [10] condenses all strongly connected components (SCCs) in the given general directed graph $G$ to obtain a DAG $G_A$, and two topological orders $X$ and $Y$ are generated for each vertex in $G_A$. Let $X_v$ and $Y_v$ denote the first and second topological order of a vertex $v$, respectively. If $u$ can reach $v$, both $X_u < X_v$ and $Y_u < Y_v$ hold. Hence, for a query $u \xrightarrow{?k} v$, we can directly return the answer *False* if $X_u > X_v$ or $Y_u > Y_v$ in FELINE index.

In *FELINE* [10], $X$ is calculated by a topological ordering algorithm, and $Y$ coordinate is assigned by applying a heuristic decision. When assigning $Y$ coordinate, let $R$ be a set storing all roots in current DAG. FELINE iteratively runs the following procedures until all vertices in $G_A$ have $Y$ coordinates.

*Step 1.* Choose the root $r$ from $R$ with largest $X_r$, assign $r$ a coordinate $Y_r$;

*Step 2.* Remove all of $r$'s outgoing edges. and some of its children may have no ancestors and become new roots. Thus, $R$ should be updated.

---

**Algorithm 1.** FELINE$^+$ Index Construction

---

**Input:** DAG $G_A$;
**Output:** Two topological orders $X$ and $Y$;
1: $X \leftarrow$ a Topological Order of $G_A$
2: $R \leftarrow$ all the roots in $G_A$ sorted w.r.t descending $X$ value
3: **while** $R$ is not empty **do**
4:    pop the first element $r$ from $R$ and assign $Y_r$
5:    $R_{tmp} \leftarrow [\,]$
6:    **for** each outgoing neighbor $t$ of $r$ **do**
7:       remove edge $(r, t)$
8:       **if** $t$ has no incoming neighbor **then**
9:          $R_{tmp} \leftarrow R_{tmp} \cup \{t\}$
10:    sort $R_{tmp}$ according to descending $X$ value
11:    insert all elements of $R_{tmp}$ in the front of $R$, while preserving the order
12: **return** $X, Y$;

---

*Example 1.* By condensing all SCCs of graph $G$ in Fig. 1(a), its corresponding DAG $G_A$ is shown in Fig. 3. After assigning $X$, we start to assign $Y$ and $R = \{a, c, f'\}$. Since $X_{f'} = 3$ is the largest one, $Y_{f'}$ is assigned to be 0, and next we assign $Y_c = 1$ and $Y_a = 2$. When all edges connecting with $b'$ are removed, we update $R = R \cup \{b'\}$ to continue assigning $Y$ coordinate to $b'$. As for online querying, for instance, vertex $a$ cannot reach vertex $c$ since $Y_a > Y_c$ in Fig. 3.

The time cost of condensing SCCs and generating $X$ coordinate is $O(|V| + |E|)$. Note that FELINE utilizes a max-heap to store all the current roots $R$, in which those roots are sorted in the descending order according to $X$. It takes $O(1)$ to pop a root $r$ from the max-heap in *Step 1*, and each vertex in $G_A$ can only be inserted into $R$ once which costs $O(log|V|)$ time. Hence, the overall time cost of building index construction for FELINE is $O(|V|log|V| + |E|)$.

In this paper, we propose an novel technique to accelerate $Y$ coordinate generation, utilizing a simple array to store all the current roots $R$ instead of a max-heap. Firstly, $R$ is initialized by putting all the roots in original $G_A$, making sure they are sorted in descending order w.r.t. $X$ value. Then the following two steps are processed iteratively until all the vertices have $Y$ coordinate.

*Step 1.* Pop the first element $r$ from the array $R$ and assign its $Y$ coordinate.

*Step 2.* Remove all of $r$'s outgoing edges. Sort those new roots w.r.t descending $X$ value, then insert them in the front of array $R$, while preserving the order.

**Theorem 1.** *The order of elements in array $R$ is always the same as the descending order of their $X$ value.*

*Proof.* At first, array $R$ is initialized with all roots in original $G_A$, which are sorted in the descending order w.r.t. $X$ value. Assume that elements in array $R$ are in the descending order of $X$ value. When we pop the first element $r$ from array $R$ to assign $Y_r$, $X_r \geq X_v$ holds for any vertex $v$ in array $R$. After removing $r$'s outgoing edges, some of its children $w$ may become new roots and $X_w > X_r$

must hold. Thus, every $w$ has larger $X$ than any $v$ in array $R$. After sorting those new roots $w$ in descending $X$ value and inserting them in the front of array $R$, all the vertices in array $R$ are still in their descending $X$ order.    □

The enhanced algorithm, denoted by FELINE$^+$, for accelerating FELINE is shown in Algorithm 1. When generating $Y$ coordinate, according to Theorem 1, the first element $r$ of array $R$ always has the largest $X_r$ value in $R$, and it actually constructs the same index as FELINE. Note that to make sure the initial roots in arrary $R$ are in descending order w.r.t. $X$ value, we only need to reverse the initial root queue of $X$ coordinate generation process, because their $X$ values are generated following the order of it. Hence, the initialization time of array $R$ is linear to the number of roots in original $G_A$. When processing each current root $r$, sorting the new roots takes $O(|w|log|w|)$, where $|w|$ is the number of new roots obtained by removing $r$'s outgoing edges. Since each vertex in $G_A$ can be a new root only once, the time cost of generating $Y$ coordinate is $O(|V|log(Deg_m^{(out)}) + |E|)$, where $Deg_m^{(out)}$ is the max number of outgoing neighbors of a vertex and $|w| \leq Deg_m^{(out)}$ always holds.

The total time cost of building index for FELINE$^+$ is $O(|V|log(Deg_m^{(out)}) + |E|)$. Theoretically, since $Deg_m^{(out)}$ is much smaller than $|V|$ in many graphs, our approach is faster than the original FELINE whose time cost is $O(|V|log|V| + |E|)$. Experiments confirm that the proposed optimization technique significantly accelerates the index construction for FELINE, as shown in Sect. 5.2.

## 3.2    Extended Spanning Tree Index for General Directed Graph

**Preliminary.** We first briefly introduce pre- and postorder index and global level for a tree, which have been used in *GRIPP* [9] and *BFSI-B* [12]. Note that *BFSI-B* applies min-post strategy, which actually has the same effect as pre- and postorders. For any vertex $v$ in the tree, $pre_v$ and $post_v$ represent the pre- and postorder index of $v$, respectively. And $level_v$ is the global level of $v$, i.e., the distance from the tree root to $v$. $pre_v$ and $post_v$ are generated during the DFS traversal, while $level_v$ is generated during the BFS traversal.

*Example 2.* Figure 4(a) illustrates the three labels. Following the visiting order in DFS, we start from root $a$ and set $pre_a$ to 0. Then we visit $b$ and $c$ and set $pre_b$ and $pre_c$ to 1 and 2, respectively. After returning from $c$, we set $post_c$ to 3. The process proceeds until all nodes have been visited. Each node is assigned both pre- and postorder index following the DFS. As for *level* index, $level_a$ is set to be 0 and we can assign *level* to other vertices following the BFS.

We say that $(pre_v, post_v) \subset (pre_u, post_u)$ iff $pre_v \geq pre_u \wedge post_v \leq post_u$. Based on the constructed index $(pre_v, post_v, level_v)$ discussed above, Theorem 2 holds in the tree, and query $u \xrightarrow{?k} v$ can be efficiently answered. For example, in Fig. 4(a) $a$ can reach $d$ in 2 hops, since $(4,5) \subset (0,11)$ and $level_d - level_a = 2$.

**Theorem 2.** *Given two vertices $u$ and $v$ in tree $T$, $u$ can reach $v$ within $k$ hops if $(pre_v, post_v) \subset (pre_u, post_u) \wedge level_v - level_u \in (0, k]$.*

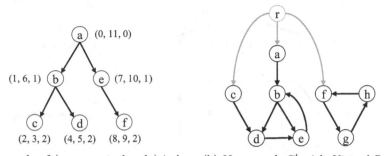

(a) Example of $(pre_v, post_v, level_v)$ index    (b) New graph $G'$ with *Virtual Root*

**Fig. 4.** Illustration of $(pre_v, post_v, level_v)$ index and *Virtual Root*

*Proof.* According to the process of pre- and postorder generation, $(pre_v, post_v) \subset (pre_u, post_u)$ indicates that $v$ is in the subtree whose root is $u$. $level_v - level_u \in (0, k]$ implies that there is a path from $u$ to $v$ which is not longer than $k$.    □

Clearly, if the input graph is a tree, both time and space cost for building the index are $O(|V| + |E|)$ and it only takes $O(1)$ for online query. However, when the input general directed graph $G$ is not a tree, to make it practical and efficient enough for answering $k$-hop reachability queries, we introduce *Virtual Root*, *Real Node* and *Virtual Node* to transform $G$ into an Extended Spanning Tree (EST). Note that our method is quite different from existing approaches like *GRIPP* [9] and *BFSI-B* [12]. *GRIPP* solves reachability queries while ignores distance information which is necessary for answering $k$-hop reachability queries, and *BFSI-B* is developed for only dealing with DAGs. However, most graphs in real life have cycles and *BFSI-B* cannot directly work on these graphs.

**Virtual Root.** Since the given graph $G$ may not be connected, e.g., the graph in Fig. 1(a), we add a virtual root $V_R$ to make sure that it can reach all vertices in $G$. We first add an edge from $V_R$ to all the vertices which have no predecessors, then explore from $V_R$ to mark all of its descendants visited. The second step is to randomly select an unvisited vertex $v$, and add an edge from $V_R$ to $v$ while all of $v$'s descendants are marked visited. We repeat the second step until all vertices have been visited. Take graph $G$ in Fig. 1(a) as an example. After adding a virtual root for it, we obtain a new graph $G'$ in Fig. 4(b).

**Real and Virtual Nodes.** When starting BFS from virtual root $V_R$, we may encounter endless loop since there may exist cycles in $G'$, or some visited vertices since they have multiple incoming edges. To solve this problem, we introduce *Real Nodes* and *Virtual Nodes*. In BFS process, if vertex $v$ has never been visited, it will be added to the spanning tree as a *Real Node* and we will continue to visit its successors. If vertex $v$ has been visited, it will be added to the tree as a *Virtual Node* while its successors will not be explored again. Following the

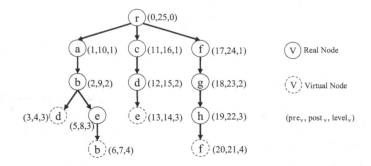

**Fig. 5.** Extended spanning tree of $G$ and $(pre_v, post_v, level_v)$ index

above definition of *Real Node* and *Virtual Node*, we can construct an extended spanning tree from graph $G'$, as shown in Example 3. Also, Theorem 3 holds.

*Example 3.* In Fig. 4(b), we start BFS from $r$ and add real nodes for $r$, $a$, $c$, $f$, $b$, $d$ and $g$. When exploring from $b$ to visit $d$, we create a virtual node for $d$ since it has been visited before. Figure 5 is the extended spanning tree of $G'$.

**Theorem 3.** *In extended spanning tree, each vertex $v$ in graph $G'$ must have exactly one real node. The total number of real and virtual nodes in this tree is equal to the number of edges in $G'$ plus 1.*

*Proof.* Since virtual root $V_R$ can reach all vertices in $G'$ and we start BFS from $V_R$ to construct the extended spanning tree, a real node is created for each vertex $v$ in $G'$ when it is visited for the first time. When $v$ is visited again, we only create a virtual node for it. Hence, each $v$ in $G'$ must have exactly one real node.

At the beginning of BFS, we create a real node for virtual root $V_R$. As for the other vertices $v$ in $G'$, a real node or virtual node will be created for $v$ only when we explore from its incoming neighbor. Hence, the number of real and virtual nodes in this tree is equal to the number of edges in $G'$ plus one, where the additional one is the real node representing virtual root $V_R$.    □

**Index Generation.** Recall that in a tree, the index of vertex $v$ consists of $pre_v$, $post_v$ and $level_v$. When constructing the extended spanning tree from graph $G'$, we have already run BFS in the tree, and *level* index will also be generated for all the nodes. Next, we explore the whole tree by DFS and assign each vertex with pre- and postorder index. Take the graph $G'$ in Fig. 4(b) as an example. The index of its extended spanning tree is shown in Fig. 5. After assigning the above index, Theorem 4 holds for all the real and virtual nodes in the tree.

**Theorem 4.** *If vertex $v$ of $G'$ has virtual nodes in the extended spanning tree, denote its unique real node as $v'_r$. For any virtual node $v'_i$ of $v$, $level_{v'_i} \geq level_{v'_r}$.*

*Proof.* When constructing the extended spanning tree by BFS, all the virtual nodes of $v$ are created after its real node is created. Hence, based on the exploration order of BFS, $level_{v'_i} \geq level_{v'_r}$.    □

Let $|V'|$ and $|E'|$ denote the number of vertices and edges in $G'$, respectively. When generating $G'$ from original graph $G$, we add a virtual root $V_R$ and at most $|V|$ edges to connect vertices in $G$. Thus, $O(|V'| + |E'|) = O(|V| + |E|)$.

The time and space comlexity of adding a virtual root is $O(|V| + |E|)$, since each vertex and edge is visited once. When constructing the extended spanning tree, each edge in $G'$ is visited once since we explore from vertex $v$ only when its unique real node is created. According to Theorem 3, it takes both time and space cost $O(|V| + |E|)$ to create all real and virtual nodes. And both BFS and DFS also take the time and space cost $O(|V| + |E|)$. Hence, the overall time and space cost for constructing the extended spanning tree and the three labels are $O(|V| + |E|)$, which indicates that it is feasible even for very large graphs.

## 3.3    Summary of Offline Indexing

The index of our proposed $ESTI$ method consists of two parts: FELINE$^+$ (Sect. 3.1) and the extended spanning tree (Sect. 3.2). The whole generation process is shown in Algorithm 2. Recall that building FELINE$^+$ index takes $O(|V| log(Deg_m^{(out)}) + |E|)$ time and $O(|V|)$ space, where $Deg_m^{(out)}$ is the maximum outgoing degree in $G_A$. And the time and space cost of constructing the extended spanning tree and three labels are both $O(|V| + |E|)$. Hence, the overall index construction time of $ESTI$ is $O(|V| log(Deg_m^{(out)}) + |E|)$, and index size is $O(|V| + |E|)$. Next, we will show how the constructed index supports efficient online $k$-hop reachability queries.

---

**Algorithm 2.** $ESTI$ Index Construction

---

**Input:** A general directed graph $G$;
**Output:** FELINE$^+$ index $X, Y$; $EST$ mapping each $v$ in $G$ to its real or virtual node
   $v'$ in extended spaning tree; $Pre, Post, Level$ index for each node $v'$ in the tree.
  1: $G_A \leftarrow$ condense SCCs in $G$
  2: $X, Y \leftarrow$ generating FELINE$^+$ index for $G_A$                    ▷ see Algorithm 1
  3: $G' \leftarrow$ add a virtual root $V_R$ and virtual edges in $G$          ▷ see Section 3.2
  4: $F \leftarrow \{(V_R, 0)\}$                                        ▷ a queue used as BFS frontier
  5: $i \leftarrow 0$
  6: **while** $F$ is not empty **do**
  7:     pop $(u, l)$ from $F$
  8:     $Level[i] \leftarrow l$
  9:     **if** $u$ has not been visited **then**
 10:         $EST[u].RealNode \leftarrow i$
 11:         **for** each out-neighbor $v$ of $u$ **do**
 12:             $F \leftarrow F \cup \{(v, l+1)\}$
 13:     **else**
 14:         add node $i$ to $EST[u].VirtualNodes$
 15:     $i \leftarrow i + 1$
 16: $Pre, Post \leftarrow$ Assign pre- and postorder for all real and virtual nodes in the tree
 17: **return** $X, Y, EST, Pre, Post, Level$;

---

---

**Algorithm 3.** Basic Query Fucntion $Query(u, v, k)$

---

**Input:** Start vertex $u$, target vertex $v$, $k$; Offline index $X,Y,EST,Pre,Post,Level$.
**Output:** $True$ or $False$.
1: **if** $X[u] > X[v] \vee Y[u] > Y[v]$ **then**
2:     **return** $False$
3: $u'_r \leftarrow EST[u].RealNode$
4: **for** each node $v'$ in $\{EST[v].RealNode\} \cup EST[v].VirtualNodes$ **do**
5:     **if** $(Pre[v'], Post[v']) \subset (Pre[u'_r], Post[u'_r]) \wedge level[v'] - level[u'_r] \leq k$ **then**
6:         **return** $True$
7: **if** $k > 1$ **then**
8:     **if** number of outgoing edges of $u \leq$ number of incoming edges of $v$ **then**
9:         **for** each outgoing neighbor $w$ of $u$ **do**
10:             **if** $Query(w, v, k - 1)$ **then**
11:                 **return** $True$
12:     **else**
13:         **for** each incoming neighbor $w$ of $v$ **do**
14:             **if** $Query(u, w, k - 1)$ **then**
15:                 **return** $True$
16: **return** $False$;

---

## 4    Online Querying

### 4.1    Basic Query Process

After constructing $ESTI$ index (Sect. 3) for the input graph $G$, we can utilize the index to answer $k$-hop reachability queries online. Given a query $u \xrightarrow{?k} v$, if $u = v$ or $k \leq 0$ we can directly return the answer. Assume that $u \neq v$ and $k > 0$, the basic query function is shown in Algorithm 3.

As discussed in Sect. 3.1, in Line 1–2, if the topological order $X$ (or $Y$) of $u$'s corresponding vertex in DAG $G_A$ is larger than $v$'s $X$ (or $Y$), we can safely return $False$. In Line 3–6, the pre- and postorders of real and virtual nodes are compared. Note that in Line 7–15, we run DFS only when $k > 1$ (Line 7) because the exploration will never return $True$ when $k \leq 1$. If $k = 1$ the answer from Line 3–6 is the final answer, and $k = 0$ is impossible since the initial input assumes that $k > 0$ while funtion $Query$ is invoked only when $k > 1$.

*Example 4.* Given the constructed index in Fig. 5, for query $c \xrightarrow{?3} b$, we invoke $Query(c, b, 3)$. The pre- and postorder of $c$'s Real Node is $(11, 16)$, but the real node of $b$ has index $(2, 9) \not\subset (11, 16)$ and its virtual node has index $(6, 7) \not\subset (11, 16)$. Then $Query(d, b, 2)$ is invoked, which results in calling $Query(e, b, 1)$. Luckily, $b$'s virtual node has index $(6, 7) \subset (5, 8)$ and the function returns $True$.

To further improve the performance of online querying, we develop three pruning strategies based on properties of the extended spanning tree.

## 4.2   Pruning Strategies

**Prune I.** For query $u \xrightarrow{?k} v$, denote $u'_r$, $v'_r$ as the real node of $u$ and $v$, respectively. Prune I strategy utilizes Theorem 5 to stop redundant exploration in advance, i.e., $Query(u, v, k)$ will directly return $False$ if $level_{v'_r} - level_{u'_r} > k$.

**Theorem 5.** *If $level_{v'_r} - level_{u'_r} > k$, $u$ cannot reach $v$ within $k$ hops.*

*Proof.* Note that as discussed above, we never invoke $Query(u, v, k)$ s.t. $k = 0$.

(Case 1). When $k = 1$, assume that $level_{v'_r} - level_{u'_r} > 1$. If $u$ can reach $v$ within 1 hop, $v$ has a real or virtual node $v'$ which is the child of $u'_r$ and $level_{v'} = level_{u'_r} + 1$. According to Theorem 4, $level_{v'} \geq level_{v'_r}$ indicates that $level_{v'_r} - level_{u'_r} \leq level_{v'} - level_{u'_r} = 1$, which contradicts the assumption.

(Case 2). When $k > 1$, in function $Query(u, v, k)$, Line 3–6 will never return $True$ since $level_{v'} \geq level_{v'_r}$ and $level_{v'} - level_{u'_r} \geq level_{v'_r} - level_{u'_r} > k$. Hence we need to invoke $Query(w, v, k-1)$ or $Query(u, w, k-1)$. For $Query(w, v, k-1)$, since the real or virtual node $w'$ is a child of $u'_r$ in the tree, the real node of $w$ satisfies $level_{w'_r} \leq level_{w'} = level_{u'_r} + 1$. Thus, we have $level_{v'_r} - level_{w'_r} \geq level_{v'_r} - level_{u'_r} - 1 > k - 1$, and $Query(w, v, k-1)$ falls into Case 1 or Case 2 again. For $Query(u, w, k-1)$, since $w'_r$ is the parent of one of the real or virtual node $v'$ in the tree, $w'_r$ satisfies $level_{w'_r} = level_{v'} - 1 \geq level_{v'_r} - 1$. Thus, we have $level_{w'_r} - level_{u'_r} \geq level_{v'_r} - level_{u'_r} - 1 > k - 1$, and $Query(u, w, k-1)$ also falls into Case 1 or Case 2 again.

Hence, if $level_{v'_r} - level_{u'_r} > k$, $u$ cannot reach $v$ within $k$ hops.          □

*Example 5.* In Fig. 5, for query $f \xrightarrow{?1} e$, both real and virtual nodes of $e$ have level 3, while the real node of $f$ has level 1. Since $3 - 1 > k = 1$, we return $False$.

**Prune II.** In Line 3–6 of Algorithm 3, we iterate all real and virtual nodes $v'$ to compare $(pre_{v'}, post_{v'})$ with $(pre_{u'_r}, post_{u'_r})$, where $u'_r$ is the unique real node of $u$. From the generation process of pre- and postorder index, $(pre_i, post_i)$ and $(pre_j, post_j)$ can never overlap for any vertex $i$ and $j$. Instead of utilizing $(pre_{v'}, post_{v'})$, we can only check whether $pre_{v'} \in (pre_{u'_r}, post_{u'_r})$. Hence, $post_{v'_i}$ index of all virtual nodes $v'_i$ will never be used in online phase, which means that we do not need to store $post$ index for all virtual nodes in offline phase.

Moreover, when vertex $v$ has lots of virtual nodes $v'_i$, checking whether $pre_{v'_i} \in (pre_{u'_r}, post_{u'_r})$ is not efficient enough. Instead of iterating them one by one for comparison, if all the virtual nodes $v'_i$ have been sorted w.r.t. their $pre_{v'_i}$ in offline phase, we can spend only $log(|v'_i|)$ to find the first virtual node whose $pre_{v'_i} > pre_{u'_r}$ and start iterating from it until $pre_{v'_i} > post_{u'_r}$, where $|v'_i|$ is the number of virtual nodes representing $v$. Note that the number of virtual nodes representing vertex $v$ is equal to its incoming degree in $G'$ minus 1, since in the extended spanning tree construction (Sect. 3.2), we create a virtual node for $v$ only when $v$ is visited again from an incoming neighbor. Hence, sorting all virtual nodes $v'_i$ w.r.t. $pre_{v'_i}$ for each vertex $v$ costs $O(|E|log(Deg_m^{(in)}))$, where $Deg_m^{(in)}$ is the maximum incoming degree of a vertex. And the overall time cost of offline indexing is $O(|V|log(Deg_m^{(out)}) + |E|log(Deg_m^{(in)}))$ if Prune II strategy is used in online phase.

---

**Algorithm 4.** *ESTI* Online Query Function $Query(u,v,k)$

---

**Input:** Start vertex $u$, target vertex $v$, $k$; Offline index $X, Y, EST, Pre, Post, Level, dist$.
**Output:** *True* or *False*.

1: **if** $X[u] > X[v] \vee Y[u] > Y[v] \vee level_{v'_r} - level_{u'_r} > k$ **then**         ▷ Prune I
2:     **return** *False*
3: $u'_r \leftarrow EST[u].RealNode$
4: $v'_i \leftarrow$ the first virtual node of $v$ s.t. $pre_{v'_i} > pre_{u'_r}$         ▷ Prune II
5: **while** $pre_{v'_i} < post_{u'_r}$ **do**
6:     **if** $level[v'] - level[u'_r] \leq k$ **then**
7:         **return** *True*
8:     $v'_i \leftarrow$ next virtual node of $v$
9: **if** $k > 1 \wedge Dist[u] < k$ **then**         ▷ Prune III
10:     **if** number of outgoing edges of $u \leq$ number of incoming edges of $v$ **then**
11:         **for** each outgoing neighbor $w$ of $u$ **do**
12:             **if** $Query(w,v,k-1)$ **then**
13:                 **return** *True*
14:     **else**
15:         **for** each incoming neighbor $w$ of $v$ **do**
16:             **if** $Query(u,w,k-1)$ **then**
17:                 **return** *True*
18: **return** *False*;

---

**Prune III.** For each real node $u'_r$ of $u$, while performing DFS traversal in offline index construction, we can find out $dist_u$ which represents the distance from $u'_r$ to the nearest virtual node $w'_i$ among all its successors in extended spanning tree. Given $dist$ index for every real node in the tree, for query $u \xrightarrow{?k} v$, if $dist_u \geq k$, we do not have to explore $u$'s successors. That is because when exploring from $u'_r$ in the tree, virtual nodes can only exists in the $k^{th}$ hop. Assume that $u$ can reach $v$ within $k$ hops. When one of $v$'s real or virtual node is in the subtree rooted at $u'_r$, the query will return *True* in Line 3–6 in Algorithm 3. When all of $v$'s real and virtual nodes are not in the subtree rooted at $u'_r$, there must exist a virtual node $w'_i$ which can jump out of the subtree to reach $v$. Note that $level_{w'_i} - level_{u'_r} < k$ holds, or it needs more than $k$ hops from $u$ to $v$. However, it contradicts $dist_u \geq k$ since the distance from $u'_r$ to $w'_i$ is smaller than $k$.

*Example 6.* In Fig. 5, for query $f \xrightarrow{?3} c$, the pre- and postorder index of $c$ is not in the interval of $f$'s index, i.e., $(11,16) \not\subset (17,24)$. Next, instead of exploring $g$ and $h$, we can safely return *False* directly since $dist_f = k = 3$.

### 4.3 Summary of Online Querying

After utilizing the three pruning strategies as discussed in Sect. 4.2, the *ESTI* query function $Query(u,v,k)$ is shown in Algorithm 4. Though in the worst case we still need to explore the whole graph, *ESTI* index still helps a lot for pruning online search space. Section 5 will demonstrate its practical efficiency.

**Table 1.** Statistics of datasets

| Graph | $|V|$ | $|E|$ | Graph | $|V|$ | $|E|$ |
|---|---|---|---|---|---|
| kegg | 3,617 | 4,395 | p2p-Gnutella31 | 62,586 | 147,892 |
| amaze | 3,710 | 3,947 | soc-Epinions1 | 75,879 | 508,837 |
| nasa | 5,605 | 6,538 | 10go-uniprot | 469,526 | 3,476,397 |
| go | 6,793 | 13,361 | 10cit-Patent | 1,097,775 | 1,651,894 |
| mtbrv | 9,602 | 10,438 | uniprotenc22m | 1,595,444 | 1,595,444 |
| anthra | 12,499 | 13,327 | 05cit-Patent | 1,671,488 | 3,303,789 |
| ecoo | 12,620 | 13,575 | WikiTalk | 2,394,385 | 5,021,410 |
| agrocyc | 12,684 | 13,657 | cit-Patents | 3,774,768 | 16,518,948 |
| human | 38,811 | 39,816 | citeseerx | 6,540,401 | 15,011,260 |
| p2p-Gnutella05 | 8,846 | 31,839 | go-uniprot | 6,967,956 | 34,770,235 |
| p2p-Gnutella06 | 8,717 | 31,525 | govwild | 8,022,880 | 23,652,610 |
| p2p-Gnutella08 | 6,301 | 20,777 | soc-Pokec | 1,632,803 | 30,622,564 |
| p2p-Gnutella09 | 8,114 | 26,013 | uniprotenc100m | 16,087,295 | 16,087,295 |
| p2p-Gnutella24 | 26,518 | 65,369 | yago | 16,375,503 | 25,908,132 |
| p2p-Gnutella25 | 22,687 | 54,705 | twitter | 18,121,168 | 18,359,487 |
| p2p-Gnutella30 | 36,682 | 88,328 | uniprotenc150m | 25,037,600 | 25,037,600 |

## 5   Experiments

We evaluate the effectiveness and efficiency of the proposed *ESTI* method by carrying extensive experiments on both small and large graphs. All the experiments are conducted on a Linux machine with an Intel(R) Xeon(R) E5-2678 v3 CPU @2.5GHz and 220G RAM, and all algorithms are implemented using C++ and complied by G++ 5.4.0 with -O3 Optimization. Each experiment has been run for 10 times and the results are consistent among 10 executions. In this section, we report the average value from 10 executions of each experiment.

### 5.1   Datasets

A variety of real graphs are used in our experiments, as shown in Table 1. *kegg, amaze, nasa, go, mtbrv, anthra, ecoo, agrocyc* and *human* are small graphs from different sources [13]. *p2p-Gnutella* graphs are 8 snapshots of Gnutella peer to peer file network, while *soc-Epinions1* is a who-trust-whom online social network [5]. As for large graphs, *10go-uniprot, go-uniprots, uniprotenc22m, uniprotenc100m* and *uniprotenc150m* come from Uniprot database. *10cit-Patent, 05cit-Patent, cit-Patents* and *citeseer* are citation networks [3]. *WikiTalk* is a Wikipedia communication network, while *soc-Pokec* and *twitter* are large-scale social networks [5,7]. *govwild* and *yago* are RDF datasets [7].

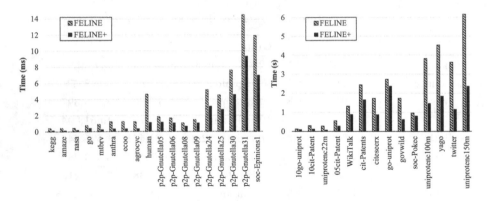

**Fig. 6.** Index construction time of FELINE and FELINE$^+$

## 5.2 Performance of FELINE$^+$

As discussed in Sect. 3.1, we propose an optimized approach named FELINE$^+$ to accelerate FELINE index generation, while obtaining exactly the same index as FELINE. Figure 6 shows the index construction time, in which FELINE$^+$ significantly speeds up the construction process in all graphs.

## 5.3 Queries with Different $k$

The efficiency of online querying is crucial for $k$-hop reachability query answering, and different values of $k$ can significantly affect the performance. We report the query time of the proposed *ESTI* method with different values of $k$ ($k = 2, 4, 8$) in Table 2, comparing it with the state-of-art $k$-reach approach [2]. For each $k$, we generate a million queries with randomly selected start and target vertices. Note that $k$-reach requires a fixed budget $b$ to construct the partial vertex cover and we set $b = 1000$, which is the same as the budget used in [2].

When the value of $k$ increases, the time cost of both $k$-reach and *ESTI* also tend to increase, because a larger $k$ indicates a larger search space when the built index cannot directly answer a query. We notice that most of queries fall into the worst case in $k$-reach, which needs traditional BFS search over the whole graph. Note that when $k = 4$ and $k = 8$, $k$-reach exceeds our time limit (4 h) in graph *soc-Pokec*. Clearly, *ESTI* is faster than $k$-reach over all graphs when $k = 2$ and $k = 4$, and it also beats $k$-reach in most graphs except for graph *WikiTalk*. Note that the diameter of *WikiTalk* is 9, which is relatively small and is quite closed to $k = 8$. In practice, $k$ will not be too large for social networks.

**Table 2.** Query time (ms) of different $k$

| Graph | k = 2 | | k = 4 | | k = 8 | |
|---|---|---|---|---|---|---|
| | k-reach | ESTI | k-reach | ESTI | k-reach | ESTI |
| kegg | 63 | **27** | 97 | **42** | 103 | **40** |
| amaze | 58 | **25** | 83 | **34** | 90 | **37** |
| nasa | 96 | **20** | 193 | **33** | 212 | **45** |
| go | 177 | **36** | 380 | **70** | 391 | **111** |
| mtbrv | 66 | **16** | 124 | **24** | 116 | **23** |
| anthra | 68 | **17** | 108 | **25** | 129 | **26** |
| ecoo | 67 | **17** | 123 | **25** | 114 | **28** |
| agrocyc | 72 | **17** | 105 | **26** | 124 | **27** |
| human | 69 | **20** | 138 | **26** | 255 | **29** |
| p2p-Gnutella05 | 630 | **95** | 8,069 | **723** | 47,595 | **9,507** |
| p2p-Gnutella06 | 624 | **93** | 9,260 | **755** | 56,856 | **9,890** |
| p2p-Gnutella08 | 656 | **73** | 5,654 | **462** | 26,063 | **5,061** |
| p2p-Gnutella09 | 529 | **74** | 5,078 | **482** | 34,501 | **7,006** |
| p2p-Gnutella24 | 467 | **74** | 4,862 | **514** | 100,276 | **15,639** |
| p2p-Gnutella25 | 509 | **65** | 4,534 | **430** | 79,991 | **14,438** |
| p2p-Gnutella30 | 668 | **74** | 6,166 | **478** | 129,051 | **24,226** |
| p2p-Gnutella31 | 806 | **78** | 5,784 | **503** | 195,265 | **34,490** |
| soc-Epinions1 | 132,712 | **596** | 999,966 | **3,747** | 765,753 | **11,932** |
| 10go-uniprot | 788 | **109** | 1,622 | **204** | 2,505 | **469** |
| 10cit-Patent | 245 | **90** | 400 | **159** | 815 | **322** |
| uniprotenc22m | 491 | **77** | 644 | **102** | 776 | **129** |
| 05cit-Patent | 458 | **130** | 722 | **222** | 1,649 | **480** |
| WikiTalk | 1,112,536 | **240** | 8,091,777 | **842** | **769,542** | 11,162,590 |
| cit-Patents | 5,259 | **382** | 36,879 | **1,605** | 306,144 | **21,195** |
| citeseerx | 927 | **205** | 2,935 | **264** | 23,154 | **763** |
| go-uniprot | 1,196 | **214** | 1,386 | **201** | 2,744 | **351** |
| govwild | 3,419 | **147** | 9,993 | **211** | 19,229 | **483** |
| soc-Pokec | 1,510,794 | **4,057** | - | **653,194** | - | **6,430,662** |
| uniprotenc100m | 744 | **95** | 987 | **113** | 1,748 | **160** |
| yago | 501 | **113** | 861 | **168** | 1,255 | **257** |
| twitter | 592 | **211** | 647 | **215** | 1,432 | **435** |
| uniprotenc150m | 938 | **103** | 1,367 | **146** | 2,019 | **205** |

**Table 3.** Index size, index construction time and query time on small graphs

| Graph | Index size (KB) | | Index time (ms) | | Query time (ms) | |
|---|---|---|---|---|---|---|
| | k-reach | ESTI | k-reach | ESTI | k-reach | ESTI |
| kegg | 129 | **101** | 80 | **1** | 107 | **44** |
| amaze | 127 | **101** | 77 | **1** | 97 | **42** |
| nasa | 229 | **139** | 78 | **1** | 200 | **43** |
| go | 284 | **212** | 81 | **3** | 298 | **68** |
| mtbrv | 345 | **233** | 76 | **2** | 120 | **31** |
| anthra | 448 | **301** | 79 | **3** | 118 | **30** |
| ecoo | 452 | **305** | 79 | **3** | 120 | **30** |
| agrocyc | 454 | **306** | 81 | **3** | 119 | **30** |
| human | 1,380 | **922** | 87 | **10** | 121 | **32** |
| p2p-Gnutella05 | 450 | **389** | 80 | **7** | 34,067 | **5,659** |
| p2p-Gnutella06 | 445 | **384** | 79 | **7** | 39,910 | **5,747** |
| p2p-Gnutella08 | 383 | **262** | 80 | **4** | 19,977 | **2,989** |
| p2p-Gnutella09 | 407 | **332** | 80 | **6** | 24,226 | **4,514** |
| p2p-Gnutella24 | 1,680 | **931** | 89 | **19** | 77,227 | **12,457** |
| p2p-Gnutella25 | 2,102 | **787** | 91 | **14** | 61,227 | **9,966** |
| p2p-Gnutella30 | 3,056 | **1,269** | 99 | **28** | 103,566 | **11,235** |
| p2p-Gnutella31 | 5,189 | **2,143** | 123 | **55** | 160,885 | **23,031** |
| soc-Epinions1 | 50,211 | **5,361** | 957 | **134** | 1,214,127 | **6,579** |

## 5.4   Comparison with the State-of-art

As discussed in Sect. 1, $k$-reach [1,2] is the only method solving $k$-hop reachablity queries on general directed graphs. We conduct experiments on both small and large graphs to compare the proposed *ESTI* method with $k$-reach. For each graph, we randomly generate a million queries while values of $k$ are generated following the distance distribution of all reachable pairs. Their index size, index construction time and query time are reported in Table 3 and 4.

The results in Table 3 shows that *ESTI* completely beats $k$-reach in all small graphs. Note that the budget of $k$-reach is also set to be 1000. *ESTI* constructs smaller index and is approximately an order of magnitude faster when building index for most small graphs. As for online querying, *ESTI* costs significantly less time. It is even more than a hundred times faster in graph *soc-Epinions1*.

For large graphs, we compare our *ESTI* method with $k$-reach in Table 4, where the budget of $k$-reach are set to be 1,000 and 50,000, respectively. Note that $k$-reach exceeds our time limit (4 h) on graph *soc-Pokec*. When answering queries online, *ESTI* method costs much less time over all large graphs. Though *ESTI* needs longer index construction time on most graphs, we believe that the efficiency of online query processing is more important than

**Table 4.** Index size, index construction time and query time on large graphs

| Graph | Index size (MB) | | | Index time (s) | | | Query time (s) | | |
|---|---|---|---|---|---|---|---|---|---|
| | k-reach (b=1k) | k-reach (b=50k) | ESTI | k-reach (b=1k) | k-reach (b=50k) | ESTI | k-reach (b=1k) | k-reach (b=50k) | ESTI |
| 10go-uniprot | **24** | 24 | 34 | 0.7 | **0.4** | 1.1 | 1.2 | 1.0 | **0.2** |
| 10cit-Patent | **26** | 39 | 31 | **0.2** | 0.6 | 0.8 | 0.4 | 0.4 | **0.1** |
| uniprotenc22m | 55 | 55 | **37** | 0.6 | **0.5** | 0.8 | 0.4 | 0.4 | **0.1** |
| 05cit-Patent | **41** | 60 | 53 | **0.3** | 1.2 | 1.7 | 0.6 | 0.7 | **0.2** |
| WikiTalk | 217 | 217 | **75** | 4.7 | 4.5 | **4.0** | 6392 | 6049 | **0.8** |
| cit-Patents | **181** | 558 | 188 | **5.2** | 22.4 | 9.2 | 94.0 | 92.5 | **9.7** |
| citeseerx | **171** | 237 | 219 | **3.5** | 11.1 | 6.4 | 2.1 | 2.2 | **0.3** |
| go-uniprot | 331 | **321** | 372 | 9.8 | **6.9** | 13.7 | 1.0 | 1.0 | **0.3** |
| govwild | **256** | 262 | 314 | 5.1 | **4.4** | 8.0 | 9.9 | 6.6 | **0.2** |
| soc-Pokec | **183** | **183** | 260 | 11.3 | **10.5** | 13.8 | - | - | 2281 |
| uniprotenc100m | 556 | 556 | **370** | 8.2 | **6.0** | 10.0 | 0.7 | 0.7 | **0.1** |
| yago | 411 | 446 | 472 | **2.9** | 4.2 | 12.3 | 0.5 | 0.6 | **0.1** |
| twitter | 609 | 609 | **443** | **5.5** | 6.9 | 11.3 | 0.6 | 0.6 | **0.3** |
| uniprotenc150m | 866 | 866 | **576** | 14.3 | **10.1** | 16.8 | 0.9 | 0.9 | **0.1** |

offline indexing. Theorectically, the overall time cost of *ESTI* offline indexing is $O(|V|log(Deg_m^{(out)}) + |E|log(Deg_m^{(in)}))$, which is a stable bound.

The index size of *ESTI* is $O(|V| + |E|)$, which is strictly linear to the size of input graph. However, $k$-*reach* with budget 1,000 has the smallest index size on some large graphs, and it also costs a lot of time to answer queries online. It seems that 1,000 is a relatively small budget, which may limit the querying performance of $k$-*reach*. But when the budget is set to be 50,000, $k$-*reach* has larger index size than *ESTI* in many graphs, while it still cost more time in online querying process. Hence, the overall query answering ability of *ESTI* method is also better over large graphs.

# 6  Related Works

## 6.1  Reachabilty Query

Before Cheng et al. [1] first proposed $k$-hop reachability problem, lots of studies about reachability query over large graphs have been carried. Reachability query is a special case of $k$-hop reachability query when $k = \infty$. Since the lack of distance information, existing reachability query methods including *BFL* [8], *IP+* [11], *GRIPP* [9], *PWAH8* [6], *GRAIL* [13] and *Path-Tree Cover* [4], etc. are not sufficient to answer $k$-hop reachability queries.

## 6.2  $k$-hop Reachabilty Query

To answer $k$-hop reachability problems, a naive idea is to process BFS or DFS in given directed graph. Both BFS and DFS don't need any pre-computed index, but they are not efficient when the graph becomes very large, since lots of search branches will be expanded while exploring in the original large graph. In contrast, storing the shortest distance between each pair of vertices helps to answer any queries within $O(1)$ time. However, in order to compute and store such distance, performing BFS from every vertex in $G$ costs $O(|V|(|V|+|E|))$ time and $O(|V|^2)$ space, which is also inefficient and even infeasible for large graphs.

**Vertex Cover Based Method.** Vertex cover is a subset of all the vertices in a given graph $G$, making sure that for each edge in $G$, at least one of the two vertices connected by this edge is contained in the vertex cover. *k-reach* [1,2] makes good use of vertex cover, and runs BFS in the subgraph constructed from vertex cover to build index. Though it is proved efficient in small graphs, when dealing with larger graphs, *k-reach* still costs infeasible index time and space.

To overcome this drawback, Cheng et al. also proposes a partial vertex cover [2] to make a trade-off between offline index and online query performance. Though it can work on very large graphs, the partial vertex cover index cannot answer a large proportion of online queries directly. In fact, traditional online BFS would be invoked for more than 95% of the queries. Hence, it is still not practical enough for answering $k$-hop reachability queries efficiently.

**Methods Work on DAGs.** To improve index efficiency, Xie et al. [12] proposed *BFSI-B* Algorithm, which uses the breadth-first spanning tree to build *BFSI* index, including *min-post* index and global BFS level $TLE$. Also, *FELINE* index [10] is adopted to filter those unreachable queries. Another method developed for DAGs is *HT* [3], which adopts the idea of partial 2-hop cover. In its indexing process, vertices with high degree are selected as hop nodes. Both backward and forward BFS are started from each hop node $u$. When visiting a new vertex $v$, current hop node's id $u$ and the distance from $u$ to $v$ will be stored as the index of $v$. Topological order is also used for filtering unreachable queries.

Though both *BFSI-B* and *HT* are more efficient than *k-reach*, they can only work for DAGs and cannot directly deal with directed graphs with cycles. Also, more efficient pruning strategies need to be utilized to further improve online querying performance.

**Algorithms for Distributed Systems.** To deal with multiple $k$-hop reachability queries concurrently on distributed infrastructures, *C-Graph* [14] focuses on improving both disk and network I/O performance when performing BFS. Compared with developing methods for a single machine, designing optimizations for distributed systems is a significantly different task.

## 7   Conclusion

We propose *ESTI* method to efficiently solve $k$-hop reachability queries for general directed graphs, which builds an extended spanning tree in offline phase and utilizes three pruning strategies to accelarte query processing. Also, an optimization named FELINE$^+$ is developed to speeds up FELINE index generation, which helps to effectively filter unreachable queries in online searching.

We also conduct extensive experiments to compare *ESTI* with the state-of-art method *k-reach*. Our experiment results confirm that on most graphs the overall performance of *ESTI* is the best, and in online querying it is significantly faster.

**Acknowledgments.** This work was supported by National Natural Science Foundation of China (Grant No. 61902074) and Science and Technology Committee Shanghai Municipality (Grant No. 19ZR1404900).

# References

1. Cheng, J., Shang, Z., Cheng, H., Wang, H., Yu, J.X.: K-reach: who is in your small world. CoRR abs/1208.0090 (2012)
2. Cheng, J., Shang, Z., Cheng, H., Wang, H., Yu, J.X.: Efficient processing of k-hop reachability queries. VLDB J. **23**(2), 227–252 (2014)
3. Du, M., Yang, A., Zhou, J., Tang, X., Chen, Z., Zuo, Y.: HT: a novel labeling scheme for k-hop reachability queries on DAGs. IEEE Access **7**, 172110–172122 (2019)
4. Jin, R., Xiang, Y., Ruan, N., Wang, H.: Efficiently answering reachability queries on very large directed graphs. In: SIGMOD 2008, pp. 595–608 (2008)
5. Leskovec, J., Krevl, A.: SNAP Datasets: Stanford large network dataset collection
6. van Schaik, S.J., de Moor, O.: A memory efficient reachability data structure through bit vector compression. In: SIGMOD 2011, pp. 913–924 (2011)
7. Seufert, S., Anand, A., Bedathur, S., Weikum, G.: Ferrari: flexible and efficient reachability range assignment for graph indexing (2013)
8. Su, J., Zhu, Q., Wei, H., Yu, J.X.: Reachability querying: can it be even faster? IEEE Trans. Knowl. Data Eng. **29**(3), 683–697 (2017)
9. Trißl, S., Leser, U.: Fast and practical indexing and querying of very large graphs. In: SIGMOD 2007, pp. 845–856 (2007)
10. Veloso, R.R., Cerf, L., Meira Jr., W., Zaki, M.J.: Reachability queries in very large graphs: a fast refined online search approach. In: EDBT, pp. 511–522 (2014)
11. Wei, H., Yu, J.X., Lu, C., Jin, R.: Reachability querying: an independent permutation labeling approach. Proc. VLDB Endow. **7**(12), 1191–1202 (2014)
12. Xie, X., Yang, X., Wang, X., Jin, H., Wang, D., Ke, X.: BFSI-B: an improved k-hop graph reachability queries for cyber-physical systems. Inf. Fusion **38**, 35–42 (2017)
13. Yildirim, H., Chaoji, V., Zaki, M.: Grail: a scalable index for reachability queries in very large graphs. VLDB J. **21**, 1–26 (2012)
14. Zhou, L., Chen, R., Xia, Y., Teodorescu, R.: C-graph: a highly efficient concurrent graph reachability query framework. In: ICPP, pp. 79:1–79:10. ACM (2018)

# NREngine: A Graph-Based Query Engine for Network Reachability

Wenjie Li[1], Lei Zou[1(✉)], Peng Peng[2], and Zheng Qin[2]

[1] Peking Universtity, Beijing, China
{liwenjiehn,zoulei}@pku.edu.cn
[2] Hunan University, Changsha, China
{hnu16pp,zqin}@hnu.edu.cn

**Abstract.** A quick and intuitive understanding of network reachability
is of great significance for network optimization and network security
management. In this paper, we propose a query engine called *NREngine*
for network reachability when considering the network security policies.
NREngine constructs a knowledge graph based on the network security policies and designs an algorithm over the graph for the network
reachability. Furthermore, for supporting a user-friendly interface, we
also propose a structural query language named *NRQL* in NREngine
for the network reachability query. The experimental results show that
NREngine can efficiently support a variety of network reachability query
services.

**Keywords:** Network reachability · Network security policies · RDF ·
Graph database

## 1 Introduction

Network reachability is an important basis for network security services, which
has attracted more and more attentions of the experts and scholars. Network
reachability is a functional characteristic of the network, which ensures smooth
communication between nodes in order for users to conveniently access the
resources of the network [1].

Considering the requirement of network security or privacy protection, users
usually configure various of security policies in network devices such as firewalls,
routers and so on. Security policies usually restrict the users' access to the network, and its function is to control network reachability. Obviously, there needs
to be a balance between ensuring normal network communication and achieving
network security or privacy protection. In another word, the network reachability needs to be maintained within a suitable range. If the network reachability
is more than the actual requirement, it may cause unnecessary communication,
or even create opportunities for the malicious attacks; and the network reachability that is less than the actual requirement will disrupt the normal network

© Springer Nature Switzerland AG 2021
C. S. Jensen et al. (Eds.): DASFAA 2021 Workshops, LNCS 12680, pp. 90–106, 2021.
https://doi.org/10.1007/978-3-030-73216-5_7

services, and even lead to huge economic losses. Therefore, the network must have suitable reachability.

In order to measure the network reachability, there are many traditional methods to check whether the network is reachable or not by using the ping, traceroute or other tools. These methods have the following two shortcomings. First, the results are dependent on the state of the devices. If some devices in the query path get offline, the query results may be always unreachable. Second, these methods send the test data packet (such as ICMP data packet) to evaluate the network reachability, which costs a lot of network resources.

With the increase of network devices and the expansion of network scale, it will become a hot and difficult point to quantify the reachability model of the whole network. Furthermore, it has important theoretical value and application prospects to validate the network reachability through an efficient network reachability query approach, and intelligently locate the defects in security policies configuration according to the query results, and optimize the security policies configuration and network performance.

After constructing the network reachability model, we can use the graph traversal search algorithms to query the reachability. Those methods require searching the graph globally for network reachability and often have low performances. Furthermore, if the network reachability model is stored in two-dimensional database tables or files, we should reconstruct the graph when querying, which greatly reduces the performance dramatically; If the network reachability model is stored in memory, once the system is offline or downtime, it cannot save the data of the network reachability model and also are limited by the memory capacity. Therefore, in this study, we build up a knowledge graph of network reachability based on network security policies and transform the network reachability query into queries over the knowledge graphs. Then, we can maintain the knowledge graph of network reachability in graph databases, like gStore [2,3], Jena [4], rdf4j [5] and Virtuoso [6], which can gain the high performances when evaluating queries over the knowledge graphs of network reachability.

## 1.1 Key Contributions

In this paper, we focus on the query of the network reachability. The main contributions of this paper are as follows:

1. We propose a novel model for the network reachability based on network security policies, and construct a knowledge graph of network reachability.
2. We extend the structured query language over knowledge graphs and propose a new structured query language called *NRQL* for network reachability. We design the efficient query algorithms for evaluating NRQL statements over the knowledge graph of network reachability.
3. We propose a novel query engine for the network reachability called *NREngine*, which implements all the above techniques.
4. To evaluate the effectiveness and efficiency of *NREngine*, we conduct extensive experiments.

## 2   Related Work

Recently, there are some effective works on the network reachability. Xie *et al.* have made a pioneering work, they define the network reachability and propose a method to model the static network reachability [1]. The key idea is to extract the configuration information of routers in the network and reconstruct the network into a graph in a formal language, so the network reachability can be calculated through classical problems such as closure and shortest path.

Zhang *et al.* propose a method to merge the IP addresses with the same reachability into the IP address sets [7]. When the network reachability is changed, the affected IP address sets can be reconstructed quickly by splitting or merging to update the network reachability in real-time. However, they do not provide an algorithm to answer whether an IP is reachable along a certain path to another IP. Benson *et al.* propose the concept of the policies unit [8]. The policies unit is a set of IP addresses affected by the same security policies. A policies unit may be distributed in many subnets, or there may be many different policies units in a subnet. They also do not provide an algorithm for the network reachability query.

Amir *et al.* propose a network reachability query scheme based on network configurations (mainly ACLs), and construct a network reachability query tool called "Quarnet" [9,10]. Its ACL model still adopts FDD model, the queries need to be split by the paths of the FDD. Chen *et al.* propose the first cross-domain privacy-preserving protocol for quantifying network reachability [11]. The protocol constructs equivalent representations of the ACL rules and determines network reachability while preserving the privacy of the individual ACLs. They do not consider the other network security policies(such as route table). Hone *et al.* propose a new method to detect IP prefix hijacking based on network reachability, which is a specific application of network reachability [12].

Recently, there are some effective works on network research based on the graph. Liang *et al.* propose an improved hop-based reachability indexing scheme 3-Hop which gains faster reachability query evaluation, which has less indexing costs and better scalabilities than state-of-the-art hop-based methods, and they propose a two-stage node filtering algorithm based on 3-Hop to answer tree pattern queries more efficiently [13]. Rao *et al.* propose a model of network reachability based on decision diagram [14]. Li *et al.* propose a verification method of network reachability based on the topology path. They transform the problem of the communication need into the verification problem of topology path reachability via SNMP and Telnet-based topology discovery and graph theory techniques [15]. Alfredo *et al.* propose a novel reachability-based theoretical framework for modeling and querying complex probabilistic graph data [16]. Hasan proposes a novel knowledge representation framework for computing sub-graph isomorphic queries in interaction network database [17].

## 3   Overview

### 3.1   Problem Definitions

The essence of network reachability query is to determine whether a certain type of network packet can reach another node from one node or from one subnet to another. Given two subnets $N_1$ and $N_2$, and two host nodes $v_1$ and $v_2$, where $v_1$ is a node of $N_1$, so $v_1 \in N_1$ holds, and $v_2$ is a node of $N_2$, so $v_2 \in N_2$ holds. The network reachability query in this paper can be divided into three categories as follows according to the query targets.

**Node to Node.** This category of query is mainly used to check the network reachability between nodes. We use $v_1 \rightarrow v_2$ to denote that $v_1$ to $v_2$ is reachable, and use $v_1 \nrightarrow v_2$ to denote that $v_1$ to $v_2$ is unreachable.

**Node to Subset.** This category of query is mainly used to check the network reachability between nodes and subsets. We use $v_1 \rightarrow N_2$ to denote that $v_1$ to $N_2$ is reachable, and use $v_1 \nrightarrow N_2$ to denote that $v_1$ to $N_2$ is unreachable. Obviously, the following formula holds.

$$v_1 \rightarrow N_2 = \{\exists v_j, \quad v_1 \rightarrow v_j\}(v_j \in N_2)$$

**Subset to Subset.** This category of query is mainly used to check the network reachability between subnets. We use $N_1 \rightarrow N_2$ to denote that $N_1$ to $N_2$ is reachable, and use $N_1 \nrightarrow N_2$ to denote that $N_1$ to $N_2$ is unreachable. Obviously, the following formula holds.

$$N_1 \rightarrow N_2 = \{\exists v_i, v_j, \quad v_i \rightarrow v_j\}(v_i \in N_1, v_j \in N_2)$$

We also can divide the network reachability query into two categories as follows according to the result of query.

1. **Boolean query.** The result of the query is a boolean value (such as yes or no). For example, "SMTP server 192.168.0.32 to host 192.168.0.54 is reachable?", and the result is "yes" or "no".
2. **Node query.** The result of the query is a set of nodes that satisfy the query condition. For example, "Which hosts in subset 192.168.0.0/24 can receive the email from the SMTP server 192.168.0.32?". The result is a set of nodes.

## 3.2   System Architecture

In this paper, we propose a query engine for network reachability based on network security policies, *NREngine*. Figure 1 shows the system architecture of NREngine. NREngine consists of two parts as follows.

In the offline part of NREngine, we collect and organize the network security policies (including ACLs and routing table). First, we remove the network security policies where the action field value is *deny* and extend OPTree [18] for maintaining the network security policies. OPTree is a homomorphic structure of network security policies, and the redundancy policies and the conflict policies can be removed when constructing OPTree. Then, we propose a network reachability model based on the network topology and the network security policies, and construct a knowledge graph based on the network reachability model. To support efficient evaluation of the network reachability query, we maintain the knowledge graph in graph databases, like gStore [2,3], Jena [4], rdf4j [5] and Virtuoso [6].

The online part of NREngine is mainly responsible for processing user query requests. In this part, we propose a structured query language, called *NRQL* (*Network Reachability Query Language*) for network reachability query. Users send the query requests to NREngine using NRQL statements. The key of NREngine query is to match the query conditions by using OPTree in the network reachability model. In order to adapt to *OPTree*, we propose a NRQL query parsing algorithm based on OPTree query algorithm. NREngine can provide three categories of queries: "node to node", "node to subnet" and "subnet to subnet". The results of those queries can be either "Yes" or "NO", or a set of nodes satisfying the query conditions.

Considering the high real-time requirement of network reachability query, the relatively low frequency of network security policies change, and the long time-consuming construction of the network reachability model based on OPTree, we construct or update the network reachability model through timing schedule in the offline part. In the online part, NREngine provides a real-time network reachability query service based on the high query efficiency of graph databases. In terms of system deployment, the online part and the offline part can be deployed independently, in which the offline part can be deployed in the intranet environment to isolate ordinary users, and the online part can be deployed in the extranet environment to provide network reachability query services to ordinary users by the GUI of NREngine.

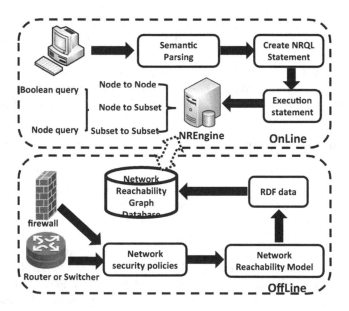

**Fig. 1.** The system architecture of NREngine

# 4    The Offline Part of NREngine

## 4.1    Network Reachability Model Based on Network Security Policies

The essence of network reachability in our paper is to determine whether network data packets can be transferred from one node to another. Many factors affect whether network data packets can be transmitted from one node to another, such as the state of the devices. The devices include firewalls, routers, switchers and hosts. If a device is not online, the data packet can not be transmitted through it. However, the state of the device is an instantaneous state, which may be man-made shutdown or the device failure. By opening or repairing the fault, the state of the device can be changed, so it cannot reflect the basic state of the network.

Another key factor affecting the network data package is the network security policies. The network security policy is not an instantaneous state, and cannot be changed due to the device off-line or the device failure. Therefore, the network security policies can well reflect the basic state of the network, and we can build a model for the network reachability based on the network security policies. Noted that, because the security policies in hosts are managed by their managers and it is hard to collect the security policies, we do not consider those security policies in our paper.

First, we formally define the network reachability model. Given a subnet $N$ and there are $n$ devices $(D_1, D_2, \ldots, D_n)$ in $N$. Here, we use the graph as the basic model of the network reachability model. The network is denoted as $G = (V, E, L)$, where $V$ denotes the set of devices like firewall, router or switcher in the subnet $N$; $E$ denotes a set of the edges between vertices and $L$ is the labels of the edges based on the network security policies. According to the properties of ACL and routing table, if $v_i$ has only one outgoing edge, $v_i$ denotes a device which has packet filtering function, such as firewall. Otherwise $v_i$ denotes a device which has packet forwarding function, such as router or switcher. Obviously, $G$ is a directed graph. Figure 2 shows an example of the network reachability model. In Fig. 2, $v_5$, $v_6$ and $v_7$ denote the network security devices which have packet filtering function, and $v_1$, $v_2$, $v_3$ and $v_4$ denote network security devices which packet forwarding function.

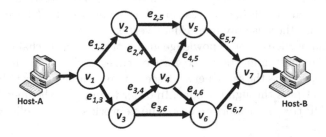

**Fig. 2.** An example of the network reachability model

The construction of network reachablity model $G = (V, E)$ for subset $N$ has two steps as follow as.

**Step 1: Creating the Vertices Set for Devices.** We create a vertex for each network security device in subset $N$. Noted that if a network security device $D_i$ not only has a packet filtering function but also has a packet forwarding function, in other words, the action field's values of the network security policies $R_i$ in $D_i$ have three categories: *Accept*, *Deny* and *NextDevice*. we should create two virtual vertexes $D_i'$ and $D_i''$ to denote $D_i$, and $D_i'$ only has a packet filtering function, and $D_i''$ only has a packet forwarding function.

**Step 2: Linking the Vertices of Devices.** According to the topological structure of the network $N$ and the flow direction of data packets, the edges between nodes are constructed. Given a vertex $v_i$ with packet filtering function, and the next node is $v_j$, we construct an edge $e_{i,j}$ that from $v_i$ to $v_j$, and construct the OPTree $T_{v_i}$ based on the network security policies which in $v_i$ , we use $L(e)$ to denote the label of the edge $e$, that is, $L(e_{i,j}) = T_{v_i}$. If $v_i$ has the packet forwarding function, then there are many next-hop nodes of $v_i$. we construct an edge $e_{i,j}$ for each next-hop node $v_j$, and use $R(i,j)$ to denote the network security policies which is in $v_i$ and the next-hop node is $v_j$, and we construct the OPTree $T_{R(i,j)}$ and $L(e_{i,j}) = T_{R(i,j)}$.

### 4.2 Knowledge Graph of Network Reachability

After constructing the network reachability model, the next step is to efficiently evaluate the network reachability query. Existing methods of maintaining the network reachability model in two-dimensional database tables or files have low performances or are limited by memory capacity. Thus, in this study, we build up a knowledge graph of network reachability based on network security policies and transform the network reachability model into edges of knowledge graphs. Figure 3 shows an example knowledge graph of network reachability. Then, we can maintain the knowledge graph of network reachability in graph databases, like gStore [2,3], Jena [4], rdf4j [5] and Virtuoso [6], which can gain the high performance of evaluating network reachability queries.

The schema of vertices in our knowledge graph of network reachability is shown in Table 1, which includes three categories of vertices ("DeviceType", "Network" and "Edge"). For the devices that have the packet forwarding function, there is more than one "Edge" vertex, and for the devices that have the packet filtering function, there is only one "Edge" vertex.

The schema of edges in our knowledge graph of network reachability is shown in Table 2. The most important category of edges is "Label". The value of a "Label" edge represents the set of network security policies which is to be matched by a data packet pass through the edge. In order to improve the query efficiency, we maintain OPTree in memory to store the network security policies.

**Table 1.** Schema of vertices in knowledge graph of network reachability

| Name | Type | Remark |
|---|---|---|
| DeviceType | Property | The device type of the vertex, such as firewall, router, switcher, and host |
| Network | Property | The network of the vertex belongs, such as $N_1$ |
| Edge | Resource | The edge of the vertex, such as $e_{1,2}$ |

**Table 2.** Schema of edges in knowledge graph of network reachability

| Name | Type | Remark |
|---|---|---|
| Label | Property | The label value of the edge. In this study, the value is an OPTree. |
| NextNode | Resource | The vertex which the edge point to |

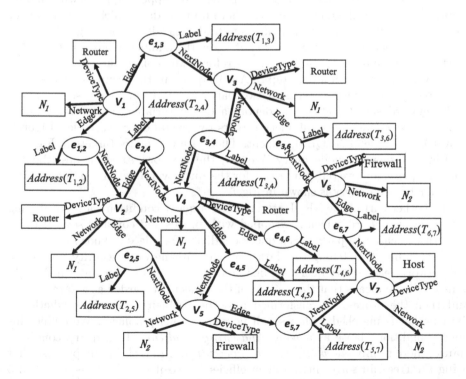

**Fig. 3.** Example knowledge graph of network reachability

```
Select ?x,?y
Where
{
   ?x  Network ?y.
   ?x  DeviceType "Host".
   ?z  NextNode ?x.
   ?z  Label ?k.
   ?k  <nr:in> <SIP=192.168.32.0/24,SP=any, DIP=192.168.24.212,DP=23,P=tcp>.
}
```

**Fig. 4.** Example point query

## 5   The Online Part of NREngine

### 5.1   Structured Query Language for Network Rechability

In this section, we extend the structured query language over knowledge graphs, SPARQL [19], for describing the user's network reachability query request over the knowledge graph of network rechability. The extended structured query language is called *NRQL* (Network Reachability Query Language).

Generally, SPARQL does not support matching operations related to the *policy matching condition*, a SPARQL statement only supports queries with limited steps. If the target of a query statement is to find the data which meet the query conditions within 3 steps, its meaning is to find data within the range of 3 edges. In network reachablility query, we do not know the number of edges between the starting node and the target node. Therefore, we replace the starting vertex recursively and perform a one step SPARQL query in the recursive process.

A NRQL statement consists of two parts: a query target clause beginning with keyword *select* and a ? to denote a query variable. Therefore, a query target clause contains several query targets by several query variables. Figure 4 shows an example of NRQL statement. There are two query targets in Fig. 4.

Figure 4 shows a node query. We use the keyword *exist* to describe the boolean query, and we only add *exist* for the query variables of the query statement for boolean query is shown in Fig. 5.

The other part of the NRQL statement is a query condition clause beginning with keyword *where*, which is wrapped with braces and contains several triple patterns. Each triple pattern is a query condition for edges in knowledge graph. Here, to support matching the data package with network security policies in the network reachability query, we define a new predicate "<nr: in>" in NRQL. This predicate means that the content of the object is regarded as a data packet, and then the process is transformed into the problem of checking whether a data packet is matched with a set of network security policies. We define this query condition clause as a *policy matching condition*. In a query condition clause, there is only one *policy matching condition*. In [18], Li *etc.* propose that using OPTree can solve this problem efficiently, so it can be transformed into finding a predicate path in OPTree, which can be solved by using OPTree search algorithm.

```
Select exist(?x)
Where
{
    ?x  Network ?y.
    ?x  DeviceType "Host".
    ?z  NextNode ?x.
    ?z  Label ?k.
    ?k  <nr:in> <SIP=192.168.32.0/24,SP=any, DIP=192.168.24.212,DP=23,P=tcp>.
}
```

**Fig. 5.** Example Boolean query

## 5.2   Execution of NRQL Statements

After we generate NRQL query statements based on the user's query request, then we execute the statements over the knowledge graph of network reachability. Unfortunately, most graph databases for knowledge graphs, like gStore [2,3], Jena [4], rdf4j [5] and Virtuoso [6], do not directly support the NRQL query statements, but they only support the SPARQL query statements. Therefore, when NRQL query statements are executed, additional parsing and pre-processing of NRQL query statements are needed.

---

**Algorithm 1: NRQLBooleanQuery$(Q, G)$**

---

**Input**: $Q$:The NRQL query statement.
$G$:The network reachability model.
**Output**: $True$:Network is reachable; $False$:Network is unreachable
1  set $v_c = v_{start}$;
2  $result = checkIsMatch(v_c, Q_{select}, R_{kg}, r_{match})$;
3  **if** $result == false$ **then**
4  |    **return** false;
5  **else**
6  |    generate a sparql statement $ask$ which the condition clause is $R_{kg} - r_{match}$.
   |    /* Execute the $ask$ by the query api of the graph database    */
7  |    **return** $graphDatabase.query(ask)$;

---

Before describing the query algorithm, we formally define the common concepts which would be used in the algorithm. We use $G = (V, E, L)$ to denote the network reachability model, and use $Q = \{Q_{select}, Q_{where}\}$ to denote a NRQL query statement, in which $Q_{select}$ denotes the query target clause and $Q_{where}$ denotes the query condition clause. According to the above description, the query condition clause contains a set of triple patterns, so $Q_{where} = R_{kg}$, where $R_{kg}$ denotes a set of edges in the knowledge graph of network reachability and $r_i$ denotes a triple tuple in the set of $R_{kg}$. The $r_{match}$ denotes the *policy matching condition*.

We design the query algorithm of the NRQL according to the categories of the query of network reachability and the categories of the query result. For boolean query, we design the boolen query algorithm, whose pseudo-code is shown as Algorithm 1.

---

**Function** Boolean checkIsMatch($v_c, Q_{select}, R_{kg}, r_{match}$)

---
1  **if** $v_c \in Q_{select}$ **then**
2    |   **return** true;
3  **else**
    |   /* get the next vertexes of $v_c$ which match with $r_{match}$      */
4    |   set $V_{next}=getNextVertexes(v_c, r_{match})$;
5    |   **if** $V_{next} \neq \emptyset$ **then**
    |   |   /* match $r_{match}$, then call the recursive function      */
6    |   |   set $flag=true$;
7    |   |   **for** $i = 0$ **to** $V_{next}.length$ **do**
8    |   |   |   set $v_c=V_{next}[i]$;
9    |   |   |   $flag=flag \parallel checkIsMatch(v_c, Q_{select}, R_{kg}, r_{match})$;
10  |   |   **if** $flag==false$ **then**
11  |   |   |   **return** false;
12  |   |   **else**
13  |   |   |   **return** true;
14  |   **else**
15  |   |   **return** false;

---

There are two key functions for boolean query in Algorithm 1. Function *checkIsMatch* is a recursive function. For example, in node to node query, we use $v_{start}$ to denote the start vertex, and use $v_{end}$ to denote the end vertex. Firstly, we set $v_c=v_{start}$ and call the function *getNextVertexes* to get the next nodes. In function *getNextVertexes*, we generate a sparql query statement to get the next vertexes by using the query api of the graph database and use $V_{next}$ to denote the vertexes which satisfy the condition clause $R_{kg}$-$r_{match}$. Secondly, we check whether each vertex $v_i$ match with $r_{match}$, if a vertex can match with $r_{match}$, we set $v_c=v_i$, and repeatedly execute the function *checkIsMatch* until every vertex in $V_{next}$ has been checked. Finally, if the result of function *checkIsMatch* is false, it means that there is no path that satisfies the query condition clause, the result of node to node query is false. Otherwise, we generate a SPARQL ask statement where the condition clause is $R_{kg}$-$r_{match}$. According to the properties of ask statement of SPARQL, the result is a boolen value.

The node query algorithm is similar to the boolean query algorithm, except that the result is a set of nodes, and the pseudo-code of the algorithm is shown in Algorithm 2.

In Algorithm 2, we finally generate a SPARQL query statement to get the vertexes which satisfy with $R_{kg}$-$r_{match}$.

### 5.3   Analysis

For Algorithm 1, we find the key process of boolean query is to match all the output edges of the starting node with the *policy matching condition*. Because the *gStore* has a high level of query efficiency, the query time of *gStore* can

**Function** Vertexes getNextVertexes($v_c, r_{match}$)

```
   /* generate the query statement based on the SPARQL         */
 1 set sparql=' select ?address,?z where { < v_c > edge ?y. ?y label ?address. ?y
   NextNode ?z.}';
 2 json=graphDatabase.query(sparql);
 3 if json==null then
 4 │   return ∅;
 5 else
 6 │   set V_list=[];
 7 │   set edgeList=json.list;
 8 │   for i = 1 to edgeList do
          /* get the Object of the OPTree according the memory address
             of the OPTree                                        */
 9 │   │   set T=Address(edgeList[i].address);
          /* use the search algorithm of the OPTree              */
10 │   │   set path=T.search(r_match);
11 │   │   if path ≠ null then
             /* It means that the edge of the vertex V_c match the r_match
                when there is a path                              */
12 │   │   │   V_list.add(edgeList[i].endNode);
13 │   │   else
14 │   │   │   continue;
15 return V_list;
```

---

**Algorithm 2: NRQLNodeQuery$(Q, G)$**

**Input**: $Q$:The NRQL query statement for the network reachability query.
$G$:The network reachability model.
**Output**: the node set which satisfies the query condition

```
 1 set v_c=v_start;
 2 result=checkIsMatch(v_c, Q_select, R_kg, r_match);
 3 if result==false then
 4 │   return null;
 5 else
 6 │   generate a sparql statement select which the query target clause is Q_select
   │   and the query condition clause is R_kg − r_match.
   │   /* Excute select by using the query api of the graph database.  */
 7 │   return graphDatabase.query(select);
```

---

be ignored. This checking process is equivalent to the searching process of
OPTree. The time complexity of OPTree search algorithm as $O(m\log)$, assuming that the number of output edges of each node is $k$, and there are $q$ nodes
between the starting node and the target nodes of the query. The best case
is that there is not a output edge $e$ of the starting node which can match the
*policy matching condition*, then we only need to execute the checking process for
the starting node once. Thus the time complexity is $O(km\log n)$. The worst case

is that we need to execute the checking process for every node between the starting node and the target node. The time complexity is $O(kqmlogn)$. Therefore, the time complexity of the boolean query algorithm is $O(kpmlogn)(1 \leq p \leq q)$.

Noted that the node query algorithm is similar to the boolean query algorithm, therefore, the time complexity of the node query algorithm is $O(kpmlogn)(1 \leq p \leq q)$.

# 6    Experiments

In this section, we perform our experiments and evaluate the effectiveness and efficiency of NREngine.

## 6.1    Setting

In this paper, we propose a knowledge graph-based query engine for network reachability, and construct the network reachability model based on the network topology and the network security policies. Therefore, in our experiments, we should generate three categories of datasets as follow.

1) **Datasets of Network Topology.** The data set includes the devices and the edges between the devices. The devices include the host, router, switcher and firewalls. The size of the devices in the generated network topology ranges from 10 to 100 with the step length of 10, and the ratio of firewalls in those devices is 30%, the ratio of routers or switchers in those devices is 70%, and the size of the forwarding ports in routers or switchers is no more than 4. Noted that in our experiment, the size of subnets is 3, and each of subnet, we generate 2 hosts.

   In order to be closer to the actual situation, we generate the random size of the forwarding ports for each router and switcher, so the size of the edges in the network reachability model is uncertain. However, the size of the edges in the network reachability model can intuitively reflect the complexity of the network. Therefore, We use the size of the edges in the network reachability model as the metrics in our experiment. Table 3 shows the generated data set of network topology in our experiments.

2) **Datasets of Network Security Policies.** We use the network security policies generation tool *ClassBench* proposed in [20], which is widely used in policies generation to generate the network security policy sets of the devices in the network. The size of the generated network security policy sets in each device range from 100 to 1000 with the step length is 100, note that each field of a network security policy generated by *ClassBench* is represented as a range, we need to transform the range value to one or more prefixes based on the properties of OPTree.

3) **Datasets of NRQL Query Statements.** We generate three categories of NRQL query statement: node to node query, node to subnet query and subnet to subnet query. We generate 100 NRQL query statements for each category. Noted that the start node is different from the end node in the generated NRQL query statements.

We perform our experiments on PC running Centos7.2 with 32 GB memory and 4 cores of Intel(R) Xeon(R) processor(3.3 GHz) and implement our prototype system using Java. The graph database used for maintaining the knowledge graph of network reachability is *gStore* [2,3].

For the offline part of NREngine, we generate the knowledge graph of network reachability and construct OPTrees for the labels of edges. We measure the execution time and memory usage of the knowledge graph.

For the online part of NREngine, we perform two categories of the network reachability query: the boolean query and the node query. To evaluate the efficiency of NREngine, we measure the query time in our experiments. In our experiments, we execute the three categories network reachability query: node to node query, node to subnet query and subnet to subnet query that use the same query condition clauses and only change the query target clauses.

In order to make the experimental results more accurate, we execute the all NRQL query statements, and then measure their average query time.

**Table 3.** Datasets

| Scale of devices | Devices | | Edges |
|---|---|---|---|
| | Routers or Switchers | Firewalls | |
| 10 | 7 | 3 | 89 |
| 20 | 14 | 6 | 503 |
| 30 | 21 | 9 | 912 |
| 40 | 28 | 12 | 1293 |
| 50 | 35 | 15 | 1723 |
| 60 | 42 | 18 | 2207 |
| 70 | 49 | 21 | 2498 |
| 80 | 56 | 24 | 2974 |
| 90 | 63 | 27 | 3319 |
| 100 | 70 | 30 | 3812 |

## 6.2   Experiments Results of Offline Part

For the offline part of NREngine, the results of experiments are shown in Table 4. Noted that the number of edges is the size of edges in network reachability model. On the one hand, the experimental results show that we can build a knowledge graph of network reachability with 3700 edges in less than three hours, and the memory usage is less than 3 GB. Because of the knowledge graph construction is an offline process, so the time cost and the space cost are acceptable.

On the other hand, the experimental results show that the time and the memory usage of OPTree construction are more than that of knowledge graph

**Table 4.** Results on knowledge graph of network reachability construction

| Number of edges | Knowledge graph generation | | OPTree construction | |
|---|---|---|---|---|
| | Ave. Time (s) | Ave. memory size (MB) | Ave. time (min) | Ave. memory size (MB) |
| 100 | 12.56 | 12.42 | 3.51 | 40.42 |
| 500 | 18.52 | 20.42 | 11.52 | 218.32 |
| 900 | 23.41 | 25.62 | 21.42 | 398.23 |
| 1300 | 26.43 | 31.24 | 35.23 | 612.42 |
| 1700 | 31.24 | 35.62 | 51.52 | 812.25 |
| 2100 | 51.42 | 39.42 | 72.51 | 978.07 |
| 2500 | 69.41 | 42.54 | 98.53 | 1234.23 |
| 2900 | 87.09 | 48.23 | 119.64 | 1592.31 |
| 3300 | 97.14 | 55.21 | 138.52 | 1892.18 |
| 3700 | 112.52 | 60.87 | 150.42 | 2132.52 |

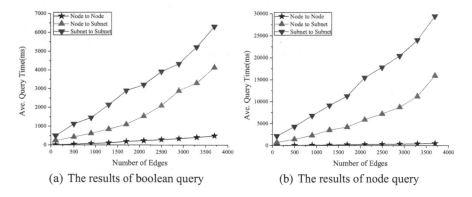

(a) The results of boolean query            (b) The results of node query

**Fig. 6.** The results of experiments

generation. The reason is that OPTree Construction is a time-consuming operation, and it includes path checking and path merging, and network reachability model can be quickly converted into knowledge graph based on pattern matching.

### 6.3  Experiments Results of Online Part

For the online part of NREngine, the results of experiments are shown in Fig. 6. The experimental results show that on the one hand the query time of network reachability is from milliseconds to seconds with the increase in the size of edges in network reachability model. On the other hand, the efficiency of node to node query is the highest, followed by node to subnet query, and the efficiency of subnet to subnet query is the lowest. The reason is that we need to traverse every node in the subnet until we find a node that satisfies the query conditions for node to subnet query and subnet to subnet query. Figure 6(a) shows the experimental result of the boolean query, and Fig. 6(b) shows the experimental result. The experimental results show that the query time of node query is similar to that of boolean query in node to node query, and the query time of node query is much longer than that of boolean query in node to subnet query and subnet to subnet query. The reason is that we need find all nodes that satisfy the query conditions in the node query, and we just find one node that satisfies the query in the boolean query.

# 7    Conclusion

In this study, we propose a model of the network reachability based on the network security policies, and propose a query engine called "NREngine" for network reachablity. In order to improve the efficiency of network reachability query, some techniques are used to construct the network as a knowledge graph of network reachability and maintain the knowledge graph in graph databases. On this basis, the knowledge graph of network reachability is proposed for network reachability query. To describe user's network reachability query requests, we propose a structured query language, which is called NRQL, and design the query algorithms for NRQL. The experimental results indicate that NREngine is effective and efficient.

**Acknowledgements.** This work is partially supported by The National Key Research and Development Program of China under grant 2018YFB1003504, the National Natural Science Foundation of China under Grant (No. U20A20174, 61772191), Science and Technology Key Projects of Hunan Province (2019WK2072, 2018TP3001, 2018TP2023, 2015TP1004), and ChangSha Science and Technology Project (kq2006029).

# References

1. Xie, G.G., Zhan, J., Maltz, D.A., Zhang, H., Greenberg, A.: On static reachability analysis of IP networks. In: IEEE INFOCOM, pp. 2170–2183 (2005)
2. Zou, L., Mo, J., Chen, L.: gStore: answering SPARQL queries via subgraph matching. VLDB Endow. **4**(8), 482–493 (2011)
3. Zou, L., Özsu, M.T., Chen, L., Shen, X., Huang, R., Zhao, D.: gStore: a graph-based SPARQL query engine. VLDB J. **23**(4), 565–590 (2014)
4. McBride, B.: Jena: implementing the RDF model and syntax specification. In: SemWeb (2001)
5. Broekstra, J., Kampman, A., van Harmelen, F.: Sesame: a generic architecture for storing and querying RDF and RDF schema. In: Horrocks, I., Hendler, J. (eds.) ISWC 2002. LNCS, vol. 2342, pp. 54–68. Springer, Heidelberg (2002). https://doi.org/10.1007/3-540-48005-6_7
6. Erling, O., Mikhailov, I.: Virtuoso: RDF support in a native RDBMS. In: de Virgilio, R., Giunchiglia, F., Tanca, L. (eds.) Semantic Web Information Management, pp. 501–519. Springer, Heidelberg (2009). https://doi.org/10.1007/978-3-642-04329-1_21

7. Zhang, B., Eugene, T.S., Wang, N.G.: Reachability monitoring and verification in enterprise networks. In: ACM SIGCOMM, pp. 459–460 (2008)
8. Benson, T., Akella, A., Maltz, D.A.: Mining policies from enterprise network configuration. In: The 9th ACM SIGCOMM Conference on Internet Measurement Conference, pp. 136–142 (2009)
9. Khakpour, A.R., Liu, A.X.: Quantifying and querying network reachability. In: the 29th International Conference on Distributed Computing Systems (ICDCS), pp. 817–826 (2010)
10. Khakpour, A.R., Liu, A.X.: Quantifying and verifying reachability for access controlled networks. IEEE/ACM Trans. Netw. (TON) 21(2), 551–565 (2013)
11. Chen, F., Bezawada, B., Liu, A.X.: Privacy-preserving quantification of cross-domain network reachability. IEEE/ACM Trans. Netw. (TON) 23(3), 946–958 (2015)
12. Hong, S.C., Ju, H., Hong, J.W.K.: Network reachability-based IP prefix hijacking detection. Int. J. Netw. Manag. 23(1), 1–15 (2013)
13. Liang, R., Zhuge, H., Jiang, X., Zeng, Q., He, X.: Scaling hop-based reachability indexing for fast graph pattern query processing. IEEE Trans. Knowl. Data Eng. (TKDE) 26(11), 2803–2817 (2014)
14. Rao, Z.C., Pu, T.Y.: Decision diagram-based modeling of network reachability. Appl. Mech. Mater. 513, 1779–1782 (2014)
15. Li, Y., Luo, Y., Wei, Z., Xia, C., Liang, X.: A verification method of enterprise network reachability based on topology path. In: The 2013 Ninth International Conference on Computational Intelligence and Security, pp. 624–629 (2013)
16. Cuzzocrea, A., Serafino, P.: A reachability-based theoretical framework for modeling and querying complex probabilistic graph data. In: IEEE International Conference on Systems, pp. 1177–1184 (2012)
17. Jamil, H.: A novel knowledge representation framework for computing sub-graph isomorphic queries in interaction network databases. In: 2009 21st IEEE International Conference on Tools with Artificial Intelligence, pp. 131–138 (2009)
18. Li, W., Qin, Z., Li, K., Yin, H., Lu, O.: A novel approach to rule placement in software-defined networks based on OPTree. IEEE Access 7(1), 8689–8700 (2019)
19. Arenas, M., Gutiérrez, C., Pérez, J.: On the semantics of SPARQL. In: de Virgilio, R., Giunchiglia, F., Tanca, L. (eds.) Semantic Web Information Management, pp. 281–307. Springer, Heidelberg (2009). https://doi.org/10.1007/978-3-642-04329-1_13
20. Taylor, D.E., Turner, J.S.: Classbench: a packet classification benchmark. IEEE/ACM Trans. Netw. 15(3), 499–511 (2007)

# An Embedding-Based Approach to Repairing Question Semantics

Haixin Zhou[1]([⊠]) and Kewen Wang[2]

[1] College of Intelligence and Computing, Tianjin University, Tianjin 300350, China
haixinzhou@tju.edu.cn
[2] School of Information and Communication Technology, Griffith University,
Brisbane, Australia

**Abstract.** A question with complete semantics can be answered correctly. In other words, it contains all the basic semantic elements. In fact, the problem is not always complete due to the ambiguity of the user's intentions. Unfortunately, there is very little research on this issue. In this paper, we present an embedding-based approach to completing question semantics by inspiring from knowledge graph completion based on our proposed representation of a complete basic question as unique type and subject and multiple possible constraints. Firstly, we propose a back-and-forth-based matching method to acknowledge the question type as well as a word2vec-based method to extract all constraints via question subject and its semantic relevant in knowledge bases. Secondly, we introduce a time-aware recommendation to choose the best candidates from vast possible constraints for capturing users' intents precisely. Finally, we present constraint-independence-based attention to generate complete questions naturally. Experiments verifies the effectiveness of our approach.

**Keywords:** Embedding · Question completion · Knowledge graph

## 1 Introduction

Question answering (QA) is a downstream task of natural language reasoning. The question answering system can answer natural language questions accurately and concisely in the statistical data set by understanding the intent of the question [12]. This data set can be a structured knowledge database (as a knowledge base) or an unstructured document collection. QA techniques have been widely used in many fields of NLP, such as chatbot, intelligent search, and recommendation [30]. Different from searching via keyword matching, a question has a complete semantics (called *complete question*), that is, it consists of all basic elements of questions if QA returns some accurate and concise answer [36]. Due to the ambiguous representation of users' intent, the questions that users often ask are not always complete [9]. For instance, *"the last Japanese metro"*, *"Parisian resident population"*, and *"actors of the movie Green Book"*

C. S. Jensen et al. (Eds.): DASFAA 2021 Workshops, LNCS 12680, pp. 107–122, 2021.
https://doi.org/10.1007/978-3-030-73216-5_8

are incomplete questions. A complete question must have a complete semantics. In practice, QA systems often complete the semantics of questions when they are under completeness [35]. For instance, to answer an incomplete question *What date is today?*, we often complete it by adding the date constraint of the current date. So it becomes very interesting to complete the question accurately and concisely for further inquiry. As we investigated, unfortunately, there is no open research work on this problem.

Knowledge graph completion (KGC) [2] as a recent popular technique in learning new entities and relations, effectively improve knowledge graphs (KGs) by filling in its missing connections. It is natural to apply KGC techniques to complete the semantics of questions since questions are often transformed to logical queries (e.g., SPARQL queries) built on triple patterns in QA systems as well as KGs are modeled in sets of triples [10]. However, there are some essential differences between questions and KGs as follows: (1) each question necessarily have unique type while an entity has multiple relations; (2) the semantics of a question is dynamically changing while the semantics of a triple in KG often is stable; (3) one question has multiple representations (as question grammar form) while the representation of triples in KG is often single. Hence it is not direct to apply KGC techniques for question completion (QC).

In this paper, we propose a novel method based on embedding. This method is based on some core technologies of mechanism map completion (KGC) to analyze and infer the semantics of incomplete questions and make an incomplete question that cannot be answered correctly. The question becomes a question with complete semantics so that some QA tasks can identify these questions and give the correct answers. The main contributions of this paper are summarized as follows:

- We recognize question type by the back-and-forth-based matching method, which is to determine it based on the type of answer found by the knowledge graph and the information entered by the users. Meanwhile, we propose a word2vec-based method to extract all constraints via question subject and its semantic relevant in KBs.
- We introduce a time-aware recommendation which overcome the contradiction between static data and dynamic question to choose the best candidates from vast possible constraints for capturing users' intents precisely.
- We present constraint-independence-based attention, which captures the deep meaning expressed by the user through the order of multiple constraints to generate complete questions naturally.

Experimental results on the datasets revised from benchmark datasets demonstrate that our approach can effectively extract semantics lost in capturing intents.

## 2    Overview of Our Approach

In this section, we introduce our proposed embedding-based model for question completion, and its overview framework is shown in Fig. 1.

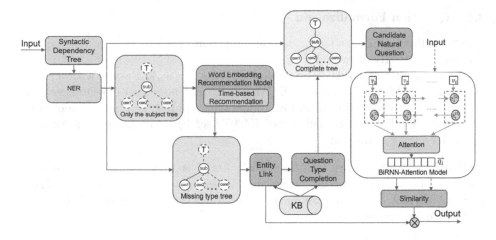

**Fig. 1.** The overview framework of our approach

Our embedding-based model consists of three modules, namely, *question representation*, *question completion*, and *question generator* as follows:

– **Question Representation** obtains a tree structure of questions. We process the input word sequence (that is, the incomplete question) to get the tree structure of the question. As the base of the question completion module, the module is used to provide the reference of candidate completed questions. In this module, we present the method based on a parsing-dependency-tree to construct a structured representation of all incomplete questions.
– **Question Completion** constructs a complete tree-structure of input questions, that is, processes the output of the question representation module. The module is used to mainly build complete tree-structures of incomplete questions for generating complete questions in the question generator. In this module, we present a back-and-forth-based matching method to learn question type and an embedding-based model to learning constraint relations.
– **Question Generator** generates questions with complete semantics for all complete tree-structures of incomplete questions. The module is mainly used to learn intents of incomplete questions based on complete tree-structures. In this module, we present a time-aware recommendation and constraint-independence-based attention for extracting complete semantics of incomplete questions.

## 3  Our Approach

In this chapter, the technical framework of the paper is divided into four departments to introduce: namely, *question formulization*, *question representation*, *question completion*, and *question generation*.

### 3.1   Question Formalization

The formalization of questions is determining what syntactic structure of a question to be correctly answered is.

Let $\Sigma$ be a set of words. A finite sequence $(w_1, \ldots, w_n)$ of words with $w_i \in \Sigma$ $(i = 1, 2, \ldots, n)$ is a *sentence* over $\Sigma$. We use $\Sigma^*$ to denote a set of sentences over $\Sigma$. A *phrase* is a collection of words. By default, a phrase is a natural sentence.

Let $V_N, V_E, V_T$ be three subsets of $\Sigma^*$. We set $V_T \cap (V_N \cup V_E) = \emptyset$, that is, $V_T$ and $V_N \cup V_E$ is disjoint.

**Definition 1 (Structure of Question).** *Let $V = V_T \cup V_N \cup V_E$. Let $q$ be a question. A* syntactic structure *(structure, for short) of $q$ over $V$ as a labeled tree $T_q = (N, E, \mathcal{L}, \lambda, \delta)$ whose children is either a labelled tree named* substructure *of $T_q$ or a leaf where*

- *$\mathcal{L} : \mathrm{roots}(T_q) \to V_T$: an injective function mapping it to one phrase;*
- *$\lambda : N \to V_N$: a function mapping each node to a phrase;*
- *$\delta : E \to V_E$: a function mapping each edge to a phrase;*

*where $\mathrm{roots}(T_q)$ is a collection of non-leaf nodes in $T_q$.*

Let $V$ be a set of phrases and $q$ be a question over $V$. $T_q = (N, E, \mathcal{L}, \lambda, \delta)$. We use $\mathrm{Root}(T_q)$ to denote the root of $T_q$ and $\mathrm{Leaf}(T_q)$ to denote all leaves of $T_q$. Let $r = \mathrm{Root}(T_q)$.

We say $\mathcal{L}(r)$ *type* of $q$. We say $\lambda(r)$ *subject* of $q$ and $\lambda(u)$ *constraint* of $q$ where $u \in \mathrm{Leaf}(T_q)$. Given an edge $(u, v) \in E$, we say $\delta(u, v)$ *modification* between $u$ and $v$. In this sense, we directly $v$ *modifying* $u$. And we say $q'$ *subquestion* of $q$ if $T_{q'}$ is a substructure of $T_q$.

A question $q$ is *complete* if its structure tree contains at least three nodes which are all mapped to some phrases and its root is labelled by a question type. Formmaly, let $T_q = (N, E, \mathcal{L}, \lambda, \delta)$, if $|N| \geq 2$ and $\mathcal{L} \neq$ and $\lambda \neq \emptyset$. Intuitively, a complete question contains at least three elements: type, subject, and constraint. See Fig. 2 Fig. 3.

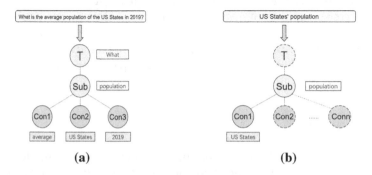

**Fig. 2.** (a)-Complete question (b)-Incomplete question

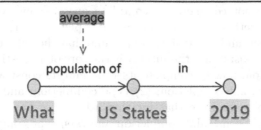

**Fig. 3.** Query graph structure

Notice that query graph or semantic graph of questions in a graph structure are substantially introduced to represent logical queries whose solutions are possible answers of questions [32]. Most approaches of QA systems mainly focus on better matching a question to candidate query graphs. Our proposed tree structure of questions is to represent the relational structure of questions themselves for determining characteristics of questions as a whole such as completeness.

### 3.2 Natural Question Representation

Given a natural question, based on the formalization of question structure, we build its structure.

Let $V$ be a set of phrases. Let $q$ be a natural question $(w_1, w_2, \ldots, w_n)$ over $V$ based on the assumption each $w_i$ is not a stopword. In other words, $\{w_1, \ldots, w_n\}^* \subseteq V$, that is, all words $w_1, \ldots, w_n$ are words occurring in $V$. $T_q = (N, E, \mathcal{L}, \lambda, \delta)$ is constructed in the following four steps:

**Type extracting.** Compute $\mathcal{L}(\text{node}_{\text{new}})$ and update $N$ by adding a new node denoted by $\text{node}_{\text{new}}$. We present a dynamic planning for entity extracting and relation learning built on indenpency-trees [27,28].

**Subject learning.** Compute $\lambda(r)$ and update $\lambda$. We present a heuristic method for subject learning by combing gammar structure with speech analysis.

**Constraint learning.** Compute $\lambda(u)$ and update $N, E$ by adding new edge from newly generated nodes and edges and further update $\delta(e)$ for all $e \in N$. We present a greedy method for identity of named entity built on *gammar indenpency-tree*. The similarity function between $\vec{w_i}$ and $\vec{r_j}$ is defined as follows:

$$\text{sim}(\vec{w_i}, \vec{r_j}) = \sum_{h-1}^{d} \frac{(w_{ih} + r_{jh})(|w_{ih} - r_{jh}|)}{\sum\limits_{h=1}^{d}(w_{ih} + w_{jh})}. \tag{1}$$

### 3.3  Question Completion

In this module, we extend a question (possibly an original sentence), whose question structure is not complete, to complete question via word2vec as an essential embedding method [23]. Recall that a complete question requires three elements: type, subject, and constraint. We assume that the subject of an incomplete question always exists since the subject is the core of a question. Now, given an incomplete question, how to complete its type and its constraints. Consider three cases: (1) constraint completion; (2) type completion; and (3) both type and constraint completion. Note that for the 3rd case of both type and constraint, we firstly apply constraint completion and then type completion. In this subsection, we mainly review the first two cases.

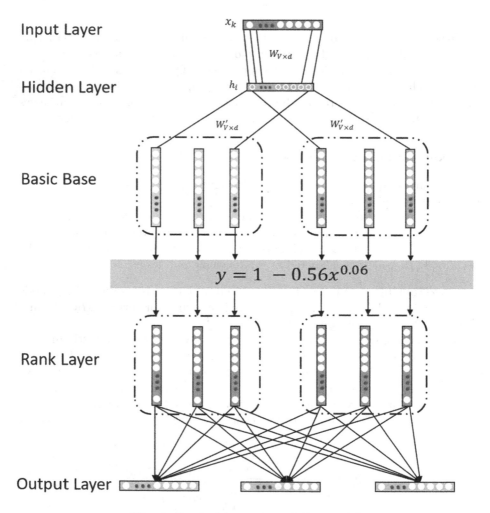

**Fig. 4.** Constraint recommendation model

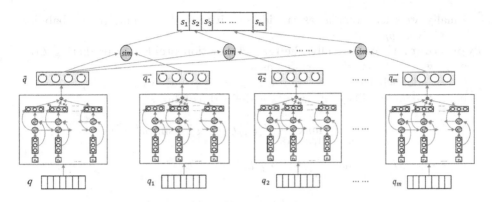

**Fig. 5.** Bi-RNN-attention

**Constraint completion.** Formally, let $V$ be a set of phrases, let $q$ be a natural imcomplete question $(w_1, w_2, \ldots, w_n)$ over $V$, $T_q = (N, E, \mathcal{L}, \lambda, \delta)$ is the structure of $q$, the constraint completion problem is predicting all possible constraints $\lambda(u)$ of $T_q$ w.r.t subject $\lambda(r)$ where $r$ is the root of $T_q$ and $u$ is a leaf of $r$.

We use $\vec{q} = (\vec{w_1}, \ldots, \vec{w_n})$ to denote the sequence of $d$-ary vectors after via word2vec. We introduce probability in two cases of $q$: basic question (the depth of its structure tree is exactly at most 2) and complex question (the depth of its structure tree is at least 3).

We use $w_i'$ to denote the predicted target word w.r.t. $w_i$. The conditional probability of $w_i'$ given $w_i$, denoted by $P(w_i'|w_i)$, is defined as follows: basic question (see Eq. (2)) and comple question (see Eq. (3)).

$$P(w_i'|w_i) = \frac{P(w_i', w_i)}{P(w_i)} = \frac{\text{count}(w_i', w_i)}{\text{count}(w_i)}; \tag{2}$$

$$P(w_i'|w_i) = \prod_{i=1}^{L(\vec{w_i'})-1} \frac{\overrightarrow{(w_{n(w_i', j)})}^{\top} \vec{w_i}}{1 + \exp\left(-\sigma(n(w_i', j))\right)}, \tag{3}$$

where $L(\vec{w_i'})$ and $n(w_i', j)$ are the length and the $j$-th node of the path from the root node to $w_i'$ respectivily. And $\sigma$ is defined as follows: let $k$ is a positive number,

$$\sigma(n(w_i', j)) = \begin{cases} 1, & \text{if } j = 2k - 1, \\ -1, & \text{if } j = 2k. \end{cases} \tag{4}$$

Let $t$ be a time. The conditional probability of $w_i'$ given $w_i$ w.r.t. $t$ via *Ebbinghaus Forgetting Curve (see Fig. 4)* is defined as follows: let $|\Sigma| = m$,

$$P(w_i'|w_i, t) = \frac{(1 - 0.56\Delta t^{0.06})P(w_i'|w_i)}{\sum\limits_{i=1}^{m}(1 - 0.56\Delta t^{0.06})P(w_i'|w_i)}. \tag{5}$$

Finally, we select some $w_i'$ as candidates with suitable conditional probability of $w_i'$ given $w_i$ w.r.t. $t$.

**Type completion.** Formally, the type completion problem is predicting $\mathcal{L}(r)$ of $T_q$ w.r.t subject $\lambda(r)$ (denoted as $s$)

Given a knowledge base $\mathcal{K}$, we use $\Omega(\mathcal{K}, s) = \{(r, o) \mid (s, r, o) \in \mathcal{K}\}$.

Now, the probability of $r_j$ w.r.t. $s$ is defined as follows:

$$P(r_j) = \max_{i=\{1,2,\dots,n\}} \{\sigma'(w_i')P(w_i'|w_i,t)\text{sim}(\overrightarrow{w_i'}, \overrightarrow{r_j})\},$$

where, let $M = \max_{\{j=1,2,\dots,n\}} \{P(w_j'|w_i,t)\}$,

$$\sigma'(w_i') = \begin{cases} 1, & \text{if } P(w_i'|w_i,t) > M - \min\{\frac{1}{n}, 0.1\}, \\ 0, & \text{otherwise.} \end{cases}$$

Finally, let $\text{Objects}(r_j, \mathcal{K}) = \{o \mid (s, r_j, o) \text{ and } P(r_j) \text{ is maximal}\}$. We assign the type $\tau$ of named entity of elements of $\text{Objects}(r_j, \mathcal{K})$ obtained identity of named entity to the value of $\lambda(r)$. That is, we obtain $\lambda(r) = \tau$.

### 3.4  Natural Question Generation

In this module, we generate a natural question from a question with a completed structure. The difficulty of question generation is to choose an optimal natural question from all candidates, maximally capturing the semantics of original questions. The process of QG consists of the three steps:

**Generating natural questions.** In this step, we generate all possible nature questions as candidates by learning stopwords via word2vec and enumerating all orders of constraints. Note that the structure of questions depends on non-stop words which have already been deleted as initialization. We learn stopwords to express a natural question comprehensively.

**Representing natural question.** In this step, we obtain the embedding representation of all words in a natural question via word2vec and then encode those embedding representation in BiRNN (see Fig. 5)as follows: let $m(q)$ be the number of natural questions and $q_j$ be a candidate of $q$, the cost function of BiRNN is as follows:

$$\text{Cost}(q) = -\frac{1}{m(q)} \sum_{j=1}^{m(q)} [q \ln q_j + (1 - q) \ln(1 - q_j)].$$

Based on the above model, we present an attention-based method to learn the weight of a word and utilize a fully-connected layer to obtain the vector representation $\overrightarrow{q_j}$ of $q$ as follows:

$$\overrightarrow{q_j} = \sum_{t=1}^{d} a_{jt} \overrightarrow{h_{jt}}, \quad a_{jt} = \frac{\exp(\overrightarrow{w_{jt}}^\top \overrightarrow{V})}{\sum_{t=1}^{d} \exp(\overrightarrow{w_{jt}}^\top \overrightarrow{V}))}, \tag{6}$$

where $V$ is the shared parameter of attention. $a_{jt}$ and $h_{jt}$ is the weights and the last hidden layer of the $t$-the word in the $q_j$,$w_{jt} = \text{sigmoid}(W\overrightarrow{h_{jt}} + \overrightarrow{b})$.

**Recommending natural question.** In this step, combining the problem similarity and the probability of the corresponding constraint, we define the Score function as follows: $\text{Score}(q_j) = P(r_j)sim(\overrightarrow{q_j}.\overrightarrow{q})$. Finally, we choose the optimal $q_j$ with the highest $\text{Score}(q_j)$, $q_j = \arg\max_{\{j=1,2,...,m\}}\{\text{Score}(q_j)\}$.

# 4 Experiments and Evaluation

In this section, we construct four sets of experiments to evaluate our approach as follows:

- The 1st experiment is to evaluate the effectiveness of our approach.
- The 2nd experiment is to compare our approach with existing approaches to QG.
- The 3rd experiment is to evaluate the effectiveness of our time-aware recommendation.
- The 4th experiment is to verify the applications of QC.

## 4.1 Experiment Setup

In our experiments, we select DBpeida[1], consisting of 5.4 million entities, 110 million triples, and 9708 predicates, as our KB. We select three representative datasets, where most of the approaches are based on them [26]:

- WebQuestions: The dataset released by [5] contains 3778 pairs for training and 1254 pairs for testing.
- QALD-7: QALD released by [24] is a series of open-domain question answering campaigns, mainly based on DBpedia.
- TREC-8 QA[2]: The dataset contains 3000 pairs mostly contributed by NIST assessors.

## 4.2 Effectiveness of Question Completion

To evaluate the accuracy of question completion, we established a QA data set (incomplete question set) obtained by WebQuestions, TrecQA, and QALD through the random selection of deletion strategies., denoted by WebQuestions*, TrecQA*, and QALD*, respectively.

In the strategy of constructing an incomplete question set, we apply the Fisher-Yates Shuffle algorithm[3] (A popular algorithm with low time complexity) randomly delete words to ensure that fair incomplete questions are generated.

---

[1] https://wiki.dbpedia.org/.
[2] https://www.aclweb.org/anthology/N13-1106.pdf/.
[3] http://hdl.handle.net/2440/10701/.

In addition, we use gAnswer [11] for answer testing. Our method uses the query parsing module of gAnswer, so any KBQA system can perform this experiment.

The experimental results are shown in Table 1, where *QC-Precision* represents the accuracy of the answer after completing the question through our method, and *Precision* represents the unfinished answer Accuracy. Through Table 1, the Precision of the three data sets is all zero, which means that any incomplete question cannot be answered. At the same time, the QC accuracy of the three data sets is 0.5450, 0.3742, and 0.3477, respectively. These results are based on the time-aware recommendation model. If there is no model, our QC accuracy is 0.5060%, 0.3322%, and 0.2877%, respectively. In other words, it can prove that our algorithm is effective.

Besides, to quantify the effectiveness of question completion, we introduce a new metric named *completion rate*, which is defined as follows:

$$QC\text{-}R = \frac{QC\text{-}Precision}{O\text{-}Precision} \tag{7}$$

Here *O-Precision* represents the accuracy obtained by processing the original datasets. The three O-Precision of WebQuestions, TrecQA, and QALD are 0.6464, 0.4555, and 0.4331, respectively.

**Table 1.** Precision of question completion

|  | Precision | QC-Precision | QC-Rate |
|---|---|---|---|
| QALD* | 0 | 0.5450(↑3.9%) | 0.8432 |
| TrecQA* | 0 | 0.3742(↑4.2%) | 0.8215 |
| WebQuestions* | 0 | 0.3477(↑6.0%) | 0.8029 |

### 4.3    Comparison to Question Generation (QG)

Though question generation (QG) (returning a question when inputting an answer) is essentially different from question generation (QC) (returning a complete question when inputting an incomplete question) as discussed previously, both QG and QC return a text when inputting a text. To distinguish QC from QG, our experiment is to check if outputs of QG and QC are different when inputting one text.

In this experiment, we employ for QG [31] and SQuAD dataset consisting of more than 100K questions posed by crowd workers on 536 Wikipedia articles. Moreover, the experimental results show that all outputs in QC and QG are totally different. Due to the limited space, we take two examples shown in Table 2. QC outputs a different result from QG for one input. Therefore the experiment verifies that QC is a different task from QG.

**Table 2.** An example of difference between QC and QG

| Input | Question Generation | Question Completion |
|---|---|---|
| Like the lombardi trophy, the "50" will be designed by tiffany & co... | Who designed the lombardi trophy? | Not A Question |
| Blank_Obama | – | Who is Blank_Obama wife? |

## 4.4 Evaluating Time-Aware Recommendation

In this experiment, we compare our QC approach based on the proposed time-aware recommendation (denoted by $QC^{tr}$) to our QC approach based on the skip-gram algorithm [23] (denoted by $QC^{sg}$) to evaluate the performance of our proposed time-aware recommendation. Note that the skip-gram algorithm is used as a mainstream tool to obtain distributed word vectors; it is widely applied in the recommended system [18].

To support time-aware recommendation, we revise the three datasets by adding timestamps in a random method[4] as follows: (1) Remove all stop words in the three datasets and unify them into lowercase letters; and (2) Add timestamps to all data randomly in a chronological order.

Note that the first step is to reduce the extra noise, which is caused by either stop words or difference between uppercase and lowercase, to improve the performance of recommendation. Moreover, the second step is to add timestamp labels.

The experimental result is shown in Table 3.

**Table 3.** Improvement of time-aware recommendation

| | Accuracy of $QC^{sg}$ | Accuracy of $QC^{tr}$ |
|---|---|---|
| TrecQA$^t$ | 68.1% | **71.0%** |
| WebQuestions$^t$ | 70.3% | **72.8%** |
| QALD$^t$ | 66.7% | **69.8%** |

By Table 3, $QC^{tr}$ achieves 1.9%, 2.5%, and 3.1% improvement of accuracy w.r.t. $QC^{sg}$ over TrecQA$^t$, WebQuestions$^t$, and QALD$^t$, respectively. Hence, we can conclude that our time-aware recommendation improves performance.

## 4.5 Applications of Question Completion

Finally, we conduct another experiment to demonstrate the effectiveness of our QC method, in which QC preprocesses the KBQA system data set to be tested.

---

[4] http://hdl.handle.net/2440/10701/.

We select the three systems NFF [11], RFF [11], Aqqu [13] where NFF and RFF are non-template systems with highest scores of QA-Task in the latest QALD[5] and Aqqu is a classical KBQA system [8]. The experimental results w.r.t. F1-score on WebQuestions and QALD are showed in Table 4.

**Table 4.** QC improving QA baselines

| | F1 | | | |
|---|---|---|---|---|
| | WebQuestions | | QALD | |
| | Baseline | Baseline+QC | Baseline | Baseline+QC |
| NFF | 0.4847 | **0.5415** | 0.7751 | **0.7892** |
| RFF | 0.3052 | **0.3825** | 0.5341 | **0.5513** |
| Aqqu | 0.4944 | **0.5532** | 0.3741 | **0.4882** |

By Table 4, all baselines with QC preprocessing achieve 0.0568–0.0773 over WebQuestions and 0.0141–0.141 over QALD. Therefore, the experiment demonstrates that QC is useful to improve the accuracy of off-the-shelf QA systems.

### 4.6    Error Analysis

We randomly analyze the major causes of errors on 100 questions with unreasonable results output due to the following major factors.

*Semantic Ambiguity and Complexity.* Due to the variety or the semantic ambiguity of incomplete questions, we hardly construct a completely reasonable tree structure for those questions practically. On the other hand, our recommendation model does not always capture all users' intentions accurately due to the ambiguity of questions. As a result, we possibly complete an incorrect question tree.

*Entity Linking Error.* This error is caused by the failure in extracting the appropriate entities for a given question. As a result, we sometimes generate incorrect question trees. In our work, we mostly fix this error by correcting those wrong entities.

*Dateset and KB Error.* This error arises due to the defects of datasets or KB. The test datasets contain many open questions, while the entity-relationship of KB does not always cover all questions.

---

[5] https://project-hobbit.eu/challenges/qald-9-challenge/.

# 5    Related Works

In this section, we discuss our approach by comparing it to existing works and some techniques.

The most relevant problem to question completion is *question generation*. Question generation (QG) [29] is a natural language problem that a question is generated given long sentences or articles [34]. The traditional approach to solve it is converting sentences into related questions based on heuristic rules [3]. An end-to-end via sequence-to-sequence learning model base on attention is present in [7] to reduce the rules of handcrafting and the constraints of sophisticated NLP. The current popular approaches of QG are based on the attention-based encoder-decoder model [6,31], where the SynNet model based on the calculation formula of conditional probability is introduced to divide the QG into answer synthesis module and problem synthesis module. Besides. There are still some works by combing QG with QA to train answers and questions within the overall system [25].

Different from QG inherently, question completion generates a natural question for any given a sentence or a phrase or even a word. QG often takes a phrase or a word as an answer.

The paper is an extension of QC [33]. Compared with the previous paper, this paper adds more technical implementation details, complete theoretical derivation, and more experiments. Compared with the previous article, the contribution discusses the improvement of the previous work. In writing, There are also considerable differences in format.

Besides, we are interested in discussing the techniques presented in our approach by comparing to existing techniques in the two aspects.

**Query graph.** Query graph (or semantic graph) is a graph-structure representing a logical query such as SPARQL BGP (basic graph pattern) query to be correctly answered in knowledge bases [28]. In general, a natural question can be structured by a Semantic Query Graph (SQG) to represent the query intent. SQG obtains semantic information such as entities and attributes from the dependency analysis tree of natural questions. In the answering question system, a question can be converted to SPARQL via SQG. These dots in SQG correspond to the entities in the knowledge graph, and each edge is associated with a relation or attribute in the knowledge base. The above is a semantic query graph of "relation (edge)-first". [11] presents a "node-first" super SQG using entity phrase, class phrase, and wh-words as nodes, and a simple path between nodes is introduced into the edge. Compared with the "relation (edge)-first" SQG, the super SQG obtains a possible relationship between nodes to retain more semantic information.

Different from the query graph, our tree-based structure of questions is a formalization of questions that are used to express the characteristics of questions such as completeness of questions.

**Question Error Detection.** Question error detection is to find a question with grammar error and then correct it for reliably answering. [12,21] presents a Bi-LSTM model modification question based on the scoring mechanism by cross-checking its accuracy of the relationship between predicted answers and subject words in a question. Different from error detection of questions treating only questions, question completion addressed in our paper is to construct a complete question from an incomplete question, which is possibly phrases, even words. Our approach is based on time-aware recommendation and density of named entity from the knowledge graph completion.

In summary, question completion is different from existing QA tasks such as question generation, and techniques developed in our approach are substantially different.

## 6  Conclusions

In this paper, we introduce a task related to natural language reasoning called question completion. Inspired by the knowledge graph completion, we propose a method based on embedding to complete questions that cannot be answered in any knowledge. Unlike question generation, our algorithm can infer the intent of incomplete questions and is used in the improvement of question answering systems. In future work, we are interested in question completion tasks in some important areas while considering more domain knowledge.

**Acknowledgments.** This work is supported by Key Research and Development Program of Hubei Province (No. 2020BAB026).

## References

1. Abujabal, A., Yahya, M., Riedewald, M., Weikum, G.: Automated template generation for question answering over knowledge graphs. In: Proceedings of WWW, pp. 1191–1200 (2017)
2. Alberto, G., Sebastijan, D., Mathias, N.: Learning sequence encoders for temporal knowledge graph completion. In: Proceedings of EMNLP, pp. 4816–4821 (2018)
3. Andrew, M., Arthur, C., Natalie, P.: Question generation from concept maps. In: Proceedings of D&D2012, pp. 75–99 (2012)
4. Bao, J., Duan, N., Yan, Z., Zhou, M., Zhao, T.: Constraint-based question answering with knowledge graph. In: Proceedings of COLING, pp. 2503–2514 (2016)
5. Berant, J., Chou, A., Frostig, R., Liang, P.: Semantic parsing on freebase from question-answer Pairs. In: Proceedings of EMNLP, pp. 1533–1544 (2013)
6. David, G., Huang, P., He, X., Deng, L.: Two-stage synthesis networks for transfer learning in machine comprehension. In: Proceedings of EMNLP, pp. 835–844 (2017)
7. Du, X., Shao, J., Claire, C.: Learning to ask: neural question generation for reading comprehension. In: Proceedings of ACL, pp. 1342–1352 (2017)
8. Denis, S., Eugene, A.: When a knowledge base is not enough: question answering over knowledge bases with external text data. In: Proceedings of SIGIR, pp. 235–244 (2018)
9. Guo, X., et al.: Learning to query, reason, and answer questions on ambiguous texts. In: Proceedings of ICLR (2017)

10. Hamid, Z., Giulio, N., Jens, L.: Formal query generation for question answering over knowledge bases Hamid Zafar1. In: Proceedings of ESWC, pp. 714–728 (2018)
11. Hu, S., Zou, L., Yu, J., Wang, H., Zhao, D.: Answering natural language questions by subgraph matching over knowledge graphs. IEEE Trans. Knowl. Data Eng. **30**(5), 824–837 (2018)
12. Huang, X., Zhang, J., Li, D., Li, P.: Knowledge graph embedding based question answering. In: Proceedings of WSDM, pp. 105–113 (2019)
13. Hannah, B., Elmar, H.: More accurate question answering on freebase. In: Proceedings of CIKM, pp. 1431–1440 (2015)
14. Jiang, T., et al.: Towards time-aware knowledge graph completion. In: Proceedings of COLING, pp. 1715–1724 (2016)
15. Luo, K., Lin, F., Luo, X., Zhu, K.Q.: Knowledge base question answering via encoding of complex query graphs. In: Proceedings of EMNLP, pp. 2185–2194 (2018)
16. Meng, Y., Anna, R., Alexey, R.: Temporal information extraction for question answering. In: Proceedings of EMNLP, pp. 887–896 (2017)
17. Mansi, G., Nitish, K., Raghuveer, C., Anirudha, R., Zachary, C.: AmazonQA: a review-based question answering task. In: Proceedings of IJCAI, pp. 4996–5002 (2019)
18. Mihajlo, G., Haibin, C.: Real-time personalization using embeddings for search ranking at airbnb. In: Proceedings of KDD, pp. 311–320 (2018)
19. Pennington, J., Socher, R., Manning, C.D.: Glove: global vectors for word representation. In: Proceedings of EMNLP, pp. 1532–1543 (2014)
20. Reddy, S., Lapata, M., Steedman, M.: Large-scale semantic parsing without question-answer Pairs. TACL 377–392 (2014)
21. Semih, Y., Izzeddin, G., Yu, S., Yan, X.: Recovering question answering errors via query revision. In: Proceedings of EMNLP, pp. 903–909 (2017)
22. Tong, Q., Young, P., Park, Y.: A time-based recommender system using implicit feedback. Expert Syst. Appl. **34**(4), 3055–3062 (2007)
23. Tomas, M., Ilya, S., Kai, C., Greg, C., Jeffrey, D.: Distributed representations of words and phrases and their compositionality. CoRR, abs /1310.4546 (2013)
24. Usbeck, R., Ngomo, A.N., Haarmann, B., Krithara, A., Röder, M., Napolitano, G.: 7th open challenge on question answering over linked data (QALD-7). In: Proceedings of ESWS, pp. 59–69 (2017)
25. Wang, T., Yuan, X., Trischler, A.: Joint model for question answering and question generation. arXiv, CoRR, abs/1706.01450 (2017)
26. Wanyun, C., Yanghua, X., Haixun, W., Yangqiu, S., Seung-won, H., Wei, W.: A joint model for question answering and question generation. CoRR, abs/1903.02419 (2019)
27. Yao, X., Van Durme, B.: Information extraction over structured data: question answering with freebase. In: Proceedings of ACL, pp. 956–966 (2014)
28. Yih, W., Chang, M., He, X., Gao, J.: Semantic parsing via staged query graph generation: question answering with knowledge base. In: Proceedings of ACL, pp. 1321–1331 (2015)
29. Yanghoon, K., Hwanhee, L., Joongbo, S., Kyomin, J.: Improving neural question generation using answer separation. In: Proceedings of AAAI, pp. 6602–6609 (2019)
30. Zhang, W., Liu, T., Qin, B., Zhang, Y.: Benben: a Chinese intelligent conversational robot. In: Proceedings of ACL, pp. 13–18 (2017)
31. Zhou, Q., Yang, N., Wei, F., Tan, C., Bao, H., Zhou, M.: Neural question generation from text: a preliminary study. In: Proceedings of NLPCC, pp. 662–671 (2017)

32. Zou, L., Huang, R., Wang, H., Yu, J.X., He, W., Zhao, D.: Natural language question answering over RDF: a graph data driven approach. In: Proceedings of SIGMOD, pp. 313–324 (2014)
33. Haixin, Z., Xiaowang, Z.: An embedding-based approach to completing question semantics. In: Proceedings of ISWC, pp. 209–213 (2020)
34. Ma, X., Zhu, Q., Zhou, Y., Xiaolin, L., Dapeng, W.: Improving question generation with sentence-level semantic matching and answer position inferring. In: Proceedings of AAAI (2019)
35. Garg, S., Vu, T., Moschitti, A.: TANDA: transfer and adapt pre-trained transformer models for answer sentence selection. In: Proceedings of AAAI (2019)
36. Peng, Q., Xiaowen, L., Mehr, L., Zijian, W.: Answering complex open-domain questions through iterative query generation. In: Proceedings of EMNLP (2019)

# Hop-Constrained Subgraph Query and Summarization on Large Graphs

Yu Liu, Qian Ge, Yue Pang, and Lei Zou$^{(\boxtimes)}$

Peking University, Beijing, China
{dokiliu,geqian,michelle.py,zoulei}@pku.edu.cn

**Abstract.** We study the problem of hop-constrained relation discovery in a graph, i.e., finding the structural relation between a source node $s$ and a target node $t$ within $k$ hops. Previously studied $s - t$ graph problems, such as distance query and path enumeration, fail to reveal the $s - t$ relation as a big picture. In this paper, we propose the $k$-hop $s - t$ subgraph query, which returns the subgraph containing all paths from $s$ to $t$ within $k$ hops. Since the subgraph may be too large to be well understood by the users, we further present a graph summarization method to uncover the key structure of the subgraph. Experiments show the efficiency of our algorithms against the existing path enumeration based method, and the effectiveness of the summarization.

**Keywords:** Hop-constrained subgraph query · $k$-hop $s - t$ subgraph · $s - t$ graph summarization

## 1 Introduction

With the advent of graph data, it has become increasingly important to manage large-scale graphs in database systems efficiently. Generally, vertices in graphs represent entities and edges the relations between them. Paths, formed by chaining together multiple edges that share vertices, can be seen as representing more complex relations between its source and destination vertices. The fundamental problem of discovering the relation between two entities has thus given rise to numerous path-finding algorithms, the majority of which aims at determining whether a relation exists between two vertices (i.e., reachability) or finding a relation between two vertices that satisfy specific properties (e.g., shortest path and top-$k$ path enumeration). However, in certain applications, focusing on one relation (path) at a time is not enough. We list two real-world scenarios in which the $s - t$ relation is demanded as a big picture.

*Motivation 1. Discovery of ownership structure.* In an equity network, vertices represent corporations, an edge points from a corporation to another if the former holds shares of the latter. An important query would be to discover the ownership structure between two corporations, characterized by chains of shareholding that may span across the whole network. The results of such queries can

© Springer Nature Switzerland AG 2021
C. S. Jensen et al. (Eds.): DASFAA 2021 Workshops, LNCS 12680, pp. 123–139, 2021.
https://doi.org/10.1007/978-3-030-73216-5_9

help gain insights into a market's dynamics, e.g., how financial risks propagate, and therefore help with risk management.

*Motivation 2. Relation discovery in social networks.* In a social network, vertices represent persons and edges their relationships, which may include *follower-of, friend-of, parent-of,* etc. A query may aim to obtain the "social group" formed with two persons of interest as the source and destination respectively, composed of other persons that act as intermediates for the former to reach the latter. Results of such queries may benefit social network analysis (e.g., for advertising) and anomaly detection (e.g., for detecting crimes and terrorism).

Unfortunately, these applications cannot be appropriately handled by existing path-finding problems and their solutions, for they call not for single paths, but a *subgraph* that merges all relevant relations between the source and the destination. In this paper, we tackle the problem of efficiently computing a subgraph that represent the relations between a source and a destination vertex. Intuitively, given a hop constraint $k$, we compute a subgraph containing the paths from $s$ to $t$ within $k$ hops, which is referred to as the $k$-hop $s - t$ subgraph. Specifically, several algorithms based on graph traversal and pruning techniques are developed to compute the subgraph. Considering the subgraph may be too large to be well understood by users (e.g., for visualization) for large graphs, we further propose a graph summarization technique to only reveal the structural skeleton of the subgraph. The main contribution of our paper is summarized as follows.

- We first propose the $k$-hop $s - t$ *graph query*, which returns a subgraph containing all paths from $s$ to $t$ within $k$ hops. Compared to existing queries such as ($k$-hop) reachability, $s - t$ distance query and path enumeration, the subgraph query reveals the $s - t$ relation as a big picture. We also propose a traversal-based algorithm which is worst-case optimal in answering subgraph queries.
- Based on the result subgraph, we further propose the notion of $s - t$ *graph summarization with hop constraint*, which contracts the subgraph into a summarized graph with only a few nodes (controlled by a user-defined parameter). We present an algorithm based on skeleton node selection and local graph clustering, and demonstrate two skeleton node selection strategies which depend on path frequency and walking probability, respectively.
- On several large graph datasets we demonstrate the efficiency of our subgraph finding algorithm against the baselines based on path enumeration, as well as the effectiveness of our algorithms in terms of $s - t$ relation discovery and subgraph summarization.

The remainder of the paper is organized as follows. Section 2 gives the formal definitions of our studied problems. We discuss related work in Sect. 3, including several baseline methods. In Sect. 4 and 5, we propose our solutions for the $k$-hop subgraph query and hop-constrained $s - t$ graph summarization, respectively. Section 6 reports the experimental results. We conclude the paper in Sect. 7.

## 2   Preliminaries

### 2.1   Problem Statement

We first give several formal definitions about paths and subgraphs. Then we describe the two studied problems.

**Definition 1 (Path and $k$-hop path).** *Given a simple and directed graph $G = (V, E)$, a path $p = (v_1 = s, v_2, \ldots, v_l = t)$ in $G$ is defined as a sequence of edges, i.e., $(v_i, v_{i+1}) \in G.E, \forall i \in [1, l)$. Note that the length of $p$ is $l - 1$, and $p$ is referred to as a $(l - 1)$-hop path.*

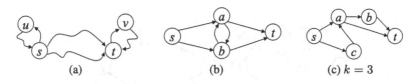

(a)            (b)            (c) $k = 3$

**Fig. 1.** An illustration of the definition of $k$-hop $s - t$ subgraph.

We say path $p$ contains a cycle if there exists some $1 \leq i < j \leq l$ such that $v_i = v_j$.

**Definition 2 (Union of paths).** *Given a set of paths $\{p_1, \ldots, p_m\}$, where each path $p_i = (v_{1_i} = s, \ldots, v_{l_i} = t)$ is from $s$ to $t$. Subgraph $G_{st} = (V_{st}, E_{st})$ is a union of paths $\{p_1, \ldots, p_m\}$, if $V_{st} = \cup_{i \in [1,m]}\{v_{1_i} \cup \ldots \cup v_{l_i}\}$, and $E_{st} = \cup_{i \in [1,m]}\{(v_{j_i}, v_{(j+1)_i}), j \in [1, l)\}$. Duplicated edges are removed during the union operation.*

**Definition 3 ($k$-hop $s-t$ subgraph).** *A subgraph $G_{st}$ ($s \neq t$) is referred to as $k$-hop $s-t$ subgraph if it is the union of all $k$-hop $s-t$ paths, such that for each path $p = (v_1 = s, \ldots, v_l = t)$, (1) $v_i \neq s, \forall i \in (1, l]$; and (2) $v_j \neq t, \forall j \in [1, l)$. We also refer to $G_{st}$ as $k$-hop subgraph, or subgraph when the context is clear.*

The definition of subgraph query aims to reveal the $k$-hop relation between $s$ and $t$ as a whole, rather than enumerating separate paths. However, we are not interested in (1) nodes only reachable to $t$ via $s$ ($u$ in Fig. 1(a)), or (2) nodes only reachable from $s$ via $t$ ($v$ in Fig. 1(a)). Intuitively, $u$ and $v$ do not contribute to the relation between $s$ and $t$, therefore we do not take them into consideration.

Nonetheless, we allow certain cycles in the $s - t$ relation. For example, after inserting two 3-hop paths $(s, a, b, t)$ and $(s, b, a, t)$ into the subgraph (Fig. 1(b)), a cycle is formed between $a$ and $b$. Such cycle may represent meaningful relations, for instance, the circulating ownership of stock in financial networks. Also note that $G_{st}$ may contain $s - t$ path longer than $k$ ($(s, c, a, b, t)$ in Fig. 1(c)), which is inevitable because of the union of different paths. We ignore these longer paths in that $G_{st}$ only focuses on the close (i.e., $k$-hop) relation between $s$ and $t$.

In this paper, we study the problems of $k$-hop $s-t$ subgraph query and $k$-hop subgraph summarization, defined as follows.

**Definition 4 (Hop-constrained subgraph query).** *Given a directed graph $G = (V, E)$, a source node $s$, a target node $t(t \neq s)$, and the hop constraint $k$, return the $k$-hop $s - t$ subgraph $G_{st}^k$.*

For simplicity, we only consider simple directed graphs in this paper. Moreover, the relation defined above points from $s$ to $t$ and is asymmetric (following out-edges). Note that the problem setting can be easily extended to other types of graphs (e.g., undirected or weighted graphs), while the relation can be defined based on a set of paths following in-edges, or allowing a mixture of outgoing and incoming edges.

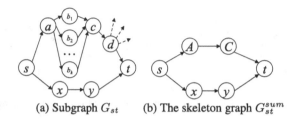

(a) Subgraph $G_{st}$     (b) The skeleton graph $G_{st}^{sum}$

**Fig. 2.** $k$-hop subgraph and its summarization.

**Remarks.** As mentioned by previous work [21,23], posing a constraint on the number of hops is reasonable, in that the strength of the relation drops dramatically with distance. Nonetheless, our studied problem is still well-defined when $k$ is set to $\infty$. Let $G_{SCC}$ be the directed acyclic graph (DAG) where each node in $G_{SCC}$ corresponds to a strongly connected component of $G$. Let $C_s$ (resp. $C_t$) be the strongly connected component containing $s$ (resp. $t$). To this end, we are essentially extracting the subgraph in $G$ which corresponds to a subgraph in $G_{SCC}$ composed by all paths from $C_s$ to $C_t$.

Although the $k$-hop $s - t$ subgraph provides a way to understand the relation between $s$ and $t$, the size of the subgraph can be extremely large, e.g., with hundreds of thousands of vertices for a reasonable $k$ (say 6) and a medium-sized graph. This prevents us from finding the underlying structure of $s - t$ relation. Therefore, we also consider the problem of subgraph summarization, which summarizes the result subgraph into a small and succinct one.

We adopt a skeleton node-based summarization method that contains two steps. First, we find a set of skeleton nodes that play important role in the underlying structure, where the number of skeleton nodes is a user-defined parameter. Then, we conduct local graph clustering from these skeleton nodes. Other nodes in the subgraph (except $s$ and $t$) is assigned to one of the communities $C_i$ corresponding to some skeleton node $v_i$. In particular, given a $k$-hop s-t subgraph $G_{st}$ and a set of skeleton nodes $V_S$ (detailed later), the skeleton graph $G_{st}^{k,h}$ is the summarization graph of $G_{st}^k$, where $G_{st}^{k,h}.V = V_S \cup \{s,t\}$, and for any $u, v \in G_{st}^{k,h}.V$, $(u, v) \in G_{st}^{k,h}.E$ if some criterion is satisfied, e.g., there exists an edge $(x, y) \in G_{st}^k.E$ s.t. $x \in C_u$ and $y \in C_v$, or the probability of node $u$ reaching node $v$ in $G_{st}^k$ is above some threshold.

**Definition 5 (Hop-constrained $s - t$ graph summarization).** *Given a directed graph $G = (V, E)$, a source node $s$, a target node $t$, the hop constraint $k$, and the number of skeleton nodes $h$, return the $k$-hop $s - t$ summarized graph $G_{st}^{k,h}$, which contains $h$ super-nodes corresponding to $h$ local communities in the $s - t$ subgraph $G_{st}^k$.*

Figure 2 demonstrates a skeleton graph of the $k$-hop subgraph. We set the number of skeleton nodes as 4. Intuitively, node $a, c$, and $d$ are more important than $b_1, \ldots, b_k$ (e.g., shell companies in financial networks) because more paths go through them. Node $a$ and $c$ are preferred over $d$ because the latter is a hot point (i.e., node of large degree, shown in dashed edges), but contributes few edges to $G_{st}$. Therefore, we compress node $a$ and $b_1, \ldots, b_k$ to a super-node $A$, $c$ and $d$ to a super-node $C$. The summarization graph highlights the structural information in the $s - t$ relation and is easier to visualize. We will discuss the strategies for finding skeleton nodes in Sect. 5.2.

## 3  Existing Work

To the best of our knowledge, there is no existing work that directly considers the problem of $s - t$ subgraph query or summarization. However, a bunch of works study similar problems that are more or less aimed at determining the relation between a pair of vertices, in which the techniques used can be extended to our problem settings. We categorize them as follows and discuss their relation to our problems.

### 3.1  Reachability/$k$-hop Reachability

The classic reachability problem studies whether there exists a path from a source vertex to a destination vertex. The majority of existing reachability algorithms [24,25,28,30] are index-based and focus on the directed acyclic graph (DAG) contraction of the input graph. Some generalized versions of reachability queries have also been proposed, including the label constrained reachability [16] and the $k$-hop reachability [7], which answers reachability with the hop constraint $k$. They can not be directly applied to our problem because too few information between $s$ and $t$ is preserved.

### 3.2  Shortest Path/$k$-shortest Paths

A plethora of works study the single-pair shortest path problem [3–5,8,11,12,14], as well as the $k$-shortest paths (KSP) problem [6,10,15,18,29] which returns the top-$k$ shortest paths between the source and the target. Though the algorithms for KSP can be used for path enumeration, it has been proved inefficient [13].

## 3.3   Path Enumeration/Top-$k$ Path Enumeration

The path enumeration problem aims to find all paths from the source to the target, possibly with additional constraints (e.g., hop constraint). To answer the subgraph query, we can first enumerate all ($k$-hop) paths and then combine them into a subgraph. We first show that depth-first search (DFS) can be easily applied to answer $k$-hop subgraph queries.

The pseudo-code is illustrated in Algorithm 1. Given a directed graph $G$, a source node $s$, a target node $t$, and the hop constraint $k$, the algorithm first initialize $G_{st}^k$ as empty graph (Line 1) and then invoke $k$-DFS (Line 2). The procedure $k$-DFS traverses the graph from $s$ in a depth-first way. Once it reaches $t$, which means a path from $s$ to $t$ is found, we insert the path into $G_{st}^k$ (Lines 6–7); then we stop traversal immediately, ignore any node only reachable from $s$ via $t$ (Line 8). Note that the traversal is limited within $k$-hops from $s$ (Lines 10–12). We have the following theorem.

**Theorem 1.** *Algorithm 1 correctly computes the $k$-hop subgraph with $O(n^k)$ worst time complexity.*

*Proof.* The correctness of Algorithm 1 can be easily derived by the property of DFS and the definition of $k$-hop subgraph. To show the algorithm runs in $O(n^k)$ in worst case, consider the following graph (Fig. 3). Since there are totally $\left(\frac{n}{k}\right)^k$ $k - hop$ paths from $s$, the time complexity of DFS is $O(k \cdot \left(\frac{n}{k}\right)^k) \sim O(n^k)$.

To improve the practical efficiency of $k$-DFS, in the theoretical paper [13], they propose T-DFS, a polynomial delay algorithm which takes $O(km)$ to find one path. Recent work [23] aims at detecting all simple cycles within $k$ hops after the insertion of an edge on dynamic graphs, by enumerating all simple paths within $k - 1$ hops between the two vertices that the new edge is adjacent to. To speed up query processing, it employs a hot point index to prevent repetitive traversals from vertices with high degrees. On the other hand, [21] employs a pruning-based algorithm to speed up $k$-hop simple path enumeration. Note that any path enumeration algorithm has at least $\Omega(k\delta)$ complexity, where $\delta$ is the number of valid paths. Since there can be tremendous numbers of paths, they are not suitable to answer top-$k$ subgraph queries of which the answer size is bounded by $O(n + m)$. Besides, current methods do not consider cycles for simplicity, whereas we include some types of cycles that also represent the relationship in the subgraph.

Other related works, such as graph summarization [9,17,22,26,27], primarily consider summarizing the whole input graph or the subgraph around a given node $s$ while preserving some properties, and are not specially tailored for the $s - t$ relation discovery.

---

**Algorithm 1.** The Baseline Algorithm

---

**Input:** Directed graph $G = (V, E)$; Source $s$; Target $t$; Hop constraint $k$
**Output:** $G_{st}^k$, the $k$-hop $s - t$ subgraph
 1: Initialize $G_{st}^k$ as empty graph and stack $\mathcal{S} = \emptyset$;
 2: $k\text{-DFS}(G, s, s, t, k, \mathcal{S}, G_{st}^k)$;
 3: **return** $G_{st}^k$;
 4: **Procedure** $k\text{-DFS}$(Graph $G$, Current node $u$, Source $s$, Target $t$, Hop constraint
    $k$, Stack $\mathcal{S}$, Partial subgraph $G_{st}^k$)
 5:    Push $u$ to $\mathcal{S}$;
 6:    **if** $u = t$ **then**
 7:        Add $p(\mathcal{S})$ to $G_{st}^k$;
 8:        **return** ;
 9:    **if** $k > 0$ **then**
10:        **for** each $v \in O(u)$ and $v \neq s$ **do**
11:            $k\text{-DFS}(G, v, s, t, k - 1, \mathcal{S}, G_{st}^k)$;

---

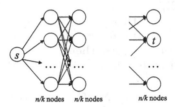

**Fig. 3.** Worst case graph for $k$-DFS.

# 4 KHSQ: The K-hop Subgraph Query Algorithm

## 4.1 Rationale

Recall that the baseline algorithm incurs $O(n^k)$ cost for $k$-hop subgraph query because each node $v$ may conduct DFS multiple times. In this section, we propose a simple yet effective algorithm that accelerates the querying speed by utilizing the distance information. Intuitively, we first conduct a $k$-hop breadth first search (BFS) to compute $d^f(s, v)$ for each $v$, which is the distance from $s$ to $v$. Similarly, we compute the distance of $v$ to $t$ (denoted by $d^b(v, t)$) by a traversal from $t$ following in-edges. Then, the distance information are used to prune most repeated or unnecessary traversals.

## 4.2 Algorithm

Our algorithm, denoted as KHSQ (<u>K</u>-<u>H</u>op <u>S</u>ubgraph <u>Q</u>uery), is demonstrated in Algorithm 2. We first invoke $k$-BFS both from $s$ and $t$ to compute the distance array (Lines 2–3). The $k$-BFS procedure (Lines 7–18) is analogous to the breadth-first search, but only traverses $k$ levels. Then we invoke DFS-SQ (Line 5), which improves the naive $k$-DFS in two aspects:

First, instead of enumerating all $k$-hop paths and union them together, we check for each edge $(u, v)$ to see if it is in $G_{st}$. To be precise, recall that

---

**Algorithm 2.** KHSQ

---

**Input:** Directed graph $G = (V, E)$; Source $s$; Target $t$; Hop constraint $k$
**Output:** $G_{st}$, the $k$-hop $s - t$ subgraph
1: Initialize $G_{st}$ as empty graph and stack $\mathcal{S} = \emptyset$;
2: $d^f(s, *) = k\text{-BFS}(G, s, t, k)$;
3: $d^b(*, t) = k\text{-BFS}(G^r, t, s, k)$;
4: Initialize $dfs(v) = false, \forall v \in V$;
5: DFS-SQ$(G, s, s, t, k, \mathcal{S}, d^f(s, *), d^b(*, t), dfs, G_{st})$;
6: **return** $G_{st}$;
7: **Procedure** $k$-BFS(Graph $G$, Source $s$, Target $t$, Hop constraint $k$)
8:    Initialize $d(s, s) = 0, d(s, v) = \infty$ for $\forall v \in V \backslash \{s\}$;
9:    Initialize queue $\mathcal{Q} = \{s\}$;
10:   Initialize $lvl = 0, numThisLvl = 1, numNextLvl = 0, visited(s) = true,$ and $visited(v) = false, \forall v \in V \backslash \{s\}$;
11:   **while** $\mathcal{Q} \neq \emptyset$ **do**
12:     **for** $i = 1$ to $numThisLvl$ **do**
13:       $u = \text{remove}(\mathcal{Q})$;
14:       **if** $u \neq t$ and $lvl < k$ **then**
15:         **for** each $v \in O(u)$ and $v \neq s$ **do**
16:           **if** $visited(v) = false$ **then**
17:             Add $v$ to $\mathcal{Q}$, $d(s, v) = d(s, u) + 1, visited(v) = true,$ $numNextLvl + +$;
18:       $lvl + +, numThisLvl = numNextLvl, numNextLvl = 0$;
19: **Procedure** DFS-SQ(Graph $G$, Current node $u$, Source $s$, Target $t$, Hop constraint $k$, Stack $\mathcal{S}$, forward distance $d^f(s, *)$ and backward distance $d^b(*, t)$, Label $dfs$, Partial subgraph $G_{st}$)
20:   Push $u$ to $\mathcal{S}$;
21:   $dfs(u) = true$;
22:   **if** $u = t$ **then**
23:     $\text{pop}(\mathcal{S})$;
24:     **return** ;
25:   **for** each $v \in O(u)$ and $v \neq s$ **do**
26:     **if** $d^f(s, u) + 1 + d^b(v, t) \leq k$ **then**
27:       Add edge $(u, v)$ to $G_{st}$;
28:     **if** $|S| = d^f(s, v)$ and $dfs(v) = false$ **then**
29:       DFS-SQ$(G, v, s, t, k, \mathcal{S}, d^f(s, *), d^b(*, t), dfs, G_{st})$;
30:   $\text{pop}(\mathcal{S})$;

---

in the definition of $k$-hop subgraph $G_{st}$, $(u, v) \in E_{st}$ if there exists some $k$-hop path containing it. If $d^f(s, u) + 1 + d^b(v, t) \leq k$, we are sure path $p = (p^*(s, u), (u, v), p^*(v, t))$ is a $k$-hop path, where $p^*(s, u)$ (resp. $p^*(v, t)$) denotes one shortest path from $s$ to $u$ (resp. from $v$ to $t$). Hence, it is sufficient to check each edge only once.

Second, we also guarantee that each node (and its out-edges) is visited only once. Each node $v$ conducts neighbor traversal only if the path in stack $\mathcal{S}$ is a shortest path to $v$, and $v$ has not conducted the traversal before. Since our algorithm checks each node and each edge at most once, the complexity is asymptotically linear of the problem inputs.

## 4.3   Analysis

The following theorem states the correctness of KHSQ.

**Theorem 2.** *Algorithm 2 correctly finds the k-hop subgraph.*

*Proof.* It is easy to see that procedure $k$-BFS correctly computes the distance from $s$ to $v$ within $k$-hops, and set the distance of other nodes as $\infty$. This also holds for the traversal from $t$ on $G^r$, which is in fact traversing in-edges of $G$. To show the correctness of procedure DFS-SQ, note that every edge $(u, v)$ in every $k$-hop path will be added to $G_{st}$ according to our checking condition (Line 27 of Algorithm 2). On the other hand, if some edge $(u', v')$ is not contained by any such path, then we have $d^f(s, u) + 1 + d^b(v, t) > k$ and it will be excluded. Path like $v \to s \to t$ (or $s \to t \to v$) and with less than $k$-hops is eliminated by setting $d^b(v, t)$ (or $d^f(s, v)$) as $\infty$ in $k$-BFS. Therefore, the subgraph returned by KHSQ algorithm is equal to the union of all valid $k$-hop paths, and the correctness follows.

We bound the time and space complexity of KHSQ as follows. Since procedure $k$-BFS only conducts breath-first search from $s$ and within $k$ hops, its time complexity is $O(n + m)$. As discussed above, we have demonstrated that procedure DFS-SQ visits each node and each edge at most once. Therefore, the cost is still bounded by $O(n + m)$. The following theorem states that KHSQ is highly efficient in answering $k$-hop subgraph queries.

**Theorem 3.** *The time complexity of KHSQ is $O(n+m)$, which is asymptotically linear with the problem inputs and worst-case optimal.*

*Proof.* Since KHSQ only invokes $k$-BFS twice and DFS-SQ from $s$ once, the time complexity can be easily derived. To see the algorithm is worst-case optimal, consider the case that $|G_{st}| = \Theta(|G|)$, where $|G| = |G.V| + |G.E| = n + m$, e.g., $G_{st} = G$. Since each node and edge must be processed with $O(1)$ cost, our claim holds.

# 5   KHGS: The $k$-hop $s - t$ Graph Summarization Algorithm

## 5.1   Problem Overview

Though the $k$-hop subgraph demonstrates the relation between $s$ and $t$ as a whole, it suffers from extremely large size for many real-world networks. For example, on a medium-sized graph, say, with millions of nodes and a reasonable $k$ (e.g., 6), the result subgraph may contain hundreds of thousand nodes, which prevents us from understanding the underlying structure of the $s - t$ relation.

Hence, we propose a skeleton node based method for subgraph summarization, which relies on a set of *skeleton nodes* (plus $s$ and $t$) while the edges and paths between them are contracted and summarized (Recall its definition in Sect. 2.1.). As long as we correctly select the most important nodes as the skeleton of $G_{st}$, the summarized graph reveals the key structure of $s - t$ relation hidden in a bunch of edges.

Our algorithm framework is shown in Algorithm 3. Given a graph $G$, a source node $s$, a target node $t$, and the hop constraint $k$, the KHGS (K-Hop Graph Summarization) algorithm first invokes KHSQ to compute the $k$-hop subgraph $G_{st}^k$. To get the summarization graph, our algorithm contains two steps. We first choose $h$ most important nodes (referred to as skeleton nodes) from $G_{st}^k$, where $h$ is a user-defined parameter. Then we contract $G_{st}^k$ into $G_{st}^{k,h}$, which can be simply implemented by $h$ distinct local traversals (e.g., clustering) from the skeleton nodes. Specifically, we present two importance measures based on path frequency and walking probability, respectively. We describe the algorithms for skeleton node selection in the following subsection.

---

**Algorithm 3. KHGS**

---

**Input:** Graph $G$; Source $s$; Target $t$; Hop constraint $k$; Number of skeleton nodes $h$
**Output:** $G_{st}^{k,h}$, the skeleton graph of the $s - t$ subgraph $G_{st}$
1: $G_{st}^k = \text{KHSQ}(G, s, t, k)$;
2: $V_S = \text{FindSkeletonNodes}(G_{st}^k, s, t, k, h)$;
3: $G_{st}^{k,h} = \text{SummarizedGraphConstruction}(G_{st}^k, s, t, k, V_S)$;
4: **return** $G_{st}^{k,h}$;

---

**Algorithm 4. FindSkeletonNodes-PathBased**

---

**Input:** Subgraph $G_{st}^k$; Source $s$; Target $t$; Hop constraint $k$; Number of skeleton nodes $h$
**Output:** $V_S$, a set of skeleton nodes, where $|V_S| \leq h$, and the subgraph $G_{st}$
1: $\{Partial(s, v), \forall v \in V_{st}\} = \text{PUSH-PATH}(G_{st}^k)$;
2: $\{Partial(v, t), \forall v \in V_{st}\} = \text{PUSH-PATH}(r(G_{st}^k))$;
3: **for** each $v \in V_{st}$ **do**
4:     $PCnt(v) = \sum_{(i,c_i) \in Partial(s,v)} \sum_{\substack{(j,c_j) \in Partial(v,t) \\ i+j \leq k}} c_i c_j$
5: Let $V_S$ be the top-$h$ nodes in $V \backslash \{s, t\}$ with largest (and non-zero) $PCnt(v)$;
6: **return** $V_S$;
7: **Procedure** PUSH-PATH($G_{st}^k$)
8:     Initialize $Partial(s, v) = \emptyset, \forall v \in V_{st}, Partial_0(s, s) = 1, Partial_0(s, v) = 0, \forall v \in V \backslash \{s\}$;
9:     **for** $l = 0$ to $k - 1$ **do**
10:         **for** each edge $(u, v) \in E_{st}$ **do**
11:             $Partial_{l+1}(s, v) += Partial_l(s, u)$;
12:         **for** each $v \in V_{st}$ **do**
13:             $Partial(s, v) = \cup_{l=0}^k Partial_l(s, v)$;

---

## 5.2  Finding Skeleton Nodes

**The Path Frequency Based Method.** According to the definition of $G_{st}^k$, which is the union of all $k$-hop paths from $s$ to $t$, a natural importance measure for a node $v \in V_{st}^k$ is the number of paths that go through $v$. We denote it as $PCnt(v)$. To be precise, we have

$$PCnt(v) = \sum_{p:s \to t, p \in G_{st}^k, |p| \leq k} I(p \text{ goes through } v), \forall v \in V_{st}, \tag{1}$$

where $I(*)$ is an indicator variable. Instead of enumerating and checking all paths, which incurs the excessive $O(n^k)$ cost, we propose an algorithm based on the *push* operation which transfers the information from $u$ to $v$ for each edge $(u, v)$ and the observation that $PCnt(v)$ can be computed by counting the number of paths from $s$ to $v$ and $v$ to $t$, respectively.

Our algorithm, denoted by FindSkeletonNodes-PathBased, takes $G_{st}^k$, source node $s$, target node $t$, hop constraint $k$, and skeleton node number $h$ as input. It invokes procedure PUSH-PATH to compute all paths from $s$ to $v$ (and from $v$ to $t$) for each $v \in V_{st}^k$ (Lines 1–2). Since every such path is a fragment of some path from $s$ to $t$, it is referred to as the *partial path*. For each node $v$, $Partial(s, v)$ contains a list of $(step, cnt)$ pairs, which indicates that there are totally $cnt$ distinct paths from $s$ to $v$ of length $step$. Once we have $Partial(s, v)$ and $Partial(v, t)$ for each $v$, we calculate $PCnt(v)$ by the following equation:

$$PCnt(v) \approx \sum_{(i,c_i) \in Partial(s,v)} \sum_{\substack{(j,c_j) \in Partial(v,t) \\ i+j \leq k}} c_i c_j, \forall v \in V_{st}^k. \tag{2}$$

Intuitively, Eq. 2 says that the number of $k$-hop paths that pass $v$ can be approximated by the number of paths from $s$ to $v$ times the number of paths from $v$ to $t$. Note that we exclude paths longer than $k$. In fact, the equation gives the exact answer when $G_{st}^k$ does not contain cycles. When the subgraph has cycles, Eq. 2 computes an upper bound of $PCnt(v)$. We illustrate it by an example as in Fig. 4. Consider the subgraph in Fig. 4(a), while we set $k = 6$. Their are totally three 6-hop paths from $s$ to $t$ (Fig. 4(b)). Similarly, we can compute all 6-hop paths from $s$ to $v$ and from $v$ to $t$, as shown in Fig. 4(c). If we concatenate these partial paths together (and eliminate paths longer than 6), the second and third path Fig. 4(b) is counted twice and three times, respectively.

Unfortunately, we are unable to eliminate duplicated counting of paths unless we can enumerate all $s - t$ paths, which is infeasible for sizable graphs. However, since $k$ is usually small in practice, which limits the repetitions in a cycle for a $k$-hop path, the over estimation of path count only has a minor effect. Therefore, our approximation achieves a good balance between efficiency and effectiveness.

Now we describe the procedure to compute $Partial(s, v)$ (and $Partial(v, t)$) for each node $v$. Take $Partial(s, v)$ as an example. Denote by $Partial_l(s, v)$ the number of path from $s$ to $v$ of length $l$, the following lemma holds. The equation for $Partial_l(v, t)$ can be defined analogously.

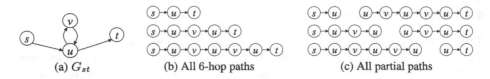

(a) $G_{st}$      (b) All 6-hop paths      (c) All partial paths

**Fig. 4.** Duplicated counting of $s - t$ paths.

**Lemma 1.**

$$Partial_l(s, v) = \begin{cases} 1, & \text{if } v = s \text{ and } l = 0, \\ 0, & \text{if } v \neq s \text{ and } l = 0, \\ \sum_{u \in I(v)} Partial_{l-1}(s, u), & \text{otherwise.} \end{cases} \quad (3)$$

*Proof.* For any node $v \neq s$, a path from $s$ to $v$ of length $l$ can be decomposed into the sub-path from $s$ to $u$ and edge $(u, v)$, where $u$ denotes the predecessor of $v$ in $p$. The length of the sub-path is exactly $l - 1$. Besides, for any $u, u' \in I(v)$ and $u \neq u'$, the path to $v$ either comes from $u$ or $u'$, and our lemma follows.

The implementation of Eq. 3 is shown in procedure PUSH-PATH. We initialize $Partial(s, v)$ in Line 8, and then proceed on $k$ iterations. During each iteration $l$, we check every edge $(u, v) \in E_{st}$, and push $Partial_l(s, u)$ to $Partial_{l+1}(s, v)$ (Lines 9–11). For the computation of $Partial(v, t)$, we invoke PUSH-PATH on the reverse graph of $G_{st}^k$ (denoted as $r(G_{st}^k)$). Notice that for practical efficiency, in each iteration $l$ we only record a set of nodes with $Partial_l(s, v) > 0$, and conduct push operation from these nodes. The following lemma and theorem are easily derived.

**Lemma 2.** *The time and space complexity of procedure PUSH-PATH is $O(k(n+ m))$ and $O(kn + m)$, respectively.*

*Proof.* Since in each iteration, each node and each edge is processed at most once, so the complexity is bounded by $O(|G_{st}^k|) = O(|V_{st}^k| + |E_{st}^k|)$, and again bounded by $O(n+m)$ because $G_{st}$ is a subgraph of $G$. The algorithm has exactly $k$ iterations, thus the time complexity is $O(k(n + m))$. For the space usage of PUSH-PATH, note that storing $Partial(s, v)$ needs $O(k)$ space. We need extra $O(|G_{st}^k|)$ space for $G_{st}^k$. In total, the space cost is bounded by $O(kn + m)$.

**Theorem 4.** *The time and space complexity of FindSkeletonNodes is $O(k^2 n + km))$ and $O(kn + m)$, respectively.*

*Proof.* Algorithm FindSkeletonNodes invokes PUSH-PATH twice, which costs $O(k(n + m))$ time. Since both $Partial(s, v)$ and $Partial(v, t)$ may contain $O(k)$ items, computing $PCnt(v)$ for each $v$ is $O(k^2)$. Consequently, the total computation cost is bounded by $O(k(n + m)) + O(k^2 n) = O(k^2 n + km))$. Since each node (resp. each edge) needs $O(k)$ (resp. $O(1)$) space cost, the space complexity is $O(kn + m)$.

In practice, we always use small $k$, e.g., $k \leq 10$, and the algorithm has near-linear time and space complexity.

**The Walking Probability Based Method.** The path frequency based definition of node importance is intuitive, but suffers from a few deficiencies owing to the graph structure inside and outside $G_{st}^k$. First, the path frequency based measure does not consider the path length, which is an indication of the closeness of the relation. Second, path frequency is vulnerable to malicious tampering of the graph structure. Take Fig. 2(a) as an example, by building more shell companies (i.e., $b_i$), more $s-t$ paths go through $a$ and $c$. Third, node $d$ is a hot point but contribute few to the $s-t$ relation, indicating that we should also consider the graph structure of the whole graph when choosing skeleton nodes. Lastly, as previously discussed, the path-based measure is also vulnerable to cycles.

Inspired by the Random Walk with Restart [20] and Personalized PageRank [19], we propose a walking probability based measure for node importance, which alleviates all drawbacks above. We only need a few modification of FindSkeletonNodes-PathBased and PUSH-PATH. Briefly speaking, instead of transfer the information of path frequency from $u$ to $v$ along each edge $(u, v)$, we transfer probability instead. We first define the walking probability as follows.

**Definition 6 (Random walk).** *A random walk from $u$ is defined as (1) for each step, with probability $\alpha$ the walk stops; (2) with $1-\alpha$ probability, $u$ randomly chooses an out-neighbor $v$ and proceeds to it. Here $\alpha$ is a decay factor in $(0, 1)$.*

**Definition 7 (Walking probability).** *The probability of node $u$ walks to $v$, denoted as $Pr(s, v)$, is defined as $Pr(s, v) = \cup_{l=0}^{k} Pr_l(s, v), \forall v \in V$, where*

$$Pr_l(s, v) = \begin{cases} 1, & \text{if } v = s \text{ and } l = 0, \\ 0, & \text{if } v \neq s \text{ and } l = 0, \\ \sum_{u \in I(v)} \alpha \cdot \frac{Pr_{l-1}(s,u)}{|O(u)|}, & \text{otherwise.} \end{cases} \quad (4)$$

It can be proved by induction that $Pr(s, v)$ is exactly the probability of a random walk from $s$ terminating at $v$. We omit the details for space constraint. By substituting the path frequency measure by the probability based one, we denote the corresponding procedures as FindSkeletonNodes-ProbBased and PUSH-PROB, respectively.

## 5.3   Summarized Graph Construction

After we have determined the skeleton node set $V_S$, we contract $G_{st}$ accordingly. The procedure is rather straightforward: for each $v \in V_S$, we conduct any off-the-shelf local clustering algorithm, and contract each cluster to a super-node. Two super-nodes have connecting edges if some nodes in the corresponding cluster are connected or the walking probability between the super-nodes is above some threshold. For example, if we employ personalized PageRank for local clustering and estimate the PPR values via a limited number of random walks, a good tradeoff between efficiency and effectiveness can be fulfilled.

Finally, we conclude with the following Theorem, which states the complexity of KHGS.

**Theorem 5.** *The time and space complexity of KHGS is $O(h(k^2 n + km))$ and $O(hn + m)$, respectively, and is near-linear when $k$ and $h$ can be viewed as constants.*

## 6   Experiments

In this section, we evaluate both the efficiency of our $k$-hop subgraph algorithm against the baseline algorithm, as well as the performance of the summarization algorithms.

### 6.1   Experimental Settings

**Dataset Details.** We employ four large directed dataset, i.e., Web-Google (WG) ($n = 875,713, m = 5,105,039$), In-2004 (IN) ($n = 1,382,908, m = 16,917,053$), Soc-LiveJournal (LJ) ($n = 4,847,571, m = 68,475,391$), and IT-2004 (IT) ($n = 41,291,594, m = 1,150,725,436$). All datasets are obtained from public sources [1,2].

**Methods.** For the $k$-hop subgraph queries, we compare the baseline algorithm (Algorithm 1) and KHSQ (Algorithm 2) in terms of efficiency. For the graph summarization problem, we report the query time of KHGS (Algorithm 3), and consider skeleton node selection via both path frequency based and walking probability based methods.

**Environments.** We randomly generate $1,000$ $s - t$ pairs, and vary $k$ from 3 to 6. We guarantee that $t$ can be reached from $s$ within $k$ hops. All experiments are conducted on a machine with a 2.6GHz CPU and 64GB memory.

### 6.2   Efficiency

Table 1 shows the query time of the baseline algorithm and KHSQ. Since the $k$-DFS procedure is extremely slow for large $k$, we set $k = 4$. Symbol '-' indicates that for some query the time cost exceeds 1,000 s. KHSQ is significantly faster than the baseline, e.g., nearly an order of magnitude faster on LJ. Moreover, for the largest dataset IT, the baseline method fails to answer the query even for $k = 4$. Figure 5 demonstrate the query speed of Baseline and KHSQ varying $k$.

In the following, we compare the result size of path enumeration and subgraph query, followed by the query time evaluation of our summarization algorithms.

**Table 1.** Query time (sec) of Baseline and KHSQ ($k = 4$).

| Method | Dataset | | | |
|---|---|---|---|---|
| | WG | IN | LJ | IT |
| Baseline | 0.013 | 0.52 | 11.12 | - |
| KHSQ | 0.005 | 0.022 | 1.59 | 1.34 |

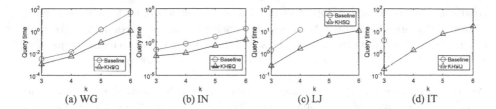

**Fig. 5.** Query time (sec) of Baseline and KHSQ, varying $k$

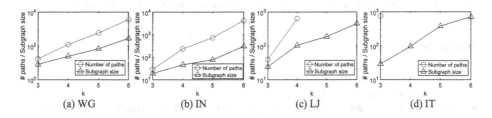

**Fig. 6.** Result size of path enumeration and subgraph query

**Table 2.** Query time (sec) of KHGS.

| Method | Dataset | | | |
|---|---|---|---|---|
| | WG | IN | LJ | IT |
| KHGS (Path-based) | 0.43 | 1.28 | 76.81 | 91.26 |
| KHGS (Probability-based) | 0.52 | 1.42 | 83.57 | 97.71 |

**Result Size.** We conduct $k$-hop path enumeration and subgraph query on four datasets and very $k$ from 3 to 6. The result is shown in Fig. 6. As $k$ increases, the number of $k$-hop paths explodes, whereas the size of $k$-hop subgraph is still limited.

**Query Time of KHGS.** We also evaluate the query efficiency of our KHGS algorithm. For all datasets, we fix the number of skeleton nodes $h$ as 8, hop constraint $k = 5$, and $\alpha = 0.6$. Note that the complexity of KHGS is linear in $h$ and quadratic in $k$. Hence their values do not have a major effect on the query efficiency. The results are shown in Table 2.

## 7   Conclusion

In this paper, we propose the problem of $k$-hop $s - t$ subgraph query and a traversal-based algorithm that answers the query in $O(n + m)$ time, which is worst-case optimal. We further introduce the notion of hop-constrained $s - t$ graph summarization, which computes a skeleton graph of the $s - t$ subgraph and provides a better understanding of the underlying structure of the $s - t$

relation. Our proposed algorithms are based on skeleton node selection with various strategies and graph traversal. Extensive experiments demonstrate that our proposed queries better reflect the $s-t$ relation compared to existing queries, while our algorithms are highly efficient even on massive graphs.

**Acknowledgements.** This work was supported by The National Key Research and Development Program of China under grant 2018YFB1003504, NSFC (No. 61932001), and Peking University Medicine Seed Fund for Interdisciplinary Research supported by the Fundamental Research Funds for the Central Universities (No. BMU2018MI015). This work was also supported by Beijing Academy of Artificial Intelligence (BAAI).

# References

1. http://konect.uni-koblenz.de/
2. http://law.di.unimi.it/webdata/
3. Abraham, I., Delling, D., Goldberg, A.V., Werneck, R.F.: Hierarchical hub labelings for shortest paths, pp. 24–35 (2012)
4. Bast, H., Funke, S., Sanders, P., Schultes, D.: Fast routing in road networks with transit nodes. Science **316**(5824), 566 (2007)
5. Bauer, R., Delling, D., Sanders, P., Schieferdecker, D., Schultes, D., Wagner, D.: Combining hierarchical and goal-directed speed-up techniques for Dijkstra's algorithm. ACM J. Exp. Algorithmics **15**(2.3) (2010)
6. Chang, L., Lin, X., Qin, L., Yu, J.X., Pei, J.: Efficiently computing top-k shortest path join. In: EDBT 2015–18th International Conference on Extending Database Technology, Proceedings (2015)
7. Cheng, J., Shang, Z., Cheng, H., Wang, H., Yu, J.X.: K-reach: who is in your small world. Proc. VLDB Endow. **5**(11), 1292–1303 (2012)
8. Delling, D., Goldberg, A.V., Pajor, T., Werneck, R.F.: Robust exact distance queries on massive networks. Microsoft Research, USA, Technical report 2 (2014)
9. Dunne, C., Shneiderman, B.: Motif simplification: improving network visualization readability with fan, connector, and clique glyphs. In: Proceedings of the SIGCHI Conference on Human Factors in Computing Systems, pp. 3247–3256. ACM (2013)
10. Eppstein, D.: Finding the k shortest paths. SIAM J. Comput. **28**(2), 652–673 (1998)
11. Geisberger, R., Sanders, P., Schultes, D., Delling, D.: Contraction hierarchies: faster and simpler hierarchical routing in road networks. In: McGeoch, C.C. (ed.) WEA 2008. LNCS, vol. 5038, pp. 319–333. Springer, Heidelberg (2008). https://doi.org/10.1007/978-3-540-68552-4_24
12. Goldberg, A.V., Harrelson, C.: Computing the shortest path: a search meets graph theory. In: Proceedings of the Sixteenth Annual ACM-SIAM Symposium on Discrete Algorithms, pp. 156–165. Society for Industrial and Applied Mathematics (2005)
13. Grossi, R., Marino, A., Versari, L.: Efficient algorithms for listing k disjoint st-paths in graphs, pp. 544–557 (2018)
14. Jiang, M., Fu, A.W., Wong, R.C., Xu, Y.: Hop doubling label indexing for point-to-point distance querying on scale-free networks. Very Large Data Bases **7**(12), 1203–1214 (2014)

15. Jiménez, V.M., Marzal, A.: Computing the $K$ shortest paths: a new algorithm and an experimental comparison. In: Vitter, J.S., Zaroliagis, C.D. (eds.) WAE 1999. LNCS, vol. 1668, pp. 15–29. Springer, Heidelberg (1999). https://doi.org/10.1007/3-540-48318-7_4

16. Jin, R., Hong, H., Wang, H., Ruan, N., Xiang, Y.: Computing label-constraint reachability in graph databases. In: Proceedings of the 2010 ACM SIGMOD International Conference on Management of Data, pp. 123–134. ACM (2010)

17. LeFevre, K., Terzi, E.: Grass: graph structure summarization. In: Proceedings of the 2010 SIAM International Conference on Data Mining, pp. 454–465. SIAM (2010)

18. Martins, E.Q., Pascoal, M.M.: A new implementation of yens ranking loopless paths algorithm. Q. J. Belgian French Italian Oper. Res. Soc. **1**(2), 121–133 (2003)

19. Page, L., Brin, S., Motwani, R., Winograd, T.: The pagerank citation ranking: bringing order to the web. Technical report, Stanford InfoLab (1999)

20. Pan, J.Y., Yang, H.J., Faloutsos, C., Duygulu, P.: Automatic multimedia cross-modal correlation discovery. In: Proceedings of the Tenth ACM SIGKDD International Conference on Knowledge Discovery and Data Mining, pp. 653–658. ACM (2004)

21. Peng, Y., Zhang, Y., Lin, X., Zhang, W., Qin, L., Zhou, J.: Hop-constrained ST simple path enumeration: Towards bridging theory and practice. Proc. VLDB Endow. **13**(4), 463–476 (2019)

22. Purohit, M., Prakash, B.A., Kang, C., Zhang, Y., Subrahmanian, V.: Fast influence-based coarsening for large networks. In: Proceedings of the 20th ACM SIGKDD International Conference on Knowledge Discovery and Data Mining, pp. 1296–1305. ACM (2014)

23. Qiu, X., et al.: Real-time constrained cycle detection in large dynamic graphs. Proc. VLDB Endow. **11**(12), 1876–1888 (2018)

24. Su, J., Zhu, Q., Wei, H., Yu, J.X.: Reachability querying: can it be even faster? IEEE Trans. Knowl. Data Eng. **29**(3), 683–697 (2016)

25. Tang, X., Chen, Z., Zhang, H., Liu, X., Shi, Y., Shahzadi, A.: An optimized labeling scheme for reachability queries. Comput Mater. Continua **55**(2), 267–283 (2018)

26. Tian, Y., Hankins, R.A., Patel, J.M.: Efficient aggregation for graph summarization. In: Proceedings of the 2008 ACM SIGMOD International Conference on Management of Data, pp. 567–580. ACM (2008)

27. Toivonen, H., Zhou, F., Hartikainen, A., Hinkka, A.: Compression of weighted graphs. In: Proceedings of the 17th ACM SIGKDD International Conference on Knowledge Discovery and Data Mining, pp. 965–973. ACM (2011)

28. Wei, H., Yu, J.X., Lu, C., Jin, R.: Reachability querying: an independent permutation labeling approach. VLDB J. Int. J. Very Large Data Bases **27**(1), 1–26 (2018)

29. Yen, J.Y.: Finding the k shortest loopless paths in a network. Manag. Sci. **17**(11), 712–716 (1971)

30. Zhu, A.D., Lin, W., Wang, S., Xiao, X.: Reachability queries on large dynamic graphs: a total order approach. In: Proceedings of the 2014 ACM SIGMOD International Conference on Management of Data, pp. 1323–1334. ACM (2014)

# Ad Click-Through Rate Prediction: A Survey

Liqiong Gu$^{(\boxtimes)}$

Tianjin Electronic Information Technician College, Tianjin 300350, China

**Abstract.** Ad click-through rate prediction (CTR), as an essential task of charging advertisers in the field of E-commerce, provides users with appropriate advertisements according to user interests to increase users' click-through rate based on user clicks. The performance of CTR models plays a crucial role in advertising. Recently, there are many approaches to improving the performance of CTR. In this paper, we present a survey to analyze state-of-art models of CTR via types of models comprehensively. Finally, we summarize some practical challenges and then open perspective problems of CTR.

**Keywords:** Click-through rate · CTR Prediction · E-commerce

## 1 Introduction

Click-through rate (CTR), defined as the probability that a specific user clicks on a displayed ad, is essential in online advertising [11,31]. To maximize revenue and user satisfaction, online advertising platforms must predict the expected user behavior of each displayed advertisement and maximize the user's expectations of clicking [28,33].

The prediction of click-through rate [9] is significant in the recommendation system (e.g., advertising system). Its task is to estimate the probability of users clicking on the recommended item. In many recommendation systems, the goal is to maximize the number of clicks, so the items returned to the user should be ranked by the estimated click-through rate. In other application scenarios, such as online advertising, increasing revenue is also substantial the ranking strategy can be adjusted. It is the bid of CTR for all candidates, where "bid" benefits the system when the user clicks on the item. In either case, the key is to estimate the CTR correctly. Hence it is crucial to improve the performance of CTR in the advertising system.

At the early phase of CTR, classical models of machine learning are applied to extract low-level features [2], such as factorization machine (FM) [23], and field-aware factorization machine (FFM) [14]. Though those models of machine learning have successfully improved the performance of CTR by extracting simple features, roughly called "*explicit*" features (e.g., independant features or combined features [18]), they are not good at capturing "*implicit*" features (e.g.,

© Springer Nature Switzerland AG 2021
C. S. Jensen et al. (Eds.): DASFAA 2021 Workshops, LNCS 12680, pp. 140–153, 2021.
https://doi.org/10.1007/978-3-030-73216-5_10

hidden features [34]) due to the sparse characteristics of advertising data [8,12]. In recent years, inspired by the success of deep learning [10,15], models of deep learning are applied to extract implicit features for further improving the performance of CTR [6,24]. Compared to models of machine learning, models of deep learning actually bring a better performance. However, it is not always acceptable to ad recommendations for the cost of learning especially online advertising systems. Though there are many approaches based on models of either machine learning or deep learning, there is few work to survey those approaches comprehensively in Ad CTR while there are some survey of current recommendation models such as sparse prediction [8].

This paper presents a survey to analyze state-of-art models of CTR via types of models, namely, machine learning model and deep learning model. For every kind of model, we technically analyze their principles via comparison. Finally, based on practical industrial applications, we summarize some practical challenges and then open perspective problems of CTR.

The rest of the paper is organized as follows. We survey CTR models based on machine learning in Sect. 2 and deep learning in Sect. 3. In Sect. 4, we discuss some challenging open problems. Finally, we conclude this paper in Sect. 5

## 2   Overview of Machine Learning CTR Prediction Models

In this section, we mainly review two popular CTR models, namely, *factorization machine (FM)* [23] and *field-aware factorization machine (FFM)* [14], which are based on linear model and non-linear models mainly extract low-level features.

Besides, [5] presents a personalized click prediction model, aiming to provide a framework for the customized click model in paid search by employing user-specific and demographic-based characteristics in reflecting the click behavior of individuals and groups.

In those models, each feature is independent without considering the implicit relationship between features [22].

### 2.1   Factorization Machine (FM)

Based on logistic regression, [18] introduces the second-order Cartesian product for characterizing a simple combination of features formalized as follows:

$$Y = \omega_0 + \sum_{i=1}^{n} \omega_i v_i + \sum_{i=1}^{n} \sum_{j=i+1}^{n} \omega_{ij} v_i v_j, \tag{1}$$

where $v_i$ and $v_j$ are feature vectors, $\omega_i$ is the weight of the feature, and $\omega_{ij}$ is the weight of the new feature combined by $v_i$ and $v_j$.

However, some shortcomings of this combination are concluded as follows:

- Causing dimensional disasters;
- Bringing negative effect on the model;
- Resulting in very sparse sample features.

Support vector machine (SVM) [3] model is a popular model applied in CTR prediction. Though non-linear SVM can perform kernel mapping on features, it can not always learn well when highly sparse features [8]. As usually, matrix decomposition is applied in the relative score matrix and then learn the implicit relationship between features and features. Since each model has its limitations for given feature scenes or inputs. In Ad CTR, FM [22,23] algorithm decouples $\omega_{ij}$ (in Eq. (1)) so that each feature can automatically learn a hidden vector.

Note that FM assumes that the two features do not appear in a sample. Indeed, however, this assumption seems rough since there is an indirect relationship [8,34]. The FM model is a machine learning model based on matrix factorization. It has a useful learning ability for sparse data. It also introduces combined features. Its objective function is formalized as follows:

$$Y = \omega_0 + \sum_{i=1}^{n} \omega_i v_i + \sum_{i=1}^{n} \sum_{j=i+1}^{n} <u_i, u_j> v_i v_j. \tag{2}$$

Note that tlogistic regression (LR) [1] model contains the first two items of Eq. (2). Compared with Eq. (1), $u_i$ represents the hidden vector of the feature $v_i$. The dot product of the invisible vectors corresponding to the two features is used to obtain the weight of the implicit relationship between the two features and obtained through model training. Moreover, FM model can make predictions in linear time.

## 2.2   Field-Aware Factorization Machine (FFM)

In advertising, one-hot variables are usually encountered, which will result in sparse data features. To solve this problem, the Field-aware Factorization Machine (FFM) [13] model improves FM model by introducing the concept of category, namely *field*. The features of the same field are individually coded by one-hot. For each dimensional feature $v_i$, for each field $f_j$ of other features, a hidden vector $u_{i,f_j}$ can be learned. This isolated vector is related to both feature and field. In other words, when a feature is associated with two different features, different implicit vectors are used. Assuming that $n$ features of the sample are divided into $f$ fields, the model is formalized as follows:

$$Y = \omega_0 + \sum_{i=1}^{n} \omega_i v_i + \sum_{i=1}^{n} \sum_{j=i+1}^{n} <u_{i,f_j}, u_{j,f_i}> v_i v_j, \tag{3}$$

where $f_i$ and $f_j$ represent the field to which the i-th feature and the $j$-th feature belong, and $u_{i,f_j}$ represents the hidden vector of the feature $u_i$ relative to field $j$. To use the FFM model, all features need to be converted into the "field$_{id}$:feat$_{id}$:value" format, where field$_{id}$ is the number of the field to which the feature belongs, feat$_{id}$ is the feature number, and the value represents the feature value.

Based on field, FFM sets the same natural features as the same field. In simple terms, it divides the numerical features generated by the same categorical feature through one-hot encoding into the same field.

In addition to the FFM model, there are many approaches to improving the FM model [13, 20] with high complexity.

Besides, Follow the Regularized Leader (FTRL) [18] algorithm popularly applied in the industry, as a linear model, improves the predictive ability by using a heavy feature engineering [6]. The gradient iteration method of LR's optimization algorithm FTRL is formalized as follows:

$$\omega_{t+1} = \text{argmin}_\omega (g_{1:t} \cdot \omega + \frac{1}{2} \sum_{s=1}^{t} \sigma_s \parallel \omega - \omega_s \parallel + \lambda \parallel \omega \parallel_1). \tag{4}$$

## 3    Overview of Deep Learning Prediction CTR Models

Neural network-based models can simultaneously extract high-order and low-order feature interactions in sparse advertising data. These neural network-based models usually extract low-level feature interactions by designing a pooling layer and then extract higher-order feature interactions through multiple hidden layers and activation units.

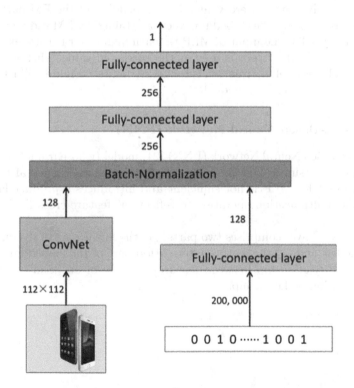

**Fig. 1.** The framework of DeepCTR [4]

### 3.1    Deep Neural Network Model (DeepCTR)

DeepCTR deep neural network model [4] (whose framework is illustrated in Fig. 1) contains three modules:

- Convnet: Take the original image $u$ as input, and then the convolutional network. The output of Convnet is the feature vector of the original image.
- Basicnet: Take the basic feature $v$ as input and applies a fully connected layer to reduce the dimensionality. Subsequently, the outputs of Convnet and Basicnet are connected into a vector and fed to two fully connected layers.
- Combnet: The last fully connected layer's output is a real value $z$. The model uses basic features and original images to predict the click-through rate of image advertisements in one step. Image features can be regarded as a supplement to the basic features.

### 3.2    Factorisation-Machine Supported Neural Network (FNN)

The Factorisation-machine supported Neural Network (FNN) [8] model uses the concept of field in FFM to attribute the original features. FNN assumes that each field has only one non-zero value. The embedding of the FM part of FNN needs to be pre-trained. The embedding vector obtained by FM is directly concat connected and used as the input of MLP to learn high-order feature expressions, and the final DNN [11] output is used as the predicted value. Therefore, FNN's presentation of low-level information is relatively limited. Figure 3 illustrates the structure of a four-layer FNN model.

### 3.3    Product-Based Neural Network (PNN)

The Product-based Neural Network (PNN) [21] model (whose framework is illustrated in Fig. 2) assumes that the cross feature expression learned after embedding to be input for MLP is not sufficient and introduces a product layer idea based on the multiplication operation to reflect the feature cross DNN network structure.

The Product Layer comprises two parts: (1) the linear part of the embedding layer on the left and (2) the feature intersection part of the embedding layer on the right. The relationship between features is more of an "and" relationship than an "addition" relationship.

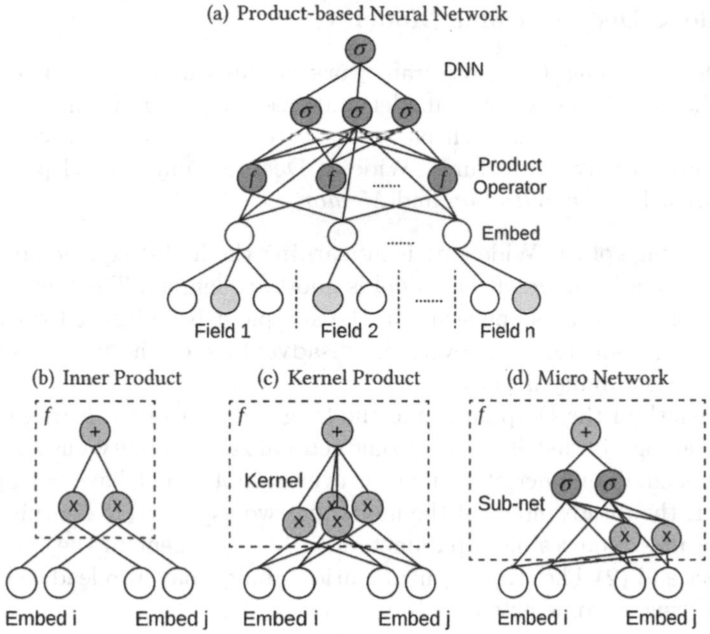

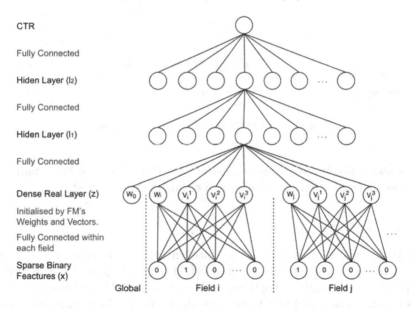

(a) Product-based Neural Network

DNN

Product
Operator

Embed

Field 1    Field 2    Field n

(b) Inner Product    (c) Kernel Product    (d) Micro Network

Kernel    Sub-net

Embed i    Embed j    Embed i    Embed j    Embed i    Embed j

**Fig. 2.** The framework of PNN [21]

CTR

Fully Connected

Hiden Layer (l2)

Fully Connected

Hiden Layer (l1)

Fully Connected

Dense Real Layer (z)

Initialised by FM's
Weights and Vectors.

Fully Connected within
each field

Sparse Binary
Feactures (x)

$w_0$    $w_i$    $v_i^1$    $v_i^2$    $v_i^3$    $w_j$    $v_j^1$    $v_j^2$    $v_j^3$

Global    Field i    Field j

**Fig. 3.** The framework of four-layer FNN [8]

## 3.4 Wide & Deep Learning Model

Wide & Deep learning [6] (whose framework is illustrated in Fig. 4) is a fusion strategy that combines a linear model and deep learning training. It designs a pooling operation to sum each pair of feature vectors to learn low-level feature interactions between features. Wide & Deep learning model presents two concepts, namely, *Generalization* and *Memory*.

- The advantage of the Wide part is in learning the high-frequency part of the sample. The advantage of the model is good at memory. The high-frequency and low-order features appearing in the sample can be learned with a small number of parameters. However, the disadvantage of the model is that the generalization ability is poor.
- The strength of the Deep part is in the long tail part of the learning sample. The advantage is that it has a strong generalization ability ensure it better support a small number of samples or even samples that have not appeared. However, the disadvantage of the model lie two aspects: (1) Learning of low-level features requires more parameters to be equivalent to the wide part of the effect and (2) The strong generalization ability may also lead to bad cases of overfitting to some extent.

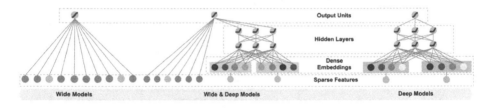

**Fig. 4.** The framework of Wide & Deep [6]

In a short, the Wide & Deep model is powerful. Since the Wide part is an LR model, however, manual feature engineering is still required.

## 3.5 DeepFM

DeepFM [9] (whose framework is illustrated in Fig. 5) combines FM and deep learning. It learns interactive features by calculating each pair of feature vectors' inner product and simply fusing them. In DeepFM, the FM and Deep parts share the embedding layer, and the parameters obtained by FM training are used as the output of the wide part and as the input of the DNN part.

Besides, Neural Factorization Machines (NFM) [12] designed a BI-Interaction pooling. The pooling layer calculates the element-wise product of two feature vectors to represent the two features' implicit relationship. Attention Factorization Machines (AFM) [32] introduces an attention mechanism based on NFM

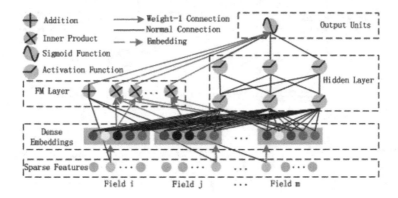

**Fig. 5.** The Wide & Deep Architecture of DeepFM [9]

to calculate the contribution of feature interaction to the target; CCPM model based on convolutional neural network (CNN) [16], which can be learned by convolution kernel Feature interaction between local features [15].

## 3.6 Deep Knowledge-Aware Network (DKN)

Deep knowledge-aware network (DKN) model [28] (whose framework is illustrated in Fig. 6) utilizes knowledge graph representation into news recommendation. DKN takes one piece of candidate news and one piece of a user's clicked news as input.

– KCNN: an extension of traditional CNN to process its title and generate an embedding vector for given news;
– Attention: an aggregator for final embedding of the user;
– DNN: a deep neural network for concatenating embedding of candidate news and the user to calculate the click predicted probability.

## 3.7 InteractionNN

InteractionNN [8,34] (whose framework is illustrated in Fig. 7) extracts information layer by layer during sparse data modeling to characterize multilevel feature interactions. InteractionNN mainly contains three modules, namely, *nonlinear interaction pooling*, *Layer-lossing*, and *embedding* as follows:

– Embedding: Extract basic dense features from sparse features of data;
– NI pooling: a hierarchical structure in constructing lowlevel feature interaction from basic dense features;
– Layer-lossing: a feed-forward neural network learning high-level feature interactions.

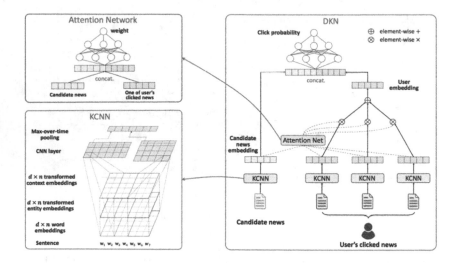

**Fig. 6.** The framework of DKN [28]

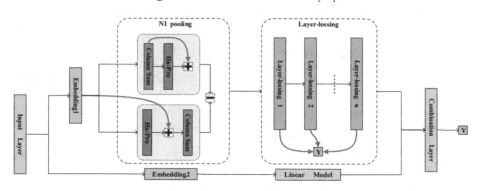

**Fig. 7.** The framework of InteractionNN [6]

## 3.8 Deep Interest Network (DIN)

*DIN.* Deep Interest Network (DIN) [35] (whose framework is illustrated in Fig. 8) designs a local activation unit to adaptively learn the representation of user interests from historical behaviors with respect to a certain Ad. This representation vector varies over different Ads, improving the expressive ability of model greatly. Besides, we develop two techniques: mini-batch aware regularization and data adaptive activation function which can help training industrial deep networks with hundreds of millions of parameters.

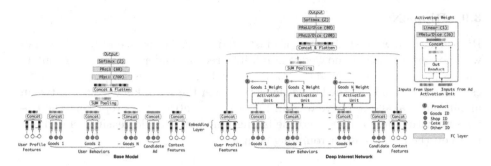

**Fig. 8.** The framework of DIN [35]

*DIEN.* Deep Interest Evolution Network (DIEN) [36] (whose framework is illustrated in Fig. 9) designs an interest extractor layer to characterize temporal interests where an auxiliary loss is introduced to supervise interest extracting each step. Moreover, an interest evolving layer is presented to capture interest evolving process via attention with considering the effects of relative interests.

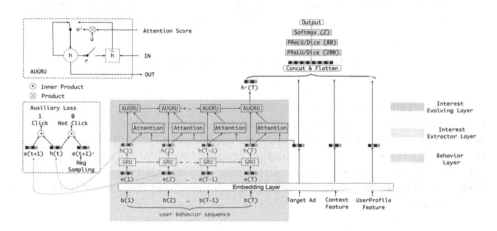

**Fig. 9.** The framework of DIEN [36]

Besides, the Deep Crossing model [24,25], based on the Residual Network (ResNet) [10], is a deep neural network for combining features to produce superior models in an automatical way. A set of individual features is input to Deep Crossing, and then crossing features are discovered implicitly. Deep Crossing Network (DCN) [30] is based on the low-order feature interaction and a cross-network (Cross Network) is proposed to learn higher-order feature interactions.

In short, the existing neural network-based models mostly use linear models to extract linear features and low order interactive features [7,36], deep learning models to extract high-order interactive features, and finally predict the final goal by fusing features.

### 3.9    RippleNet

RippleNet [27] (whose framework is illustrated in Fig. 10) is an end-to-end framework for incorporating knowledge graph into recommender systems to stimulate the propagation of user preferences. RippleNet extends a user's potential interests along with links in the knowledge graph over the set of knowledge entities in an automatical and iterative way. In Fig. 10, concentric circles represent the ripple sets with different hops, and the fading blue indicates decreasing relatedness between the center and surrounding entities.

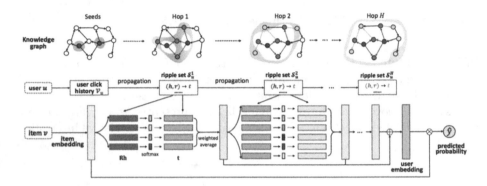

**Fig. 10.** The framework of RippleNet [27]

Besides, knowledge graph convolutional networks (KGCN) [29] is an end-to-end framework for characterizing inter-item relatedness via their associated attributes to be mined. For each entity, its neighbors to be sampled are used to discover high-order structure information and semantic information. KGQR [37] presents a CTR model for capturing rich side information for recommendation decision making by leveraging knowledge graph processing the rules of reinforcement learning.

## 4    Challenges of CTR Prediction

In this section, based on reviews above, we discuss some crucial challenges of CTR prediction as follows:

- **Judgment of Click Traffic Attributes.** That is, how to maximize the matching degree between traffic and advertisers. In other words, how to recommend the most suitable Ads for all types of users. For instance, since the types of advertisers are also roughly divided into the brand, effect, or vague e-commerce, it is not suitable to recommend effects of advertising to some users without purchasing power, such as elementary school students/middle school students. Besides, older users, such as the elderly, may not be interested in some promotional activities of young brands instead of some health-care advertisements. Besides, older users, such as the elderly, may also promote some young brands. Generally, the problem of maximizing the effectiveness of traffic is a combinatorial optimization problem. It is challenging to find solutions in practice, although there are some third-party DMP handling this problem.

- **User Value Prediction of Media Side.** Advertising opportunities are generated during the user's using APP. The lowest value judgment is made when the user's advertising display opportunities are sold in the market based on the current user attributes at the present time, which can be used to determine whether the current user is suitable for advertising exposure this time. If the current user can have exposure, then the current user's minimum value is predicting the user experience damage value of the current user if the advertisement is displayed.

- **Traffic Distribution Problem.** For a user's advertising opportunity, when there are $N$ advertisers' needs ($N > 1$), then how should the advertising display opportunity be allocated so that APP can maximize advertising revenue within a certain period. When the advertiser's advertising budget How to allocate when there are no restrictions, how to allocate when the advertiser's budget is limited, how to allocate when the advertiser has both a restricted APP and a minimum selling standard, the goal is to make it in restricted or unrestricted situations To maximize the revenue of APP within a certain period of time, if possible, to maximize the revenue of the user of APP during the life cycle. In this way, the maximum and longest revenue of APP can be guaranteed.

- **Quotation Prediction Problem.** In the case of limited data and limited feedback, the advertiser's quotation prediction is carried out. For example, some advertisers will be the next day or inconvenient to obtain feedback data. Predict the money that the advertiser will pay from the media's perspective to maximize the overall revenue of the media. At the same time, an AB scheme is designed to make the profit experiment with limited feedback.

Last but important, since the high-efficiency of prediction is still vital to practical online advertising, the optimization of deep learning models becomes more and more critical when deep learning models exhibit the great success of Ad CTR prediction.

# 5    Conclusions

This paper gives a comprehensive survey of the existing approaches of Ad CTR prediction based on machine learning models and deep learning models. Based on reviews, we open challenging problems raised from practical advertising businesses. We think that this survey can inspire new methods and develop new optimization, as well as it also provides an introduction for beginners.

**Acknowledgments.** We thank three anonymous reviewers for their constructive comments to revise this paper and technical discussions with engineers in Bayescom.com (http://www.bayescom.com/). This work is supported by the Joint Program of Bayescom.com.

# References

1. Asar, Y., Arashi, M., Wu, J.: Restricted ridge estimator in the logistic regression model. Commun. Stat. Simul. Comput. **46**(8), 6538–6544 (2017)
2. Bocca, F., Rodrigues, L.: The effect of tuning, feature engineering, and feature selection in data mining applied to rainfed sugarcane yield modelling. Comput. Electron. Agric. **128**, 67–76 (2016)
3. Chapelle, O., Keerthi, S.S.: Efficient algorithms for ranking with SVMs. Inf. Retrieval **13**(3), 201–215 (2010)
4. Chen, J., Sun, B., Li, H., Hua, X.S.: Deep CTR prediction in display advertising. In: Proceedings of MM 2016, pp. 811–820 (2016)
5. Cheng, H., Erick, C.P.: Personalized click prediction in sponsored search. In: Proceedings of WSDM 2010, pp. 351–360 (2010)
6. Cheng, H., et al.: Wide & deep learning for recommender systems. In: Proceedings of DLRS@RecSys 2016, pp. 7–10 (2016)
7. Feng, Y., Lv, F., Shen, W., Wang, M., Sun, F., Yang, K.: Deep session interest network for click-through rate prediction. In: Proceedings of IJCAI 2019, pp. 2301–2307 (2019)
8. Gao, Q.: Sparse prediction based on feature implicit relationship. Master's Thesis. Tianjin University, China (2020)
9. Guo, H., Tang, R., Ye, Y., Li, Z., He, X.: DeepFM: a factorization-machine based neural network for CTR prediction. In: Proceedings of the IJCAI 2017, pp. 1725–1731 (2017)
10. He, K., Zhang, X., Ren, S., Sun, J.: Deep residual learning for image recognition. In: Proceedings of CVPR 2016, pp. 770–778 (2016)
11. He, X., et al.: Practical lessons from predicting clicks on Ads at Facebook. In: Proceedings of ADKDD@KDD 2014, pp. 5:1–5:9 (2014)
12. He, X., Chua, T.: Neural factorization machines for sparse predictive analytics. In: Proceedings of SIGIR 2017, pp. 355–364 (2017)
13. Juan, Y., Lefortier, D., Chapelle, O.: Field-aware factorization machines in a real-world online advertising system. In: Proceedings of WWW 2017, pp. 680–688 (2017)
14. Juan Y., Zhuang Y., Chin W.S., Lin C.J.: Field-aware factorization machines for CTR prediction. In Proc. of RecSys'16, pp. 43–50 (2016)
15. Krizhevsky, A., Sutskever, I., Hinton, G.: ImageNet classification with deep convolutional neural networks. In: Proceedings of NIPS 2012, pp. 1106–1114 (2012)
16. Liu Q., Yu F., Wu S., Wang L.: A convolutional click prediction model. In Proc. of CIKM'15, pp. 1743–1746 (2015)
17. Lyu, Z., Dong, Y., Huo, C., Ren, W.: Deep match to rank model for personalized click-through rate prediction. In: Proceedings of AAAI 2020, pp. 156–163 (2020)
18. McMahan, H., et al.: Ad click prediction: a view from the trenches. In: Proceedings of KDD 2013, pp. 1222–1230 (2013)
19. Ouyang, W., Zhang, X., Ren, S., Li, L., Liu, Z., Du, Y.: Click-through rate prediction with the user memory network. CoRR abs/1907.04667 (2019)

20. Pan, J., Xu, J., Ruiz, A.L., Zhao, W., Pan, S., Lu, Q.: Field-weighted factorization machines for click-through rate prediction in display advertising. In: Proceedings of WWW 2018, pp. 1349–1357 (2018)
21. Qu, Y., et al.: Product-based neural networks for user response prediction over multi-field categorical data. ACM Trans. Inf. Syst. **37**(1), 5:1–5:35 (2018)
22. Rendle, S.: Factorization machines. In: Proceedings of ICDM 2010, pp. 995–1000 (2010)
23. Rendle, S.: Factorization machines with libFM. ACM Trans. Intell. Syst. Technol. **3**(3), 1–22 (2012)
24. Shan, Y., Hoens, T., Jiao, J., Wang, H., Yu, D., Mao, J.C.: Deep crossing: web-scale modeling without manually crafted combinatorial features. In: Proceedings of KDD 2016, pp. 255–262 (2016)
25. Shi, S.T., et al.: Deep time-stream framework for click-through rate prediction by tracking interest evolution. In: Proceedings of AAAI 2020, pp. 5726–5733 (2020)
26. Wang, C., Zhang, M., Ma, W., Liu, Y., Ma, S.: Make it a Chorus: knowledge- and time-aware item modeling for sequential recommendation. In: Proceedings of SIGIR 2020, pp. 109–118 (2020)
27. Wang, H., et al.: RippleNet: propagating user preferences on the knowledge graph for recommender systems. In: Proceedings of CIKM 2018, pp. 417–426 (2018)
28. Wang, H., Zhang, F., Xie, X., Guo, M.: DKN: deep knowledge-aware network for news recommendation. In: Proceedings of WWW 2018, pp. 1835–1844 (2018)
29. Wang, H., Zhao, M., Xie, X., Li, W., Guo, M.: Knowledge graph convolutional networks for recommender systems. In: Proceedings of WWW 2019, pp. 3307–3313 (2019)
30. Wang, R., Fu, B., Fu, G., Wang, M.: Deep & Cross network for Ad click predictions. In: Proceedings of ADKDD@KDD'17, pp. 12:1–12:7 (2017)
31. Wang, X., Li, W., Cui, Y., Zhang, R., Mao, J.: Click-through rate estimation for rare events in online advertising. Online Multimedia Advertising: Techniques and Technologies, pp. 1–12 (2010)
32. Xiao, J., Ye, H., He, X., Zhang, H., Wu, F. Chua, T.: Attentional factorization machines: learning the weight of feature interactions via attention networks. In: Proceedings of IJCAI 2017, pp. 3119–3125 (2017)
33. Zhang, W., Du, T., Wang, J.: Deep learning over multi-field categorical data. In: Ferro, N., et al. (eds.) ECIR 2016. LNCS, vol. 9626, pp. 45–57. Springer, Cham (2016). https://doi.org/10.1007/978-3-319-30671-1_4
34. Zhang, X., Gao, Q., Feng, Z.: InteractionNN: a neural network for learning hidden features in sparse prediction. In: Proceedings of IJCAI 2019, pp. 4334–4340 (2019)
35. Zhou, G., et al.: Deep interest network for click-through rate prediction. In: Proceedings of KDD 2018, pp. 1059–1068 (2018)
36. Zhou, G., et al.: Deep interest evolution network for click-through rate prediction. In: Proceedings of AAAI 2019, pp. 5941–5948 (2019)
37. Zhou, S., et al.: Interactive recommender system via knowledge graph-enhanced reinforcement learning. In: Proceedings of SIGIR 2020, pp. 179–188 (2020)

# An Attention-Based Approach to Rule Learning in Large Knowledge Graphs

Minghui Li[1], Kewen Wang[2(✉)], Zhe Wang[2], Hong Wu[2], and Zhiyong Feng[1]

[1] College of Intelligence and Computing, Tianjin University, Tianjin, China
[2] School of Information and Communication Technology, Griffith University,
Brisbane, Australia
k.wang@griffith.edu.au

**Abstract.** This paper presents a method for rule learning in large knowledge graphs. It consists of an effective sampling of large knowledge graphs (KGs) based on the attention mechanism. The attention-based sampling is designed to reduce the search space of rule extraction and thus to improve efficiency of rule learning for a given target predicate. An implementation ARL (Attention-based Rule Learner) of rule learning for KGs is obtained by combining the new sampling with the advanced rule miner AMIE+. Experiments have been conducted to demonstrate the efficiency and efficacy of our method for both rule learning and KG completion, which show that ARL is very efficient for rule learning in large KGs while the precision is still comparable to major baselines.

**Keywords:** Knowledge graph · Rule learning · Link prediction

## 1 Introduction

In recent years, many large knowledge graphs (KGs), such as DBpedia [1], Freebase [3], NELL [5], Wikidata [19] and YAGO [16], have been manually or automatically created. These KGs store millions of entities and facts in the form of RDF triples, and the facts are not isolated, but are connected to other facts through shared entities, thus forming a graph representation of knowledge of interest. Knowledge graphs provide flexible organisation of often huge amounts of data integrated from heterogeneous sources (including the Web) and are often coupled with ontological rules that describe domain knowledge or business rules [2]. For instance, if we know that a rule $bornInState(x, y) \leftarrow bornInCity(x, z), cityInState(z, y)$ in a given KG, it can be useful for deriving a new fact $bornInState(x, y)$ when both facts $bornInCity(x, z)$ and $cityInState(z, y)$ are already in the KG.

However, it is challenging to produce rules from large KGs by hand and thus the problem of automatically extracting rules from a KG received extensive attention in the past few years. In the case of rule learning in a large KG, no negative examples are explicitly given, which makes it difficult to effectively employ rule learning algorithms that have been developed in Inductive Logic

© Springer Nature Switzerland AG 2021
C. S. Jensen et al. (Eds.): DASFAA 2021 Workshops, LNCS 12680, pp. 154–165, 2021.
https://doi.org/10.1007/978-3-030-73216-5_11

Programming (ILP) [7,11,12,24]. Also, it is a challenge for them to scale to the current KGs with large data volume and growing scale. Large data volume leads to a large rule search space, which makes the current rule learning systems unable to process or the processing speed is very slow. This brings about a problem of how to shrink search space. To solve this problem, some rule learning systems have proposed new strategies to reduce the search space.

Several methods for rule learning in KGs have been proposed in the literature, such as AMIE+ [8], RLvLR [15] and ScaLeKB [6]. ScaLeKB proposes an ontological path finding method consisting of pruning and eliminating imperfect or low-efficiency rules, a series of parallel algorithms, and splitting the entire learning task into several independent subtasks to learn first-order inference rules from a given KG. AMIE+ [8] adopts a series of accelerated operations for rule refinement and speeds up confidence evaluation of a rule based on statistics. DistMult [22] uses the embedding learned by the representation learning method based on matrix factorization to express the relations in a KG as a diagonal matrix to reduce computational complexity. At the same time, predicates in a rule body are split into different types by using relational domain restrictions. Such optimisation techniques can significantly reduce the search space and thus improve the efficiency of rule mining. RLvLR [15] sampling subgraphs related to target the predicate, and proposes the so-called co-occurrence score function to rank all possible candidate rules. R-Linker [21] proposes a novel subgraph sampling method and an embedding-based score function for rule selection. However, it is still challenging for efficiently learning rules in KGs with large data volume and growing scale.

Different from the traditional AMIE+ system, many recently proposed rule miners use representation learning methods in rule learning. For example, Dist-Mult uses tensor decomposition and RLvLR uses RESCAL [13,14] to construct embeddings of entities and predicates.

Inspired by the self-attention mechanism [18], we propose an attention-based sampling method to select relevant predicates for each target predicate $P_t$ for processing rule learning on large-scale knowledge graphs. The usefulness is manifested in the importance of the selected predicates to the $P_t$. In this paper, we present an attention-based predicate sampling method to improve the efficiency of rule learning on large-scale knowledge graphs. In order to evaluate the effectiveness of the sampling method, we performed experiments on the two tasks of rule learning and link prediction. The experimental results show that our ARL method is very efficient while keeping a comparative precision for both rule learning and KG completion. The maximum length of CP rules including head atom learned by ARL can be up to 5, which is longer than those rules learned by most of the state-of-the-art rule learners to the best of our knowledge.

The remainder of this paper is structured as follows: Sect. 2 introduces some basic concepts related to the knowledge graph. Section 3 introduces the sampling method and the rule learning system ARL in detail. Section 4 presents experimental results and evaluation. Section 5 summarizes our work.

## 2   Preliminaries

In this section, we briefly introduce some concepts and define some notations to be used later, which are related to knowledge graphs and representation learning methods.

### 2.1   Knowledge Graphs and Rules

A knowledge graph can be represented by $G = (E, P, F)$, where $E$ and $P$ denote the set of entities and predicates respectively, and $F$ is the fact set, for instance, $bornInCity(Mary, Beijing)$. Following the convention in knowledge representation, a fact can be defined as $P_t(e, e')$, where $e$ and $e'$ represent the subject and object entities of the predicate or relation $P_t$, indicating the two nodes $e, e' \in E$ are connected by a labeled directed edge $P_t \in P$.

Like many other works on rule learning in KGs, we focus on mining a special class of Horn rules called *closed path rules* or *CP rules*, which are in the following form:

$$r : P_1(x, z_1) \wedge P_2(z_1, z_2) \wedge ... \wedge P_n(z_{n-1}, y) \rightarrow P_t(x, y),$$

where $x, y$ and $z_i$ are variables, which can be instantiated by entities. $P_t(x, y)$ is the head atom of rule $r$, denoted $head(r)$; while $P_1(x, z_1) \wedge P_2(z_1, z_2) \wedge ... \wedge P_n(z_{n-1}, y)$ is the conjunction of body atoms, denoted $body(r)$. Intuitively, if there are facts in KG that make the $body(r)$ hold at the same time, then $P_t(x, y)$ holds too. Besides, CP rules have some parameter constraints, that is, each variable in rule $r$ appears at least twice and allowing the predicate of head atom to appear in the body.

Following the major rule learning methods AMIE+ [8] and RLvLR [15], we use standard confidence (SC) and head coverage (HC) to measure the quality of the learned rules. Take the above rule $r$ as an example, the degrees of SC and HC are defined as follows:

$$supp(r) = \#(e, e') : body(r)(e, e') \wedge P_t(e, e'),$$

$$SC(r) = \frac{supp(r)}{\#(e, e') : body(r)(e, e')}, \quad HC(r) = \frac{supp(r)}{\#(e, e') : P_t(e, e')}.$$

Here $supp(r)$ is the support degree of $r$, which is the number of facts that satisfy $body(r)$ and $head(r)$ in the knowledge graph. $\#(e, e') : body(r)(e, e')$ denotes the number of entity pairs that only satisfy the body. For example, if there are entities $e, e_1, ..., e'$ and facts $P_1(e, e_1), P_2(e_1, e_2), ..., P_n(e_{n-1}, e')$ in the given KG, then it is said that the entity pair $(e, e')$ satisfies the body of $r$. Similarly, the denominator of HC represents the number of entity pairs in KG that satisfy the head atom.

### 2.2   Representation Learning

The triple facts in a KG are discrete symbolized knowledge. While representation learning of knowledge bases is to learn distributed representations of entities and relationships (embeddings) by encoding the graph structure into a continuous, low-dimensional vector space through a certain pattern.

Two most classic knowledge graph embedding methods are translation-based and tensor decomposition-based. The tensor decomposition models regard the knowledge graph as a n-dimensional tensor and decompose this tensor into the product of several embeddings, such as RESCAL [13,14], ComplEx [17], ANALOGY [10], while translation-based models utilize the spatial structure of the graph to model from the head entity to the tail entity, such as TransE [4], TransH [20], TransR [9].

The rule learner RLvLR employs RESCAL to obtain entity and predicate embeddings to guide the candidate rule search. In this paper, we choose TransE to guide the selection of preferred predicates in the sampling method as it is a more recent method and widely used in the literature.

TransE embeds each entity and relation (or predicate) into a d-dimensional vector. For each given fact $(h, r, t)$, the following scoring function is computed to measure the possibility of a fact as true:

$$\phi(h, r, t) = ||\mathbf{h} + \mathbf{r} - \mathbf{t}||_{L_1/L_2},$$

where $\mathbf{h}$, $\mathbf{r}$ and $\mathbf{t}$ denote the embedding of $h, r$ and $t$, respectively.

## 3  Attention-Based Rule Learning (ARL)

The framework ARL for rule learning is shown in Fig. 1, which consists of two independent parts: attention-based sampling and rule search. The attention-based sampling method is the core of ARL, including hop-based sampling and attention-based predicate selection, we will introduce it in detail below. After sampling the preferred predicates, we can input them into two different kinds of rule learners for rule search: an embedding-based rule miner like RLvLR or an ILP-based method for searching rules like AMIE+.

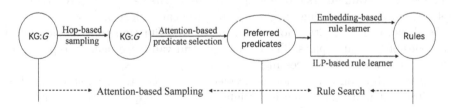

**Fig. 1.** An overview of ARL.

### 3.1  Hop-Based Sampling

Given a target predicate $P_t \in P$, we aim to learn CP rules of different lengths whose head predicate is $P_t$. For each $P_t$, it aims to obtain relevant samples $G'$ of

$P_t$, i.e., a subgraph of $G$. The hop-based sampling is necessary, since many facts irrelevant to $P_t$ will not be useful, so we can use it for preliminary sampling.

The hops for mining different lengths of rules are different. For a fact $(h, P_t, t)$, $h, t \in E$, we start from the nodes $h$ and $t$ respectively, perform a breadth-first traversal of specific hops in $G$ to obtain a subgraph $G'$. In particular, let the length of rules be $n(n \leq 5)$, we can generate the sampling entity sets $E_0, ..., E_{\lfloor n/2 \rfloor}$ by $\lfloor n/2 \rfloor$ hops, where $E_0$ is the entity set directly connected to the $P_t$ in the knowledge graph $G$, such as $P_t(e, e')$. $E_i$ $(0 < i \leq \lfloor n/2 \rfloor)$ is the entity directly connected to $E_{i-1}$ through a predicate $P_i$, such as $P_i(e', e'')$ or $P_i(e, e'')$.

The union entity set is generated by different hops of entity sets: $E' = \bigcup_{i=0}^{\lfloor n/2 \rfloor} E_i$. And the union fact set is $F' = \{P(e_1, e_2) \mid e_1, e_2 \in E', P(e_1, e_2) \in F\}$. Since too many facts are sampled, it will be difficult for TransE to deal with. Thus, we tried two ways to limit the number of samples: limit the number of facts, or limit the number of facts and entities.

*Limit the number of facts.* We limit the number of one-hop facts of $P_t$ in a certain percentage, a hyper-parameter *Ratio*. For other-hops, we collect the remaining facts by filtering out the triples for too high-frequency or too low-frequency predicates.

*Limit the numbers of entities.* We limit the number of one-hop facts of $P_t$ by allowing at most $N_1$ entities. For other-hops, while we filter out facts with too high-frequency or too low-frequency predicates, we also require that each remaining predicate be connected to at most $N_2$ facts. Here $N_1$ and $N_2$ are hyper-parameters.

### 3.2   Attention-Based Predicate Selection

The core of attention-based sampling method is the mechanism for selection preferred predicate. The scaled dot-product attention [18] maps a set of queries and a set of key-value pairs into a weighted sum of the values. Inspired by it, we apply the way of calculating weights to our predicate selection.

As shown in Fig. 2, attention is equivalent to a function that maps a query to a series of key-value pairs. The query, key and value are usually represented as a sequence vector for processing.

General attention calculation is divided into three steps, as shown in Fig. 2(a). First, calculate the similarity weight between the query and each key in a certain way; second, use softmax function to normalize the weight; finally, the weighted key and the corresponding value are summed to obtain the attention.

The scaled dot-product attention that we use is a part of self-attention in Fig. 2(b). The calculation process is the same as that in Figure (a). While the $f(\mathbf{Q}, \mathbf{K}_i)$ is a dot-product, the scaling factor $1/\sqrt{d_k}$ is added to the general dot-product, where $d_k$ is the dimension of queries. The reason for using a scaling factor is to avoid too large values when calculating the dot-product operation,

1. $f(\mathbf{Q}, \mathbf{K}_i) = \begin{cases} \mathbf{Q}^T \mathbf{K}_i & \text{Dot product} \\ \mathbf{Q}^T \mathbf{W}_a \mathbf{K}_i & \text{Matrix multiplication} \\ \mathbf{W}_a[\mathbf{Q}; \mathbf{K}_i] & \text{Cascade} \\ v_a{}^T tanh(\mathbf{W}_a \mathbf{Q} + \mathbf{U}_a \mathbf{K}_i) & \text{Perceptron} \end{cases}$

2. $a_i = softmax(f(\mathbf{Q}, \mathbf{K}_i)) = \dfrac{exp(f(\mathbf{Q}, \mathbf{K}_i))}{\sum_j exp(f(\mathbf{Q}, \mathbf{K}_j))}$

3. $Attention(\mathbf{Q}, \mathbf{K}, \mathbf{V}) = \sum_i a_i \mathbf{V}_i$

(a)  The general calculation process of attention          (b)  Scaled Dot-Product Attention

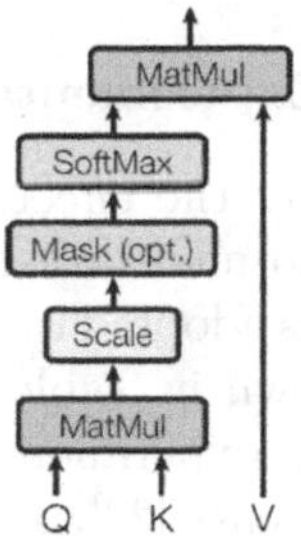

**Fig. 2.** The attention mechanism.

but resulting in a smaller gradient during model training. The formula for the attention mechanism is as follows.

$$Attention(\mathbf{Q}, \mathbf{K}, \mathbf{V}) = softmax(\frac{\mathbf{Q} \cdot \mathbf{K}^T}{\sqrt{d_k}}) \cdot \mathbf{V}$$

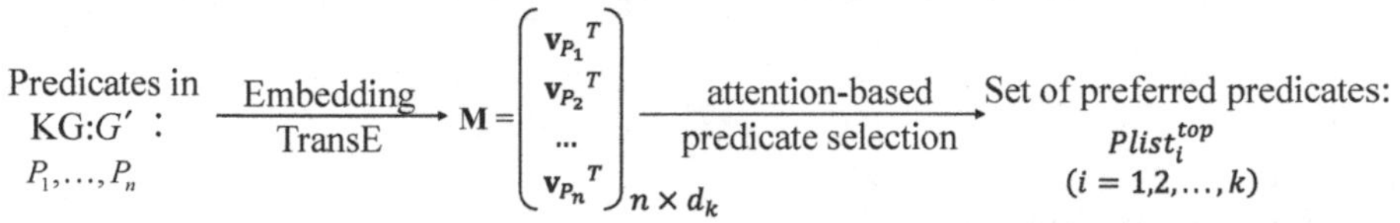

Predicates in KG:$G'$ : $P_1, ..., P_n$ $\xrightarrow{\text{Embedding TransE}}$ $\mathbf{M} = \begin{pmatrix} v_{P_1}{}^T \\ v_{P_2}{}^T \\ \cdots \\ v_{P_n}{}^T \end{pmatrix}_{n \times d_k}$ $\xrightarrow{\text{attention-based predicate selection}}$ Set of preferred predicates: $Plist_i^{top}$ $(i = 1, 2, ..., k)$

**Fig. 3.** Attention-based predication selection.

As shown in Fig. 3, the attention-based predicate selection consists of three major steps. Firstly, we choose the simple and effective TransE to obtain the matrix representation $\mathbf{M} \in \mathbf{R}^{n \times d_k}$ of predicates for the sampled $G'$. Based on this, we initialize the three matrices $\mathbf{Q}, \mathbf{K}, \mathbf{V}$ with $\mathbf{M}$, and use the formula $Attention(\mathbf{Q}, \mathbf{K}, \mathbf{V})$ above to calculate a weight matrix $\mathbf{W} \in R^{n \times n}$, where $d_k$ is the embedding dimension, $n$ is the total number of $G'$ predicates. Each row of the matrix $\mathbf{W}$ corresponds to a predicate $P$, and the value in the row vector indicates the importance of other predicates to $P$.

Next, we perform the predicate selection based on weight matrix $\mathbf{W}$. For each target predicate $P_t$, we first take the corresponding weight row of $P_t$ in $\mathbf{W}$, rank it and get the top $n \times rate_1$ important predicate list $Plist_t^{top}$ for $P_t$. For predicates $P_i \in Plist_t^{top}$, we can consider that they are in the environment $S$ of $P_t$, where the predicates are important or related to $P_t$. Besides, there may be also several predicates that are important to those predicates in $S$. So they may also be important to $P_t$. Similarly, we obtain the top $n \times rate_1$ important predicate list $Plist_i^{top}$ for $P_i$. Finally, the intersection of all $Plist_i^{top}$ is the final predicate set we preferred.

# 4    Result and Discussion

## 4.1    Experimentation

To verify the effectiveness of our attention-based predicate sampling method, we performed two sets of experiments on rule learning and link prediction. The datasets adopted in our experiments are widely used benchmarks, with the statistics shown in Table 1. The first three benchmarks are commonly used for rule learning experiments [15], and the last two are often used for link prediction experiments [22].

Our rule learning experiments was run on a Linux Ubuntu 18.04 server with 3.5 GHz CPU and 64 GB of memory, and the link prediction experiments was run on a Linux Ubuntu 16.04 server with 3.4 GHz CPU and 64 GB of memory.

**Table 1.** Benchmark specifications.

| KG | #Facts | #Entities | #Predicates |
| --- | --- | --- | --- |
| YAGO2s | 4.12M | 2.26M | 37 |
| Wikidata | 8.40M | 3.08M | 430 |
| DBpedia 3.8 | 11.02M | 3.10M | 650 |
| FB75K | 0.32M | 0.07M | 13 |
| FB15K-237 | 0.31M | 0.01M | 237 |

## 4.2    Rule Learning

We have conducted three experiments to evaluate the performance of ARL and to validate the following statements:

1. ARL can learn quality rules faster than RLvLR [15].
2. The efficiency and effectiveness are achieved through the attention-based mechanism.
3. The sampling method can be used in other rule learners, like AMIE+ [8], to enhance their scalability.
4. The rules learned by ARL have good quality for link prediction.

We compared the (average) numbers of learned rules (#R, with a threshold $SC \geqslant 0.1$ and $HC \geqslant 0.01$ as in [15]) and those of learned quality rules (#QR, with $SC \geqslant 0.7$ and $HC \geqslant 0.01$). The SC and HC measures are calculated over the whole datasets, not just on the samples.

*Experiment 1.* Our system is most similar to RLvLR, so we first compare our system with RLvLR to verify the benefit of our sampling algorithm on enhancing the efficiency of rule learning. We randomly selected 20 predicates as target predicates from each of the datasets YAGO2s, Wikidata, and DBPedia 3.8. We

**Table 2.** Comparison with RLvLR on rule learning.

| Model | DBpedia3.8 | | Wikidata | | YAGO2s | |
|-------|-----|------|--------|------|--------|------|
|       | #R  | #QR  | #R     | #QR  | #R     | #QR  |
| RLvLR | 99.36 | 12.71 | 165.03 | **24.39** | 23.14 | 6.18 |
| ARL   | 61.97 | **23.51** | 34.08 | 19.41 | 10.91 | **6.37** |

compared the numbers of rules and quality rules learned per hour, and the results are shown in Table 2.

Note that despite the numbers of rules learned by ARL per hour are often smaller than those of RLvLR, the numbers quality rules learned per hour are larger than those of RLvLR. This shows our predicate sampling method can greatly reduce the search space of rule learning without missing quality rules.

*Experiment 2.* The previous experiment demonstrates the efficiency of our sampling method in learning quality rules, and in the following experiment, we evaluate the effectiveness of our attention-based mechanism in sampling. To this end, we replaced our attention-based sampling method with a random predicate selection method, denoted ARL(R), which randomly select the same number of predicates as the original ARL method.

To compare ARL with ARL(R) on rule learning, we randomly selected 10 target predicates for each of the three datasets, and recorded the total numbers of rules and quality rules learned by both systems and their total times.

**Table 3.** Random sampling vs. Attention-based sampling.

| KG | DBpedia3.8 | | | Wikidata | | | YAGO2s | | |
|----|-----|------|------|------|------|------|------|------|------|
|    | #R  | #QR  | Time | #R   | #QR  | Time | #R   | #QR  | Time |
| ARL(R) | 63 | 13 | 9.09 | 128 | **72** | 7.83 | 7 | 2 | 1.56 |
| ARL | **71** | **21** | **0.33** | **130** | 71 | **0.39** | **15** | **11** | **0.25** |

Table 3 shows the results. As can be seen, our attention-based sampling method clearly outperforms the random method in both the numbers of (quality) rules learned and surprisingly, the learning efficiency, which shows the effectiveness of our attention-based mechanism.

*Experiment 3.* The third experiment is to evaluate the benefit of our sampling method in enhancing the scalability of rule learning, and we used our sampling method as a pre-processing module for AMIE+ and compare the performance with AMIE+ itself. As the search space increase exponentially in the rule length, the existing rule learners all face the challenge of learning long rules. In the original evaluation of AMIE+, the maximum rule length is 4 (for Wikidata and DBpedia 3.8, the maximum rule length is 3). As a 'stress' test, we aim to learn rules with the maximum length of 5.

For this experiment, we did not specify any target predicate, that is, to learn rules whose heads may contain any predicates in the KG, which is a default setting of AMIE+. We ran AMIE+ (with and without our sampling) with its native predicate selection heuristics (if any). To verify sampling indeed allows more quality rules to be learned in a reasonable time, we first ran AMIE+ with our sampling and recorded the number of rules with time spent, and then we made AMIE+ to run without sampling for the same time and recorded the number of rules learned. The experimental result are summarised in Table 4.

**Table 4.** AMIE+ with and without sampling over the same timeframes.

| KG | w. sampling | | | w.o. sampling | |
|---|---|---|---|---|---|
| | #R | #QR | Time (h) | #R | #QR |
| Wikidata | 4084 | 651 | 4.20 | 81 | 30 |
| DBpedia 3.8 | 18070 | 1590 | 10.32 | 255 | 38 |

Table 4 clearly shows the benefit of our sample method in enhancing the scalability of AMIE+, allowing up to 8 times more quality rules to be learned.

### 4.3 Link Prediction

The link prediction experiment further evaluates the quality of rules learned by ARL in an important task of KG completion. Given a KG, a predicate $P_t$, and an entity $e$, the task of link prediction is to predict another entity $e'$ such that $P_t(e, e')$ or $P_t(e', e)$ is valid in the KG [15,22].

We set the threshold SC to be 0.005 and HC to be 0.001. We adopted the standard link prediction metrics, Mean Reciprocal Rank (MRR) and Hits@10. MRR is the average of the reciprocal ranks of the desired entities, Hits@10 is the percentage of desired entities being ranked among top 10.

We compared the performance of ARL with several major systems on three benchmark datasets FB75K, and a more challenging version FB15K-237. Each dataset is divided into training set (70%) and test set (30%). Note that we use rules with a maximum length of 4 to infer new facts.

Tables 5 shows the results on respective datasets, the performance of the other systems are obtained from [15].

From the results, although the number of rules learned by ARL is about one-tenth of those learned by RLvLR, it obtains a higher Hits@10 on FB15K-237 compared to RLvLR. This shows ARL has advantages in learning quality rules with higher efficiency.

**Table 5.** Comparison on link prediction.

|  | FB75K | | FB15K-237 | |
|---|---|---|---|---|
|  | MRR | Hits@10 | MRR | Hits@10 |
| ARL | 0.27 | 42.0 | 0.23 | 40.3 |
| RLvLR | 0.34 | 43.4 | 0.24 | 39.3 |
| Neural LP [23] | 0.13 | 25.7 | 0.24 | 36.1 |
| R-Linker [21] | - | - | 0.24 | 38.1 |
| DISTMULT [22] | - | - | 0.25 | 40.8 |

## 5 Conclusion

In this paper, we have proposed a sampling method and then implemented a system ARL (attention-based rule learner) for rule learning in large KGs. ARL employs attention mechanism to model the importance of predicates in a given KG with respect to each target predicate. As a result, our sampling method significantly reduces the search space of rules, especially for large KGs, while keeping a comparable quality of learned rules. Our attention-based sampling method can be used as a preprocessing for any rule learners while the implementation is based on AMIE+. Our experimental results show that the efficiency of AMIE+ can be significantly improved. In particular, longer rules can be learned by ARL. We are working on further improving the accuracy of ARL while keeping the scalability and efficiency.

**Acknowledgements.** This work was partially supported by the National Natural Science Foundation of China under grant 61976153.

## References

1. Auer, S., Bizer, C., Kobilarov, G., Lehmann, J., Cyganiak, R., Ives, Z.: DBpedia: a nucleus for a web of open data. In: Aberer, K., et al. (eds.) ASWC/ISWC -2007. LNCS, vol. 4825, pp. 722–735. Springer, Heidelberg (2007). https://doi.org/10.1007/978-3-540-76298-0_52
2. Bellomarini, L., Gottlob, G., Pieris, A., Sallinger, E.: Swift logic for big data and knowledge graphs. In: Tjoa, A.M., Bellatreche, L., Biffl, S., van Leeuwen, J., Wiedermann, J. (eds.) SOFSEM 2018. LNCS, vol. 10706, pp. 3–16. Springer, Cham (2018). https://doi.org/10.1007/978-3-319-73117-9_1
3. Bollacker, K., Evans, C., Paritosh, P., Sturge, T., Taylor, J.: Freebase: a collaboratively created graph database for structuring human knowledge. In: Proceedings of the 2008 ACM SIGMOD International Conference on Management of Data, pp. 1247–1250 (2008)
4. Bordes, A., Usunier, N., Garcia-Duran, A., Weston, J., Yakhnenko, O.: Translating embeddings for modeling multi-relational data. In: Advances in Neural Information Processing Systems, pp. 2787–2795 (2013)

5. Carlson, A., et al.: Toward an architecture for never-ending language learning. In: Fox, M., Poole, D. (eds.) Proceedings of the Twenty-Fourth AAAI Conference on Artificial Intelligence, AAAI 2010, Atlanta, Georgia, USA, 11–15 July 2010. AAAI Press (2010)

6. Chen, Y., Wang, D.Z., Goldberg, S.: Scalekb: scalable learning and inference over large knowledge bases. VLDB J. **25**(6), 893–918 (2016)

7. Fürnkranz, J., Gamberger, D., Lavrač, N.: Foundations of Rule Learning. Springer, Heidelberg (2012). https://doi.org/10.1007/978-3-540-75197-7

8. Galárraga, L., Teflioudi, C., Hose, K., Suchanek, F.M.: Fast rule mining in ontological knowledge bases with amie+. VLDB J. **24**(6), 707–730 (2015)

9. Lin, Y., Liu, Z., Sun, M., Liu, Y., Zhu, X.: Learning entity and relation embeddings for knowledge graph completion. In: Bonet, B., Koenig, S. (eds.) Proceedings of the Twenty-Ninth AAAI Conference on Artificial Intelligence, 25–30 January 2015, Austin, Texas, USA, pp. 2181–2187. AAAI Press (2015)

10. Liu, H., Wu, Y., Yang, Y.: Analogical inference for multi-relational embeddings. In: Precup, D., Teh, Y.W. (eds.) Proceedings of the 34th International Conference on Machine Learning, ICML 2017, Sydney, NSW, Australia, 6–11 August 2017, pp. 2168–2178. PMLR (2017)

11. Muggleton, S.: Inverse entailment and progol. New Gen. Comput. **13**(3–4), 245–286 (1995)

12. Muggleton, S.: Learning from positive data. In: Muggleton, S. (ed.) ILP 1996. LNCS, vol. 1314, pp. 358–376. Springer, Heidelberg (1997). https://doi.org/10.1007/3-540-63494-0_65

13. Nickel, M., Rosasco, L., Poggio, T.A.: Holographic embeddings of knowledge graphs. In: Schuurmans, D., Wellman, M.P. (eds.) Proceedings of the Thirtieth AAAI Conference on Artificial Intelligence, 12–17 February 2016, Phoenix, Arizona, USA, pp. 1955–1961. AAAI Press (2016)

14. Nickel, M., Tresp, V., Kriegel, H.: A three-way model for collective learning on multi-relational data. In: Getoor, L., Scheffer, T. (eds.) Proceedings of the 28th International Conference on Machine Learning, ICML 2011, Bellevue, Washington, USA, 28 June - 2 July 2011, pp. 809–816. Omnipress (2011)

15. Omran, P.G., Wang, K., Wang, Z.: Scalable rule learning via learning representation. In: Lang, J. (ed.) Proceedings of the Twenty-Seventh International Joint Conference on Artificial Intelligence, IJCAI 2018, 13–19 July 2018, Stockholm, Sweden, pp. 2149–2155 (2018)

16. Suchanek, F.M., Kasneci, G., Weikum, G.: Yago: a core of semantic knowledge. In: Proceedings of the 16th International Conference on World Wide Web, pp. 697–706 (2007)

17. Trouillon, T., Welbl, J., Riedel, S., Gaussier, É., Bouchard, G.: Complex embeddings for simple link prediction. In: Proceedings of the International Conference on Machine Learning (ICML), pp. 2071–2080 (2016)

18. Vaswani, A., et al.: Attention is all you need. In: Advances in Neural Information Processing Systems 30: Annual Conference on Neural Information Processing, Long Beach, CA, USA, pp. 5998–6008 (2017)

19. Vrandečić, D., Krötzsch, M.: Wikidata: a free collaborative knowledgebase. Commun. ACM **57**(10), 78–85 (2014)

20. Wang, Z., Zhang, J., Feng, J., Chen, Z.: Knowledge graph embedding by translating on hyperplanes. In: Brodley, C.E., Stone, P. (eds.) Proceedings of the Twenty-Eighth AAAI Conference on Artificial Intelligence, Québec, Canada, pp. 1112–1119. AAAI Press (2014)

21. Wu, H., Wang, Z., Zhang, X., Omran, P.G., Feng, Z., Wang, K.: A system for reasoning-based link prediction in large knowledge graphs. In: Proceedings of the ISWC Satellites, pp. 121–124 (2019)
22. Yang, B., Yih, W., He, X., Gao, J., Deng, L.: Embedding entities and relations for learning and inference in knowledge bases. In: Bengio, Y., LeCun, Y. (eds.) Proceedings of the 3rd International Conference on Learning Representations, ICLR 2015, San Diego, CA, USA, 7–9 May (2015)
23. Yang, F., Yang, Z., Cohen, W.W.: Differentiable learning of logical rules for knowledge base reasoning. In: Guyon, I., von Luxburg, U., Bengio, S., Wallach, H.M., Fergus, R., Vishwanathan, S.V.N., Garnett, R. (eds.) Advances in Neural Information Processing Systems 30: Annual Conference on Neural Information Processing Systems, December 4–9, 2017, pp. 2319–2328. Long Beach, CA, USA (2017)
24. Zeng, Q., Patel, J.M., Page, D.: Quickfoil: scalable inductive logic programming. Proc. VLDB Endow. **8**(3), 197–208 (2014)

# The 1st International Workshop on Machine Learning and Deep Learning for Data Security Applications

# Multi-scale Gated Inpainting Network with Patch-Wise Spacial Attention

Xinrong Hu[1,2], Junjie Jin[1,2(✉)], Mingfu Xiong[1,2], Junping Liu[1,2], Tao Peng[1,2], Zili Zhang[1,2], Jia Chen[1,2], Ruhan He[1,2], and Xiao Qin[3]

[1] Engineering Research Center of Hubei Province for Clothing Information, Wuhan, China
junjie.jin@qq.com
[2] School of Mathematics and Computer Science, Wuhan Textile University, Wuhan, China
[3] Department of Computer Science and Software Engineering, Aubern University, Auburn, USA

**Abstract.** Recently, deep-model-based image inpainting methods have achieved promising results in the realm of image processing. However, the existing methods produce fuzzy textures and distorted structures due to ignoring the semantic relevance and feature continuity of the holes region. To address this challenge, we propose a detailed depth generation model (GS-Net) equipped with a Multi-Scale Gated Holes Feature Inpainting module (MG) and a Patch-wise Spacial Attention module (PSA). Initially, the MG module fills the hole area globally and concatenates to the input feature map. Then, the module utilizes a multi-scale gated strategy to adaptively guide the information propagation at different scales. We further design the PSA module, which optimizes the local feature mapping relations step by step to clarify the image texture information. Not only preserving the semantic correlation among the features of the holes, the methods can also effectively predict the missing part of the holes while keeping the global style consistency. Finally, we extend the spatially discounted weight to the irregular holes and assign higher weights to the spatial points near the effective areas to strengthen the constraint on the hole center. The extensive experimental results on Places2 and CelebA have revealed the superiority of the proposed approaches.

**Keywords:** Image inpainting · Feature reconstruction · Gated mechanism · Spacial attention · Semantic relevance

## 1 Introduction

The goal of image completion is the task to fill the missing pixels in an image in a way that the corresponding restored image to have a sense of visual reality. The restored area needs continuity and consistency of texture while seeking semantic consistency between the filled area and any surrounding area. Image completion techniques are widely adopted in photo recovery, image editing, object deletion and other image tasks [1, 5]. At present, the existing methods have focused on the restoration of the rectangular areas near the image centers [6, 7]. This kind of regular hole restoration could result in the model over-fitting accompanied by poor migration effect [8]. The overarching objective of this

C. S. Jensen et al. (Eds.): DASFAA 2021 Workshops, LNCS 12680, pp. 169–184, 2021.
https://doi.org/10.1007/978-3-030-73216-5_12

work is to propose an image-restoration model, which is sufficiently robust to repair regular and irregular holes. Our proposed technique produces semantically meaningful predictions to ensure that the repaired parts are perfectly integrated with other portions without any expensive post-processing.

Traditional image restoration methods mainly exploit the texture synthesis technology to address the challenge of hole fillings. These methods assume that the missing regions should contain a pattern similarity to those of background regions. And they use the certain statistics of the remaining image to restore the damaged image region [1–4]. As one of the most advanced techniques used in the past, PatchMatch [1] can quickly find the nearest neighbor matching to replace the repaired hole area through the stochastic algorithm. Although it usually produces the smooth results especially in background rendering tasks, it is limited by the available image statistics and just considers the low-level structures without any high-level semantics or global structures for captured images. In addition, the traditional diffusion-based and block-based methods assume that missing blocks can be found in the background image and they cannot generate new image content for complex and non-repetitive structural regions (e.g. human faces) [9].

Nowadays, the deep-learning-based methods are constantly explored to overcome the aforementioned obstacles of the above methods by training a large amount of data [8–10, 12, 13]. In particular, deep convolutional neural networks (CNNs) and generative adversarial networks (GANs) have been introduced to implement the image complement tasks [9, 14, 15]. Broadly speaking, image inpainting tasks equipped with deep module mainly can be divided into two categories. The first ones uses global spatial attention to fill holes by building the similarity between a missing area and the other areas [6, 7, 19, 31]. Although this group of methods can ensure a consistency between generated information and context semantics, there often exist pixel discontinuity and semantic gaps [12]. The second family of schemes is to attach different levels of importance to the valid pixels of the original image to predict the missing pixels [8, 14]. These methods correctly handle irregular vulnerabilities correctly, but the generated content still suffers from semantic errors and boundary artifacts [12]. The above methods work poorly due to ignoring the semantic relevance and the feature continuity of the generated contents, which are related to the continuity of local pixels.

Inspired by the human mind coupled with the partial convolution [8], we propose a Multi-Scale Gated Inpainting module (MG) and a Patch-wise Spacial Attention module (PSA) are proposed to fill an unknown area of the feature map with similar method as a part of our model. The MG module first fills each unknown feature patch in an unknown region with the most similar feature patch in the known regions. Subsequently, the selection of information in the filled area is controlled by a two-scale gating strategy. As a result, the global semantic consistency is guaranteed by the first step, and the local feature consistency is optimized by the second step optimization. In addition to controlling the style relation under local features, the PSA module handles the repaired features with block-level attention.

Technically, our model uses the U-Net [20] architecture as a baseline to propagate the consistency of global and local styles and detailed texture information to the missing areas. On the whole, this model continuously collects the features of an effective region through partial convolution. At a higher level of the encoding phase, we develop a distinctive Multi-Scale Gated Inpainting module (MG) to carry out two phases. First, MG revises a current hole area for its global style alignment through Contextual Attention [6]. Second, MG brings a resulting feature fix into alignment with the overall style through a multi-scale gated mechanism. In the decoding stage, PSA divides the feature channel into multiple patch blocks to further optimize the consistency of local styles so that the network learns more effective local features. Finally, the repaired image is delivered to VGG16 [29] to gauge the style loss and perception loss. Which help generate details consistent with the global style. In addition, our model is finally down-sampled to the size of $4 \times 4$ in order to obtain a higher level of semantic consistency. The experiments driven by the two standard datasets (Places2 [25] and CelebA [26]) reveal that the proposed methods produce higher quality results than those of the existing competitors. The main contributions of this work are summarized as follows:

- We develop a new Multi-Scale Gated Inpainting module (MG) applied to the model structure. MG combines feature maps generated by gated modules of different proportions to obtain structural information of features at different scales, thereby flexibly leveraging background information to balance the image requirements.
- We extend the spatial attention module by adding the minimization feature of patchwise to ensure that the pixels generating holes area are true and locally stylized.
- We introduce the concepts of style loss and the perception loss to construct the proposed loss function, which yield a consistent style. The proposed new spatial discounted loss of irregular holes helps to strengthen hole-center constraints, thus promoting texture consistency.
- The experiments with two standard datasets (Places2 [25] and CelebA [26]) demonstrate the superiority of our approaches over the most advanced methods found in the literature.

## 2   Related Work

### 2.1   Image Inpainting

Traditional non-learning methods propagate and reproduce information by calculating the similarity with the other background regions [2, 4]. PatchMatch [1] can well synthesize surface textures through the nearest neighbor matching algorithm, which is an excellent patch matching algorithm. However, these methods do not semantically originate meaningful contents, neither can the methods deal with large missing areas. For the nonexistent detailed texture features, these schemes are unable to generate new features while exhibiting poor recovery effect.

In recent years, the methods based on deep learning have become a significant symbol of the image restoration. Context Encoder [15] tries to restore the central area $(64 \times 64)$ of $128 \times 128$ images. This technique is the first deep network model to handle the inpainting tasks, which provides reasonable results for the holes semantic filling. Unfortunately, it has a poor inpainting ability at fine textures. Shortly thereafter, Iizuka *et al.* extends the context encoder by proposing local and global discriminators to improve repaired quality for the image consistency [10]. This extension overlooks the consistent relation between holes and the other areas as a whole. Therefore, there exist more obvious color differences. The Context Encoder [15] is trained to act as the constraint of global content [13], local texture constraints are constructed by using the local patches similarity of the missing part and the known regions to obtain high-resolution prediction.

### 2.2  Feature Matching and Feature Selection

Global spatial attention mechanisms have also been deployed to address image inpainting challenges by the virtue of similarity relation. CA [6] creates a rough prediction for the hole area through similarity calculation. The Multi-Scale Contextual-Attention [7] patch is located in the missing region. Ultimately, the re-weight for both is located on the Squeeze-And-Excitation [16] module, which improves generalization ability of the model. SCA [12] build a patch-wise continuity relationship between adjacent features of the missing area to enhance the continuity of features inside the holes area. Shift-net [19] selects a specific encode-decode layer of the same level for similarity measures, encoder features of the known region are shifted to serve as an estimation of the missing parts. The RFR [31] harvests remote information and progressively infers a boundary of the hole by the KCA module, thereby gradually strengthening the constraints on the hole's center.

The above methods adopt the similar treatment for the corruption areas and non-corruption area, thereby leading to artifacts such as color discrepancy and blurriness. Only the effective features of each layer are processed by partial convolution [8]. By updating the mask of each layer and normalizing the convolution weights and mask values, which ensures that the convolution filter focuses on the effective information of the known regions to deal with irregular holes. Partial convolution is regarded as a kind of hard mask [14], which confronts roadblocks learn specific mask information. Furthermore, it introduces automatic learning soft mask by using gated convolution and combines with SN-Patch GAN discriminator to achieve optimized predictions. When it comes to feature normalization, the above methods do not consider the influence of mask areas, which limits the trainings of network repair. Treating the damaged areas and the undamaged areas separately [11], the mean value and variance deviation are solved to continuously improve network performance.

Unlike the leading-edge strategies proposed in the literature [6, 8, 14], our solution is tailored for process images where backgrounds are misleading or lacks similarity. Our technique has an edge over the existing methods, because ours leverages the multi-scale gated module to control the degree of feature extraction while dynamically screening useful features to alleviate the problem of information redundancy and loss. In order to enrich repaired details, an extended spatial attention module [28] performs the patch-wise division on the channel to dynamically extract local features. In this way, our model is adept at generalizing scenes and understanding styles as well as picture details.

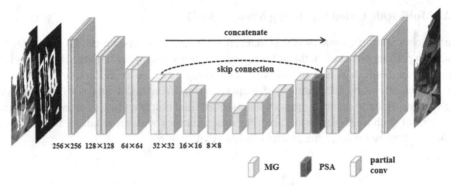

**Fig. 1.** The architecture of our GS-Net model. We augment the MG and PSA layer at the resolution of 32 × 32 in the network.

## 3 Approach

We describe the entire model structure from top to bottom, and then introduce the MG module and PSA module in details. Some extensions to the model are also expressed to allow optional user guidance.

### 3.1 An Overview of Our Model (GS-Net)

Our model is a one-stage and end-to-end image inpainting model, thereby making our approach simpler and easier to implement than other methods. More specifically, U-Net [20] is used as the baseline structure and the partial convolution [8] is stacked as the basic modules for deep feature extraction in GS-Net (see also Fig. 1). More formally, we denote W as convolution layer filter weights, b as the bias, M as the mask, and X as the feature values for the current convolution window. The partial convolution is expressed as:

$$x' = \begin{cases} W^T (X \otimes M) \frac{sum(1)}{sum(M)} + b, & if\ sum(M) > 0 \\ 0, & otherwise \end{cases} \tag{1}$$

In each convolution window with effective feature values, partial convolution layer assigns greater weight to the convolution result with fewer feature values through the above operation. After each partial convolution operation is accomplished, whether the mask has a valid pixel update mask through the convolution region. This process is expressed as:

$$m' = \begin{cases} 1, & if\ sum(M) > 0 \\ 0, & otherwise \end{cases} \tag{2}$$

After the feature map passes through partial convolutional layer, the missing area is filled with the surrounding effective feature area and becomes smaller. Therefore, all the features areas of the holes will be completely filled after sufficient successive applications of the partial convolution layer.

## 3.2  Multi-scale Gated Inpainting Module (MG)

Partial convolution in our model is stacked with layers to update masks and feature maps. In partial convolution, the holes region gradually disappear with the deepening of convolution depth, which is conducive to extracting effective depth features. However, directly interpolating the features of empty regions from the features of non-empty regions during an up-sampling process leads to the final blurry texture of missing regions. The root of such a problem is to directly extract features without repairing the features of holes, thereby ignoring the spatial continuity of features.

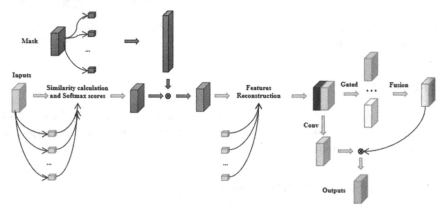

**Fig. 2.** In the MG Module, an input feature is transferred to a global attention module to fill a hole feature area, which is concatenated into the input feature to obtain different gated scores through two convolution kernels of multiple sizes (3 × 3 and 1 × 1). In the end, element-wise multiplications are performed specifically by multiplying the convoluted concatenate by the two gated scores of fusion as an output feature map.

To address the challenge of blurred image content and distorted structure, we propose a repair network with MG module in Fig. 2. First of all, The MG module uses CA [6] method to fill the holes in the high-level feature. Partial convolution is still the interpolation of depth features on the hole area, and the inability to match the optimal patch leads to information loss and confusion. Therefore, CA algorithm is used to construct similarity matrix, and deconvolution operation ensures the trainability of interpolation process.

Given an input feature map $\phi_{in}$, we firstly replace fore-ground feature map using attention mechanism. For each attention map, we use similar strategy to calculate scores as [6] the calculation of attention score could be implemented as convolution calculation.

$$s_{x,y,x',y'} = \ <\frac{f_{x,y}}{||f_{x,y}||}, \frac{b_{x',y'}}{||b_{x',y'}||}> \tag{3}$$

where, $f_{x,y}$, $b_{x',y'}$ are fore-ground patches and background patches respectively. $s_{x,y,x',y'}$ is similarity matrix between all the patches. To compute the weight of each patch, softmax is applied on the channel of score map to obtain softmax scores. Since any change in the foreground patch is more related to the similar change in the background patch. CA adopts a left-right propagation followed by a top-down propagation with kernel size of k, and then propagate score to better merge patch.

$$\hat{s}_{x,y,x',y'} = \sum_{i\in\{-k,...,k\}} s^*_{x+i,y,x'+i,y'} \tag{4}$$

where $s^*$ is channel-level softmax applied to feature mapping. Finally, the multichannel mask multiplication is used to preserve the current information and then deconvolution operation are responsible for restoring a missing feature map. However, the misleading highly similar regions and the existence of the hole regions could lead to the disappearance of effective features in deconvolution, which is detrimental to feature restoration. Fortunately, we are inspired by gated convolution to ensure the dynamic selection of effective features by constructing a soft mask mechanism, in which the expected output features are learned under fixed gateway scale. This learning mechanism is dynamic, general and accurate. When the hole feature is repaired, other inappropriate features are incorrectly filled. And the gate control mechanism dynamically adjusts the gate value to construct an appropriate output feature.

Moreover, it is nontrivial to determine the appropriate patch match size for various image to reveal images details. In general, larger patch size helps ensure style consistency while smaller patch size is more flexible on using background feature map. Patch matching on a single fixed scale seriously limits the capability to fit the model into different scene [7]. To this end, we devise a novel MG module that helps to make use of background content flexibly based on the overall image style. The MG integrate feature selection at two different scales by convolution opteration. In order to better distinguish the importance of the two, we simply use a learnable parameter $\lambda$ as the dynamic threshold. Formally, the gated feature of the output is written as:

$$output_{gated} = \lambda output_1 + (1 - \lambda)output_2 \tag{5}$$

where, $output_1$ and $output_2$ are two gated values at two different scales. Therefore, the value output by the MG module has comprehensive information at multiple scales.

Through the MG module's information, global features become continuous thanks to global-feature-hole fillings and the multi-scale gated selection mechanism. Therefore, the partial convolution layer has no need to distinguish between the region of holes and non-holes and; thus, the mask is set to 1. The MG module is adroit at capturing background information on high-level semantics while producing contents with elegant details.

### 3.3 Patch-Wise Spacial Attention Module (PSA)

A vital feature of a human visual system is that people have no intent to process an entire scene at once. Instead, humans take advantage of a series of local glimpses and selectively focus on salient parts in order to capture visual structure in a swift manner. Although the attention mechanism is widely used in image classification, the mechanism has no appearance in image inpainting. An important reason is that when it comes to incomplete image, there may still be hole information in high-level features. At this time, the traditional s spatial attention mechanism gives rise to structure dispersion and texture loss in generated images. In order to solve the above challenges and take advantage of the attention mechanism, we extend the CBAM technique [28] to devise a patch-wise attention mechanism depicted in Fig. 3.

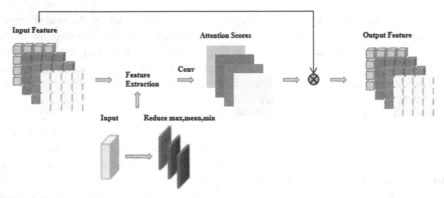

**Fig. 3.** PSA Module. 3D features are divided into $3 \times 3 \times 4$ small blocks, and the maximum, average and minimum values of channel-wise are calculated for each small block, and the spatial attention of each block is obtained through convolution.

Instead of using the channel-wise reduction directly, we choose a $3 \times 3 \times 4$ feature block as a unit of attention restoration in order to ensure the consistency relationship of local features and maximize the effect of the restoration of the hole area. We first apply maximum pool, average pool, and minimum pool operations on the channel axis and concatenate them to produce a valid feature descriptor. On the feature descriptor of the connection, convolution layer is applied to generate the spatial attention graph and extract the information features of each patch through the convolution operation. Different from classification and identification tasks, the minimum pool operation obtains the characteristics of possible hole repair to hold hole is emphasized so as to ensure its information flow in the network.

We aggregate channel information of a feature map by using three pooling operations, generating three 2D maps in $i^{th}$ patch (The $i^{th}$ patch here represents the channel between $(i - 1) * 4$ and $i * 4$): $F_{max}^i, F_{avg}^i, F_{min}^i \in R^{1 \times H \times W}$. Each denotes max-pooled features, average-pooled features and min-pooled features across the channel. Those features are then concatenated and convolved by a standard convolution layer, producing our 2D patch-wise spatial attention map. In short, the spatial attention is computed as:

$$M_i(F) = \sigma(f^{3 \times 3}([F_{max}^i, F_{avg}^i, F_{min}^i]))$$
(6)

where $\sigma$ denotes the sigmoid function and $f^{3 \times 3}$ represents a convolution operation with the filter size of $3 \times 3$.

# 4   Loss Function

Similar to the design of PC [8], the style consistency and detail level are also taken into the consideration of our loss function. For the model learning process can fully pay attention to the texture details and structural information, we consider a pre-trained VGG16 [29] as a fixed model to extract high-level features. The perceptual loss [30] and style loss compare the difference between the deep feature map of the generated image and the ground truth under different descriptors. All parameter symbols are described as follows. $\phi_i$ denotes feature maps $i^{th}$ pooling layer. H, W, C refer to the height, weight and channels number for a feature map, respectively. And N is the number of feature maps generated by the VGG16 feature extractor. The perceptual loss can be expressed as follows:

$$L_{perceptual} = \sum_{i=1}^{N} \frac{1}{H_i W_i C_i} |\phi_i^{gt} - \phi_i^{out}|_1 \tag{7}$$

Although perceptual loss helps to capture high level structures, the perceptual loss lacks the ability to preserve style consistency. To address this drawback, we advocate for the style loss ($L_{style}$) as an integral apart of our loss function. With the help of the style loss, our model is adroit at learning color and overall style information from backgrounds.

$$L_{style} = \sum_{i=1}^{N} |\frac{1}{H_i W_i C_i} (\phi_i^{style_{gt}} - \phi_i^{style_{out}})|_1 \tag{8}$$

$$\phi_i^{style} = \phi_i \phi_i^T \tag{9}$$

Total variation (TV) loss $L_{tv}$, the smoothing penalty [30] on R, is introduced into the loss function. Here R is the area of a sliding window that contains missing pixels. However, the cost of directly applying TV losses to a hole area is to promote texture blurring of the hole area. More unfortunately, in the case of large losing areas, this approach leads to a failure to repair void areas -- the hole areas remain void areas. In order to address the problem of huge amount, we benefit from two cognitions: a hole area has a certain similarity with the TV of surrounding areas; the edge of the hole area maintains a certain continuity with the surrounding area. The TV loss is expressed as follows:

$$L_{tv} = L_{row} + L_{col} \tag{10}$$

$$L_{row} = \sum_{(i,j)\in R,(i,j+1)\in R} \frac{||I_R^{i,j+1} - I_R^{i,j}||_1}{N_{I_R}}, L_{col} = \sum_{(i,j)\in R,(i+1,j)\in R} \frac{||I_R^{i+1,j} - I_R^{i,j}||_1}{N_{I_R}} \tag{11}$$

where $I_R^{i,j}$ represents an image pixel point, $N_{I_R}$ is defined as the number of elements in the hole's region. Especially for large holes, boundaries are sometimes still artifacts, which may be the lack of constraints on the center of the holes. Similar to the spatial discounted loss of the CA algorithm, the closer the hole region is to the known region, the more attention should be given to it. However, the distance between a hole point and a surrounding effective region is difficult to calculate when it comes to irregular holes. To simplify the computation, we traverse the symmetric mask value near each hole point and undertake a bit operation to quickly obtain the hole length. This process is formally articulated as follows:

$$Sym_{i,j,length} = mask_{i,j}^{i-length,j} | mask_{i,j}^{i+length,j} \qquad (12)$$

where $mask_{i,j}^{i-length,j}$, $mask_{i,j}^{i+length,j}$ are the fields of length at the coordinates of point (i, j). Finally, our target becomes the maximum length value of Sym = [0].

$$discounted_{left-right} = \max_{length}\{Sym_{i,j,length} = [0]\} \qquad (13)$$

where [0] represents the matrix with all values of 0. The upper and lower relation of the hole region is solved by the same strategy and denoted as $discounted_{top-bottom}$. Therefore, the total spatial discounted weight is formalized as follows:

$$discounted = \gamma^{(discounted_{left-right}+discounted_{top-bottom})/2} \qquad (14)$$

where, $\gamma$ represents a weighting factor. Futher, $L_{valid}$ and $L_{hole}$ which calculate L1 differences in the unmasked area and masked area respectively. The total loss $L_{total}$ is the combination of all the above loss functions. Thus, we have

$$L_{total} = \lambda_{vaild}L_{valid} + \lambda_{hole}(L_{hole} \odot discounted) + \lambda_{perceptual}L_{[erceptual} + \lambda_{style}L_{style} + \lambda_{tv}L_{tv} \qquad (15)$$

## 5   Experiments

### 5.1   Datasets and Experimental Details

In this section, we evaluate our model on two datasets: the Places2 [25] dataset and the CelebA [26] dataset. The Places2 dataset is a garden scene selected from the Places365-Standard dataset, which embraces 9069 images. The dataset is divided into the train, validate, and test subsets with a ratio of 8:1:1. The CelebA dataset contains 162,770 training images, 19,867 validation images, and 19,962 test images. We use both the training set and validation set for training purpose, whereas the test set is dedicated for testing. In the end, we use the mask dataset of partial convolution, which contains 55,116 masks for the training and 24,866 masks for testing. The size of these masks is $512 \times 512$. After resizing these masks to $256 \times 256$, we place the masks into our network model.

For all the parameter settings similar to those elaborated in the literature [8], the tradeoff parameters are set as $\lambda_{valid} = 1$, $\lambda_{hole} = 6$, $\lambda_{perceptual} = 0.05$, $\lambda_{style} = 120$ and $\lambda_{tv} = 0.1$. Our model is initialized the weights using the initialization method described in [9] and use Adam [27] for optimization with a learning rate of 0.0001, and train on a single NVIDIA V100 GPU (32 GB) with a batch size of 6. The Places2 models are

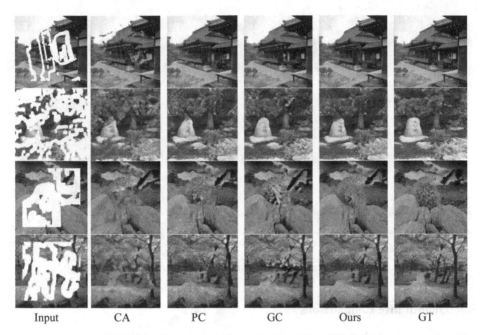

| Input | CA | PC | GC | Ours | GT |

**Fig. 4.** A Comparison of test results on Places2 images.

trained for two days, whereas the CelebA models are trained for approximately one week.

We compare the proposed MG and PSA algorithm with the following three state-of-the-art methods: CA [6]: Contextual Attention, PC [8]: Partial Convolution, GC [14]: Gated Convolution.

To make fair comparisons with the CA and GC approaches, we retrain the CA and GC models on the same datasets. Both CA and GC methods are trained using a local discriminator available in a local boundary box of the hypothetical hole, which makes no sense for the shape of masks [8]. As such, we directly use CA and GC released pretrained models. And PC is trained under the same conditions as those in our experimental setup until the PC model is converged.

## 5.2 Qualitative Comparisons

Figure 4 unveils the comparison results among our method and the three most advanced approaches processing in the Places2 dataset. All images are displayed at the same resolution ($256 \times 256$). The CA approach is effective at semantic inpainting, but the results shown above appear to be abnormally blurry and artifact. The PC method fills the hole areas with the corresponding styles, but PC loses some of the detail textures. The GC method exhibits a strong inpainting ability in local details and overall styles. Unfortunately, GC suffers from the local overshine problems. Compared with the other methods, our solution has an edge under large hole conditions by originating inpainting results that alleviate artificial traces. Figure 5 unravels that our model is able to generate fully detailed, semantically plausible, and authentic images with superb performance.

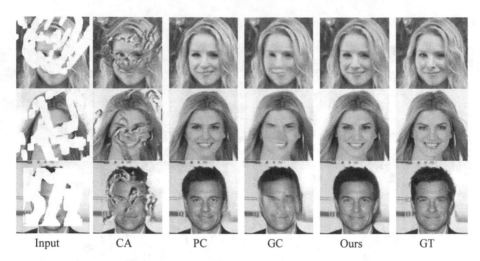

| Input | CA | PC | GC | Ours | GT |

**Fig. 5.** Comparison of test results on CelebA images.

## 5.3 Quantitative Comparisons

Now we quantitatively evaluate our model on the two datasets, using three quality methods, namely, the structural similarity (SSIM), peak signal-to-noise ratio (PSNR), and mean L1 loss assessment image similarity. Because the image restoration application in the application scenario will not stick to the above mask structure. To make a fair numerical comparison, we apply the mask generation method of GC [14] to compare the mask repair effects under the three different proportions in Fig. 6. One thousand masks and their corresponding random pictures are elected in the tests, the results of which are recapped in Table 1.

10%-20%                30%-40%                50%-60%

**Fig. 6.** Some test masks for each hole-to-image area ratio category.

Table 1 illustrates that our method produces the decent results with the best SSIM, PSNR and mean l1 loss on the Places2 dataset and the CelebA faces dataset. Similar to the aforementioned results, our MG and PSA algorithm is a front runner in terms of numerical performance on the Places2 and CelebA datasets. When it comes to repairing large holes, the performance improves of our algorithm over the existing techniques become more pronounced.

**Table 1.** Numerical comparison on two datasets.

| Dataset | | Places2 | | | CelebA | | |
|---------|----|---------|---------|---------|---------|---------|---------|
| Mask ratio | | 10%–20% | 30%–40% | 50%–60% | 10%–20% | 30%–40% | 50%–60% |
| Mean l1(%) | CA | 3.0941 | 5.7280 | 7.4529 | 3.2433 | 6.0052 | 8.4367 |
| | PC | 3.2495 | 4.4537 | 5.3843 | 1.8712 | 2.5208 | 3.2301 |
| | GC | 2.0385 | 3.5036 | 4.7996 | 1.2488 | 2.1232 | 2.9248 |
| | Ours | 2.1274 | 3.1917 | 4.2045 | 0.9542 | 1.5228 | 2.0834 |
| PSNR | CA | 21.5031 | 18.1033 | 17.2827 | 20.8873 | 17.5012 | 16.0160 |
| | PC | 24.7846 | 22.1610 | 21.1155 | 29.3626 | 26.7636 | 24.9999 |
| | GC | 24.7426 | 21.5232 | 20.1670 | 28.6721 | 25.5052 | 23.9649 |
| | Ours | 25.7142 | 23.1374 | 22.0227 | 32.0948 | 28.9088 | 27.0451 |
| SSIM | CA | 0.8327 | 0.7042 | 0.6080 | 0.8337 | 0.7067 | 0.6015 |
| | PC | 0.8296 | 0.7307 | 0.6476 | 0.9050 | 0.8567 | 0.8074 |
| | GC | 0.8638 | 0.7623 | 0.6758 | 0.9180 | 0.8589 | 0.8050 |
| | Ours | 0.8650 | 0.7762 | 0.6917 | 0.9463 | 0.9061 | 0.8638 |

## 5.4 Ablation Study and Discussion

GS-Net, being carried out on partial convolution, is equivalent to the superposition processing of partial convolution layer excluding our proposed two modules. To clearly present the effectiveness of these operations, we compare various indicators by respectively removing the MG and PSA modules in Places2 dataset. Figure 7 and Table 2 reveal that compared to the results yielded by our algorithm, the results from the non-MG and non-PSA models exhibit more artifacts and distortions. At the same time, the MG module is superior to the PSA module in terms of performance index.

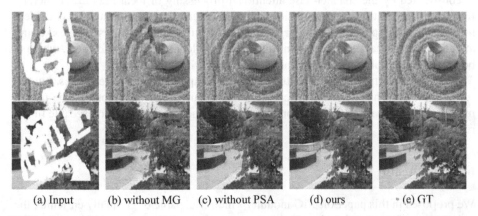

(a) Input    (b) without MG    (c) without PSA    (d) ours    (e) GT

**Fig. 7.** Comparison results for different attention manners. From the left to the right are: (a) Input, (b) Without MG, (c) Without PSA, (d) MG + PSA, (e) Ground Truth

**Table 2.** Numerical comparison on Places2 dataset.

| Method | Mean l1 loss (%) | PSNR | SSIM |
|---|---|---|---|
| Without MG | 3.8346 | 21.2940 | 0.7283 |
| Without PSA | 3.5739 | 22.5376 | 0.7615 |
| With MG and PSA | 3.1411 | 23.2501 | 0.7808 |

Apart from delivering strong capabilities in terms of recovery, GS-Net can be widely applied to intelligent face modification or face synthesis. Figure 8 shows two faces with different detail textures.

Input                    Ours                    GT

**Fig. 8.** In face of effect.

Features will be input into the PSA module after a globally filled hole area of the MG module. The information flowing through the MG module is well repaired, this PSA module is focused on controlling the relationship among local feature blocks. The PSA is constructed by the channel-wise attentional processing of local 3D blocks, thereby forming local relations such as local maximum, average, and minimum. It is evident that each patch repaired may be larger than unrepaired feature values. Thus, exerting an attention will pay more attention to the repaired local 3D region features, which is beneficial to the subsequent upsampling process.

Our model outperforms the cutting-edge techniques in most tested cases, but the repair effect still has a certain difference under a pure color background. The reason may be caused by partial convolution, which will be addressed in our foreseeable future research pathway.

## 6   Conclusion

We proposed in this paper the MG module, which is capable of gradually enriching the information of mask regions by offering semantically consistent embedding results . We developed the PSA module to further promote the enrichment of local texture details.

We conducted extensive qualitative and quantitative comparisons against the leading-edge solutions. The validity analysis and ablation learning demonstrate that our GS-Net outperforms the existing solutions over the Places2 and CelebA datasets.

# References

1. Barnes, C., Shechtman, E., Finkelstein, A., Goldman, D.B.: PatchMatch: a randomized correspondence algorithm for structural image editing. TOG **28**(3), 24:1–24:11 (2009)
2. Ballester, C., Bertalmio, M., Caselles, V., Sapiro, G., Verdera, J.: Filling-in by joint interpolation of vector fields and gray levels. IEEE Trans. Image Process. **10**(8), 1200–1211 (2018)
3. Criminisi, A., Pérez, P., Toyama, K.: Region filling and object removal by exemplar-based image inpainting. IEEE TIP **13**(9), 1200–1212 (2004)
4. Wilczkowiak, M., Brostow, G. J., Tordoff, B., Cipolla, R.: Hole filling through photomontage. In: Proceedings of the British Machine Vision Conference (BMVC), pp. 492–501. British Machine Vision Association, Oxford (2005)
5. Shetty, R., Fritz, M., Schiele, B.: Adversarial scene editing: automatic object removal from weak supervision. In: Thirty-second Conference on Neural Information Processing Systems, pp. 7717–7727. Curran Associates, Montréal Canada (2018)
6. Yu, J., Lin, Z., Yang, J., Shen, X., Lu, X., Huang, T. S.: Generative image inpainting with contextual attention. In: Proceedings of the IEEE/CVF Conference on Computer Vision and Pattern Recognition (CVPR), pp. 5505–5514 (2018)
7. Wang, N., Li, J., Zhang, L., Du, B.: Musical: multi-scale image contextual attention learning for inpainting. In: Proceedings of the Twenty-Eighth International Joint Conference on Artificial Intelligence (IJCAI), pp. 3748–3754 (2019)
8. Liu, G., Reda, F.A., Shih, K.J., Wang, T.-C., Tao, A., Catanzaro, B.: Image inpainting for irregular holes using partial convolutions. In: Ferrari, V., Hebert, M., Sminchisescu, C., Weiss, Y. (eds.) ECCV 2018. LNCS, vol. 11215, pp. 89–105. Springer, Cham (2018). https://doi.org/10.1007/978-3-030-01252-6_6
9. Zhou, T., Ding, C., Lin, S., Wang, X., Tao, D.: Learning oracle attention for high-fidelity face completion. In: Proceedings of the IEEE/CVF Conference on Computer Vision and Pattern Recognition, pp. 7680–7689 (2020)
10. Iizuka, S., Simo-Serra, E., Ishikawa, H.: Globally and locally consistent image completion. ACM TOG **36**(4), 1–4 (2017)
11. Yu, T., et al.: Region normalization for image inpainting. In: Proceedings of the AAAI Conference on Artificial Intelligence, pp. 12733–12740 (2020)
12. Liu, H., Jiang, B., Xiao, Y., Yang, C.: Coherent semantic attention for image inpainting. In: ICCV, pp. 4170–4179 (2019)
13. Yang, C., Lu, X., Lin, Z., Shechtman, E., Wang, O., Li, H.: High-resolution image inpainting using multi-scale neural patch synthesis. In: The IEEE Conference on Computer Vision and Pattern Recognition (CVPR), pp. 6721–6729 (2017)
14. Yu, J., Lin, Z., Yang, J., Shen, X., Lu, X., Huang, T.S.: Free-form image inpainting with gated convolution. In Proceedings of ICCV, pp. 4471–4480 (2019)
15. Pathak, D., Krahenbuhl, P., Donahue, J., Darrell, T., Efros, A.A.: Context encoders: feature learning by inpainting. In: The IEEE Conference on Computer Vision and Pattern Recognition (CVPR), pp. 2536–2544 (2016)
16. Hu, J., Shen, L., Sun, G.: Squeeze-and-excitation networks. In: The IEEE Conference on Computer Vision and Pattern Recognition (CVPR), pp. 7132–7141 (2018)

17. Nazeri, K., Ng, E., Joseph, T., Qureshi, F., Ebrahimi, M.: Edgeconnect: structure guided image inpainting using edge prediction. In Proceedings of ICCV Workshops (2019)
18. Xiong, W., et al.: Foreground-aware image inpainting. In: The IEEE Conference on Computer Vision and Pattern Recognition (CVPR), pp. 5840–5848 (2019)
19. Yan, Z., Li, X., Li, M., Zuo, W., Shan, S.: Shift-net: image inpinting via deep feature rearrangement. In: Proceedings of ECCV, pp. 3–19 (2018)
20. Ronneberger, O., Fischer, P., Brox, T.: U-Net: convolutional networks for biomedical image segmentation. In: Navab, N., Hornegger, J., Wells, W.M., Frangi, A.F. (eds.) MICCAI 2015. LNCS, vol. 9351, pp. 234–241. Springer, Cham (2015). https://doi.org/10.1007/978-3-319-24574-4_28
21. Levin, A., Zomet, A., Weiss, Y.: Learning how to inpaint from global image statistics. In: Proceedings of International Conference on Computer Vision (ICCV), pp. 305–312 (2003)
22. Ding, D., Ram, S., Rodríguez, J.J.: Image inpainting using nonlocal texture matching and nonlinear filtering. IEEE Trans. Image Process. 28(4), 1705–1719 (2018)
23. Snelgrove, X.: High-resolution multi-scale neural texture synthesis. In: SIGGRAPH Asia Technical Briefs, pp. 1–4 (2017)
24. Goodfellow, I.J., Pouget-Abadie, J., Mirza, M.: Generative adversarial networks. In: NIPS, pp. 2672–2680 (2014)
25. Zhou, B., Lapedriza, A., Khosla, A., Oliva, A., Torralba, A.: Places: A10 million image database for scene recognition. IEEE TPAMI 40(6), 1452–1464 (2018)
26. Liu, Z., Luo, P., Wang, X., Tang, X.: Deep learning face attributes in the wild. In: Proceedings of the IEEE International Conference on Computer Vision (ICCV), pp. 3730–3738 (2014)
27. Kingma, D.P., Ba, J.: Adam: a method for stochastic optimization. arXiv preprint arXiv:1412.6980 (2014)
28. Woo, S., Park, J., Lee, J.-Y., Kweon, I.S.: CBAM: convolutional block attention module. In: Ferrari, V., Hebert, M., Sminchisescu, C., Weiss, Y. (eds.) ECCV 2018. LNCS, vol. 11211, pp. 3–19. Springer, Cham (2018). https://doi.org/10.1007/978-3-030-01234-2_1
29. Simonyan, K., Zisserman, A.: Very deep convolutional networks for large-scale image recognition. arXiv preprint arXiv:1409.1556 (2014)
30. Johnson, J., Alahi, A., Fei-Fei, L.: Perceptual losses for real-time style transfer and super-resolution. In: Leibe, B., Matas, J., Sebe, N., Welling, M. (eds.) ECCV 2016. LNCS, vol. 9906, pp. 694–711. Springer, Cham (2016). https://doi.org/10.1007/978-3-319-46475-6_43
31. Li, J., Wang, N., Zhang, L., Du, B., Tao, D.: Recurrent feature reasoning for image inpainting. In: The IEEE Conference on Computer Vision and Pattern Recognition (CVPR), pp. 7760–7768 (2020)
32. Zheng, C., Cham, T. J., Cai, J.: Pluralistic image completion. In: CVPR, pp. 1438–1447 (2019)

# An Improved CNNLSTM Algorithm for Automatic Detection of Arrhythmia Based on Electrocardiogram Signal

Jingyao Zhang[1] , Fengying Ma[1] , and Wei Chen[2](✉)

[1] School of Electrical Engineering and Automation,
Qilu University of Technology (Shandong Academy of Sciences), Jinan 250353, China
[2] School of Mechanical Electronic and Information Engineering,
China University of Mining and Technology-Beijing, Beijing 100083, China
http://dqxy.qlu.edu.cn/2019/1229/c8150a141516/page.htm,
http://cs.cumt.edu.cn/fc/b7/c11272a392375/page.htm

**Abstract.** Arrhythmia is one of the most common types of cardiovascular disease and poses a significant threat to human health. An electrocardiogram (ECG) assessment is the most commonly used method for the clinical judgment of arrhythmia. Using deep learning to detect an ECG automatically can improve the speed and accuracy of such judgment. In this paper, an improved arrhythmia classification method named CNN-BiLSTM, based on convolutional neural network (CNN) and bidirectional long short-term memory (BiLSTM), is proposed that can automatically identify four types of ECG signals: normal beat (N), premature ventricular contraction (V), left bundle branch block beat (L), and right bundle branch block beat (R). Compared with traditional CNN and BiLSTM models, CNN-BiLSTM can extract the features and dependencies before and after data processing better to achieve a higher classification accuracy. The results presented in this paper demonstrate that an arrhythmia classification method based on CNN-BiLSTM achieves a good performance and has potential for application.

**Keywords:** CNN · BiLSTM · ECG · Arrhythmia · Classification

## 1 Introduction

Cardiovascular disease is a serious problem with a high fatality rate and can easily lead to multiple complications, posing a significant threat to human health [1–3]. Arrhythmia is one of the most common types of cardiovascular disease;

This work was supported by the National Natural Science Foundation of China (Approval Number: 61903207), Shandong University Undergraduate Teaching Reform Research Project (Approval Number: M2018X078), and the Shandong Province Graduate Education Quality Improvement Program 2018 (Approval Number: SDYAL18088). The work was partially supported by the Major Science and Technology Innovation Projects of Shandong Province (Grant No. 2019JZZY010731).

C. S. Jensen et al. (Eds.): DASFAA 2021 Workshops, LNCS 12680, pp. 185–196, 2021.
https://doi.org/10.1007/978-3-030-73216-5_13

therefore, the automatic diagnosis of arrhythmia has attracted attention from researchers. An ECG is widely used in the diagnosis of arrhythmia owing to its noninvasive nature and because it provides a rich heart rhythm, thereby making the diagnostic process convenient for medical workers [4,5]. However, an ECG signal is nonlinear, and small changes might get ignored when an ECG is viewed by the naked eye; moreover, an accurate diagnosis of arrhythmia manually requires a 24 h holter recording process, which is a cumbersome and lengthy process [6–8]. Therefore, it is necessary to use computer algorithms to diagnose arrhythmia. Additionally, the use of such algorithms can improve the accuracy and robustness of the diagnosis and reduce the diagnosis time and workload.

With the development of information technology, many arrhythmia classification methods that use computer algorithms have emerged currently, some of which are traditional ECG algorithms. First, the ECG signals are extracted from features [9,10] and then put into a support vector machine and random forest in the classifier [11–14]. However, because the features are manually extracted, the obtained information may not fully reflect the true ECG signal, leading to the loss of important features. Therefore, to determine an arrhythmia, it is difficult to obtain the best results based solely on the use of machine learning.

Compared with machine learning, deep learning provides greater advantages. In deep learning, all hidden features are noticed, and no manual feature extraction is required [15]. In terms of ECGs, deep learning has also been applied to many studies on ECG signals. Xiong et al. proposed a 16-layer one-dimensional CNN to classify ECGs [16], and Acharya et al. proposed an 11-layer deep CNN network as an ECG computer-aided diagnosis system to develop four different types of automatic arrhythmia classification method [6]. Fujita et al. used a CNN combined with raw data or a continuous wavelet transform for classification of the four types of ECG signals [17]. Oh et al. used a CNN and a long short-term memory(LSTM) model to diagnose N sinus rhythms, left bundle branch blocks, right bundle branch blocks, premature atrial beats, and premature ventricular beats. The ECG signal has achieved good classification results [7]. Zheng et al. converted a one-dimensional ECG signal into a two-dimensional gray image and used the combined model of a CNN-LSTM to detect and classify the input data [18]. Several researchers have shown that the application of deep learning to the classification of ECG signals significantly improves the performance of the system. Various neural networks can extract complex nonlinear features from the original data without manual intervention, thereby making the classification results more ideal. However, learning the thinking mechanism of the ECG signal features with the high accuracy required for monitoring remains a difficult task. CNNs and BiLSTM models [19] have their own advantages in terms of feature extraction and dependency learning and they can be used for arrhythmia monitoring, improving the accuracy and stability of automatic arrhythmia detection.

In this paper, we proposed an end-to-end arrhythmia detection method to utilize the advantages of the CNN and BiLSTM networks completely. The innovative ECG classification algorithm is called CNN-BiLSTM, and it can identify

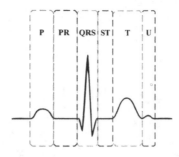

**Fig. 1.** A normal ECG signal

and classify abnormal signals from ECG signals. The contributions of this study are as follows.

I. The CNN provides advantages in terms of image processing, and the BiL-STM model can compensate for the shortcomings of the CNN in terms of context sequences. Therefore, the end-to-end network of the CNN-BiLSTM can effectively improve the accuracy of arrhythmia detection.
II. Adaptive segmentation and resampling are adopted to align the heartbeats of patients with various heart rates. Multi-scale signals that represent electrocardiographic characteristics can be used as the input of the network to extract multi-scale features.
III. Using a small amount of data as the input of the network reduces the computing resources and yields good experimental results. This improves the generalization of the network model and provides a high-precision classification method to meet the needs of automatic detection.

Section 1 of this paper introduces the current research background on automatic arrhythmia detection and the related research algorithms that have been implemented. Section 2 describes the operations conducted prior to the experiment and introduces the data and related network structure needed for the experiment. Section 2.3 introduces the experimental details and results. Section 3 provides some concluding remarks and areas of future research.

## 2 Materials and Methods

### 2.1 Description of Dataset

In this study, an ECG signal was obtained from the MIT-BIH Arrhythmia Database, which is an internationally recognized open-source database [20]. It includes 48 ECG records of 47 subjects. Each record contains a 30 min ECG signal, digitized at a rate of 360 samples per second within a range of 10 millivolts with an 11-bit resolution. Each record has an annotation file for the computer to read. A complete normal ECG signal is shown in Fig. 1.

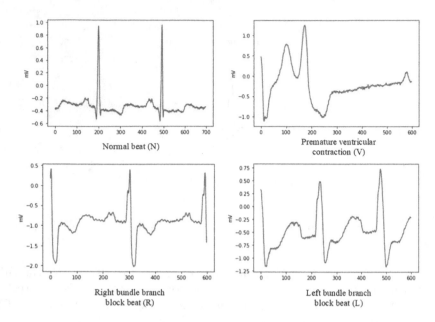

**Fig. 2.** N, V, R, L signal performance

This study used some of the signals in the database, including the following four types: normal heartbeat(N), ventricular premature beat(V), left bundle branch block heartbeat(L), and right bundle branch block heartbeat(R). The performances of the four signals are depicted in Fig. 2. Ventricular premature beats often manifest as QRS waves with wide deformities, and the direction of T waves is opposite to the direction of the QRS waves. An L often shows that the QRS wave becomes a broad R wave, and the time limit is extended. An R often shows that the QRS wave is M-shaped, and the R wave is wide and has notches.

Normal pulsation dominates the dataset; thus, we select a portion of a normal pulsation and simultaneously balance the data of the remaining three types of beats to avoid bias in the experimental results. When the ECG signal is evenly distributed, the neural network exhibits a better convergence. Therefore, we normalized the experimental data. In this study, 80% of the data were used for training, and 20% were used for testing.

## 2.2  Networks

**Convolutional Neural Network.** CNN is one of the most commonly used neural networks in the field of image processing [21–23]. It mainly includes an input layer, a convolution layer, a pooling layer, and an output layer. Among them, the convolutional layer and pooling layer are the core of the CNN. The CNN used in this study contains four convolutional layers and four pooling layers. Its architecture is shown in Fig. 3. The fully connected layer is not connected

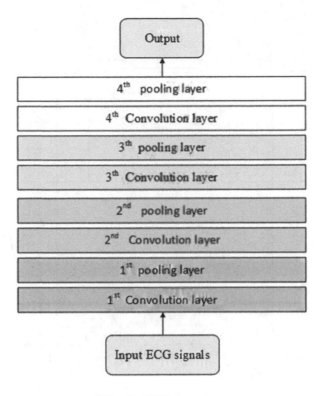

**Fig. 3.** CNN structure

here, and the output data are input into the subsequent BiLSTM network to continue the training.

The CNN is connected to the input layer through the convolution kernel. The convolution kernel performs point multiplication through a sliding window to achieve multi-scale feature extraction. Simultaneously, the weight sharing mechanism of the convolutional layer makes it more effective for feature extraction, thereby significantly reducing the number of free variables that need to be learned. The pooling layer follows the convolutional layer and performs a downsampling to reduce the feature size [24]. After going through several convolution and pooling layers, the features obtained are converted into a single one-dimensional vector for classification.

**Bidirectional Long and Short-Term Memory Network.** An LSTM network is an improved model of a cyclic neural network. It not only transmits forward information but also processes the current information. An LSTM network mainly includes three control gate units, i.e., an input gate, a forget gate, and an output gate. The input gate controls how much input information needs to be kept at the current moment, whereas the forget gate controls how much information needs to be discarded at the previous moment. The output gate

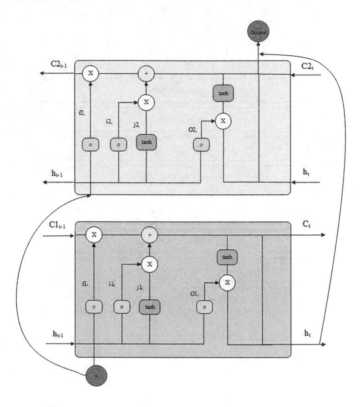

**Fig. 4.** BiLSTM hidden layer structure diagram

controls the amount of information that needs to be output to the hidden state at the current moment. The hidden layer structure is shown in Fig. 4.

Assuming that a given input sequence is represented by $X_t$, its update state has the following formula:

$$f_t = \delta\left(W_f\left[h_{t-1}, X_t\right] + b_f\right) \tag{1}$$

$$i_t = \delta\left(W_i\left[h_{t-1}, X_t\right] + b_i\right) \tag{2}$$

$$j_t = tanh(W_c\left[h_{t-1}, X_t\right] + b_c) \tag{3}$$

$$O_t = \delta\left(W_o\left[h_{t-1}, X_t\right] + b_o\right) \tag{4}$$

$$h_t = O_t * tanh(C_t) \tag{5}$$

Here, $C_t$ is the state information of the memory unit, $j_t$ is the accumulated information at the current moment, W is the weight coefficient matrix, b is the

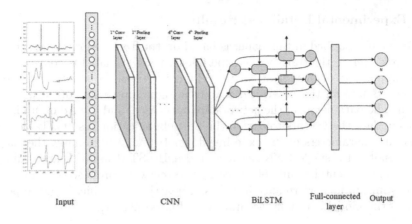

**Fig. 5.** CNN-BiLSTM network structure

bias term, sigma is the sigmoid activation function, and tanh is the hyperbolic tangent activation function.

As shown in Fig. 4, the first layer is a forward LSTM, and the second layer is a backward LSTM. The final output is calculated using the following formula:

$$h_t = \alpha h_t^f + \beta h_t^b \tag{6}$$

In this formula, $h_t^f$ is the output of the forward LSTM layer, which will be is from $x_1$ to $x_t$ as input; $h_b^t$ is the output of a backward LSTM layer, which is from $x_t$ to $x_1$; $\alpha$ and $\beta$ are the sequence control of the forward LSTM and backward LSTM factors ($\alpha + \beta = 1$),; respectively; and $h_t$ is the element sum of two unidirectional LSTM factors at time t.

**Proposed Architecture.** In the aforementioned networks, CNNs have significant advantages in terms of image processing. The CNN model extracts local features in the input signal through a sliding convolution kernel, and the dependence of the data is difficult to learn. BiLSTM can learn the forward and backward information of the feature vector extracted by the CNN by controlling the gate unit; thus, the feature extraction is more perfect. In this paper, an ECG signal classification model based on CNN-BiLSTM is proposed. The ECG signal is preprocessed and input into the model. The CNN obtains the local features of the ECG signal through the convolutional and pooling layers and then places these features into the BiLSTM. The hidden layer obtains the best feature information. The learning rate used by the network is 0.01, and the batch size is 16. Finally, the data are divided into four categories, i.e., N, V, R, and L, through the fully connected layer and the softmax function. The network structure of CNN-BiLSTM is shown in Fig. 5.

## 2.3   Experimental Detail and Results

The research described in this paper is based on the TensorFlow neural network framework. Before the start of the experiment, the data label is converted into the corresponding one-key heat carrier. This study uses the same amount of N, V, R, and L data for the experiments, and the signal processing of the dataset is random. The parameters of the network are optimized, and the Adam updater is used to update the weights to obtain the best classification results. Table 1 lists the relevant parameters of the experimental network. Additionally, the classification results of a single CNN network and a BiLSTM network are compared with the experimental results of the composite network proposed herein.

To estimate the performance in terms of heartbeat classification, the performance of the model is usually accurately evaluated [24–27].

**Table 1.** The parameters of the model

| Network layer type | Filters | Kernel size | Strides |
|---|---|---|---|
| Input layer | – | – | – |
| Con1d | 4 | 3 | 1 |
| pooling1d | – | 5 | 5 |
| Con2d | 8 | 5 | 1 |
| pooling2d | – | 5 | 5 |
| Con3d | 16 | 7 | 1 |
| pooling3d | – | 5 | 5 |
| Con4d | 32 | 9 | 1 |
| pooling4d | – | 5 | 5 |
| BiLstm | – | – | – |
| Full-connected layer | – | – | – |
| Softmax | – | – | – |

Figure 6 presents the loss function curve of the experiment using three different networks, i.e., a CNN, BiLSTM, and CNN-BiLSTM, under the same data. It can clearly be observed that the convergence effect of the proposed CNN-BiLSTM network is better than that of the two single networks.

Figure 7 presents the overall accuracy of the three networks, CNN, BiLSTM, and CNN-BiLSTM, and the classification accuracy for the four types of data. The data volume of the specific classification is provided in Tables 2, 3, 4. It can be observed that the classification accuracy of the CNN-BiLSTM network is higher than CNN and BiLSTM.

As mentioned previously, the CNN-BiLSTM model achieves an overall classification accuracy of 99.69% on the test set, where N is 99.75%, V is 99.56%, R is 99.92%, and L is 99.52%.

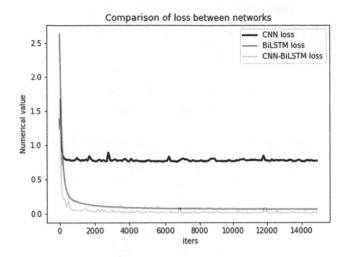

**Fig. 6.** CNN, BiLSTM, CNN-BiLSTM loss function curve comparison

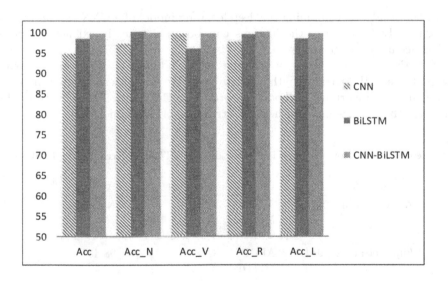

**Fig. 7.** The accuracy of CNN, BiLSTM and CNN-BiLSTM

**Table 2.** Classification results of CNN network

|   | N | V | R | L |
|---|---|---|---|---|
| N | 3921 | 84 | 26 | 2 |
| V | 2 | 3975 | 9 | 7 |
| R | 60 | 37 | 3904 | 5 |
| L | 0 | 622 | 2 | 3344 |

**Table 3.** Classification results of BiLSTM

|   | N | V | R | L |
|---|---|---|---|---|
| N | 3980 | 2 | 1 | 1 |
| V | 12 | 3864 | 42 | 110 |
| R | 0 | 19 | 3997 | 5 |
| L | 0 | 66 | 1 | 3900 |

**Table 4.** Classification results of CNN-BiLSTM

|   | N | V | R | L |
|---|---|---|---|---|
| N | 3996 | 9 | 1 | 0 |
| V | 1 | 4041 | 2 | 15 |
| R | 0 | 2 | 3944 | 1 |
| L | 1 | 17 | 1 | 3969 |

In this study, by combining the deep learning model of a CNN and an LSTM to extract ECG features, the features can be automatically extracted, and a higher accuracy can be achieved.

Table 5 presents a series of scientific studies based on ECG signals with regard to the MIT-BIH Arrhythmia database. We can observe that, compared with other deep learning methods, the proposed CNN-BiLSTM network model improves the input signal and network structure of the model.

**Table 5.** Comparison with previous work on the MIT-BIH Arrhythmia database

| Author (year) | Database | Classifier | Accuracy (%) |
|---|---|---|---|
| Acharya et al. (2016) [6] | AFDB MITDB CUDB | CNN | Net A:92.5 Net B:94.9 |
| Fujita et al. (2019) [17] | AFDB MITDB | CNN | 98.45 |
| Oh et al. (2018) [7] | MITDB | CNN and LSTM | 98.1 |
| Zheng et al. (2020) [18] | MITDB | CNN and LSTM | 99.01 |
| Proposed model (2020) | MITDB | CNN-BiLSTM | 99.69 |

## 3   Conclusion

At present, arrhythmia is one of the most common types of cardiovascular disease, and it seriously endangers human health. In this paper, an automatic system for arrhythmia classification was proposed based on a CNN-BiLSTM. This network can automatically extract and classify ECG signal features, thereby significantly reducing the workload of doctors. The network includes four convolutional layers, four pooling layers, a BiLSTM layer, and a fully connected layer and has achieved good classification performance. This classification method reduces the computing resources required and achieves a high accuracy; thus, it can be used as an auxiliary diagnostic method for clinical arrhythmia detection.

# References

1. Weiwei, C., Runlin, G., Lisheng, L., Manlu, Z., Wen, W., Yongjun, W., et al.: Outline of the report on cardiovascular diseases in China, 2014. Eur. Heart J. Suppl. J. Eur. Soc. Cardiol. **18**(Suppl F), F2 (2016)
2. Ince, T., Kiranyaz, S., Gabbouj, M.: A generic and robust system for automated patient-specific classification of ECG signals. IEEE Trans. Biomed. Eng. **56**(5), 1415–1426 (2009)
3. Rahman, Q.A., Tereshchenko, L.G., Kongkatong, M., Abraham, T., Abraham, M.R., Shatkay, H.: Utilizing ECG-based heartbeat classification for hypertrophic cardiomyopathy identification. IEEE Trans. Nanobiosci. **14**(5), 505–512 (2015)
4. Bie, R., Zhang, G., Sun, Y., Xu, S., Li, Z., Song, H.: Smart assisted diagnosis solution with multi-sensor Holter. Neurocomputing **220**, 67–75 (2017)
5. Zhang, Y., Sun, L., Song, H., Cao, X.: Ubiquitous WSN for healthcare: recent advances and future prospects. IEEE Internet Things J. **1**(1), 311–318 (2014)
6. Acharya, U.R., Fujita, H., Lih, O.S., et al.: Automated detection of arrhythmias using different intervals of tachycardia ECG segments with convolutional neural network. Inf. Sci. **405**, 81–90 (2017)
7. Oh, S.L., Ng, E.Y., San Tan, R., et al.: Automated diagnosis of arrhythmia using combination of CNN and LSTM techniques with variable length heart beats. Comput. Biol. Med. **102**, 278–287 (2018)
8. Chen, C., Hua, Z., Zhang, R., et al.: Automated arrhythmia classification based on a combination network of CNN and LSTM. Biomed. Signal Process. Control **57**, 101819 (2020)
9. Osowski, S., Hoai, L.T., Markiewicz, T.: Support vector machine-based expert system for reliable heartbeat recognition. IEEE Trans. Biomed. Eng. **51**, 582–589 (2004)
10. Rodriguez, J., Goni, A., Illarramendi, A.: Real-time classification of ECGs on a PDA. IEEE Trans. Inf. Technol. Biomed. **9**, 23–34 (2005)
11. Melgani, F., Bazi, Y.: Classification of electrocardiogram signals with support vector machines and particle swarm optimization. IEEE Trans. Inf. Technol. Biomed. **12**, 667–677 (2008)
12. Gnecchi, J.A.G., Archundia, E.R., Anguiano, A.D.C.T., et al.: Following the path towards intelligently connected devices for on-line, real-time cardiac arrhythmia detection and classification. In: IEEE International Autumn Meeting on Power. IEEE (2016)
13. Huang, H., Liu, J., Zhu, Q., Wang, R., Hu, G.: A new hierarchical method for interpatient heartbeat classification using random projections and RR intervals. Biomed. Eng. Online **13**(1), 90 (2014). https://doi.org/10.1186/1475-925X-13-90
14. Lin, C.C., Yang, C.M.: Heartbeat classification using normalized RR intervals and wavelet features. In: International Symposium on Computer, vol. 2014, pp. 650–653 (2014)

15. Krizhevsky, A., Sutskever, I., Hinton, G.: ImageNet classification with deep convolutional neural networks. Adv. Neural. Inf. Process. Syst. **25**, 84–90 (2012)
16. Xiong, Z., Stiles, M.K., Zhao, J.: Robust ECG signal classification for detection of atrial fibrillation using a novel neural network. In: Proceedings of the 2017 Computing in Cardiology (CinC), Rennes, France, 24–27 September 2017, pp. 1–4 (2017)
17. Fujita, H., Cimr, D.: Computer aided detection for fibrillations and flutters using deep convolutional neural network. Inf. Sci. **486**, 231–239 (2019)
18. Zheng, Z., Chen, Z., Hu, F., et al.: An automatic diagnosis of arrhythmias using a combination of CNN and LSTM technology. Electronics **9**(1), 121 (2020)
19. Schuster, M., Paliwal, K.K.: Bidirectional recurrent neural networks. IEEE Trans. Signal Process. **45**(11), 2673–2681 (1997)
20. Mark, R., Moody, G.: MIT-BIH arrhythmia database directory. Massachusetts Institute of Technology, Cambridge (1988)
21. Schmidhuber, J.: Deep learning in neural networks: an overview. Neural Netw. **61**, 85117 (2015)
22. Ma, F., Zhang, J., et al.: An automatic system for atrial fibrillation by using a CNN-LSTM model. Discret. Dyn. Nat. Soc. (2020)
23. Qian, Y., Bi, M., Tan, T., et al.: Very deep convolutional neural networks for noise robust speech recognition. IEEE-ACM Trans. Audio Speech Lang. **24**, 2263–2276 (2016)
24. Dmitrievich, L.A.: Deep learning in information analysis of electrocardiogram signals for disease diagnostics. The Ministry of Education and Science of The Russian Federation Moscow Institute of Physics and Technology (2015)
25. Ma, F., Zhang, J., Liang, W., et al.: Automated classification of atrial fibrillation using artificial neural network for wearable devices. Math. Prob. Eng. **2020**, Article ID 9159158, 6 p. (2020)
26. Powers, D.M.W.: Evaluation: from precision, recall and F-factor to ROC, informedness, markedness correlation. J. Mach. Learn. Technol. **2**, 3763 (2011)
27. Lin, C.C., Yang, C.M.: Heartbeat classification using normalized RR intervals and wavelet features. In: Proceedings of the 2014 International Symposium on Computer, Consumer and Control, Taichung, Taiwan, June 2014

# Cross-Domain Text Classification Based on BERT Model

Kuan Zhang, Xinhong Hei⑩, Rong Fei$^{(\boxtimes)}$ ⑩, Yufan Guo, and Rui Jiao

Xi'an University of Technology, Xi'an, China
annyfei@xaut.edu.cn

**Abstract.** Diversity of structure and classification are difficulties for information security data. With the popularization of big data technology, cross-domain text classification becomes increasingly important for the information security domain. In this paper, we propose a new text classification structure based on the BERT model. Firstly, the BERT model is used to generate the text sentence vector, and then we construct the similarity matrix by calculating the cosine similarity. Finally, the k-means and mean-shift clustering are used to extract the data feature structure. Through this structure, clustering operations are performed on the benchmark data set and the actual problems. The text information can be classified, and the effective clustering results can be obtained. At the same time, clustering evaluation indicators are used to verify the performance of the model on these datasets. Experimental results demonstrate the effectiveness of the proposed structure in the two indexes Silhouette coefficient and Calinski-Harabaz.

**Keywords:** Text classification · BERT model · K-means

## 1 Introduction

Information security is related to the survival and core interests of individuals, enterprises and even a country. The issue of information security is without worry, and national security and social stability are also guaranteed. An important aspect of information security technology is the efficient processing and classification of the existing massive information. After sorting out the data, it is convenient for managers to search and check regularly. The traditional text classification method, which usually takes words as the basic unit of text, is not only easy to cause the lack of semantic information, but also easy to lead to the high dimension and sparsity of text features. At present, the application of text classification technology is mostly machine learning. This method usually extracts TF-IDF (Term Frequency – Inverse Document Frequency, word frequency, inverse document frequency) or word bag features, and then trains LR (Logistic Regression) model. There are many models, such as Bayes, SVM (Support Vector Machine), etc. However, the generalization ability of text classifiers based on traditional methods tends to decrease when processing text data with diverse features. In recent years, deep learning technology has developed rapidly and has been applied in various fields with remarkable results. Deep learning has been widely used with its unique network structure, which

© Springer Nature Switzerland AG 2021
C. S. Jensen et al. (Eds.): DASFAA 2021 Workshops, LNCS 12680, pp. 197–208, 2021.
https://doi.org/10.1007/978-3-030-73216-5_14

can improve the effect of text preprocessing and solve the current problems of text classification.

Text classification [1] is a basic task in natural language processing and the most important step to solve the problem of text information overload, which can sort out and classify the existing massive text resources. The processing of text classification can be divided into text preprocessing, text feature extraction, classification model construction and so on. Classic text classification algorithms such as NB (Naïve Bayes), KNN (K-Nearest Neighbor), DTree (Decision Tree), AAC (Arithmetical Average Centroid), SVM (Support Vector Machine), etc. Text classification technology has an early origin and has experienced a development process from the expert system to machine learning and then to deep learning. The development of deep learning improves the high dimension and sparsity of traditional machine learning in text classification, which leads to more effective text representation methods. After Bengioetal (2003) proposed the forward Neural Network (feed-forward neural network, FNN) language model, its lexical vector measures the semantic correlation between words. In 2013, Mikolov proposed the Word2vec framework, which mainly uses the deep learning method to map words to low-dimensional real vector space, and captures semantic information of words by calculating the distance between word vectors. The deep learning-based text classification method uses lexical vectors to express the semantic meaning of words, and then obtains the semantic representation of text through semantic combination. The methods of a semantic combination of neural network mainly include convolutional neural network, cyclic neural network and attention mechanism, etc. These methods rise from semantic representation of words to semantic representation of text through different combination methods.

The BERT [2] (Bidirectional Encoder Representations from Transformers) model is the language presentation model released by Google in October 2018, and the replacement of the Word2Vec model has become a major breakthrough in NLP (Natural Language Processing) technology. BERT model in the top level machine reading comprehension test SQuAD1.1 showed remarkable achievements: two indexes are better than others, and it also makes the best grades in 11 different NLP tests, including pushing GLUE benchmark to 80.4% absolute improvement (7.6%), MultiNLI accuracy reached 86.7% (absolute improvement rate 5.6%). BERT model is actually a language encoder, which can transform the input sentence or paragraph into feature vectors. Word embedding is a way to map words to Numbers, but a simple real number contains too little information, generally we map to a numerical vector. In the process of natural language processing, it is necessary to retain some abstract features of the language itself, such as semantics and syntax. Throughout the development history of NLP, many revolutionary achievements are the development achievements of word embedding, such as Word2Vec, ELMo and BERT, which have well preserved the features of natural language in the transformation process.

This paper uses the BERT model in the process of generating word vectors, and then calculates the cosine similarity to generate the similarity matrix. In order to solve the problem of massive information management and improve the efficiency of managers, we have designed a set of research plans. First, we extract the information in the dataset and combine the BERT model to complete the construction of the text word vector, and then create the text similarity matrix. On the premise of ensuring accuracy, reduce the time loss in the community discovery process, and then perform clustering operations through k-means and mean-shift, and finally achieve the purpose of classifying papers.

## 2  Related Work

### 2.1  BERT Model

BERT [3] is a language representation model based on deep learning. The emergence of BERT technology has changed the relationship between pre-trained word vectors and downstream specific tasks. The word vector model is an NLP tool that transforms abstract text formats into vectors that can be used to perform mathematical computations on which NLP's task is to operate. In fact, NLP technology can be divided into two main aspects: training text data to generate word vectors and operating these word vectors in the downstream model. At present, the technology to generate word vectors mainly includes word2VEc, ELMo, BERT and other models. The core module of BERT model is transformer, and the core of transformer is the attention mechanism. The attention mechanism draws lessons from human visual attention, which enables the neural network to focus on a part of the input, that is, it can distinguish the influence of different parts of the input on the output. BERT's network architecture uses a multi-tier Transformer structure, which abandons the traditional RNN and CNN. The transformer is an Encoder-Decoder structure, which consists of several encoders and decoders stacked.

### BERT Model Input

As shown in Fig. 1 below, BERT's input adds "CLS" at the beginning of the first sentence as the beginning of the text, and "SEP" as the end.

Token Embeddings: To represent the word vector, each word is converted into a vector by establishing a word-to-word scale, which is used as the input representation of the model.

Segment Embeddings: This part is used to distinguish two sentences, because the pre-training needs to do the task of classifying two sentences.

Position Embeddings: This part is obtained through model training.

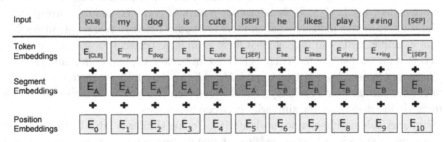

**Fig. 1.** BERT model input.

**BERT Model Pre-training Tasks**

I. *Masked Language Model (MLM)*

In order to train in-depth both transformer representation, BERT [4] model adopts a simple approach: That is, on the model input text, the input words are randomly screened, and then the rest of the words are used to predict which part of the input is screened. For the masked text, a special symbol [MASK] is used 80% of the cases, a random word is used in 10% of the cases, and the original word is left unchanged in 10% of the cases. After this operation, the model does not know whether the words at the corresponding position are correct or not when predicting a word, which makes the model have to rely more on the context information to predict the word, and at the same time, the model has a certain ability to correct errors.

II. *Next Sentence Prediction*

The work of the Next Sentence Prediction is as follows: In any two sentences of a given text, determine whether the second sentence follows the first sentence in the original dataset. In other words, when inputting sentences A and B, B is 50% likely to be the next sentence of A, and 50% may be from anywhere in the text. Consider only two sentences and decide if they are the preceding and following sentences in an article. In the actual pre-training process, 50% correct sentence pairs and 50% wrong sentence pairs are randomly selected from the text corpus for training. Combined with the Masked LM task, the model can more accurately describe the semantic information of the sentence and even the text level.

The BERT [5] model performs joint training on the Masked LM task and the Next Sentence Prediction task, so that the vector representation of each word output by the model can describe the overall semantic information of the input text (single sentence or sentence pair) as comprehensively and accurately as possible. Provide better initial values of model parameters for subsequent fine-tuning tasks.

## 2.2 Matrix Preprocessing

Given graph $H = (K, L)$, $K = \{k_1, k_2, \ldots, k_n\}$. $K$ represents the node set in the graph, and $L$ represents the edge set in the graph. $N(u)$ is the set of neighbor nodes of node $u$. Suppose matrix $A = [a_{ij}]_{n \times n}$ is the adjacency matrix of graph $H$, and the corresponding elements of the matrix indicate whether there is an edge between two points in graph $H$.

For example, $a_{ij}$ is 1 to indicate that there is an edge between $k_i$ and $k_j$ in graph $H$. If $a_{ij}$ is 0, it means two points. There is no edge in between $k_i$ and $k_j$ in graph $H$.

Similarity matrix: For a network graph $H = (K, L)$, the similarity matrix $S = [s_{ij}]_{n \times n}$ is calculated through the similarity of nodes between two points in the graph $H$. The elements in the matrix are the similarity between two nodes.

Similarity calculation-cosine similarity [6]: Cosine similarity, also known as cosine distance, uses the cosine of the angle between two vectors in a vector space as a measure of the difference between two individuals. The closer the cosine value is to 1, the closer the angle is to $0°$, that is, the more similar the two vectors are. This is called "cosine similarity" [7].

We denote the angle between two vectors, vector a and vector $b$ as $\theta$, then the law of cosine is used to get similarity.

$$\cos\theta = \frac{A \cdot B}{\|A\|\|B\|} = \frac{\sum_{i=1}^{n} A_i \times B_i}{\sqrt{\sum_{i=1}^{n} (A_i)^2} \times \sqrt{\sum_{i=1}^{n} (B_i)^2}}. \tag{1}$$

## 2.3 K-means Clustering

K-means [8] clustering is an unsupervised learning algorithm. Assuming that the existing training samples are $\{x^{(1)}, \cdots, x^{(m)}\}$, and $x^{(i)} \in R^n$, The main steps of the K-means algorithm are as follows.

Step 1.  Randomly select k points as cluster centers, denoted as: $u_1, u_2, \cdots u_k \in R^n$;
Step 2.  Traverse all the data and divide each data into the nearest center point, thus forming k clusters;
Step 3.  Calculate the average value of each cluster as the new center point;
Step 4.  Repeat Step 2 and Step 3 until the position of the cluster center no longer changes or the iteration reaches a certain number of times.

In order to solve the inaccuracy problem of clustering of a small number of samples, the K-means algorithm adopts an optimized iterative operation, and iteratively corrects and prunes the clusters that have been obtained to determine the clustering of some samples, which optimizes the places where the initial supervised learning sample classification is unreasonable. At the same time, it can reduce the total clustering time complexity for some small samples.

## 2.4 Mean-Shift Clustering

The core of mean-shift [9] algorithm can be understoodcd by name, mean (mean), shift (offset), in short, there is a point, there are many points around it, we calculate the point and move to each point. The sum of the required offsets is averaged, and the average offset is obtained. (The direction of the offset is the direction where the surrounding points are densely distributed) The offset includes the size and direction. Then the point moves in the direction of the average offset, and then uses this as a new starting point to iterate continuously until a certain condition is met.

The algorithm flow of mean-shift [10] is shown as follows.

Step 1.  Select the center point $x$ and make a high-dimensional sphere with radius h (if we are in image or video processing, it is a 2-dimensional window, not limited to a sphere, it can be a rectangle), and mark all points that fall into the window as $x_i$.
Step 2.  Calculate the mean-shift vector. If the value is less than the threshold or the number of iterations reaches a certain threshold, stop the algorithm, otherwise update the dot and continue to Step 1.

## 3 Model Specification

The specific algorithm design of this paper is as follows.

Step 1.  Use BERT pre-training model to build BERT server and client on the device.
Step 2.  Randomly select from the two original datasets to get two new datasets C and E.
Step 3.  Input C and E into the BERT model to generate the corresponding vector file.
Step 4.  Calculate the generated vector to get the cosine similarity, and at the same time, create the similarity matrix M.
Step 5.  Clustering by k-means and mean-shift. The input parameters are the similarity matrix and the number of communities, and the clustering results of the corresponding communities are obtained, namely RES.
Step 6.  silhouetteScore(M, RES) → $IND_{silhouetteScore}$: Obtain the value of silhouette coefficient.
Step 7.  calinski_harabaz_index(M, RES) → $IND_{CalinskHarabaz}$: Obtain the value of Calinski-Harabaz Index.

| Algorithm: |
| --- |
| **Input:** information dataset C and E |
| **Output:** clustering result set RES1 and RES2 |
| Evaluation index result: $IND_{silhouetteScore}$ and $IND_{CalinskiHarabazIndex}$ |
| 1    BERT(C, E) → $\{C_v, E_v\}$ |
| 2    Create$_{similarity\,matrix(C_v, E_v)}$ → $\{M_C, M_E\}$ |
| 3    **For** i: |
| 4        Mean − shift$(M_C)$ → $res1_i$ and K − means$(M_E)$ → $res2_i$ |
| 5        RES1.append($res1_i$) and RES2.append($res2_i$) |
| 6    **End for** |
| 7    silhouetteScore(M, RES) → $IND_{silhouetteScore}$ |
| 8    calinski_harabaz_index(M, RES) → $IND_{CalinskHarabaz}$ |

## 4 Experimental Analysis

### 4.1 Experimental Design

I.  Experimental environment
    Processor: Inte(R) Core(TM) i5-8300H CPU@2.30GHz; RAM: 8GB; Operating system: windows server 2012 R2 standard:IDE: pycharm.
II.  DataSet
    This experiment uses two datasets (toutiao_cat_data and CORD-19-research-challenge) for testing. In this experiment, for the above two datasets, 1,024 and 2,048 data were randomly selected to obtain the experimental results.

The URLs of datasets are listed as follows.
    The URL of toutiao_cat_data is addressed as https://github.com/aceimnorstuvwxz/toutiao-multilevel-text-classfication-dataset.
    The URL of CORD-19-research-challenge is addressed as https://www.kaggle.com/allen-institute-for-ai/CORD-19-research-challenge.

## 4.2  Evaluation Indices

Silhouette coefficient [11]: Silhouette coefficient is an index used to measure the effectiveness of clustering. It can describe the sharpness of each category after clustering. The profile coefficient contains two factors: cohesion and separation.

Cohesion: reflects the closeness of a sample point to the elements in the class.

Separation: reflects the closeness of a sample point to elements outside the class.

The formula of the silhouette coefficient is as follows.

$$S(i) = \frac{b(i) - a(i)}{max\{a(i), b(i)\}}. \tag{2}$$

Among them, $a(i)$ represents the cohesion degree of the sample points, and the calculation formula is as follows.

$$a(i) = \frac{1}{n-1} \sum \text{distance}(i, j). \tag{3}$$

Where $j$ represents other sample points in the same class as sample $i$, and distance represents the distance between $i$ and $j$. So the smaller $a(i)$, the tighter the class. The calculation method of $b(i)$ is similar to that of $a(i)$. The value range of the silhouette coefficient $S$ is $[-1, 1]$. The larger the silhouette coefficient, the better the clustering effect.

Calinski-Harabaz(CH) Index [12]: The clustering model is an unsupervised learning model. Generally speaking, the clustering result is that the closer the data distance between the same categories, the better, and the farther the data distance between different categories, the better. The CH index measures the tightness within a class by calculating the sum of the squares of the distances between each point in the class and the center of the class, and measures the separation of the dataset by calculating the sum of the squares of the distances between various center points and the center of the dataset. The CH index is obtained from the ratio of separation and compactness. Therefore, the larger the CH, the tighter the cluster itself, the more dispersed the clusters, that is, the better clustering results. The CH index is calculated as follows.

$$s(k) = \frac{tr(B_k)}{tr(W_k)} \frac{m-k}{k-1}. \tag{4}$$

Where $m$ is the number of training set samples, $k$ is the number of categories, Bk is the covariance matrix between categories, $W_k$ is the covariance matrix of the data within the category, and tr is the trace of the matrix. It can be seen from the above formula that the smaller the covariance of the data within the category, the better, the larger the covariance between categories, the better, so the Calinski-Harabaz score will be higher.

## 4.3  Analysis of Experimental Results

Perform clustering operations on the two datasets respectively, and combine the BERT model to complete the construction of the text word vector matrix to obtain the similarity matrix. Under the premise of ensuring accuracy, reduce the time loss in the community discovery process, and finally achieve the purpose of paper classification. The analysis results are discussed as follows.

**CORD-19-research-challenge Dataset**

On this dataset, we selected 2,048 thesis topics, input the BERT model to generate word vectors for the topics, calculate the cosine similarity between vectors, and generate the similarity matrix that is 2,048*2,048. Then the mean-shift clustering is used for classification. Finally, use Silhouette coefficient and Calinski-Harabaz Index evaluate the performance of the process. The specific results are as follows (Fig. 2):

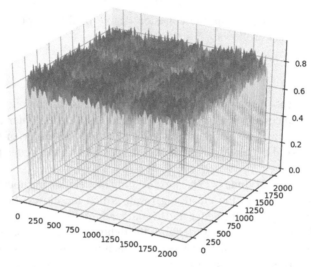

**Fig. 2.** Three-dimensional similarity matrix of the CORD-19-research-challenge dataset.

The conclusion is obtained by using Mean-Shit clustering: Fig. 3 uses the contour coefficient to evaluate the clustering effect. It can be seen from the figure that when the number of communities decreases, the clustering effect is obviously better, and the effect of categorizing papers is also the best at this time; Fig. 4 uses the Calinski-Harabaz index to analyze the data. When the community increases, the CH indicator gradually becomes smaller. According to the nature of the evaluation index, it can be known that the steeper the curve, the better the classification effect. From the general trend of the curve in the figure, it can be estimated that then the number of classification communities is about 2, the classification effect is the best and the research significance is the greatest.

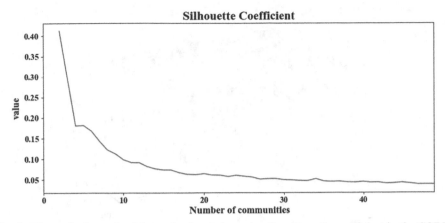

**Fig. 3.** The analysis result of the evaluation index using the silhouette coefficient in the CORD-19-research-challenge dataset.

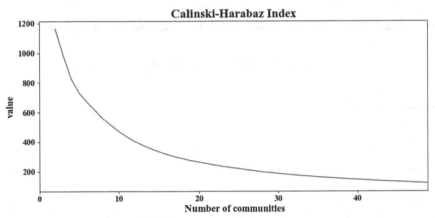

**Fig. 4.** The analysis results of the evaluation indicators using the Calinski-Harabaz Index in the CORD-19-research-challenge dataset.

**Toutiao_cat_data Dataset**

The original dataset contains 382,688 pieces of information, which are distributed in 15 categories. We selected 1,024 pieces of data for experimentation. The operation is the same as the previous dataset, except that when clustering the data, we use K-means instead of mean-shift. Similarly, at the end of the experiment, Silhouette coefficient and Calinski-Harabaz Index are used to evaluate the classification effect of the process. The specific results and analysis are as follows (Fig. 5).

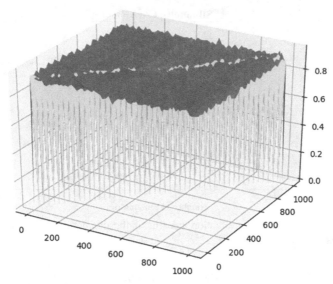

**Fig. 5.** Three-dimensional similarity matrix of the toutiao_cat_data dataset.

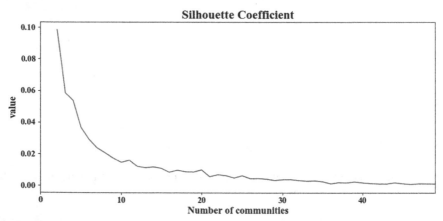

**Fig. 6.** The analysis result of the evaluation index using the silhouette coefficient in the toutiao_cat_data dataset.

The conclusion is obtained by using K-means clustering: Fig. 6 also uses the contour coefficient to evaluate the clustering effect. From the curve in the figure, it can be known that the smaller the number of communities, the better the clustering effect. At this time, the data is also The effect of centralized information classification is the best; Fig. 7 uses the Calinski-Harabaz index to evaluate the clustering results. When the number of communities increases, the CH indicator gradually becomes smaller, and the steeper the curve, the better the classification effect. It can be seen from the figure that when the number of classification communities is about 3, the classification effect is the best.

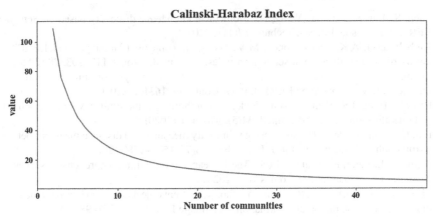

**Fig. 7.** The analysis results of the evaluation indicators using the Calinski-Harabaz Index in the toutiao_cat_data dataset.

## 5 Conclusions

This paper uses a relatively new word vector model to process the text data and achieves a more optimistic classification effect through a series of operations. This method improves the accuracy of classic clustering algorithms, such as K-means, mean-shift, and so on. Firstly, the BERT model is used to extracts the title of the paper in the CORD-19-research-challenge dataset and the information in the dataset--toutiao_cat_data to complete the generation of the word vector, and at the same time, construct the cosine similarity matrix. Then on the premise of ensuring accuracy, the time loss in the community detection process is reduced. Finally, K-means and mean-shift are used for clustering on the two datasets, and finally, the purpose of text classification is achieved. Through the analysis of the clustering effect of the dataset, for the CORD-19-research-challenge dataset, the experimental results show that when the number of communities is 2, the clustering effect is the best. For the dataset-- toutiao_cat_data, it can be concluded that when the number of communities is 3, the clustering effect is the best. we can achieve the experimental purpose well, which provides an operational basis for information security technology to a certain extent, and need to improve the clustering operation, so as to achieve the best classification effect of the results.

**Acknowledgments.** This research was partially supported by National Key Research and Development Program of China (No. 2018YFB1201500) and Natural Science Foundation of China (No. 61773313).

## References

1. Zhang, Q., Havard, M., Ford, E.C.: Categorizing incident learning reports by narrative text clustering to improve safety. Int. J. Radiat. Oncol. Biol. Phys. **108**(3), S55 (2020)
2. Amer, A.A., Abdalla, H.I.: A set theory based similarity measure for text clustering and classification. J. Big Data **7**(1), 1–43 (2020). https://doi.org/10.1186/s40537-020-00344-3

3. Shen, S., Liu, X., Sun, H., Wang, D.: Biomedical knowledge discovery based on Sentence-BERT. Proc. Assoc. Inf. Sci. Technol. **57**(1) (2020)
4. Ambalavanan, A.K., Devarakonda, M.V.: Using the contextual language model BERT for multi-criteria classification of scientific articles. J. Biomed. Inform. **112**, 103578 (2020)
5. Zheng, J., Wang, J., Ren, Y., Yang, Z.: Chinese sentiment analysis of online education and Internet buzzwords based on BERT. J. Phys. Conf. Ser. **1631**(1) (2020)
6. Park, K., Hong, J.S., Kim, W.: A methodology combining cosine similarity with classifier for text classification. Appl. Artif. Intell. **34**(5), 396–411 (2020)
7. Liu, D., Chen, X., Peng, D.: Some cosine similarity measures and distance measures between q -rung orthopair fuzzy sets. Int. J. Intell. Syst. **34**(7), 1572–1587 (2019)
8. Ahmed, M., Seraj, R., Islam, S.M.S.: The k-means algorithm: a comprehensive survey and performance evaluation. Electronics **9**(8) (2020)
9. Li, J., Chen, H., Li, G., He, B., Zhang, Y., Tao, X.: Salient object detection based on meanshift filtering and fusion of colour information. IET Image Proc. **9**(11), 977–985 (2015)
10. Du, Y., Sun, B., Lu, R., Zhang, C., Wu, H.: A method for detecting high-frequency oscillations using semi-supervised k-means and mean shift clustering. Neurocomputing **350**, 102–107 (2019)
11. Nidheesh, N., Abdul Nazeer, K.A., Ameer, P.M.: A hierarchical clustering algorithm based on silhouette index for cancer subtype discovery from genomic data. Neural Comput. Appl. **32**(15), 11459–11476 (2019). https://doi.org/10.1007/s00521-019-04636-5
12. Putri, D., Leu, J.-S., Seda, P.: Design of an unsupervised machine learning-based movie recommender system. Symmetry **12**(2), 185 (2020). https://doi.org/10.3390/sym12020185

# Surface Defect Detection Method of Hot Rolling Strip Based on Improved SSD Model

Xiaoyue Liu and Jie Gao[✉]

North China University of Science and Technology, Tangshan 063210, Hebei, China

**Abstract.** In order to reduce the influence of surface defects on the performance and appearance of hot-rolled steel strip, a surface defect detection method combining attention mechanism and multi-feature fusion network was proposed. In this method, the traditional SSD model was used as the basic framework, and the ResNet50 network after knowledge distillation was selected as the feature extraction network. The low-level features and high-level features were fused and complementary to improve the accuracy of detection. In addition, channel attention mechanism was introduced to filter and retain important information, which reduced the network computation and improves the network detection speed. The experimental results showed that the accuracy of RAF-SSD model for surface defect detection of hot rolled steel strip was significantly higher than that of traditional deep learning models, and the detection speed was 12.9% higher than that of SSD model, which can meet the real-time requirements of industrial detection.

**Keywords:** Hot rolled strip · Surface defect · SSD model · ResNet50 · Feature fusion · Channel attention mechanism

## 1 Introduction

Hot rolled steel strip is an important material in industrial production, widely used in aerospace, machinery manufacturing, construction and other fields. Its surface quality is very important to the aesthetics, performance and durability of products, and will directly affect the evaluation of product quality level [1]. However, due to the poor actual production environment and complex technological process, hot rolled steel strip is prone to be affected by rolling equipment, technology, raw materials and external environment in the production process, thus forming various types of defects on the surface [2].

The traditional surface defect detection methods of hot rolled steel strip are divided into frequency-flash detection method and manual detection method, but these two methods are non-automatic detection methods, mainly rely on human eyes for detection. However, in the production process, human eyes are very easy to produce fatigue, and it is impossible to accurately detect the type and grade of defects, resulting in high false detection rate and false detection rate, low detection efficiency and other problems. The more advanced detection methods of hot rolled steel strip include infrared detection method and computer vision detection method. Although the former detection speed is

© Springer Nature Switzerland AG 2021
C. S. Jensen et al. (Eds.): DASFAA 2021 Workshops, LNCS 12680, pp. 209–222, 2021.
https://doi.org/10.1007/978-3-030-73216-5_15

faster and the accuracy is higher, it cannot realize the accurate classification of the surface defects of hot rolled steel strip. The latter has strong defect recognition ability, but the algorithm flow is complex and the robustness is poor [3]. For this reason, the rapid and accurate detection of hot rolled steel strip has become the focus of many domestic and foreign research institutions.

In recent years, with the continuous development and progress of deep learning, a lot of target detection algorithms have emerged. At present, target detection networks based on deep learning can be structurally divided into two categories [4]: one-stage models and two-stage models. The core idea of the two-stage models represented by RCNN, Fast-RCNN and Faster-RCNN is based on the method of candidate region. The candidate box of the target may exist, and then the target detection is further carried out. The one-stage models represented by SSD and YOLO directly predict and classify target locations on the basis of network feature extraction. These algorithms have been improved by researchers and applied to defect detection in various fields, and good results have been achieved. Based on SSD model, Jiang Jun combined void convolution and feature enhancement algorithm to construct AFE-SSD model, which significantly improved the detection accuracy of small targets [5]. Zhu Deli used MobileNet as the feature extractor of SSD model, which improved the feature extraction capability and the robustness of the model [6]. Wu Shoupeng improved the Faster-RCNN model by using the bidirectional feature pyramid to enhance the detection capability of multi-scale defects [7]. Although the above methods effectively improve the detection accuracy, they also increase the number of parameters for the model, which makes it difficult for the detection speed to meet the real-time detection requirements.

From the detection accuracy and speed of the surface defect of hot rolled steel belt, considering two aspects in this paper, on the basis of the SSD model, use after knowledge distilling ResNet50 replaced VGG - 6 as a network of feature extraction, and the characteristics of shallow and deep characteristics of fusion, make the characteristics of shallow semantic information is not enough rich up, then get the characteristics of the image through the channel attention mechanism for information filtering, so as to realize the effective use of important information, and reduce the computational complexity. The results showed that the average accuracy of RAF-SSD model was 71.4%, which was 1.6% and 1.8% higher than SSD model and YOLO-V3 one-stage detection model, respectively, and 1.0% higher than the two-stage detection model Faster RCNN model. The detection speed of RAF-SSD model was 12.9% higher than that of the fastest SSD model among the three models. This indicates that the improved method in this paper can effectively improve the extraction and utilization of feature information of the model, and improve the accuracy and detection speed of the model for surface defects of hot rolled steel strip, which can meet the requirements of the industry for real-time and accurate detection.

## 2   SSD Model and its Improved Method

### 2.1   SSD Model

SSD target detection model [8] is a multi-classification single-order target detection model proposed on the basis of drawing on the Anchor generation method of Faster

RCNN [9] and the mesh division idea of YOLO [10]. This model maintains the detection speed of the one-stage model and is equivalent to the two-stage deep learning target detection model in detection accuracy. It is one of the mainstream deep learning target detection model at present.

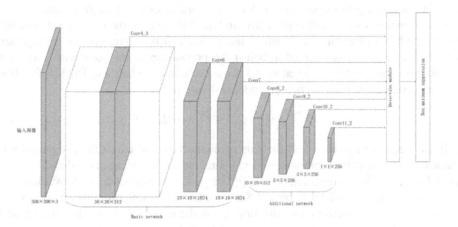

**Fig. 1.** SSD model structure diagram

The structure of SSD model is shown in Fig. 1, which takes VGG-16 as the basic network for feature extraction of input images, and adds an additional 4-level convolutional layer. The input image will generate a series of characteristic maps through the network. The characteristic maps of Conv4_3, Conv7, Conv8_2, Conv9_2, Conv10_2, Conv11_2 are mainly used for the final prediction. Finally, through detection module, classification and regression calculation are carried out on each feature image respectively to obtain the type of detection target and the position of prediction box in the feature image. Finally, the results obtained from the detection module are integrated, and the final detection results are obtained by the method of non-maximum suppression.

## 2.2 Knowledge Distillation Algorithm

Knowledge distillation is an effective model compression method by transferring learning ideas in neural networks and obtaining smaller models that are more suitable for reasoning through trained larger models [11]. The model introduced by knowledge distillation algorithm can be divided into teacher model and student model, among which the teacher model has large number of parameters, strong feature extraction ability and high accuracy. The number of student model parameters is relatively small, but the accuracy of individual training is insufficient, so it is difficult to meet the actual needs. The training method of knowledge distillation is to guide the training process of students' network through teacher network, so as to extract the soft knowledge of large teacher model into small student model. The knowledge distillation algorithm uses the teacher model's soft output to punish the student model because the soft output can provide more information from the native network. Therefore, the soft output of the teacher network is used to train and compress the student model in the form of a soft label. Different from the traditional hard label, which only uses "0" and "1" to label the data, the soft label uses the data between 0 and 1 to label the picture, which can well represent the distance between classes.

The principle of knowledge distillation algorithm is shown in Fig. 2. In the knowledge distillation training of the student model, a network model with more parameters and higher accuracy should be trained firstly, and the soft label of the training set obtained by the model and the real label should be used as the distillation training object together, and the parameter α is selected to adjust the proportion of the parameter. After the training is completed, the student model is used to make predictions. The loss function of knowledge distillation of student model is shown in Eq. (1):

$$L_{KD} = \alpha T^2 D_{KL}(Q_s^\tau, Q_t^\tau) + (1 - \alpha)L_{ce}(Q_s, y_{ture}) \tag{1}$$

$\alpha$ is the weight of two parts loss function parameter, used to adjust the two partial loss values of weights of back-propagation gradient; $T$ is the temperature coefficient; $D_{KL}$ is the KL divergence loss function; $Q_t^\tau$ and $Q_s^\tau$ corresponds to the model of the teachers and students respectively model improved Softmax functions softening the output results; $L_{ce}$ stands for cross entropy loss function; $Q_s$ is the output of the student model after passing the Softmax function; $y_{true}$ is the true tag value.

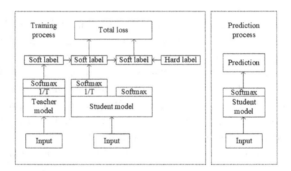

**Fig. 2.** Schematic diagram of knowledge Distillation Algorithm

Equation (1) contains two components. One is the KL divergence loss function of the student model and the teacher model, which is used to calculate the distance between the output of the student model and the teacher model. The other part is the cross-entropy loss value between the student model and the real label, and the weighted sum of these two parts is used as the loss function of distillation training.

In the training process of knowledge distillation, a temperature coefficient T can be introduced into the Softmax function to obtain a softer label. The improved Softmax function is shown in Eq. (2).

$$S_j = \frac{e^{z_j/T}}{\sum_j e^{z_j/T}} \tag{2}$$

Where, $z_j$ is the output value of the jth neuron; $S_j$ is the Softmax output value of $z_j$. The size of T can change the distribution of the output results of the neural network. The lager the value of T, the smoother the probability distribution. Therefore, in the training of knowledge distillation, the timely adjustment of T value can control the softening degree of the output results of teacher model and student model in the loss function.

## 2.3  Residual Network

Although the deepening of the network can improve the ability of feature extraction, the simple stacking of network layers will lead to gradient explosion, gradient disappearance and network degradation, which is reflected in the decline of training accuracy and test accuracy.

He proposed Residual Network (ResNet) to solve the problem of network degradation [12]. The principle of each residual learning module in the residual network is shown in Fig. 3.

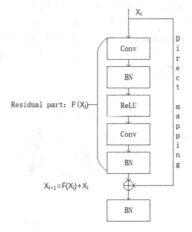

**Fig. 3.** Flow chart of attention mechanism algorithm

$X_1$ represents the residual block input, and the output of the real for $F(X_1)$, expect to get the output of $X_{1+1}$, each module will also enter superimposed on the output in the form of direct mapping, but in the need to use the output of the ReLU function activated. After superposition, the output into $F(X_1)+X_1$, network learning content into $F(X_1) = X_{1+1} - X_1$ residual form. If the network layer number more than the optimal number of residual network will map the extra layer of training for $F(X) = 0$, namely the equal identity map layer to the input and output, so it can avoid the phenomenon of network degradation. Moreover, the residual network simplifies the learning objective and makes the network training converge more quickly.

The commonly used ResNet models are ResNet50, ResNet101, ResNet152, etc. Using knowledge distilling algorithm mentioned in the previous section, ResNet50 as teacher model, ResNet101 as student model, to students in the model and the number of smaller at the same time, through the supervision and training process, the teacher model transfer from "knowledge" ResNet101 network to ResNet50 network, improve the accuracy of the ResNet50, and get the KD-ResNet50 distillation after training the student model.

## 2.4  Feature Fusion

SSD model is based on forward propagation convolutional neural network, which is hierarchical and can extract feature graphs of different scales and different semantic information. In general, the feature map extracted by shallow layer network has a high resolution, but the receptive field corresponding to each feature point is small, and the semantic information is poor, so it is suitable for predicting small objects. After multi-layer convolutional pooling operation, feature images extracted from deep network have low resolution, large receptive field corresponding to each feature point and rich semantic information, which are suitable for prediction of large objects [13].

Most of the surface defects of hot rolled steel strip are small area defects, and the proportion of the defects in pixels is low, which belongs to small target detection. Small object detection requires that the features extracted by the network have high resolution and rich semantic information, while SSD mdoel uses multi-scale feature map for target detection, which leads to its unsatisfactory effect in small object detection. In view of this situation, the fusion of shallow and deep feature images is considered to improve the detection accuracy of small targets.

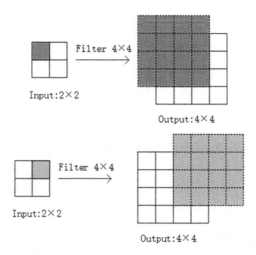

**Fig. 4.**  Deconvolution principle diagram

However, the resolution of the feature images extracted by different convolution layers may be different, so it is necessary to expand and shrink the convolution feature images with different resolutions first, and then carry out feature fusion after unified resolution. Upsampling is a method of converting low-resolution images to high-resolution images. In the convolutional neural network, the commonly used upsampling method is transpose convolution, also known as deconvolution, but it only realizes the adjustment of size, rather than the reduction of value in the mathematical sense. Deconvolution principle as shown in Fig. 4, assume that the input dimension of $2 \times 2$, in the process of the deconvolution using convolution kernels of size is 4 * 4, set the padding value is 1, as convolution kernels to 2 for pace in the input image, get four output window, the size of 4 * 4 different output window overlapping part of overlay, then remove the outermost layer of padding, the resulting output image size is 4 * 4, amplification for the input size 2 times.

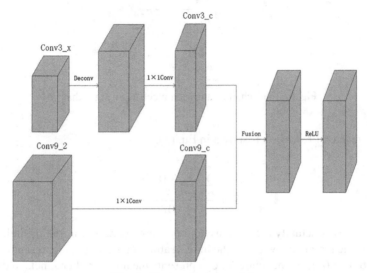

**Fig. 5.** Feature fusion module

This paper refers to the deconvolution module of DSSD model [14], whose structure is shown in Fig. 5. First, the up-sampling of high-level and low-resolution images is carried out, and then the channel number is unified through $1 \times 1$ convolution layer. After that, the adjusted deep and shallow feature maps are fused, and finally the ReLU function is used for activation.

### 2.5 Attention Mechanism

The working principle of the attention mechanism [15] is to establish a new layer of weight, and after learning and training, make the network learn more important areas in each training image, and strengthen the weight of these areas, thus forming the so-called attention. This paper uses the channel attention mechanism module including three parts, namely extrusion, incentive and attention. The algorithm flow after adding attention mechanism is shown in Fig. 6.

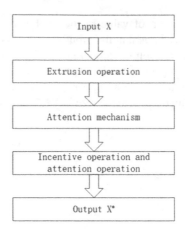

**Fig. 6.** Flow chart of attention mechanism algorithm

The principle of extrusion is shown in Eq. (3).

$$S = \frac{\sum_{i=1}^{H} \sum_{j=1}^{W} X(i,j)}{H * W} \tag{3}$$

Equation (3) is actually a global average pooling operation. In the formula, $H$ and $W$ represent the length and width of the input feature image $X$; $(i,j)$ represent the points at the position of $(i,j)$ on the figure $X$; $C$ represents the number of channels in the figure $X$; all the eigenvalues in each channel are averaged and summed up to obtain a one-dimensional array $S$ of length $C$. After the output $S$ is obtained, the correlation between channels is modeled, that is, the process of excitation. The principle is shown in Eq. (4).

$$R = \mathrm{Sigmoid}(W_2 \cdot \mathrm{ReLU}(W_1 S)) \tag{4}$$

The dimension of $W_1$ is $C_1 \times C_1$, and the dimension of $W_2$ is $C_2 \times C_2$, of which $C_2 = C_1/4$, are these two weights trained by ReLU function and Sigmoid function, and a one-dimensional excitation weight is obtained to activate each channel, and the obtained $R$ is dimension $C_1 \times 1 \times 1$. Finally, the attention calculation is carried out.

$$X^* = X \cdot R \tag{5}$$

The feature graph $X^*$ obtained by the attention module is used to replace the original input $X$, and then sent into the improved algorithm model for detection. In other words, this process is actually a process of scaling and scaling, in which different weights are multiplied by different channel values, so as to enhance the attention to important channels.

# 3   RAF-SSD Model

## 3.1   RAF-SSD Model Structure

The improved RAF-SSD model structure is shown in Fig. 7.

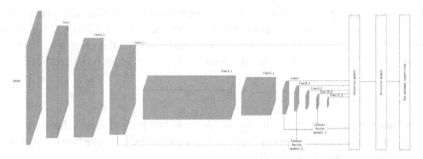

**Fig. 7.** RAF-SSD model structure diagram

Firstly, KD-ResNet50 is used as the backbone network of RAF-SSD. After knowledge distillation, KD-ResNet50 still has the same structure as ResNet50, which is mainly composed of five residual learning modules. The convolution layer in the middle of each residual block first reduces the calculation amount through a convolution layer, then another convolution layer is used for reduction, which not only ensures the accuracy, but also reduces the amount of calculation. In addition, the structure of residual blocks can deepen the network depth while avoiding the gradient explosion and gradient disappearance and other problems caused by VGG16 as the main network.

In view of the poor performance of traditional SSD model in small target detection, two feature fusion modules were added to the model, which fused the features of Conv3_x and Conv8_2, Conv7 and Conv10_2 respectively. The obtained feature map has both strong resolution and rich semantic information, which is more suitable for small target detection task.

But the added feature fusion module adds new parameters to the model and increases the computation. Therefore, in order to guarantee the detection speed of the model, the feature images extracted from the backbone network and feature fusion module are sent to the channel attention module to screen and retain important information and reduce the computing space occupied by unnecessary feature information. Finally, the feature map processed by the channel attention mechanism is applied to the detection task, and the detection results are obtained by the way of maximum suppression.

## 3.2   Priority Box Setting and Matching

The prior box is set from two main aspects: scale and length-width ratio. The scale follows the principle of linear increase, and its size changes are shown in Eq. (6).

$$S_p = S_{\min} + \frac{S_{\max} - S_{\min}}{m - 1}(p - 1), p \in [1, m] \tag{6}$$

Where, $S_p$ is the ratio of the prior box to the image size; $S_{max}$ and $S_{min}$ are the maximum and minimum values of the ratio respectively; p is the current feature graph; m is the number of feature graphs. During training, set $S_{min}$ to 0.3 and $S_{max}$ to 1.0. In the first feature graph, the scale scale of the prior box is set as $S_{min}/2$, and the subsequent feature graph is linearly increased according to the principle in Eq. (6). Prior to training frame aspect ratio is set to $\{1, 2, 1/2, 1/3\}$.

The matching principle is: first of all, each target box in the picture selects a priori box with the largest intersection ratio as the matching object, and this priori box is identified as the priori box of positive sample.IOU is the intersection of the prior box. By setting the value of IOU, the match between the prior box and the target box is realized.

### 3.3  Loss Function

The loss function of the model is mainly divided into two parts: location loss and regression loss, and the detection effect of the network is evaluated by these two parts together. The overall loss is shown in Eq. (7):

$$L(x, c, w, h) = \frac{1}{N}(\mathrm{L}_{conf}(x, c)) + \beta L_{loc}(x, w, h) \tag{7}$$

In the first part of the formula, $\frac{1}{N}(L_{conf}(x, c))$ represents the position loss, where $N$ represents the number of matching boxes with the complete target box, and c represents confidence. The second part $\beta L_{loc}(x, w, h)$ represents return loss, where x is the location of the center of the target frame, w is the width of the box, and h is the height of the box.

## 4  Experiment

### 4.1  Experimental Environment

Windows10 64-bit operating system, Intel(R)Core(TM) i7-10170U CPU and NVIDIA GeForce 1070 graphics card are used in this lab. Use Python language on Tensorflow deep learning framework combined with opencv library.

### 4.2  Experimental Data

The data used in this paper are NEU-DET data set of surface defects of hot-rolled steel strip, an open source data set of a certain university, which includes the following six main surface defects of hot-rolled steel strip:

Rolled-in scale: In the rolling process, the Rolled iron sheet is pressed into the surface of the steel plate, generally in the shape of strip, block or fish scale, and the color is brown or black.

Crazing: it is a serious surface defect. Cracks of different shapes, depths and sizes will form on the surface of the steel strip, which will cause serious damage to the mechanical properties of the steel plate.

Surface Inclusion: it is divided into metal Inclusion, non-metal Inclusion and mixed Inclusion. The surface of the steel strip presents brownish-red, yellow-brown or black embedded structures, and the inclusions are randomly distributed in different shapes.

Pitted surface: Local or patchy rough surface formed on the surface of a steel plate, resulting in pits of varying shades and shapes.

Scratches: mostly produced in the conveying process of rolling steel strip, Scratches under high temperature appear brown or light blue, irregular shape, generally long.

Patches: Approximately circular bright spots that appear as dense Patches. Which are iron sheet pressure, cracks, surface inclusions, pitting, scratches, and surface spots (Fig. 8).

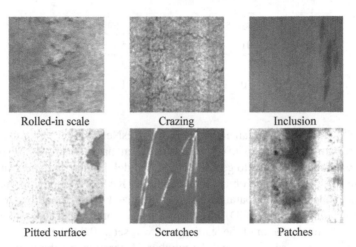

| Rolled-in scale | Crazing | Inclusion |
| Pitted surface | Scratches | Patches |

**Fig. 8.** Sample image of NEU-DET dataset

In order to facilitate training and testing, 0 to 5 are used to represent the above five defect types. Each type of defect in the original data set contains 300 samples and a total of 1800 sample images. In order to prevent the occurrence of overfitting and improve the generalization ability of the model after training, images of the data set were amplified to 13400 by flipping, random cropping, adding noise and other image enhancement methods. However, there were only six surface defect classification labels in the original data set, which was difficult to meet the requirements of deep network training. Therefore, based on the amplified 13400 samples, the defect positions in each image were marked with a rectangular frame to obtain a new data set, NEU-DETX.

### 4.3 Experimental Results

Firstly, the NEU-DET dataset with only classification label is divided into training data set and detection data set in a 4:1 ratio, which is used to verify the effect of knowledge distillation. In this paper, knowledge distillation is carried out with RESNET50 as student model and ReSNet101 as teacher model. Table 1 lists the three kinds of model number, number of floating point arithmetic (Floating point operations, FLOPs) and classification accuracy of the data, it can be seen that the teacher model ResNet101 in NEU-DET data set on the classification accuracy of 98.5%, without distillation training, learning model ResNet50 classification accuracy of 97.2%, after knowledge distilling the training of students model KD-ResNet50 classification accuracy of 98.1%, compared to before this compared with the distillation training model, classification accuracy of 0.9%. It can be seen that under the supervision and training of the teacher model, the student model achieves better classification accuracy, and compared with the teacher network, the number of parameters in the student network is only half that of the teacher network.

**Table 1.** ResNet classification effect comparison

| Model | Layers | Parameters/M | Accuracy/% | FLOPs/G |
|-------|--------|--------------|-----------|---------|
| ResNet101 | 101 | 45.37 | 98.5 | 7.6 |
| ResNet50 | 50 | 24.71 | 97.2 | 3.8 |
| KD-ResNet50 | 50 | 24.71 | 98.1 | 3.8 |

The R-SSD model was obtained by using KD-ResNet 50 after distillation training as the feature extraction network of SSD network. Then, based on the R-SSD model, the feature fusion module is added to get the RF-SSD model. Finally, attention modules are added to the RF-SSD model to obtain the final model RAF-SSD model in this paper.

After the model was built, parameters were initialized firstly. Small batch stochastic gradient descending method (SGD) was selected as the optimizer. The learning rate was set as 0.0001, the decay factor of the learning rate was set as 0.92, the number of training samples per batch was set as 32, and the number of iterations of the training data set was set as 20,000. Then, the NEU-DETX data set was divided into training data set and detection data set in the same 4:1 ratio, and the six models of SSD, R-SSD, RF-SSD, RAF-SSD, Faster RCNN and YOLO-V3 were trained.

According to the experimental results, Frame Per Second (FPS) and Mean Average Precision (mAP) are used as evaluation indexes to compare the performance of six trained models.

**Table 2.** Performance comparison of six models

| | SSD | R-SSD | RF-SSD | RAF-SSD | Faster-RCNN | YOLO-V3 |
|---|-----|-------|--------|---------|-------------|---------|
| rolled-in scale | 68.1 | 68.7 | 70.1 | 70.2 | 68.8 | 68.2 |
| crazing | 68.2 | 68.2 | 68.4 | 68.5 | 69.1 | 67.9 |
| inclusion | 72.4 | 73.4 | 73.9 | 73.7 | 74.3 | 71.5 |
| pitted surface | 68.4 | 68.9 | 70.1 | 70.2 | 69.2 | 68.4 |
| scratches | 67.8 | 67.8 | 69.9 | 70.8 | 68.1 | 68.3 |
| patches | 73.7 | 73.9 | 74.8 | 74.9 | 72.9 | 73.2 |
| mAP/% | 69.8 | 70.2 | 71.2 | 71.4 | 70.4 | 69.6 |
| FPS | 54 | 60 | 58 | 61 | 41 | 52 |

As can be seen from Table 2, the mAP value of R-SSD is 70.0%, which is an increase of 0.4% compared with the traditional SSD model, indicating that the KD-ResNet50 network trained by distillation can better extract image feature information than the VGG16 network. Because the number of parameters of KD-ResNet50 is also smaller than VGG16, the detection speed of R-SSD model is also improved by 11.1% compared with SSD model. Then the MAP value of the RF-SSD model is 1.0% higher than that of the R-SSD model, which indicates that through the addition of feature fusion module, the information of shallow feature mAP and deep feature mAP can be effectively fused, and it is more suitable for small target detection. However, because the feature fusion module adds new parameters to the model, the detection speed of the RF-SSD model is reduced by 3.3% compared with the R-SSD model. And finally to the RAF-SSD model mAP value and detection speed, respectively 71., 4% and 61, compared with the RF-SSD model, mAP value increased by 0.2%, the detection rate of 5.1%, this shows that attention so module characteristic figure of the important contents for higher weight, filtered out the important content, the characteristic information more efficiently, also reduce the computational complexity of the network.

Comparing RAF-SSD model with SSD model and YOLO-V3 one-step model, the mAP value increased by 1.6% and 1.8%, and the detection speed increased by 12.9% and 17.3%. Compared with Faster-RCNN, the RAF-SSD model improves the mAP value by 1.0% and the detection speed by 48.8% (Fig. 9).

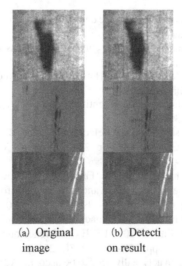

(a) Original    (b) Detecti
image    on result

**Fig. 9.** Experimental detection effect drawing

From the inspection effect diagram, the defect area in the original image is marked out by a rectangular box, and the type and confidence of the defect are also shown. It shows that the improved algorithm proposed in this paper is accurate and effective in the detection of surface defects of hot rolled steel strip, and can realize the detection of surface defects of different kinds of hot rolled steel strip.

# 5   Conclusion

Based on the SSD model, this paper uses knowledge distillation to train ResNet50 network, and uses the obtained KD-ResNet50 to replace VGG16 as the backbone network. On this basis, the feature fusion module is added to make the shallow feature and the deep feature fusion to improve the detection accuracy of small targets. Finally, the attention module is added to screen the important information, which can effectively use the feature information and reduce the computation. The experimental results show that the quasi-accuracy of RAF-SSD model is higher than that of SSD model, Faster RCNN model and YOLO-V3 model, and the detection speed is 12.9% higher than that of the fastest SSD model among the three models, which can effectively meet the real-time detection requirements for surface defects of hot rolled steel strip. However, the data set used in this paper has a small number of samples, so the next step will be to enhance the data, expand the sample size, and then improve the performance of the model.

# References

1. Xu, K., Wang, L., Wang, J.: Surface defect recognition of hot-rolled steel plates based on tetrolet transform. Instrum. Sci. Technol. **52**(4), 13–19 (2016)
2. Wu, P., Lu, T., Wang, Y.: Nondestructive testing technique for strip surface defects and its applications. Nondestr. Test. **22**(7), 312–315 (2000)
3. Zheng, J.: Research on strip surface defect detection method. Xian University of architecture and technology (2006)
4. Tao, X., Hou, W., Xu, D.: A survey of surface defect detection methods based on deep learning. Acta Automatica SinicamMonth (2020)
5. Jiang, J., Zhai, D.: Single-stage object detection algorithm based on atrous convolution and feature enhancement. Comput. Eng. (2020)
6. Zhu, D., Lin, Z.: Corn silk detection method based on MF-SSD convolutional neural network. J. South China Agricul. Univ. (2020)
7. Wu, S., Ding, E., Yu, X.: Foreign body identification of belt based on improved FPN. Saf. Coal Mines **50**(12), 127–130 (2019)
8. Liu, W., et al.: SSD: single shot MultiBox detector. In: Leibe, B., Matas, J., Sebe, N., Welling, M. (eds.) ECCV 2016. LNCS, vol. 9905, pp. 21–37. Springer, Cham (2016). https://doi.org/10.1007/978-3-319-46448-0_2
9. Ren, S.Q., He, K.M., Girshick, R., Sun, J.: Faster R-CNN: towards real-time object detection with region proposal networks. In: Advances in Neural Information Processing Systems (NIPS), Montreal, Quebec, Canada, pp. 91–99. MIT Press (2015)
10. Redmon, J., Divvala, S., Girshick, R., Farhadi, A.: You only look once: unified, real-time object detection. In: Proceedings of 2016 IEEE Conference on Computer Vision and Pattern Recognition, Las Vegas, USA, pp. 779–788. IEEE (2016)
11. Ba, L.J., Caruana, R.: Do deep nets really need to be deep. In: Advances in Neural Information Processing Systems, pp. 2654–2662 (2013)
12. He, K.M., Zhang, X.Y., Ren, S.Q., et al.: Deep residual learning for image recognition. In: 2016 IEEE Conference on Computer Vision and Pattern Recognition, pp. 770–778 (2016)
13. Lei, P., Liu, C., Tang, J., et al.: Hierarchical feature fusion attention network for image super-resolution reconstruction. J. Image Graph. **9**, 1773–1786 (2020)
14. Fu, C.Y., Liu, W., Ranga, A., et al.: DSSD: deconvolutional single shot detector. arXiv:1701.06659v1 (2017)
15. Hu, J., Li, S., Gang, S.: Squeeze-and-excitation networks. arXiv preprint arXiv:1709.01507 (2017)

# Continuous Keystroke Dynamics-Based User Authentication Using Modified Hausdorff Distance

Maksim Zhuravskii, Maria Kazachuk, Mikhail Petrovskiy[✉],
and Igor Mashechkin

Lomonosov Moscow State University, Vorobjovy Gory, Moscow 119899, Russia
paperlark@icloud.com, {mkazachuk,michael,mash}@cs.msu.ru

**Abstract.** Continuous keystroke dynamics-based user authentication methods are one of the most perspective means of user authentication in computer systems. Such methods do not require specialized equipment and allow detection of user change anytime during a user session. In this paper, we explore new approaches to solving the problem based on Hausdorff distance and its modification, including a new method, the sum of maximum coordinate deviations. We compare proposed methods to existing ones that are based on distance functions defined in feature space, statistical criteria, and neural networks. Based on the experiments, we observe that the proposed method based on the sum of maximum coordinate deviations with $k$ nearest feature vector selection reports the highest accuracy of all reviewed methods.

**Keywords:** Outlier detection · Continuous authentication · Keystroke dynamics · Hausdorff distance · SMCD

## 1 Introduction

In recent decades, cloud services have gained much attention. They are used for accomplishing both personal and enterprise tasks. Thus, it is crucial to ensure the security of information processed by the cloud services. One of the key tasks in ensuring information security is authentication, a process by which subjects, normally users, establish their identity to a system [4]. For authentication purposes, subjects provide an identifier to the system.

Authentication methods differ by the nature of the identifier used in the process. Although methods that employ secret knowledge, such as a password, or an identification object, such as a magnetic card, are easy to implement, they are vulnerable to identifier compromise. In this regard, the most promising methods are those that use user's biometrics for identification. Biometrics is divided into two categories: physiological samples and behavioral ones.

Physiological samples represent the physiological characteristics of a person which remain with him throughout his life. These include fingerprints, iris, and

C. S. Jensen et al. (Eds.): DASFAA 2021 Workshops, LNCS 12680, pp. 223–236, 2021.
https://doi.org/10.1007/978-3-030-73216-5_16

facial geometry. A significant drawback of such methods is that they depend on specialized equipment.

Behavioral patterns represent the behavioral characteristics of a person, such as voice, gait, and handwriting. These methods do not require specialized equipment. However, existing approaches are less stable than those based on physiological samples.

As for authentication frequency, continuous authentication is preferred as it eliminates the possibility of an attacker gaining access to the system sometimes after a legitimate user passed authentication.

Thus, the most promising task is continuous user authentication based on behavioral biometric characteristics. As such, we can consider keystroke dynamics, i.e. characteristics of a person's dynamics when working with a computer keyboard. Since the keyboard is one of the primary means of human interaction with a computer, keystroke dynamics can be effectively used for continuous user authentication.

In this paper, we discuss continuous keystroke dynamics-based user authentication as an outlier detection problem [9]. The goal of the research was to improve authentication effectiveness when a small dataset is used to train a user's model. With this goal in mind, we propose a new method, the sum of maximum coordinate deviations, or SMCD. Along with this method, we explore new approaches based on the Hausdorff distance [7] and its modifications. Our hypothesis was that these approaches would achieve higher efficiency when compared to other ones.

This article has the following structure. Section 2 provides an overview of existing continuous keystroke dynamics-based authentication methods. Section 3 describes methods based on the Hausdorff distance and the sum of maximum coordinate deviations. Section 4 is devoted to an experimental study of the proposed methods and their comparison with existing approaches to solving the problem. Section 5 summarises the obtained results.

## 2    Related Work

When considering authentication methods, we should pay attention to what is considered features of keystroke dynamics in a specific method, how these features are preprocessed, which algorithm is used to decide the user's legitimacy, and what accuracy the method has. In this section, we discuss each of these steps in detail.

### 2.1    Feature Extraction and Preprocessing

Most studies use single keystroke characteristics [2,3,10,12,13,17] and digraph characteristics [2,3,8,10,13,17] as features of user's keystroke dynamics. Here, a digraph is a combination of two consecutive user keystrokes. The duration of keystrokes is considered as a characteristic of single clicks. As for digraphs, all possible intervals between digraph keystrokes are used as its features.

Here and further to facilitate notation we will denote a set $\{i \in N | a \leq i \leq b\}$ where $a, b \in N$ with $\overline{a, b}$. With this in mind, let us denote $t_i^{down}, i = \overline{1, 2}$ the time when the $i$-th digraph key is pressed, $t_i^{up}, i = \overline{1, 2}$ the time when the $i$-th digraph key is released (see Fig. 1). Then, we can define four digraph characteristics as follows:

1. DD-time, or duration of the interval between keypresses: $t_2^{down} - t_1^{down}$;
2. DU-time, or digraph input duration: $t_2^{up} - t_1^{down}$;
3. UU-time, or duration of the interval between key releases: $t_2^{up} - t_1^{up}$;
4. UD-time, or duration of the jump between keystrokes: $t_2^{down} - t_1^{up}$.

To construct feature vectors a sliding window can be used [8]. In this case, we select a window of a certain size from the sequence of user keystrokes and calculate features based on the keystrokes contained in the window. When the user clicks, the sliding window shifts, and authentication is performed again (see Fig. 2). This means that the user is authenticated continuously while he is using the keyboard.

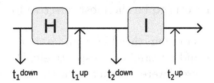

$$t_1\text{down} \qquad t_1\text{up} \qquad t_2\text{down} \qquad t_2\text{up}$$

**Fig. 1.** Digraph features.

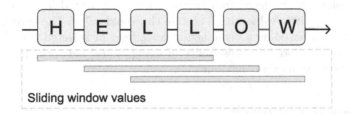

Sliding window values

**Fig. 2.** Sliding window operation.

Using a sliding window allows to use aggregate feature values, such as their empirical means [3,8,10,12,17] and standard deviations [3,8,12,17]. This increases the stability of the method to possible outliers in individual feature values. It is worth noting, however, that a smaller sliding window makes a method more responsive to a possible user change. Hence, smaller window size is preferable.

Modern keyboards contain more than $n = 70$ different keys. As a result, the dimension of the feature space of all possible digraphs and all possible single keystrokes exceeds $n^2 + n = 4970$. To reduce the dimensionality of the feature space, in studies [8,10,17] it is proposed to split keyboard keys into groups based on their physical location or functional purpose. In this case, features of digraphs or individual keystrokes are calculated only within the group and between different groups. As a result, we can decrease the number of considered features, and increase the number of their values observed in the sliding window.

## 2.2   User Model Construction

As mentioned earlier, we consider the problem of continuous user keystroke dynamics-based authentication an outlier detection problem [9]. In study [8], various classification methods are considered in application to the problem of deciding the user's legitimacy based on their keystroke dynamics. The study covers methods based on statistical criteria, distance functions defined in features space, and machine learning methods, such as SVM, $k$ nearest neighbors, and SVDD. According to the results obtained, the method based on the Kolmogorov-Smirnov statistics [15] reports to be the most effective for small sliding window size.

This method considers random variables $\xi$ and $\eta$ that correspond to some feature of a legitimate user and a tested one respectively. Based on the feature values obtained during two users were interacting with the keyboard, we construct empirical cumulative distribution functions $F_\xi(x)$ and $F_\eta(x)$. The Kolmogorov-Smirnov statistic is:

$$D_{\xi,\eta} = \sup_{x \in \mathbb{R}} \mid F_\xi(x) - F_\eta(x) \mid .  \tag{1}$$

The decision on the legitimacy of the user being tested is made based on comparing the statistic value with a certain threshold value. The threshold value can be calculated based on a given significance level or selected experimentally.

Some works explored the possibility of using neural networks for creating a model of legitimate user's behaviour [11].

# 3   Proposed Approach

This section describes the proposed approaches to continuous keystroke dynamics-based user authentication.

## 3.1   Feature Extraction

To build features based on the available data, we use the sliding window method described in Sect. 2. We use sliding windows of size from 100 to 300 events, where each event describes either a key press or a release. The event description consists of its type: press or release, a key code, and a timestamp of the moment when the event occurred.

Feature values that correspond to long pauses when the user stops typing are abnormal for the legitimate user model. Their presence in the sliding window degrades the quality of the authentication method. Thus, to prevent this we split the window if the pause between consecutive keystrokes exceeds the 40-s threshold. The optimal values of the sliding window parameters were selected experimentally.

It is worth noting that the space of keystroke dynamics features is sparse. This is because only a small subset of all possible digraphs and single keystrokes can occur in the sliding window. To select only the most significant features from the set of all features and reduce the dimensionality of feature space, in [10] we proposed to reduce the dimension of the feature space by selecting a fixed number of single keystrokes and digraphs that are most common in the observed window. This way, in each window we select 37 most frequently encountered keys and 100 most frequently encountered digraphs. For selected keystrokes, we calculate their keystroke duration, and for selected digraphs, we calculate the average duration of the UU and DU intervals. The optimal ratio of single keypress and digraph features was found during a preliminary series of experiments.

Along with the features described above we also consider the following features: average keystroke duration by groups (see Fig. 3), average frequency of keystrokes in the sliding window, average frequency of command keys used by the user in the sliding window.

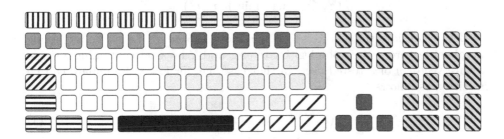

**Fig. 3.** Proposed key groups for feature extraction.

### 3.2 Quantile Discretization

To preprocess the obtained feature values we propose to use quantile discretization [10].

Let $\mathbb{X}$ be a training sample of values of some feature $\eta$. Using sample $\mathbb{X}$ we calculate empirical quantiles $q_i$ of orders $\frac{i}{k}, i = \overline{1, k-1}$, where number $k$ is an algorithm hyper-parameter. Then we replace every value $\hat{\eta}$ of feature $\eta$ with the number $j$ such that:

$$j = \begin{cases} 1 & \text{if } \hat{\eta} \in (-\infty, q_1]; \\ k & \text{if } \hat{\eta} \in (q_{k-1}, +\infty); \\ t & \text{if } t \in \overline{2, k-1} \text{ and } \hat{\eta} \in (q_{t-1}, q_t]. \end{cases} \quad (2)$$

Quantile discretization of features allows mitigating small fluctuations in values of continuous features by discretizing their values. In addition to that, it was shown in [10] that the distribution of the considered features is multimodal. By applying quantile discretization it is possible to smoothen feature distributions and thereby improve the quality of machine learning methods designed to work with more homogeneous data.

Based on the results of a series of experiments, the optimal number of sampling intervals is $k = 7$.

### 3.3  Hausdorff Distance

We propose to use the Hausdorff distance [7] for outlier detection. Let $\mathbb{X} = \{x^1, \dots, x^m\}$ and $\mathbb{Y} = \{y^1, \dots, y^l\}$ be two sets of vector from a metric space $\mathbb{M}$ with the distance function $\rho$. The Hausdorff distance between sets of vectors in $\mathbb{X}$ and $\mathbb{Y}$ is defined as:

$$\begin{aligned} \rho_H(\mathbb{X}, \mathbb{Y}) &= \max\{h(\mathbb{X}, \mathbb{Y}), h(\mathbb{Y}, \mathbb{X})\}; \\ h(\mathbb{X}, \mathbb{Y}) &= \max_{i \in \overline{1,m}} \min_{j \in \overline{1,l}} \rho(x^i, y^j); \\ h(\mathbb{Y}, \mathbb{X}) &= \max_{j \in \overline{1,l}} \min_{i \in \overline{1,m}} \rho(x^i, y^j). \end{aligned} \quad (3)$$

Let us consider the feature space as a metric space $\mathbb{M}$. Let $\mathbb{X}$ be a training sample of feature vectors of a legitimate user and $\mathbb{Y}$ be a set containing a single feature vector $y$ of a tested user. In that case, i.e. for $l = 1$, Formula 3 can be rewritten as follows:

$$\rho_H(\mathbb{X}, \mathbb{Y} \mid l = 1) = \max_{i \in \overline{1,m}} \rho(x^i, y). \quad (4)$$

To decide the legitimacy of the tested user, we compare the value of Hausdorff distance $\rho_H(\mathbb{X}, \mathbb{Y})$ to a threshold value set a priori. If $\rho_H(\mathbb{X}, \mathbb{Y})$ exceeds the threshold, the hypothesis about the legitimacy of the current user is rejected, and the user's session is suspended. Otherwise, the user continues working in the system.

Let us also consider a method based on a modified Hausdorff distance [5]. The modified Hausdorff distance, or MHD, is defined as:

$$\rho_{\text{MHD}}(\mathbb{X}, \mathbb{Y}) = \max\{h_M(\mathbb{X}, \mathbb{Y}), h_M(\mathbb{Y}, \mathbb{X}));$$

$$h_M(\mathbb{X}, \mathbb{Y}) = \frac{1}{m} \sum_{i=1}^{m} \min_{j=\overline{1,l}} \rho(x^i, y^j);$$

$$h_M(\mathbb{Y}, \mathbb{X}) = \frac{1}{l} \sum_{j=1}^{l} \min_{i=\overline{1,m}} \rho(x^i, y^j). \tag{5}$$

The modified Hausdorff distance is less sensitive to the presence of outliers in the sets $\mathbb{X}$ and $\mathbb{Y}$. For $l = 1$, Formula 5 is equivalent to:

$$\rho_{\text{MHD}}(\mathbb{X}, \mathbb{Y} \mid l = 1) = \frac{1}{m} \sum_{i=1}^{m} \rho(x^i, y). \tag{6}$$

We also investigate the method based on interpolated modified Hausdorff distance [14]. Interpolated modified Hausdorff distance, or IMHD, is defined as follows:

$$\rho_{\text{IMHD}}(\mathbb{X}, \mathbb{Y}) = \max\{h_I(\mathbb{X}, \mathbb{Y}), h_I(\mathbb{Y}, \mathbb{X}));$$

$$h_I(\mathbb{X}, \mathbb{Y}) = \frac{1}{l} \sum_{j=1}^{l} \min_{i=\overline{2,m}} \rho(\frac{x^i + x^{i-1}}{2}, y^j);$$

$$h_I(\mathbb{Y}, \mathbb{X}) = \frac{1}{m} \sum_{i=1}^{m} \min_{j=\overline{2,l}} \rho(\frac{y^j + y^{j-1}}{2}, x^i). \tag{7}$$

For $l = 1$, Formula 7 is equivalent to:

$$\rho_{\text{IMHD}}(\mathbb{X}, \mathbb{Y} \mid l = 1) = \min_{i=\overline{2,m}} \rho(\frac{x^i + x^{i-1}}{2}, y). \tag{8}$$

### 3.4   Sum of Maximum Coordinate Deviations

As a modification of the method based on the Hausdorff distance, we propose a method based on the sum of maximum coordinate deviations.

Let us denote $\mathbb{M}$ a $n$ – dimensional feature space and let $\mathbb{X} = \{x^1, \ldots, x^m\}$ be a set of $m$ vectors from space $\mathbb{M}$ where $x^i = (x_1^i, \ldots, x_n^i), i = \overline{1, m}$. The sum of the maximum coordinate deviations, or SMCD, between the vector $y = (y_1, \ldots, y_n) \in \mathbb{M}$ and the set $\mathbb{X}$ is defined as:

$$\rho_S(y, \mathbb{X}) = \frac{1}{n} \sum_{i=1}^{n} \max_{j \in \overline{1,m}} |x_i^j - y_i|. \tag{9}$$

This means that the greater the value of the sum of the maximum coordinate deviations is the greater is the difference between the vector $y$ and the vectors from the set $\mathbb{X}$.

Let us introduce a feature space $\mathbb{M}$, a training sample of the feature vectors $\mathbb{X}$, and a feature vector $y$ of the tested user in the same way as they were introduced in the previous subsection. As before, the decision on the legitimacy of the tested user can be made based on the result of comparing the value of the sum of the maximum coordinate deviations $\rho_S(y, \mathbb{X})$ with a threshold value set a priori. The current user's legitimacy hypothesis is rejected if the value $\rho_S(y, \mathbb{X})$ exceeds the specified threshold value.

Note that the sum of the maximum coordinate deviations is related to the Hausdorff distance. Let $\mathbb{M}_{l1}$ be an $n$-dimensional linear metric space with norm $\|a\| = \frac{1}{n}\sum_{i=1}^{n}|a_i|, a \in \mathbb{M}_{l1}$ and $\mathbb{X} = \{x^1, ..., x^m\}$, $\mathbb{Y} = \{y\}$ be two sets of vectors from $\mathbb{M}_{l1}$. Then, taking into account Formula 4, the following is true:

$$\rho_H(\mathbb{X}, \mathbb{Y}) = \frac{1}{n} \max_{j=\overline{1,m}} \sum_{i=1}^{n} |x_i^j - y_i|. \tag{10}$$

In a linear metric space $\mathbb{M}_{l1}$ for an arbitrary set of vectors $\mathbb{Z} = \{z^1, \ldots, z^m\}$ is the following true:

$$\max_{j=\overline{1,m}} \sum_{i=1}^{n} |z_i^j| = \sum_{i=1}^{n} |z_i^{j_0}| \le \sum_{i=1}^{n} \max_{j=\overline{1,m}} |z_i^j|, \tag{11}$$

where $z^{j_0} \in \mathbb{Z}$ is the element of the set $\mathbb{Z}$ that provides the maximum in the left part of the equation.

Combining Formulas 9, 10, and 11 we conclude that the following is true:

$$\rho_H(\mathbb{X}, \mathbb{Y}) \le \rho_S(y, \mathbb{X}). \tag{12}$$

Thus, sum of maximum coordinate deviations serves as an upper boundary of the Hausdorff distance in the linear space $\mathbb{M}_{l1}$.

### 3.5    K Nearest Neighbors

Along with other methods, an outlier detection method based on the $k$ nearest neighbors search method is considered in study [8].

Let us introduce the feature space $\mathbb{M}$, a training sample of feature vectors $\mathbb{X}$, and the feature vector $y$ of the tested user as in the previous subsection. Let $\rho(.,.)$ be a distance function defined in space $\mathbb{M}$. Then, for the vector $y$ and a given value $k$ there is a subset $\hat{\mathbb{X}} = \{\hat{x}^1, \ldots, \hat{x}^k\} \subset \mathbb{X}$ containing $k$ nearest to the vector $y$ of vectors from the set $\mathbb{X}$ based on the distance function $\rho(.,.)$. The decision on the legitimacy of the tested user is made based on comparing the average distance from the vector $y$ to the vectors from $\hat{\mathbb{X}}$:

$$d(y, \mathbb{X}) = \frac{1}{k} \sum_{j=1}^{k} \rho(y, \hat{x}^j). \tag{13}$$

As before, if the value is $d(y, \mathbb{X})$ exceeds a certain threshold set a priori, and the hypothesis about the legitimacy of the tested user is rejected.

In study [8], the L2 distance function is defined in feature space. However, other distance functions can also be used. In this regard, we also considered a variation of this method that defines a cosine distance in feature space.

Also, we propose a modification of this method. As before, we find a sub-sample $\hat{\mathbb{X}}$ of $k$ vectors closest to the vector $y$ in the sample $\mathbb{X}$. To decide the legitimacy of the tested user, we will use the value of the Hausdorff distance $\rho_H(\mathbb{X}, \{y\})$ (see Formula 3), modified Hausdorff distance $\rho_{\mathrm{MHD}}(\mathbb{X}, \{y\})$ (see Formula 5), interpolated modified Hausdorff distance $\rho_{\mathrm{IMHD}}(\mathbb{X}, \{y\})$ (see Formula 7), or the sum of maximum coordinate deviations $\rho_S(y, \hat{\mathbb{X}})$ (see Formula 9).

## 4   Results

We tested the proposed methods on a dataset used in studies [12,16]. It contains collected data for 144 users while they were working with the computer keyboard. The dataset contains over 1,345 registered clicks for each user as well as the following information about each user: platform used by the user: desktop or laptop, user's gender, user's age group, user's dominant hand, user's awareness of the data collection.

In our study we use only the following information about user's actions: code of the pressed key, keypress start timestamp, keypress end timestamp.

When testing each method, we used 80% of the legitimate user's sample to fit the method, and the remaining 20% as well as other users' samples to evaluate the method.

To assess each method we only used those users for whom we obtained more than 10 feature vectors during feature extraction. We do so to mitigate the fact that a smaller training sample may result in poor model fit and, as a result, poor accuracy of the tested authentication method.

To compare the effectiveness of the methods under consideration, we used ROC AUC, which is equal to the value of the area under the ROC curve. It can be interpreted as the probability that the model ranks a randomly chosen positive instance higher than a randomly chosen negative instance [6]. Hence, it can be used to evaluate the quality of classification without setting an exact threshold for deciding the user's legitimacy.

Along with ROC AUC, we assessed methods' effectiveness based on equal error rate (EER) [8] and average precision (AP) [18].

As a baseline model, we consider a one-class SVM with an RBF core [10]. This model is often used in outlier detection problems. We also consider the Fuzzy method [10] and neural network models.

We considered different neural network models based on the autoencoder architecture [1], including fully connected, recurrent, and fully convolutional. These models are trained to encode feature vectors of the legitimate user into vectors of reduced dimensionality. The decision on the legitimacy of the user being tested is made based on comparing the Euclidean distance between the initial and restored vectors with a certain threshold value set a priori.

Based on the results of preliminary experiments a fully convolutional autoencoder (see Fig. 4). It is worth noting that deep neural networks require a larger dataset to get an optimal model. Hence, to assess the method based on the fully convolutional autoencoder we only used those users for whom we obtained more than 100 feature vectors. This drawback of neural network models is the reason why we did not focus our research on these models.

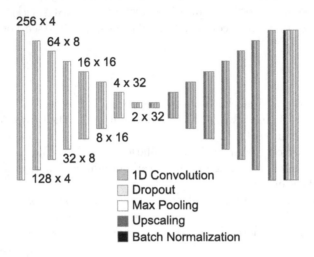

**Fig. 4.** Fully convolutional autoencoder model.

In this study we evaluated the existing methods based on Kolmogorov-Smirnov statistic, a one-class SVM with RBF core, the $k$ nearest neighbors algorithm, the Fuzzy method, the approach based on a fully convolutional autoencoder, the proposed approaches based on Hausdorff distances, and the proposed methods based on SMCD. Every algorithm was tested both with and without quantile discretization. Table 1 contains descriptions of algorithms with their optimal hyper-parameter values that were found using grid search method. Table 2 shows the experimental results.

According to the results, the proposed methods of continuous authentication based on SMCD are more effective than all other methods. Meanwhile, methods based on the composition of the $k$ nearest neighbors method and the sum of maximum coordinate deviations report the best accuracy. It is worth noting that the proposed method based on the sum of maximum coordinate deviations surpasses methods based on other considered modifications of the Hausdorff distance.

**Table 1.** Description of tested algorithms.

| Authentication method | Method parameter | Parameter value |
|---|---|---|
| One-class SVM | Feature preprocessing | Quantile discretization |
| | Kernel function | RBF |
| | Kernel width ($\gamma$) | 0.000837 |
| | Error fraction in training set ($\nu$) | 0.00144 |
| Kolmogorov-Smirnov statistic | Feature preprocessing | – |
| Hausdorff distance | Feature preprocessing | – |
| | Distance function | L1 |
| Modified Hausdorff distance | Feature preprocessing | Quantile discretization |
| | Distance function | L1 |
| Interpolated modified Hausdorff distance | Feature preprocessing | Quantile discretization |
| | Distance function | L1 |
| SMCD | Feature preprocessing | Quantile discretization |
| $k$ nearest neighbors | Feature preprocessing | Quantile discretization |
| | Distance function | L2 |
| | $k$ | 74 |
| Hausdorff distance with $k$ nearest neighbors selection | Feature preprocessing | Quantile discretization |
| | Distance function | L1 |
| | $k$ | 72 |
| MHD with $k$ nearest neighbors selection | Feature preprocessing | Quantile discretization |
| | Distance function | L1 |
| | $k$ | 73 |
| IMHD with $k$ nearest neighbors selection | Feature preprocessing | Quantile discretization |
| | Distance function | L1 |
| | $k$ | 73 |
| SMCD with $k$ nearest neighbors selection | Feature preprocessing | Quantile discretization |
| | Distance function | L2 |
| | $k$ | 58 |
| Fuzzy | Feature preprocessing | Quantile discretization |
| | Kernel function | RBF |
| | Kernel width ($\gamma$) | 1e−06 |
| | Error fraction in training set ($k$) | 0.001 |
| | Affiliation level decrease rate ($m$) | 12.5 |
| Fully convolutional autoencoder | Feature preprocessing | Quantile discretization |
| | Optimization algorithm | Adam |
| | Loss function | LogCosh |
| | Learning rate | 0.01 |
| | Epochs | 300 |

**Table 2.** Experiments results.

| Authentication method | Median ROC AUC | IQR ROC AUC | Median AP | Median EER |
|---|---|---|---|---|
| One-class SVM | 0.961 | 0.103 | 0.186 | 0.111 |
| Kolmogorov-Smirnov statistic | 0.558 | 0.181 | 0.351 | 0.500 |
| Hausdorff distance | 0.837 | 0.242 | 0.005 | 0.333 |
| Modified Hausdorff distance | 0.857 | 0.246 | 0.011 | 0.214 |
| Interpolated modified Hausdorff distance | 0.935 | 0.152 | 0.077 | 0.158 |
| SMCD | 0.950 | 0.091 | 0.305 | 0.064 |
| $k$ nearest neighbors | 0.977 | 0.129 | 0.315 | 0.089 |
| Hausdorff distance with $k$ nearest neighbors selection | 0.929 | 0.188 | 0.090 | 0.151 |
| MHD with $k$ nearest neighbors selection | 0.893 | 0.235 | 0.046 | 0.173 |
| IMHD with $k$ nearest neighbors selection | 0.960 | 0.156 | 0.176 | 0.130 |
| SMCD with $k$ nearest neighbors selection | **0.987** | **0.030** | **0.503** | **0.065** |
| Fuzzy | 0.972 | 0.079 | 0.183 | 0.102 |
| Fully convolutional autoencoder | 0.908 | 0.128 | 0.095 | 0.185 |

## 5    Conclusion

Continuous keystroke dynamics-based user authentication is one of the most promising areas of authentication methods development. We propose new approaches to this problem that are based on the Hausdorff distance and its modifications as well as new methods based on the Hausdorff distance modification, the sum of maximum coordinate deviations.

According to the results of experimental research, the proposed method based on the sum of the maximum coordinate deviations with $k$ nearest keyboard vectors selection reports the median value of the ROC AUC equal to 0.987. Thus, this method surpasses all other considered approaches, including the existing method based on Kolmogorov-Smirnov statistics, which is considered one of the most effective methods of one-class user authentication using keystroke dynamics, and a method based on a fully convolutional autoencoder.

Despite favorable experimental results, there are ways to extend the research. First of all, to guarantee language independence, the proposed method's performance should be evaluated on datasets in other languages. Secondly, with the proposed approach the size of the feature vector strongly depends on the sliding window size, i.e. in case of a smaller window fewer significant features can be extracted. Thus, to use a smaller sliding window different methods of feature extraction should be explored. Also, we plan to make a more detailed comparison with statistical and deep learning methods.

# References

1. Badrinarayanan, V., Kendall, A., Cipolla, R.: SegNet: a deep convolutional encoder-decoder architecture for image segmentation. IEEE Trans. Pattern Anal. Mach. Intell. **39**(12), 2481–2495 (2017). https://doi.org/10.1109/TPAMI.2016. 2644615
2. Bicakci, K., Salman, O., Uzunay, Y., Tan, M.: Analysis and evaluation of keystroke dynamics as a feature of contextual authentication. In: 2020 International Conference on Information Security and Cryptology (ISCTURKEY), pp. 11–17 (2020). https://doi.org/10.1109/ISCTURKEY51113.2020.9307967
3. Bours, P., Mondal, S.: Performance evaluation of continuous authentication systems. IET Biometr. **4**(4), 220–226 (2015). https://doi.org/10.1049/iet-bmt.2014. 0070
4. Butterfield, A., Ngondi, G.E., Kerr, A. (eds.): A Dictionary of Computer Science. Oxford University Press, Oxford (2016). https://doi.org/10.1093/acref/ 9780199688975.001.0001
5. Dubuisson, M.P., Jain, A.: A modified Hausdorff distance for object matching. In: Proceedings of 12th International Conference on Pattern Recognition, vol. 1, pp. 566–568 (1994). https://doi.org/10.1109/ICPR.1994.576361
6. Fawcett, T.: An introduction to ROC analysis. Pattern Recogn. Lett. **27**(8), 861–874 (2006). https://doi.org/10.1016/j.patrec.2005.10.010
7. Huttenlocher, D.P., Klanderman, G.A., Rucklidge, W.J.: Comparing images using the Hausdorff distance. IEEE Trans. Pattern Anal. Mach. Intell. **15**(9), 850–863 (1993). https://doi.org/10.1109/34.232073
8. Kang, P., Cho, S.: Keystroke dynamics-based user authentication using long and free text strings from various input devices. Inf. Sci. **308**, 72–93 (2015). https:// doi.org/10.1016/j.ins.2014.08.070
9. Kazachuk, M., Petrovskiy, M., Mashechkin, I., Gorokhov, O.: Outlier detection in complex structured event streams. Moscow Univ. Comput. Math. Cybern. **43**(3), 101–111 (2019). https://doi.org/10.3103/S0278641919030038
10. Kazachuk, M., et al.: One-class models for continuous authentication based on keystroke dynamics. In: Yin, H., et al. (eds.) IDEAL 2016. LNCS, vol. 9937, pp. 416–425. Springer, Cham (2016). https://doi.org/10.1007/978-3-319-46257-8_45
11. Kim, J., Kang, P.: Recurrent neural network-based user authentication for freely typed keystroke data. CoRR abs/1806.06190 (2018). http://arxiv.org/abs/1806. 06190. Withdrawn
12. Monaco, J.V., Bakelman, N., Cha, S., Tappert, C.C.: Developing a keystroke biometric system for continual authentication of computer users. In: 2012 European Intelligence and Security Informatics Conference, pp. 210–216 (2012). https://doi. org/10.1109/EISIC.2012.58
13. Raul, N., Shankarmani, R., Joshi, P.: A comprehensive review of keystroke dynamics-based authentication mechanism. In: Khanna, A., Gupta, D., Bhattacharyya, S., Snasel, V., Platos, J., Hassanien, A.E. (eds.) International Conference on Innovative Computing and Communications. AISC, vol. 1059, pp. 149–162. Springer, Singapore (2020). https://doi.org/10.1007/978-981-15-0324-5_13
14. Shao, F., Cai, S., Gu, J.: A modified Hausdorff distance based algorithm for 2-dimensional spatial trajectory matching. In: 2010 5th International Conference on Computer Science Education, pp. 166–172 (2010). https://doi.org/10.1109/ICCSE. 2010.5593666

15. Shiryayev, A.N.: 15. On the empirical determination of a distribution law. In: Shiryayev, A.N. (ed.) Selected Works of A. N. Kolmogorov. Mathematics and Its Applications (Soviet Series), vol. 26, pp. 139–146. Springer, Dordrecht (1992). https://doi.org/10.1007/978-94-011-2260-3_15
16. Tappert, C.C., Villani, M., Cha, S.H.: Keystroke biometric identification and authentication on long-text input. In: Behavioral Biometrics for Human Identification: Intelligent Applications, pp. 342–367. IGI global (2010)
17. Villani, M., Tappert, C., Ngo, G., Simone, J., Fort, H., Cha, S.: Keystroke biometric recognition studies on long-text input under ideal and application-oriented conditions. In: 2006 Conference on Computer Vision and Pattern Recognition Workshop (CVPRW 2006), p. 39 (2006). https://doi.org/10.1109/CVPRW.2006.115
18. Zhang, E., Zhang, Y.: Average Precision. Springer, Boston (2009). https://doi.org/10.1007/978-0-387-39940-9_482

# Deep Learning-Based Dynamic Community Discovery

Ling Wu[1], Yubin Ouyang[1], Cheng Shi[2], and Chi-Hua Chen[1(✉)]

[1] College of Mathematics and Computer Science, Fuzhou University, Fuzhou, China
[2] School of Computer Science and Engineering, Xi'an University of Technology, Xi'an, China

**Abstract.** Recurrent neural networks (RNNs) have been effective methods for time series analyses. The network representation learning model and method based on deep learning can excellently analyze and predict the community structure of social networks. However, the node relationships of complex social networks in the real world often change over time. Therefore, this study proposes a dynamic community discovery method based on a recurrent neural network, which includes (1) spatio-temporal structure reconstruction strategy; (2) spatio-temporal feature extraction model; (3) dynamic community discovery method. Recurrent neural networks can be used to obtain the time features of the community network and help us build the network time feature extraction model. In this study, the recurrent neural network model is introduced into the time series feature learning of dynamic networks. This research constructs a network spatiotemporal feature learning model combining RNN, convolutional neural networks (CNN), and auto-encoder (AE), and then uses it to explore the dynamic community structure on the spatiotemporal feature vector. The experiment chose the Email-Enron data set of the Stanford Network Analysis Platform (SNAP) website to evaluate the method. The experimental results show that the proposed method has higher modularity than Auto-encoder in the dynamic community discovery of the real social network data set. Therefore, the dynamic community discovery method based on the recurrent neural network can be applied to analyze social networks, extract the time characteristics of social networks, and further improve the modularity of the community structure.

**Keywords:** Dynamic community discovery · Social network · Recurrent neural network · Deep learning

## 1  Introduction

The static community discovery algorithm can explore the network topology community structure at a certain moment. However, the relationship between nodes in the network always changes with time. The characteristics of dynamic communities can be classified as follows: time-varying, unstable, and short-term smoothness. Based on these three characteristics, the existing community discovery and community evolution research in dynamic social networks can be roughly divided into three categories which

© Springer Nature Switzerland AG 2021
C. S. Jensen et al. (Eds.): DASFAA 2021 Workshops, LNCS 12680, pp. 237–248, 2021.
https://doi.org/10.1007/978-3-030-73216-5_17

include instantaneous optimal community discovery methods, time-varying balanced community discovery methods, and cross-time community discovery methods.

The instantaneous optimal community discovery method is implemented in two stages. The first stage explores the static community structure for network clustering at each moment, and the second stage matches the community structure on different time slices to calculate and analyze the community evolution. In essence, the instantaneous optimal community discovery method separates the relationship between time and network space topology changes. The time-varying balanced community discovery method only considers the relationship between historical moments and network space topology changes, and only the cross-time community discovery method comprehensively considers historical and future moments for network space topology changes. However, the cross-time community discovery method cannot handle real-time community detection as each new step requires recalculation of all historical steps when the network is modified. Therefore, it is of high theoretical and practical significance to design an effective model that learns the evolutionary characteristics and laws of dynamic networks by comprehensively considering the influence of historical and future time slice networks on community discovery.

Existing research proposes to use deep learning methods to analyze the temporal characteristics of social network structure and predict future changes in the social network structure. These studies make community discovery based on the structure of future social networks [1–5]. It is an important and meaningful topic to treat community discovery as a problem related to time and space and to analyze the impact of the interaction of time and space factors on community discovery.

This study will utilize the recurrent neural network to extract the time-dependent features and network evolution features of network data at different times to achieve the purpose of dimension adjustment. This study proposes to establish a combined neural network model based on a convolutional neural network, a recurrent neural network, and an auto-encoder for extracting the spatial and temporal characteristics of the network. This research obtains the dynamic community structure and evolution of the network by clustering the temporal and spatial feature vectors of the network. The performance of the algorithms will be analyzed and compared through simulation experiments.

The main contributions of this study are as follows:

(1) **Constructing a model of time feature extraction of a dynamic network**. Based on the dynamics of the dynamic network, this study takes the auto-encoder as the main model framework and integrates the recurrent neural network to generate a combined model of the recurrent neural network and the auto-encoder. This model is used for network time feature extraction.

(2) **Constructing a spatio-temporal feature extraction model for dynamic networks**. Based on the large-scale and dynamic nature of the network, this research constructs a combined model based on a recurrent neural network, a convolutional neural network, and an auto-encoder to extract the temporal and spatial features of the network. This study proposes a dynamic community discovery algorithm based on a recurrent neural network; this method is entitled "Deep Recurrent Auto-Encoding neural network based on Reconstructive matrix for dynamic community detection" (DRAER). The overall model framework of this study is shown in Fig. 1.

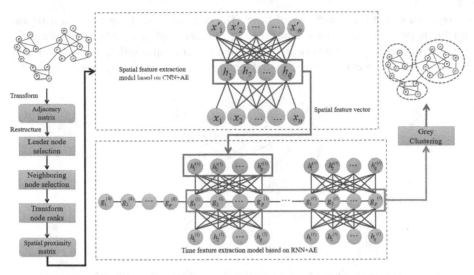

**Fig. 1.** Deep Recurrent Auto-Encoding neural network based on Reconstructive matrix for dynamic community detection (DRAER)

The remainder of the paper is organized as follows. Section 2 proposes a spatio-temporal feature extraction method for dynamic community discovery, and Sect. 3 introduces the process of dynamic community discovery method. In Sect. 4, the actual experimental results are analyzed and discussed. Section 5 summarizes the conclusions and future research directions.

## 2 Spatio-Temporal Feature Extraction Method

The method of dynamic community discovery based on recurrent neural network needs could be used to extract the time-dependent characteristics and network evolution characteristics of network data at different times. As an important tool for time analyses, a recurrent neural network could be combined with an auto-encoder method to extract time features. This study proposes a spatio-temporal feature extraction model that builds convolutional neural networks and recurrent neural networks, whose input layer and output layer have the same variables. These neural networks use convolutional layers and recurrent layers as the hidden layers between the input layer and the output layer to extract the spatio-temporal features through convolutional and recurrent operations and to represent the original input variables for encoding into vectors with spatio-temporal features.

### 2.1 Principle Explanation

A simple case study is given to explain the principle of the reconstruction adjacency matrix variable with the community matrix [6] as four nodes (i.e., $x_1$, $x_2$, $x_3$, and $x_4$), which contains four neurons in the input layer and a filter with the size of $1 \times 3$ (i.e.,

$\alpha_1$, $\alpha_2$, and $\alpha_3$) in the convolutional layer. The convolutional layer has two neurons, and the output layer have four neurons (i.e., $x_1'$, $x_2'$, $x_3'$, and $x_4'$). The simple schematic structure of the deep neural network combining the convolutional neural network and the auto-encoder for extracting the spatial features (i.e., $h1$ and $h2$) is shown in Fig. 2.

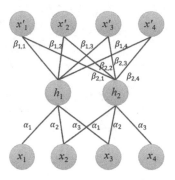

**Fig. 2.** The combination of the convolutional neural network and the auto-encoder

In the case study in Fig. 2, a recurrent neural network is adopted to analyze the spatial features at the first time point $\mathbf{H}^{(1)} = \left\{h_1^{(1)}, h_2^{(1)}\right\}$ and the spatial features at the second time point $\mathbf{H}^{(2)} = \left\{h_1^{(2)}, h_2^{(2)}\right\}$. The spatio-temporal features (i.e., $g^{(1)}$ and $g^{(2)}$) could be extracted by the combination of the recurrent neural network and the auto-encoder (shown in Fig. 3).

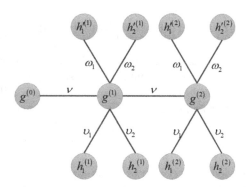

**Fig. 3.** The combination of the recurrent neural network and the auto-encoder

In this case study, the weights of the input layer are $\{\upsilon_1, \upsilon_2\}$; the adjustment variable of the neuron in the hidden layer is $\{b_{1,1}\}$; the weight between the recurrent layer is $\{v\}$; the weights between the hidden layer and the output layer are $\{\omega_1, \omega_2\}$; the adjustment variables of the two neurons in the output layer are $\{b_{2,1}, b_{2,2}\}$. The values of neurons in the recurrent layer (i.e., $\{g^{(0)}, g^{(1)}, g^{(2)}\}$) are shown in Eq. (1), Eq. (2), and Eq. (3); the values of neurons in the output layer (i.e., $\left\{h_1'^{(1)}, h_2'^{(1)}, h_1'^{(2)}, h_2'^{(2)}\right\}$) are shown in Eq. (4)

and Eq. (5); the loss function can be presented in Eq. (6).

$$g^{(0)} = 0. \tag{1}$$

$$g^{(1)} = vg^{(0)} + \sum_{i=1}^{2} v_i h_i^{(1)} + b_{1,1}. \tag{2}$$

$$g^{(2)} = vg^{(1)} + \sum_{i=1}^{2} v_i h_i^{(2)} + b_{1,1}. \tag{3}$$

$$h_i'^{(1)} = \omega_i g^{(1)} + b_{2,i}. \tag{4}$$

$$h_i'^{(2)} = \omega_i g^{(2)} + b_{2,i}. \tag{5}$$

$$E = \sum_{i=1}^{2} \frac{1}{2} \left( h_i'^{(1)} - h_i^{(1)} \right)^2 + \frac{1}{2} \left( h_i'^{(2)} - h_i^{(2)} \right)^2 = \sum_{i=1}^{2} \frac{1}{2} \left( \sigma_i'^{(1)} \right)^2 + \frac{1}{2} \left( \sigma_i'^{(2)} \right)^2. \tag{6}$$

For optimization, the gradient descent method is used, and the updated values of weights and adjustment variables are shown in Eq. (7), Eq. (8), Eq. (9), Eq. (10), and Eq. (11). The spatio-temporal features of the two input time points are cyclically extracted as $\{g^{(1)}, g^{(2)}\}$ when the training is completed.

$$\frac{\partial E}{\partial \omega_i} = \sum_{j=1}^{2} \frac{\partial E}{\partial \sigma_i'^{(j)}} \frac{\partial \sigma_i'^{(j)}}{\partial h_i'^{(j)}} \frac{\partial h_i'^{(j)}}{\partial \omega_i} = \sum_{j=1}^{2} \sigma_i'^{(j)} \times 1 \times g^{(1)} = \sum_{j=1}^{2} \sigma_i'^{(j)} \times g^{(1)}. \tag{7}$$

$$\frac{\partial E}{\partial b_{2,i}} = \sum_{j=1}^{2} \frac{\partial E}{\partial \sigma_i'^{(j)}} \frac{\partial \sigma_i'^{(j)}}{\partial h_i'^{(j)}} \frac{\partial h_i'^{(j)}}{\partial b_{2,i}} = \sum_{j=1}^{2} \sigma_i'^{(j)} \times 1 \times 1 = \sum_{j=1}^{2} \sigma_i'^{(j)}. \tag{8}$$

$$\frac{\partial E}{\partial v} = \sum_{i=1}^{2} \sum_{j=1}^{2} \frac{\partial E}{\partial \sigma_i'^{(j)}} \frac{\partial \sigma_i'^{(j)}}{\partial h_i'^{(j)}} \frac{\partial h_i'^{(j)}}{\partial g_i^{(j)}} \frac{\partial g_i^{(j)}}{\partial v} = \sum_{i=1}^{2} \sum_{j=1}^{2} \sigma_i'^{(j)} \times 1 \times \omega_i \times g_i^{(j-1)}.$$

$$= \sum_{i=1}^{2} \sum_{j=1}^{2} \sigma_i'^{(j)} \times \omega_i \times g_i^{(j-1)} \tag{9}$$

$$\frac{\partial E}{\partial v_k} = \sum_{i=1}^{2} \sum_{j=1}^{2} \frac{\partial E}{\partial \sigma_i'^{(j)}} \frac{\partial \sigma_i'^{(j)}}{\partial h_i'^{(j)}} \frac{\partial h_i'^{(j)}}{\partial g_i^{(j)}} \frac{\partial g_i^{(j)}}{\partial v_k} = \sum_{i=1}^{2} \sum_{j=1}^{2} \sigma_i'^{(j)} \times 1 \times \omega_i \times h_k^{(j)}.$$

$$= \sum_{i=1}^{2} \sum_{j=1}^{2} \sigma_i'^{(j)} \times \omega_i \times h_k^{(j)} \tag{10}$$

$$\frac{\partial E}{\partial b_{1,1}} = \sum_{i=1}^{2} \sum_{j=1}^{2} \frac{\partial E}{\partial \sigma_i'^{(j)}} \frac{\partial \sigma_i'^{(j)}}{\partial h_i'^{(j)}} \frac{\partial h_i'^{(j)}}{\partial g_i^{(j)}} \frac{\partial g_i^{(j)}}{\partial b_{1,1}} = \sum_{i=1}^{2} \sum_{j=1}^{2} \sigma_i'^{(j)} \times 1 \times \omega_i \times 1.$$

$$= \sum_{i=1}^{2} \sum_{j=1}^{2} \sigma_i'^{(j)} \times \omega_i \tag{11}$$

## 2.2  General Description

In this study, $q$ spatial features during $t$ time points are extracted as the input of a recurrent neural network. The recurrent neural network is combined with an auto-encoder neural network, which includes $q$ neurons in the input layer, a recurrent layer with $p$ neurons, and $q$ neurons in the output layer. The neural network structure is shown in Fig. 4. In the optimization process, the mean square error between the output layer and the input layer is adopted as the loss function, and the gradient descent method is used to correct each weight. In the operation stage, the trained recurrent neural network can be combined with an auto-encoder to extract the spatiotemporal features $\mathbf{G}^{(i)}$ of the network at the $i$ time point, which is shown in Eq. (12).

$$\mathbf{G}^{(i)} = \left[ g_1^{(i)} \ g_2^{(i)} \ \cdots \ g_p^{(i)} \right]. \tag{12}$$

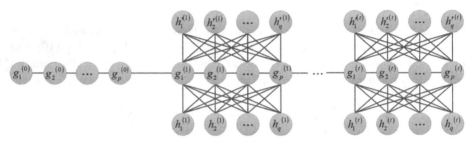

**Fig. 4.** The combination of the recurrent neural network and the auto-encoder for a general case

## 3  Dynamic Community Discovery Method

In this study, the K-means algorithm is used to group data for achieving dynamic community discovery. After extracting the spatiotemporal features of the network, a total of $n$ feature vectors of $t \times p$ dimensions can be obtained, and the $n$ data are grouped by using the K-means algorithm.

### 3.1  Algorithm

The steps of dynamic community discovery based on recurrent neural network include the following steps.

(1)  **Randomly selecting $k$ data from $n$ data as the cluster center of $k$ clusters in the initialization phase.**
(2)  **Updating the feature vector of each group center**. For instance, the feature vector $\mathbf{L}_i$ of the $i$-th group center is expressed in Eq. (13). In this study, the Euclidean distance is used to calculate the correlation. After obtaining the correlation between each data and each cluster center, the data is grouped into the cluster in accordance

with the highest correlation. The calculation of the correlation coefficient between the $j$-th data $\mathbf{G}_j$ (shown in Eq. (14)), and the correlation coefficient $\varphi(i, j)$ of the $i$-th group center is shown in Eq. (15). The calculation results could be used for grouping.

$$\mathbf{L}_i = \left[ l_{i,1}^{(1)} \ l_{i,2}^{(1)} \ \cdots \ l_{i,p}^{(1)} \ \cdots \ l_{i,1}^{(t)} \ \cdots \ l_{i,p}^{(t)} \right]. \tag{13}$$

$$\mathbf{G}_j = \left[ g_{j,1}^{(1)} \ g_{j,2}^{(1)} \ \cdots \ g_{j,p}^{(1)} \ \cdots \ g_{j,1}^{(t)} \ \cdots \ g_{j,p}^{(t)} \right]. \tag{14}$$

$$\varphi(i, j) = \sqrt{\sum_{k=1}^{p} d(i, j, k)^2}, \text{ where } d(i, j, k) = g_{j,k}^{(b)} - l_{i,k}^{(b)}. \tag{15}$$

(3)  **Determining the feature vector of each group center.** If each group center has no changes, the algorithm has reached convergence; if more than one of group centers has changes, Steps (2) and (3) are repeated.

## 3.2  Time Complexity Analyses

The neural networks include the following parameters: $n$ records are in the dataset; the number of neurons in the $i$-th hidden layer is $m_i$; the total number of hidden layers is $l$; there are $t$ time points; the number of training times is $r$. The time complexities of methods (i.e., (1) an auto-encoder model, (2) a CNN model, (3) the combination of an auto-encoder and a CNN model, and (4) the combination of an auto-encoder, a CNN, and RNN model) are shown in Eq. (16), Eq. (17), Eq. (18), and Eq. (19), respectively. Therefore, the time complexity of the proposed dynamic community discovery method is $O(trnm^2)$.

$$O\left( trn \sum_{1}^{l-1} m_i m_{i+1} \right) = O\left( trnm^2 \right). \tag{16}$$

$$O\left( trn \sum_{1}^{l-1} m_i m_{i+1} \right) = O\left( trnm^2 \right). \tag{17}$$

$$O\left( trn \sum_{1}^{l-1} m_i m_{i+1} \right) = O\left( trnm^2 \right). \tag{18}$$

$$O\left( trn \sum_{1}^{l-1} m_i m_{i+1} \right) = O\left( trnm^2 \right). \tag{19}$$

# 4  Experimental Results and Discussions

This section consists of two subsections: (1) experimental environments and (2) experimental results and analyses.

## 4.1 Experimental Environments

This subsection presents data sets, algorithm performance indicators, run configuration and algorithm comparisons.

(1) **Datasets**. This study uses the Enron Email-Enron dataset [7] from the SNAP website to verify the performance of the proposed algorithm for a real network. The statistical characteristics of Email-Enron dataset are shown in Table 1.

**Table 1.** The statistical characteristics of Email-Enron dataset.

| Maximum number of nodes | Maximum number of edges | Number of timestamps | Increasing the number of edges at a moment | Reducing the number of edges at a moment | Description |
|---|---|---|---|---|---|
| 251 | 5923 | 8 | 216 | 126 | The Enron Mail Data Set records the email communication data of the employees of Enron in the United States |

(2) **Evaluation Factor**. For evaluation, the extension modularity (EQ) [7] is used as an evaluation factor to evaluate the performance of the proposed method. EQ is defined in Eq. (20), where $O_i$ represents the number of communities to which node $i$ belongs.

$$EQ = \frac{1}{2m} \sum_{k=1}^{c} \sum_{i,j \in C_k} \frac{1}{O_i O_j} \left[ A_{ij} - \frac{k_i k_j}{2m} \right]. \tag{20}$$

## 4.2 Experimental Results and Analyses

After finding the most influential observation leader at time $t_1$ [6] and using the strategy of space structure reconstruction [6], the dataset of Email-Enron's real network was used to sequentially reconstruct the eight-time network of Email-Enron. The unreconstructed matrix and reconstructed matrix at each time are shown in Fig. 5, Fig. 6, Fig. 7, Fig. 8, Fig. 9, Fig. 10, Fig. 11, and Fig. 12.

In Table 2, the better modularity of the algorithm is highlighted in bold. Table 2 shows that the AE algorithm has achieved better modularity at $t_1$ and $t_2$, and the proposed method has achieved better modularity from $t_3$ to $t_8$.

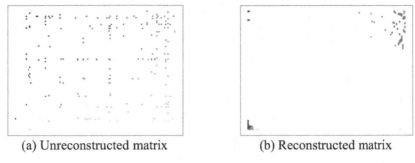

(a) Unreconstructed matrix                    (b) Reconstructed matrix

**Fig. 5.** The unreconstructed matrix and the reconstructed matrix at time $t_1$.

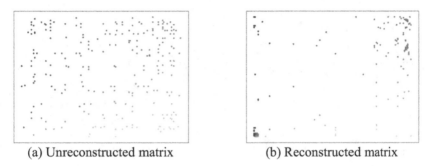

(a) Unreconstructed matrix                    (b) Reconstructed matrix

**Fig. 6.** The unreconstructed matrix and the reconstructed matrix at time $t_2$.

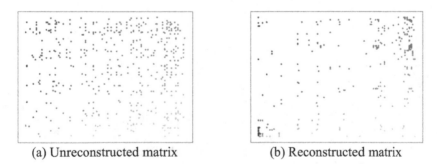

(a) Unreconstructed matrix                    (b) Reconstructed matrix

**Fig. 7.** The unreconstructed matrix and the reconstructed matrix at time $t_3$.

The learning effect of the proposed method shows insufficient accuracy in the initial stage. The modularity of the proposed method is lower than the benchmark algorithm AE at time $t_1$ and $t_2$. This is mainly due to the fact that the proposed method uses RNN to learn the time-dependent characteristics of the network. The initial two moments have fewer network information nodes and fewer connections, and the network information at a single moment is relatively insufficient. More importantly, there is only one time-series network at the dependent moment. The number of time-series network input is insufficient, which makes the proposed method unable to learn the time features of the

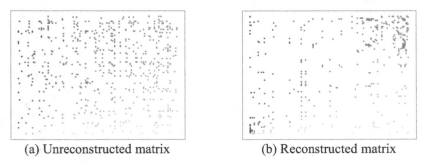

(a) Unreconstructed matrix          (b) Reconstructed matrix

**Fig. 8.** The unreconstructed matrix and the reconstructed matrix at time $t_4$.

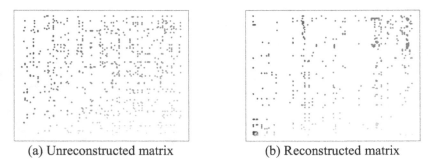

(a) Unreconstructed matrix          (b) Reconstructed matrix

**Fig. 9.** The unreconstructed matrix and the reconstructed matrix at time $t_5$.

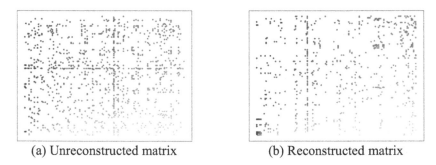

(a) Unreconstructed matrix          (b) Reconstructed matrix

**Fig. 10.** The unreconstructed matrix and the reconstructed matrix at time $t_6$.

sequence network, so that the modularity on the $t_1$ time slice and $t_2$ time slice is low. The learning effect of the proposed method in the initial stage has an insufficient precision.

From $t_3$ to $t_8$, the modularity of the proposed method is higher than that of the AE algorithm. Since from $t_3$, there are already 2 groups of networks at the time of dependent time. The RNN has been able to learn the evolution of the time series dynamic network from these 2 groups. Therefore, the accuracy of the algorithm is stable from this moment and higher than the AE algorithm.

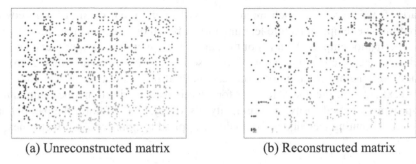

(a) Unreconstructed matrix                (b) Reconstructed matrix

**Fig. 11.** The unreconstructed matrix and the reconstructed matrix at time $t_7$.

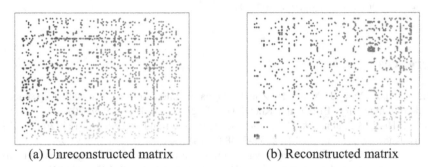

(a) Unreconstructed matrix                (b) Reconstructed matrix

**Fig. 12.** The unreconstructed matrix and the reconstructed matrix at time $t_8$.

**Table 2.** Modularity comparison between the proposed method and the AE method.

| Time $t$ | Auto-encoder | The proposed method |
|---|---|---|
| $t1$ | **0.086** | 0.075 |
| $t2$ | **0.063** | 0.035 |
| $t3$ | 0.190 | **0.297** |
| $t4$ | 0.213 | **0.345** |
| $t5$ | 0.210 | **0.287** |
| $t6$ | 0.165 | **0.267** |
| $t7$ | 0.211 | **0.274** |
| $t8$ | 0.213 | **0.384** |

## 5   Conclusions and Future Work

Firstly, the network of each time slice is reconstructed through the space structure reconstruction strategy, and the space learning model is used to obtain the spatial features at each moment. Secondly, the space feature learning model based on convolutional auto-encoder is subtly integrated into the temporal feature learning model in this study. Then, the spatial feature vector at each moment is used as the input of the neural network and input into the temporal feature learning model based on a recurrent neural network. Finally, the spatio-temporal features of the network are extracted. Experimental results showed that when the dynamic network time series network is greater than 2 groups, the proposed dynamic community discovery method based on a recurrent neural network can extract spatio-temporal features for effectively detecting dynamic communities and improving the modularity of community discovery. In the future, the parallel and distributed computing techniques [8] could be applied to improve the effectiveness of method.

**Acknowledgements.** This work was supported by the National Natural Science Foundation of China (Nos. 62002063, 61902313, and 61906043) and Natural Science Foundation of Fujian Province (Nos. 2018J07005 and 2020J05112).

# References

1. Xu, H., Yu, Z., Yang, J., Xiong, H., Zhu, H.: Dynamic talent flow analysis with deep sequence prediction modeling. IEEE Trans. Knowl. Data Eng. **31**(10), 1926–1939 (2018)
2. Liu, S., Hooi, B., Faloutsos, C.: A contrast metric for fraud detection in rich graphs. IEEE Trans. Knowl. Data Eng. **31**(12), 2235–2248 (2018)
3. Sumith, N., Annappa, B., Bhattacharya, S.: influence maximization in large social networks: heuristics, models and parameters. Futur. Gener. Comput. Syst. **89**, 777–790 (2018)
4. Zhang, G., Xu, L., Xue, Y.: The time dependency predictive model on the basis of community detection and long-short term memory. Concurr. Comput. Pract. Exp. **29**(4), article no. e4184 (2017)
5. Fang, H., Wu, F., Zhao, Z., Duan, X., Zhuang, Y., Ester, M.: Community-based question answering via heterogeneous social network learning. In: Proceeding of the 30th AAAI Conference on Artificial Intelligence, pp. 122–128 (2016)
6. Wu, L., Zhang, Q., Chen, C.H., Guo, K., Wang, D.: Deep learning techniques for community detection in social networks. IEEE Access **8**, 96016–96026 (2020)
7. McAuley, J., Leskovec, J.: Learning to discover social circles in ego networks. In: Proceedings of the 26th Annual Conference on Neural Information Processing Systems, pp. 548–556 (2012)
8. Chen, H., Jin, H., Wu, S.: Minimizing inter-server communications by exploiting self-similarity in online social networks. IEEE Trans. Parallel Distrib. Syst. **27**(4), 1116–1130 (2016)

# 6th International Workshop on Mobile Data Management, Mining, and Computing on Social Network

# Deep Attributed Network Embedding Based on the PPMI

Kunjie Dong, Tong Huang, Lihua Zhou$^{(\boxtimes)}$, Lizhen Wang, and Hongmei Chen

School of Information, Yunnan University, Kunming 650091, China
kunjiedong@qq.com, huangtong@mail.ynu.edu.cn,
{lhzhou,hmchen}@ynu.edu.cn, lzhwang2005@126.com

**Abstract.** The attributed network embedding aims to learn the latent low-dimensional representations of nodes, while preserving the neighborhood relationship of nodes in the network topology as well as the similarities of attribute features. In this paper, we propose a deep model based on the positive point-wise mutual information (PPMI) for attributed network embedding. In our model, attribute features are transformed into an attribute graph, such that attribute features and network topology can be handled in the same way. Then, we perform the random surfing and calculate the PPMI on the attribute/topology graph to effectively maintain the structural characteristics and the high-order proximity information. The node representations are learned by a shared Auto-Encoder. Besides, the local pairwise constraint is used in the shared Auto-Encoder to improve the quality of node representations. Extensive experimental results on four real-world networks show the superior performance of the proposed model over the 10 baselines.

**Keywords:** Attributed network embedding · Random surfing · Positive point-wise mutual information · Auto-encoder

## 1 Introduction

Network embedding (NE) aims to learn the latent low-dimensional representations of nodes in a network while preserving the intrinsic essence of the network [8,15,19], which can provide precise service and higher efficiency in practical applications, such as targeted detection and personalized recommendation [22,28]. Therefore, NE has aroused many researchers' interests under the drive of great requirements in recent years.

Supported by the National Natural Science Foundation of China (61762090, ,62062066, 61966036, and 61662086), the Natural Science Foundation of Yunnan Province (2016FA026), the Program for Innovation Research Team (in Science and Technology) in University of Yunnan Province (IRTSTYN), and the National Social Science Foundation of China (18XZZ005).
K. Dong and T. Huang—Both authors have contributed equally to this work.

C. S. Jensen et al. (Eds.): DASFAA 2021 Workshops, LNCS 12680, pp. 251–266, 2021.
https://doi.org/10.1007/978-3-030-73216-5_18

In a network consisting of nodes and edges, nodes represent objects and edges describe the interactive relationships amongst nodes. For example, in a cite network, nodes represent papers and edges describe the cite relationship amongst papers. In general, the interactive relationships amongst nodes are referred to as network topology, which plays a vital role in network analysis tasks. Network topology, typically in the form of node adjacency matrix, is the most common form of network representation. An important goal of NE is to preserve the neighborhood relationship of nodes in the network topology. To this end, various NE methods, such as DeepWalk [15] used the random walks based on the sampling strategies to convert a general graph structure into a large collection of linear sequences, and then utilized the skip-gram model [13] to learn low-dimensional representations for nodes from such linear sequences. This is one effective way to express graph structural information, because the sampled node sequences characterize the connections amongst nodes in a graph. However, the procedure involves a slow sampling process, and the hyper-parameters (such as walk length and total walks) are not easy to determine, especially for the large graphs. Because the sampled sequences have finite lengths, furthermore, it is difficult to capture the correct contextual information for nodes that appear at the boundaries of the sampled sequences, such that some relationships amongst nodes cannot be captured accurately and completely. To make up for the shortcomings of random walk, DNGR [4] adopts a random surfing model to capture graph structural information directly, instead of using the sampling-based method for generating linear sequences. The random surfing model first randomly orders the nodes in a graph, and then directly yield a probabilistic co-occurrence (PCO) matrix that capturing the transition probabilities amongst different nodes. Based on the PCO matrix, the positive point-wise mutual information (PPMI) can be computed, which avoids the expensive sampling process. As an explicit representation of a graph, the PPMI can effectively maintain the structural characteristics of a graph and contain the high-order similarity information of the nodes [4], so the PPMI representations of nodes can more accurately capture potentially complex, non-linear relations amongst different nodes. But DNGR used network topology alone while did not take the attribute features affiliated to nodes into consideration.

The attribute features affiliated to nodes, such as authors, research themes and keywords associated with papers in a citation network, describe the individual profile of nodes in a micro-perspective. This information often carries orthogonal and complementary knowledge beyond node connectivity and network topology, so incorporating semantic information is expected to significantly enhance NE based on network topology alone. A network whose nodes are associated with attribute features referred to as an attributed network [2]. The embedding of an attributed network (ANE) aims to learn the latent low-dimensional representations of nodes while preserving the neighborhood relationship of nodes in the network topology as well as the semantics of attribute features. This is not a trivial task, because network topology and attribute features are two heterogeneous information, although they describe the same network from two different perspectives [5,9]. How to integrate two heterogeneous information and preserve the intrinsic essence contained in network topology and attribute features simultaneously is a key issue in ANE. Some existing approaches, such as TADW [24], ASNE [12], CANE [20], first converted network topology into the feature representations, then which were used to embed into a low-dimensional space. Meanwhile, attribute features were also used to derive low-dimensional embedding on node semantics. The two low-dimensional representations of all these NEs are concatenated to joint learn the final embedding. Due to converting a network topology into a feature representation may lose or may not faithfully represent non-linear relationship amongst the nodes [11], and the individual feature vector only contains individual information without inter-individual association relationships, combining topological feature vector and attribute feature vector together may unsatisfactory to explore and exploit the complementary relationship between these two types of information. Since, UWMNE [11] maintained network topology in the graph form and built attribute graph to represent semantic information, and then used deep neural networks to integrate the topological and semantic information in these graphs to learn a unified embedding representation.

Inspired by the DNGR [4] and the UWMNE [11], we propose a deep model based on the PPMI for ANE in this paper. The model is referred to as DANEP. Specifically, we first transform attribute features into an attribute graph, which is homogeneous with topology graph, so we can deal with them in the same way. Next, we carry out random surfing on the attribute/topology graph respectively to generate a attribute/topology probabilistic co-occurrence (PCO) matrix, and then calculate the attribute/topology PPMI based on the attribute/topology PCO matrix. After that, using a shared Auto-Encoder to learn low-dimensional node representations. The advantages of DANEP lie in: the attribute graph describes the geometry of potential non-linear manifolds under attribute features information more clearly, the uniformed graph representation of attribute features and network topology contributes to integrating the complementary relationship between two types of information; the random surfing captures graph structural information concerning attribute/topology, the PPMIs calculated from the attribute/topology PCO matrixes effectively maintain both the structural characteristics and the high-order proximity information of attribute/topology graph; and the shared Auto-Encoder learns high-level abstractions from low-level features as well as captures highly non-linear information conveyed by the graph via non-linear projections. Besides, the local pairwise constraint is further designed in shared Auto-Encoder to improve the quality of node representations. We also conduct extensive experiments on four real-world networks and compare our approach with 10 baselines. The experimental results demonstrate the superiority of our approach.

It is needed to note that our DANEP model is different from the DNGR [4] and the UWMNE [11]. In our DANEP model, we apply the deep learning method on PPMIs of attribute features and network topology, but the DNGR just apply matrix factorization on PPMI of network topology, and the UWMNE directly use network topology and attribute graph as the input of an Auto-Encoder.

The rest of the paper is arranged as follows. Section 2 offers a brief overview of related work. The details of DANEP are presented in Sect. 3. Section 4 provides extensive experiments and results, and in Sect. 5, conclusions are given.

## 2 Related Work

### 2.1 Network Embedding

Many network embedding approaches only utilized network topology to learn the latent low-dimensional representations. DeepWalk [15] first employed the truncated random walks to capture the local information and then learn the latent embedding result by making use of the local information. Node2vec [8] proposed a biased random walk method to explore various neighborhoods. Line [19] considered to preserve the first-order and second-order proximity of network topology into the learned embedding representation. SDNE [21] proposed a semi-supervised model to jointly preserve the first-order and second-order similarity of network topology. Struc2vec [17] utilized a weighted random walk to obtain a similar node sequence and conceived a hierarchical structure strategy to capture node proximity at different scales. GraRep [3] integrated global structural information learned from different models into the embedding representation. DNGR [4] first adopted a random surfing model to capture graph structural information, and then used a stacked denoising Auto-Encoder to learn low-dimensional vertex representations.

### 2.2 Attributed Network Embedding

In recent years, many researchers learned representations of nodes by integrating network topology and attribute features of nodes. This brings new opportunities and development for embedding learning. In detail, AANE [10] considered the proximity of the attribute features into embedding learning and adopted a distributed manner to accelerate the learning process. TADW [24] proposed the text-associated DeepWalk model to integrate node's text features into embedding learning by matrix factorization. ASNE [12] adopted a deep neural network to model the complex interrelations between attribute features and network topology. DANE [6] employed two symmetrical Auto-Encoders to capture the consistency and complementary information between attribute features and network topology, where the two symmetrical Auto-Encoders are allowed to interact with each other. ANRL [27] utilized a neighbor enhancement Auto-Encoder with attribute-aware skip-gram to extract the correlations of attribute features and the network topology. NANE [14] considered the local and global information in the embedding process by a pairwise constraint. Based on the observations that nodes with similar topology may be dissimilar in their attribute features and vice versa,which are referred to as the partial correlation, PRRE [29] taken the partial correlation of nodes into account in the learning process.

# 3    The Proposed Model

In this section, we first present the definition of ANE and then develop a deep attributed network embedding model based on the positive pointwise mutual information.

## 3.1    Problem Definition

Given an attributed network with $n$ nodes and $m$ edges $G = (V, E, \mathbf{A})$, wherein $V = \{v_1, \cdots, v_n\}$ and $E = \{e_{ij}\}_{i,j=1}^n$ represent the sets in items of nodes and edges, respectively, and $\mathbf{A} \in R^{n \times m}$ represent the attribute matrix affiliated to the nodes, whose row vector $\mathbf{a}_i \in R^m$ corresponds to the attribute features of the node $v_i$. Let $\mathbf{S} \in R^{n \times n}$ be the adjacent matrix affiliated to the edges, whose the element $s_{ij}$ corresponds to the relationship of the edge between nodes $v_i$ and $v_j$, i.e., $s_{ij} = 1$ indicates there exists an edge linked $v_i$ to $v_j$, and $s_{ij} = 0$ indicates the edge is nonexistent. The goal of the ANE is to find a map function $f(\mathbf{A}, \mathbf{S}) \rightarrow \mathbf{H}$ that map attribute features $\mathbf{A}$ and network topology $\mathbf{S}$ into a unified low-dimensional representation $\mathbf{H} \in R^{n \times d}(d \ll n, d \ll m)$ while preserving the proximities existing in both the attribute of the nodes and the topology of the network. More precisely, nodes with similar attribute and topology in the original network should be closer in the embedding space.

## 3.2    The Architecture of Proposed Model

The architecture of DANEP is shown in Fig. 1. DANEP first constructs an attribute graph based on the attribute features $A$, such that the attribute graph and the topology graph are homogeneous. Based on the homogeneous representations of the attribute graph and topology graph, the random surfing is first conducted to obtain the attribute/topology probabilistic co-occurrence (PCO) matrix, and then the PPMIs concerning the attribute graph and topology graph are calculated, represented as **PPMI_AF** and **PPMI_NT**, respectively. The row vectors of **PPMI_AF** and **PPMI_NT** depict the profile and the neighborhood relationships of node $v_i$ with respect to attribute features and network topology. After that, a shared Auto-Encoder equipped with the local enhancement of graph regulation is applied to learn the unified low-dimensional representation for each node from the PPMIs concerning the attribute graph and topology graph.

**The Construction of the Attribute Graph.** In this subsection, we construct an attribute graph based on the attribute features $\mathbf{A}$. Let $\mathbf{B} \in R^{n \times n}$ be the attribute similarity matrix, whose elements $b_{ij} \in \mathbf{B}$ can be measured by the similarity of attribute vectors $\mathbf{a}_i \in \mathbf{A}$ and $\mathbf{a}_j \in \mathbf{A}$, such as the cosine similarity can be calculated by Eq. (1), where "·" signifies the dot product of the two vectors, "$|| \cdot ||$" denotes $L2$ norm, and "$\times$" indicates the product of two scalars.

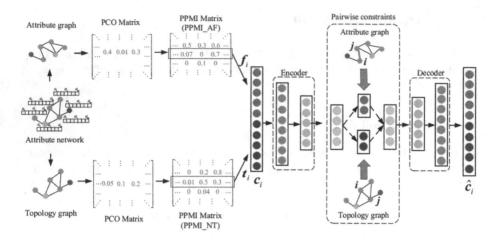

**Fig. 1.** The architecture of DANEP.

$$b_{ij} = \frac{\mathbf{a}_i \cdot \mathbf{a}_j}{\|\mathbf{a}_i\| \times \|\mathbf{a}_j\|} \tag{1}$$

Intuitively, the distance between two nodes is closer, the more intimate relationship they should have. Therefore, we apply the k-nearest neighbor method [11,18] on the $\mathbf{B}$ to construct the attribute graph with $n$ nodes, where each node $v_i$ is connected to $k$ nodes with top-k similarities in $b_i$. Let $\mathbf{B}^{new} \in R^{n \times n}$ be the adjacent matrix of the constructed attribute graph, then the element $b_{ij}^{new} = 1$ indicates there exists an edge linked $v_i$ to $v_j$, and $b_{ij}^{new} = 0$ indicates the edge is nonexistent.

**The Calculation of PPMIs.** Motivated by DNGR [4], we adopt the random surfing model on the topology graph $\mathbf{S}$/attribute graph $\mathbf{B}^{new}$ to obtain the attribute/topology probabilistic co-occurrence (PCO) matrix through k-step iterative. The iterative process can be represented by Eq. (2), where $\mathbf{p}_0$ is the initial one-hot vector with $i$-th value is 1 and the other values are 0, coefficient $\alpha$ and $1 - \alpha$ represent the probabilities with respect to the node jumps to the next node and returns to original vertex (restart), respectively.

$$\mathbf{p}_k = \alpha \cdot \mathbf{p}_{k-1} + (1 - \alpha)\mathbf{p}_0 \tag{2}$$

Based on the attribute/topology PCO matrix, the pointwise mutual information (PMI) can be calculated by Eq. (3), where $p(v_i, v_j)$ represents the number of co-occurrences that nodes $v_i$ and $v_j$ are in the same context, $|D| = \sum_{v_i} \sum_{v_j} p(v_i, v_j)$, $p(v_i)$ and $p(v_j)$ represents the number of occurrences of nodes $v_i$ and $v_j$, respectively.

$$\mathbf{PMI}_{v_i, v_j} = \log\left(\frac{p(v_i, v_j) \cdot |D|}{p(v_i) \cdot p(v_j)}\right) \tag{3}$$

Then, PPMI can be calculated by Eq. (4) [23], which means that negative values in attribute/topology PMI are assigned to zeros.

$$\mathbf{PPMI}_{v_i,v_j} = \max(\mathbf{PMI}_{v_i,v_j}, 0) \tag{4}$$

**The Design of the Shared Auto-Encoder.** In general, an Auto-Encoder consists of an encoder and a decoder which can extract inherent essence and non-linear information of a network. In DANEP, we designed a shared Auto-Encoder with $2K - 1$ layers to incorporate attribute features and network topology. The input of the Auto-Encoder is the concatenation of the row vectors of **PPMI_AF** and **PPMI_NT**, i.e. $\mathbf{c}_i = (\mathbf{f}_i, \mathbf{t}_i) = (f_{i1}, \cdots, f_{in}, t_{i1}, \cdots, t_{in})$, where $\mathbf{C} = [\mathbf{F}, \mathbf{T}] \in R^{n \times 2n}$, $\mathbf{c}_i$, $\mathbf{f}_i$ and $\mathbf{t}_i$ are the $i$-th row vector of $\mathbf{C}$, $\mathbf{F}$, and $\mathbf{T}$, respectively. Let $\mathbf{y}_{i,k}(k = 1, \cdots, K)$ and $\hat{\mathbf{y}}_{i,k}(k = 1, \cdots, K)$ be the desired embedding representation and the reconstructed representation of the Auto-Encoder, then $\mathbf{y}_{i,k}(k = 1, \cdots, K)$ and $\hat{\mathbf{y}}_{i,k}(k = 1, \cdots, K)$ can be computed by Eq. (5)–(9).

$$\mathbf{y}_{i,1} = f(\mathbf{W}_1 \mathbf{c}_i + \mathbf{b}_1) \tag{5}$$

$$\mathbf{y}_{i,k} = f(\mathbf{W}_k \mathbf{y}_{i,k-1} + \mathbf{b}_k)(k = 2, \cdots, K-1) \tag{6}$$

$$\mathbf{y}_i = \hat{\mathbf{y}}_{i,1} = \mathbf{y}_{i,K} = f(\mathbf{W}_K \mathbf{h}_{i,K-1} + \mathbf{b}_K) \tag{7}$$

$$\hat{\mathbf{y}}_{i,k} = f(\mathbf{W}_{K+k-1} \hat{\mathbf{y}}_{i,k-1} + \mathbf{b}_{K+k-1})(k = 2, \cdots, K-1) \tag{8}$$

$$\hat{\mathbf{y}}_{i,K} = f(\mathbf{W}_{2K-1} \hat{\mathbf{y}}_{i,K-1} + \mathbf{b}_{2K-1}) \tag{9}$$

Where $f(\cdot)$ represents the non-linear activation function, and $\theta = \{\mathbf{W}_k, \mathbf{b}_k\}(k = 1, \cdots, 2K - 1)$ are weight and bias parameters of the shared Auto-Encoder.

Let $\hat{\mathbf{C}}$ be the output of the decoder, where $\hat{\mathbf{c}}_i = \hat{\mathbf{y}}_{i,K} = f(\mathbf{W}_{2K-1} \hat{\mathbf{y}}_{i,K-1} + \mathbf{b}_{2K-1})$. The goal of Auto-Encoder is to minimize the reconstruction loss between the $\mathbf{C}$ and $\hat{\mathbf{C}}$, so the loss function is defined as:

$$\mathcal{L}_{rec} = \sum_{i=0}^{n} ||\hat{\mathbf{c}}_i - \mathbf{c}_i||_2^2 \tag{10}$$

To further improve the quality of node representation of the shared Auto-Encoder, we designed the local pairwise constraint, which is used to reinforce the consistency and complementary information contained in attribute features and network topology. Given the adjacent matrix $\mathbf{S}/\mathbf{B}^{new}$ of attribute/topology graph, the local pairwise constraint is defined as:

$$\mathcal{L}_{local} = \frac{1}{2} \sum_{i=1}^{n} \sum_{j=1}^{n} s_{ij} ||\mathbf{y}_i - \mathbf{y}_j||_2^2 + \frac{1}{2} \sum_{i=1}^{n} \sum_{j=1}^{n} b_{ij}^{new} ||\mathbf{y}_i - \mathbf{y}_j||_2^2 \tag{11}$$
$$= tr((\mathbf{Y}^C)^T \mathbf{L}_1 \mathbf{Y}^C) + tr((\mathbf{Y}^C)^T \mathbf{L}_2 \mathbf{Y}^C)$$

where $\mathbf{L}_1 = \mathbf{D}' - \mathbf{S}$, $\mathbf{L}_2 = \mathbf{D}'' - \mathbf{B}^{new}$, both $\mathbf{D}' = [d'_{ij}] \in R^{n \times n}$ and $\mathbf{D}'' = [d''_{ij}] \in R^{n \times n}$ are diagonal matrices, $D'_{ii} = \sum_{j=1}^{n} s_{ij}$, $D''_{ii} = \sum_{j=1}^{n} b_{ij}^{new}$.

Thus, the objective function of the DANEP is defined as:

$$\mathcal{L} = \alpha\mathcal{L}_{local} + \beta\mathcal{L}_{rec} \tag{12}$$

Where $\alpha$ and $\beta$ are the hyper-parameter to balance the weights among different losses.

## 4    Experiments and Results

In this section, we conduct extensive experiments on the four real-world networks by adopting three widely used applications, i.e., node classification, node clustering, and network visualization, to evaluate the effectiveness of our proposed method DANEP.

### 4.1    Datasets

In experiments, four publicly available networks with class labels are used, i.e., Cora, Citeseer, BlogCatelog and Flicker networks, where the first two datasets are academic papers citation network, and the last two datasets are the social networks. In Cora/Citeseer networks, nodes and edges represent academic papers and the citation relationships amongst those papers, respectively, each paper can be represented as a bag-of-words vector with 1433/3703-dimensions, and papers are divided into 7/6 categories, such as Genetic algorithm, Neural Networks and Reinforcement Learning. In BlogCatelog/ Flicker networks, nodes and edges represent the users and relationships amongst those users, respectively, each user can be represented as a bag-of-words vector with 8189/12047-dimensions, and those users are divided into 6/9 categories based on social preferences. The statistics for each network are summarized in Table 1.

**Table 1.** The statistics of networks.

| Dataset | Nodes | Edges | Features | Classes |
|---|---|---|---|---|
| Cora | 2708 | 5278 | 1433 | 7 |
| Citeseer | 3312 | 4660 | 3703 | 6 |
| BlogCatalog | 5196 | 171743 | 8189 | 6 |
| Flicker | 7575 | 239738 | 12047 | 9 |

### 4.2    Baselines

To verify the effectiveness of DANEP model, we select 10 approaches as the baselines, including: 4 "Topology-only" algorithms, i.e., DeepWalk [15], Node2Vec [8], GraRep [3], DNGR [4], and 6 "Topology +Attribute" algorithms, i.e., AANE [10], TADW [24], ASNE [12], DANE [6], ANRL [27], NANE [14]. The details of these baselines are illustrated as follows:

**"Topology-Only" Algorithms:** DeepWalk [15]: It employed the truncated random walks to capture the local topology information, and then learned the latent embedding representation by making full use of the captured local information.

Node2Vec [8]: It proposed biased random walks to project node into a low-dimensional space while preserving the network essence by exploring and preserving network neighborhoods of nodes.

GraRep [3]: It developed a model to learn the node representation for the weighted graph by integrating global structural similarity in the learning process.

DNGR [4]: It adopted a random surfing model to capture topology information, and then utilized the stacked denoising Auto-Encoder to extract meaningful information into the low-dimensional vector representation.

**"Topology +Attribute" Algorithms:** AANE [10]: AANE considered and integrated the proximity of attribute features into the embedding learning and adopted a distributed manner to accelerate the learning process.

TADW [24]: It employed a matrix factorization method based on DeepWalk to learn low-dimensional representations of text and network topology, and then concatenate them to form the final representation.

DANE [6]: DANE allowed neighborhood topology obtained by random walks and attribute features to interact with each other to preserve the consistent and complementary information during the learning process.

ANRL [27]: It designed a neighbor enhancement Auto-Encoder model with an attribute-aware skip-gram to integrate the attribute features and network topology proximities in the learning process simultaneously.

ASNE [12]: ASNE integrated the adjacent matrix of network topology and attribute matrix on the input layer, and allowed them to interact with each other for capturing the complex relationships and the more serviceable information.

NANE [14]: It cascaded the adjacent matrix of network topology and cosine similarity of attribute features into the unified representation to capture the local information and non-linear correlation in the network.

## 4.3   Parameter Settings

To get a fair comparison, we set the embedding dimension $d$ of all datasets to be 128 for all baselines. For DeepWalk and Node2Vec, we set the window size as 10, the walk length as 80, and the number of walks per node as 10. For GraRep, the maximum transition step is set to 5. For TADW, we set the regularization parameter to 0.2. Besides, the default values of the other parameters for these methods were set the same as the open-source codes released by the original authors.

## 4.4   Node Classification

In this subsection, we randomly select 10%, 30%, 50% nodes as the training set and the remained nodes as the testing set, apply the linear SVM as the classifier

**Table 2.** The performance evaluation of node classification.

| Metrics | Methods | 10% | | 30% | | 50% | |
|---|---|---|---|---|---|---|---|
| | | Mi-F1 | Ma-F1 | Mi-F1 | Ma-F1 | Mi-F1 | Ma-F1 |
| Cora | DeepWalk | 0.7341 | 0.7180 | 0.7905 | 0.7796 | 0.8233 | 0.8136 |
| | Node2Vec | 0.7059 | 0.6880 | 0.7747 | 0.7627 | 0.8013 | 0.7913 |
| | GraRep | 0.7375 | 0.7207 | 0.7750 | 0.7561 | 0.7835 | 0.7631 |
| | DNGR | 0.6493 | 0.6380 | 0.7071 | 0.6968 | 0.7335 | 0.7164 |
| | AANE | 0.6582 | 0.6177 | 0.7195 | 0.6891 | 0.7305 | 0.7015 |
| | TADW | 0.7945 | 0.7777 | 0.8360 | 0.8220 | 0.8436 | 0.8298 |
| | DANE | 0.7751 | 0.7545 | 0.8188 | 0.8033 | 0.8302 | 0.8143 |
| | ANRL | 0.7487 | 0.7224 | 0.7659 | 0.7430 | 0.7731 | 0.7538 |
| | NANE | 0.5114 | 0.4660 | 0.5816 | 0.5470 | 0.6459 | 0.6113 |
| | ASNE | 0.457 | 0.4353 | 0.5347 | 0.5099 | 0.5723 | 0.5466 |
| | DANEP | **0.8215** | **0.8078** | **0.8398** | **0.8259** | **0.8575** | **0.8435** |
| Citeseer | DeepWalk | 0.5037 | 0.4723 | 0.5888 | 0.5502 | 0.6199 | 0.5790 |
| | Node2Vec | 0.4875 | 0.4508 | 0.5621 | 0.5223 | 0.5851 | 0.5401 |
| | GraRep | 0.5063 | 0.4644 | 0.5421 | 0.4881 | 0.5542 | 0.4957 |
| | DNGR | 0.4581 | 0.4228 | 0.4970 | 0.4547 | 0.5331 | 0.4895 |
| | AANE | 0.6549 | 0.6008 | 0.6834 | 0.6293 | 0.6930 | 0.6436 |
| | TADW | 0.6621 | 0.6181 | 0.7182 | 0.6520 | **0.7421** | **0.6972** |
| | DANE | 0.6415 | 0.5960 | 0.6950 | 0.6520 | 0.7163 | 0.6722 |
| | ANRL | **0.6795** | **0.6371** | 0.7270 | 0.6747 | 0.7397 | 0.6880 |
| | NANE | 0.4488 | 0.4171 | 0.5783 | 0.5334 | 0.6298 | 0.5749 |
| | ASNE | 0.3261 | 0.3028 | 0.4110 | 0.3695 | 0.4385 | 0.3874 |
| | DANEP | 0.6494 | 0.5970 | **0.7357** | **0.6834** | 0.7349 | 0.6787 |
| Flickr | DeepWalk | 0.4389 | 0.4352 | 0.5223 | 0.5144 | 0.5483 | 0.5385 |
| | Node2Vec | 0.3899 | 0.3863 | 0.4896 | 0.4804 | 0.5171 | 0.5049 |
| | GraRep | 0.4908 | 0.4829 | 0.5422 | 0.5327 | 0.5558 | 0.5466 |
| | DNGR | 0.4656 | 0.4027 | 0.4653 | 0.4552 | 0.4768 | 0.4656 |
| | AANE | 0.5865 | 0.6068 | 0.6151 | 0.6323 | 0.6244 | 0.6369 |
| | TADW | 0.6117 | 0.6026 | 0.7020 | 0.6940 | 0.7218 | 0.7143 |
| | DANE | 0.6453 | 0.6439 | 0.7160 | 0.7144 | 0.7395 | 0.7380 |
| | ANRL | 0.2740 | 0.1984 | 0.2947 | 0.2268 | 0.2978 | 0.2250 |
| | NANE | 0.3733 | 0.3690 | 0.4993 | 0.4931 | 0.5290 | 0.5224 |
| | ASNE | 0.4366 | 0.4316 | 0.5218 | 0.5116 | 0.5500 | 0.5413 |
| | DANEP | **0.8457** | **0.8431** | **0.8585** | **0.8565** | **0.8690** | **0.8670** |
| BlogCatalog | DeepWalk | 0.5781 | 0.5733 | 0.6624 | 0.6549 | 0.6899 | 0.6811 |
| | Node2Vec | 0.5296 | 0.5258 | 0.6283 | 0.6215 | 0.6592 | 0.6509 |
| | GraRep | 0.6890 | 0.6851 | 0.7272 | 0.7230 | 0.7450 | 0.7413 |
| | DNGR | 0.5896 | 0.5850 | 0.6515 | 0.6423 | 0.6682 | 0.6585 |
| | AANE | 0.8565 | 0.8539 | 0.8846 | 0.8828 | 0.8920 | 0.8897 |
| | TADW | 0.8199 | 0.8175 | 0.8610 | 0.8799 | 0.8789 | 0.8772 |
| | DANE | 0.8436 | 0.8400 | 0.8180 | 0.8752 | 0.8876 | 0.8856 |
| | ANRL | 0.8073 | 0.8004 | 0.8285 | 0.8225 | 0.8333 | 0.8274 |
| | NANE | 0.6806 | 0.6778 | 0.7764 | 0.7739 | 0.8043 | 0.8016 |
| | ASNE | 0.5838 | 0.5823 | 0.6651 | 0.6615 | 0.6759 | 0.6695 |
| | DANEP | **0.8749** | **0.8739** | **0.9126** | **0.9117** | **0.9226** | **0.9219** |

**Table 3.** The performance evaluation of node clustering.

| Metrics | Methods | Cora | Citeseer | BlogCatalog | Flickr | Average |
|---------|---------|------|----------|-------------|--------|---------|
| ACC | DeepWalk | 0.5609 | 0.4029 | 0.3646 | 0.3163 | 0.4112 |
|     | Node2Vec | 0.6128 | 0.4208 | 0.3553 | 0.3209 | 0.4275 |
|     | GraRep | 0.5027 | 0.3238 | 0.3710 | 0.2934 | 0.3727 |
|     | DNGR | 0.5948 | 0.3929 | 0.3599 | 0.2750 | 0.4057 |
|     | AANE | 0.3904 | 0.5379 | 0.4320 | 0.1338 | 0.3735 |
|     | TADW | 0.6686 | 0.5689 | 0.6848 | 0.3640 | 0.5716 |
|     | DANE | **0.7187** | 0.4884 | 0.4879 | 0.2445 | 0.4849 |
|     | ANRL | 0.5129 | 0.5730 | 0.4720 | 0.2137 | 0.4429 |
|     | NANE | 0.1503 | 0.2430 | 0.5837 | 0.2861 | 0.3158 |
|     | ASNE | 0.3889 | 0.4088 | 0.3896 | 0.2283 | 0.3539 |
|     | DANEP | 0.7179 | **0.6097** | **0.7048** | **0.6886** | **0.6803** |
| NMI | DeepWalk | 0.4021 | 0.1371 | 0.1966 | 0.1694 | 0.2263 |
|     | Node2Vec | 0.4396 | 0.2227 | 0.2060 | 0.1801 | 0.2621 |
|     | GraRep | 0.3749 | 0.1673 | 0.2040 | 0.1482 | 0.2236 |
|     | DNGR | 0.4424 | 0.2017 | 0.1836 | 0.1462 | 0.2435 |
|     | AANE | 0.2206 | 0.2774 | 0.2759 | 0.0901 | 0.216 |
|     | TADW | **0.5515** | 0.3550 | 0.4352 | 0.1833 | 0.3813 |
|     | DANE | 0.5494 | 0.2975 | 0.3277 | 0.1165 | 0.3228 |
|     | ANRL | 0.3812 | 0.3619 | 0.3417 | 0.1004 | 0.2963 |
|     | NANE | 0.2096 | 0.2637 | 0.3583 | 0.1329 | 0.2411 |
|     | ASNE | 0.2096 | 0.1221 | 0.2165 | 0.1290 | 0.1693 |
|     | DANEP | 0.5510 | **0.3792** | **0.5207** | **0.6002** | **0.5128** |

and use 5-fold cross-validation to train the classifier in the learning process. This process is repeated 10 times and the average performance in terms of both Macro-F1 and Micro-F1 [25] is reported as the classification results. The detailed results are shown in Table 2, where the bold numbers indicate the best results. From Table 2, we have the following observations and analyses:

(1) DANEP obtains the best performance with respect to the Micro-F1 and Macro-F1 on the Cora, Flickr and BlogCatalog datasets when the training rates are 10%, 30% and 50%, respectively. The improved performance concerning the Micro-F1 and Macro-F1 is significantly on different datasets, such as DANEP achieves the 20.04%, 19.92%, 14.25%, 14.21%, 12.95 and 12.9% than the best baseline DANE on Flicker dataset when the training rates are 10%, 30% and 50%, respectively. Those results demonstrated the superiority of DANEP with random surfing and PPMI schemes.

(2) ANRL and TADW achieve the highest values on the Citeseer dataset when the training rates are 10% and 50%, respectively, which indicate that the ncighbor enhancement mechanism and text-associated matrix factorization have some the ability to capture the essence of the network, but they are still obviously inferior to DANEP.

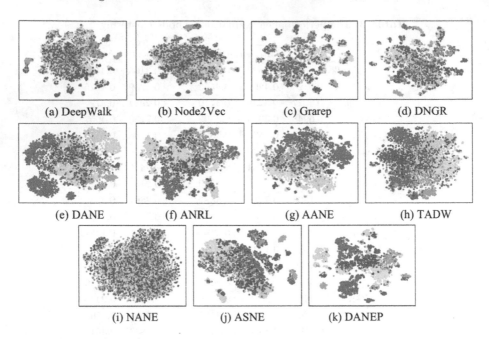

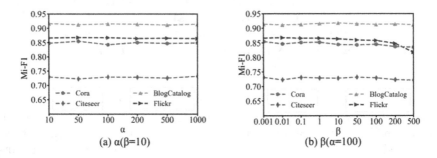

**Fig. 2.** The visualization result of different methods on the BlogCatalog dataset

**Fig. 3.** The sensitivity of DANEP w.r.t. different $\alpha$ and $\beta$ for node classification

### 4.5 Node Clustering

Node clustering is an unsupervised downstream task of network analysis based on the learned node representation. In this study, we use k-means [1] as the clustering algorithm, accuracy (ACC) [6] and normalized mutual information (NMI) [14] as metrics to evaluate the clustering performance. Similarly, this process is repeated 10 times and the average performance in terms of both ACC and NMI is reported as the clustering results. The final results for each baseline are shown in Table 3. From Tables 3, we have the following observations and analyses:

(1) DANEP acquires the best clustering performance on the Citeseer, BlogCatalog and Flickr datasets against all the baselines. The promotion of performance is significantly on different datasets, such as, DANEP with 32.46% and 41.69% than the best baseline TADW on the Flicker dataset. Besides, DANEP ranked the second on the Cora dataset, but only with the slightly inferior in ACC than DANE, i.e., −0.008, and in NMI than TADW, i.e., −0.005, respectively. Those results indicated that DANEF based on the graph representation and PPMI has a good clustering performance than all baselines.

(2) From the perspective of average performance, TADW obtains better clustering results than the other baselines, but TADW is seriously inferior to DANEP. In detail, DANEP averagely improves 10.87% in ACC and 13.11% in NMI than TADW, which demonstrated attribute graph and PPMI matrix have powerful assistance in node clustering.

### 4.6   Network Visualization

To verify whether the learned node representations have the discriminative essence features, we use the t-SNE [16] to project the learned embedding representation for each node into the 2D space. The color of a point indicates the class label. The desired embedding layout should be that nodes with the same color (label) to closer each other and different colors (label) to distant each other with the obvious boundary. Due to the space limitation, we only show the visualization result on the BlogCatalog dataset in Fig. 2, and the visualization results on other datasets are similar.

From Fig. 2, we can see that the DANEP, i.e., sub-figure (k), performs the best result with the nodes of the same color are close to each other and the boundaries amongst the different colors are discernible. Besides, DANE, sub-figure (e), performs the suboptimal result that the separation of boundaries is inferior to DANEP. Nevertheless, the visualization results of the DeepWalk, Node2Vec, Grarep, DNGR, ANRL, AANE, TADW, NANE and ASNE, i.e., sub-figure (a), (b), (c), (d), (f), (g), (h), (i) and (j), are mixed with different color nodes.

### 4.7   Sensitivity Analysis of Parameters

The hyper-parameters $\alpha$ and $\beta$ are used to balance the weights between the pairwise constraint loss and reconstruction loss of the DANEP. In this subsection, we analyze the sensitivity of hyper-parameters of DANEP via node classification and node clustering tasks. Experimental results of Micro-F1 of node classification and ACC of node clustering are presented in Fig. 3 and Fig. 4, respectively. The trends of Macro-F1 and NMI with respect to $\alpha$ and $\beta$ are similar to that of Micro-F1 and ACC, so we do not present them due to the space limitation.

From Fig. 3, we can observe that the tendencies of the Micro-F1 value of node classification are stable under different hyper-parameters and different datasets,

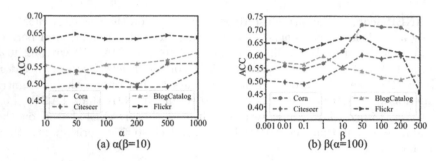

**Fig. 4.** The sensitivity of DANEP w.r.t. different $\alpha$ and $\beta$ for node clustering

which indicates DANEP has stable performance for node classification. In Fig. 4, the fluctuation of ACC of node clustering is obvious with the various hyper-parameter $\beta$ than hyper-parameter $\alpha$, which indicates that reconstruction loss plays a vital role in the node clustering process.

## 5 Conclusion

In this study, we develop the DANEP model to integrate attribute features and network topology into a unified graph format and encode each node into a low-dimensional embedding representation. In our model, the k-nearest neighbor graph can reveal some potential non-linear manifold under the attribute features, the random surfing model and PPMI can capture the structural characteristics and high-order proximity information of the attribute/topology graph, and the pairwise constraint can improve the quality of node representation. Experiment results on four real-life datasets in node classification, node clustering and visualization tasks indicated that the performance of the DANEP outperformed 10 representative baselines, including "Topology-only" algorithms and "Topology+Attribute" algorithms.

The DANEP is designed to handle the homogeneous networks with single-typed nodes and edges. However, real-world networks are usually with multiple-typed nodes and edges, which contain richer semantic information and more complex network topology for network representation learning [7,26]. Therefore, extending the DANEP to heterogeneous networks and improving the stability of clustering are our future works.

## References

1. Arthur, D., Vassilvitskii, S.: k-means++: the advantages of careful seeding. In: SODA, pp. 1027–1035 (2007)
2. Cai, H., Zheng, V.W., Chang, K.C.: A comprehensive survey of graph embedding: problems, techniques, and applications. IEEE Trans. Knowl. Data Eng. **30**(9), 1616–1637 (2018)

3. Cao, S., Lu, W., Xu, Q.: GraRep: learning graph representations with global structural information. In: CIKM, pp. 891–900 (2015)
4. Cao, S., Lu, W., Xu, Q.: Deep neural networks for learning graph representations. In: AAAI, pp. 1145–1152 (2016)
5. Dong, K., Zhou, L., Kong, B., Zhou, J.: A dual fusion model for attributed network embedding. In: Li, G., Shen, H.T., Yuan, Y., Wang, X., Liu, H., Zhao, X. (eds.) KSEM 2020, Part I. LNCS (LNAI), vol. 12274, pp. 86–94. Springer, Cham (2020). https://doi.org/10.1007/978-3-030-55130-8_8
6. Gao, H., Huang, H.: Deep attributed network embedding. In: IJCAI, pp. 3364–3370 (2018)
7. Gao, X., Chen, J., Zhan, Z., Yang, S.: Learning heterogeneous information network embeddings via relational triplet network. Neurocomputing 412, 31–41 (2020)
8. Grover, A., Leskovec, J.: node2vec: scalable feature learning for networks. In: Knowledge Discovery and Data Mining, pp. 855–864 (2016)
9. Huang, T., Zhou, L., Wang, L., Du, G., Lü, K.: Attributed network embedding with community preservation. In: DSAA, pp. 334–343 (2020)
10. Huang, X., Li, J., Hu, X.: Accelerated attributed network embedding. In: SIAM, pp. 633–641 (2017)
11. Jin, D., Ge, M., Yang, L., He, D., Wang, L., Zhang, W.: Integrative network embedding via deep joint reconstruction. In: IJCAI, pp. 3407–3413 (2018)
12. Liao, L., He, X., Zhang, H., Chua, T.: Attributed social network embedding. IEEE Trans. Knowl. Data Eng. 30(12), 2257–2270 (2018)
13. Mikolov, T., Sutskever, I., Chen, K., Corrado, G.S., Dean, J.: Distributed representations of words and phrases and their compositionality. In: NIPS, pp. 3111–3119 (2013)
14. Mo, J., Gao, N., Zhou, Y., Pei, Y., Wang, J.: NANE: attributed network embedding with local and global information. In: Hacid, H., Cellary, W., Wang, H., Paik, H.-Y., Zhou, R. (eds.) WISE 2018, Part I. LNCS, vol. 11233, pp. 247–261. Springer, Cham (2018). https://doi.org/10.1007/978-3-030-02922-7_17
15. Perozzi, B., Al-Rfou, R., Skiena, S.: DeepWalk: online learning of social representations. In: KDD, pp. 701–710 (2014)
16. Rauber, P.E., Falcão, A.X., Telea, A.C.: Visualizing time-dependent data using dynamic t-SNE. In: EuroVis - Short Papers, pp. 73–77 (2016)
17. Ribeiro, L.F.R., Saverese, P.H.P., Figueiredo, D.R.: struc2vec: learning node representations from structural identity. In: Knowledge Discovery and Data Mining, pp. 385–394 (2017)
18. Ruan, J., Dean, A.K., Zhang, W.: A general co-expression network-based approach to gene expression analysis: comparison and applications. BMC Syst. Biol. 4, 8 (2010)
19. Tang, J., Qu, M., Wang, M., Zhang, M., Yan, J., Mei, Q.: LINE: large-scale information network embedding. In: WWW, pp. 1067–1077 (2015)
20. Tu, C., Liu, H., Liu, Z., Sun, M.: CANE: context-aware network embedding for relation modeling. In: ACL, Volume 1: Long Papers, pp. 1722–1731 (2017)
21. Wang, D., Cui, P., Zhu, W.: Structural deep network embedding. In: Knowledge Discovery and Data Mining, pp. 1225–1234 (2016)
22. Wang, Z., Liu, H., Du, Y., Wu, Z., Zhang, X.: Unified embedding model over heterogeneous information network for personalized recommendation. In: IJCAI, pp. 3813–3819 (2019)
23. Weigend, A.S., Rumelhart, D.E., Huberman, B.A.: Generalization by weight-elimination with application to forecasting. In: NIPS, pp. 875–882 (1990)

24. Yang, C., Liu, Z., Zhao, D., Sun, M., Chang, E.Y.: Network representation learning with rich text information. In: IJCAI, pp. 2111–2117 (2015)
25. Yang, Y., Chen, H., Shao, J.: Triplet enhanced autoencoder: model-free discriminative network embedding. In: IJCAI, pp. 5363–5369 (2019)
26. Yu, G., Wang, Y., Wang, J., Domeniconi, C., Guo, M., Zhang, X.: Attributed heterogeneous network fusion via collaborative matrix tri-factorization. Inf. Fusion **63**, 153–165 (2020)
27. Zhang, Z., et al.: ANRL: attributed network representation learning via deep neural networks. In: IJCAI, pp. 3155–3161 (2018)
28. Zhou, L., Lü, K., Yang, P., Wang, L., Kong, B.: An approach for overlapping and hierarchical community detection in social networks based on coalition formation game theory. Expert Syst. Appl. **42**(24), 9634–9646 (2015)
29. Zhou, S., et al.: PRRE: personalized relation ranking embedding for attributed networks. In: CIKM, pp. 823–832 (2018)

# Discovering Spatial Co-location Patterns with Dominant Influencing Features in Anomalous Regions

Lanqing Zeng, Lizhen Wang[⊠], Yuming Zeng, Xuyang Li, and Qing Xiao

School of Information Science and Engineering, Yunnan University, Kunming 650091, China
lzhwang@ynu.edu.cn

**Abstract.** As one of the important exogenous factors that induce malignant tumors, environmental pollution poses a major threat to human health. In recent years, more and more studies have begun to use data mining techniques to explore the relationships among them. However, these studies tend to explore universally applicable pattern in the entire space, which will take a high time and space cost, and the results are blind. Therefore, this paper first divides the spatial data set, then combined with the attenuation effect of pollution influence with increasing distance, we proposed the concept of high-impact anomalous spatial co-location region mining. In these regions, industrial pollution sources and malignant tumor patients have a higher co-location degree. In order to better guide the actual work, the pollution factors that have a decisive influence on the occurrence of malignant tumors in the pattern is explored. Finally, a highly targeted new method to explore the dominant influencing factors when multiple pollution sources act on a certain tumor disease at the same time is proposed. And extensive experiments have been conducted on real and synthetic data sets. The results show that our method greatly improves the efficiency of mining while obtaining effective conclusions.

**Keywords:** Spatial data mining · Anomalous region · Co-location patterns with dominant influencing features

## 1 Introduction

Malignant tumors are caused by a variety of human carcinogenic risk factors including chemical, physical, biological, etc. As one of the important exogenous factors that cause malignant tumors, environmental pollution factors will not only cause damage to the ecological environment, but also pose a great threat to human health. Existing studies have shown that various types of pollution can cause humans to produce different tumor diseases, such as: increased levels of PM2.5 in air pollution will increase the risk of lung cancer [1], the gastric cancer mortality rate of residents in areas with high levels of organic pollutants and heavy metals in water is significantly higher than that in low-polluted areas [2].

To control environmental pollution and protect human health, it is necessary to find the correlation between pollution and human disease in order to propose better measures to protect the environment.

© Springer Nature Switzerland AG 2021
C. S. Jensen et al. (Eds.): DASFAA 2021 Workshops, LNCS 12680, pp. 267–282, 2021.
https://doi.org/10.1007/978-3-030-73216-5_19

Data mining refers to the process of obtaining information hidden in a large amount of data from a large amount of data according to a certain mining algorithm. At present, there have been studies using data mining technology to explore the potential connections between diseases and pathogenic factors [3]. However, these researches often explore universally applicable pattern in the entire space or data set, which will take a high time and space cost, and will produce a lot of useless results. And because of the greater interference from useless information, the final mining results are blind. Therefore, it is necessary to propose a targeted and instructive and efficient mining method to explore the correlation between tumor diseases and pollution sources.

Spatial data mining is the process of discovering potential connections between data in spatial database. The spatial co-location pattern is a set of spatial features with higher spatial proximity relations. The application and promotion of spatial co-location pattern mining methods in different fields has become a hot spot for scholars all over the world. Huang et al. proposed a join-based co-location pattern mining method [4]. Priya et al. studied the similarities and differences between the co-location mining problem and the classic association rule mining problem, and formalized the co-location pattern mining problem [5]. Manikandan and Srinivasan use R-tree index to mine spatial co-location patterns to reduce spatial data search time [6]. Manikandan and Srinivasan proposed a new co-location pattern mining algorithm [7], this method uses Prim algorithm to reduce the calculation amount of traditional algorithms without losing co-location pattern instances. Considering the differences in the contribution of different features to the pattern, a method to measure the feature differences in the co-location pattern is proposed by Fang et al. [8].

All the above methods explore universally applicable patterns in the whole data set and ignore the existence conditions of the patterns, which will undoubtedly make the obtained results blind. Moreover, they consider all spatial features equally and ignore the interaction between features.

Anomalous spatial co-location region refers to the regions in the space with spatial co-location intensity that higher or lower than expected. In the anomalous spatial co-location region, we can often obtain more interesting spatial associations. Recently, Cai et al. proposed an adaptive anomalous spatial region mining algorithm based on the co-location intensity between different features in the space [9]. This algorithm only considers whether the instances are adjacent to each other when measuring the spatial co-location intensity, but ignores the influence of distance on co-location intensity.

Based on anomalous spatial co-location pattern mining technology, combined with the attenuation effect of pollution influence with increasing distance, we propose a mining algorithm for mining high-impact anomalous regions. In high-impact anomalous regions, industrial pollution sources and patients with malignant tumors have a higher degree of spatial co-location. In this case, malignant tumors are more likely to be caused by the influence of pollution.

Then, in order to better guide the actual work, we proposed the concept of mining co-location patterns with dominant influencing features to discover the decisive factor of pollution over malignant tumors. To sum up, a new method of mining spatial co-location patterns with dominant influencing features in high-impact anomalous regions is proposed. The method is implemented using Java, and a large number of experiments are performed on real and synthetic data sets. The results show that the proposed method is practical and efficient.

The main contributions of this paper can be summarized as follows: (1) Proposed a definition of spatial co-location patterns with dominant influencing features, by dividing spatial features into primary features and the influencing features. (2) The concepts of influencing clique, influencing degree and difference degree of features are defined for mining co-locations with dominant influencing features. (3) A method to obtain high-impact anomalous regions that have a higher of co-location is provided. (4) Designed an algorithm for finding spatial co-location patterns with dominant influencing features in high-impact anomalous regions.

The organization of this paper is as follows: Sect. 2 defines the basic con-cept. Section 3 presents the mining algorithm. Section 4 gives the evaluation of the experimental results. Section 5 shows the conclusion and future works.

## 2  Basic Concepts

For a spatial instance set $O = \{o_1, o_2, \ldots, o_s\}$, if the Euclidean distance between two spatial instances $o_j$ and $o_k$ is less than or equal to a specified distance threshold $d$, it is considered that there is a proximity relationship $R$ between them, which can be expressed as:

$$R(o_j, o_k) \leftrightarrow (distance(o_j, o_k) \leq d) \tag{1}$$

When there is a spatial proximity relationship R between the spatial instances, we think that the two instances are adjacent to each other.

**Definition 1 (Primary Features (PF) and Influencing Features (IF)).** Spatial feature set $F = \{f_1, f_2, \ldots, f_k\}$ is a collection of different kinds of things. The primary feature set $PF = \{f_1^p, f_2^p, \ldots, f_m^p\}$ refers to a collection of features whose occurrence probability and distribution will be affected by other features. It is a subset of F, $PF \subseteq F$. The influencing feature set $IF = \{f_1^i, f_2^i, \ldots, f_n^i\}$ is the set of features that will affect other features. It is the complement of the primary feature set in the spatial feature set $F$, $IF = F - PF$. Figure 1 shows an example of the primary features and influencing features in a spatial dataset.

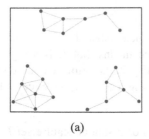
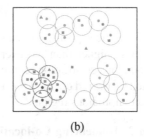

(a)                                      (b)

**Fig. 1.** An example of spatial dataset, (a) An example of primary feature instances (red dots), (b) Multiple different types of influencing feature instances (blue squares) within its proximity threshold range (circle) (Color figure online)

**Definition 2 (Influencing Cliques (IC)).** For a spatial instance set $O$, including the primary feature instance set $PO$ and the influencing feature instance set $IO$, if there is a spatial instance set $IC = \{o_j^p, o_p^i, \ldots, o_q^i\}$, and any influencing feature instance in it has a proximity relationship with the primary feature instance $o_j^p$, namely $\{R(o_j^p, o_k^i)|p \leq k \leq q, o_j^p \in PO, o_k^i \in IO\}$, then $IC$ can be called an influencing clique.

For an influencing clique $IC = \{o_j^p, o_p^i, \ldots, o_q^i\}$, for each influencing feature instance $o_i^i$, its co-location effectiveness $p$ decreases as the increase of distance between it and the primary feature instance $o_j^p$. In order to better simulate this attenuation process, we refer to the literature [10] and propose the following formula:

$$p_i = 1 * e^{-\frac{distance(o_j^p, o_i^i)^2}{d^2 * 5}} \tag{2}$$

where, $d$ value is the distance threshold specified by us. According to the formula, $p_i$ will be close to 0 when the distance is equal to $d$.

The influencing co-location intensity $ICI$ of the primary feature instance $o_j^p$ in $IC$ is defined the accumulation of $p$-values of all impact feature instances in $IC$, expressed as:

$$ICI_j = \sum_{i=p}^{q} p_i \tag{3}$$

For region A that has multiple primary feature instances, we use statistics $G*$, which proposed in [11] to measure the influencing co-location intensity of the region, expressed as:

$$G * (A) = \frac{\sum_{o_j^p \in A} ICI_j - n * \overline{ICI}}{S * \sqrt{\frac{N*n - n^2}{N-1}}} \tag{4}$$

Wherein $n$ is the number of primary feature instances contained in region A, $N$ is the total number of primary feature instances in all regions, $ICI_i$ is the number of influencing feature instances that have a neighboring relationship with the primary feature instance, $\overline{ICI}$ is the average of $ICI_i$, and $S$ is expressed as:

$$S = \sqrt{\frac{\sum_{j=1}^{N} ICI_j^2}{N} - ICI^2} \tag{5}$$

If the $G*$ value is positive, it means that the influencing spatial co-location intensity of region A is higher than the average level, which means that A is a high-impact spatial anomalous co-location region. As shown in Fig. 2, the primary feature instance (represented by the red dot) and the adjacent influencing feature instances constitute the high-impact anomalous regions.

**Definition 3 (Influencing Co-location Patterns).** For a spatial feature set $F$, which including the primary feature set $PF$ and the influencing feature set $IF$, influencing co-location pattern $c = \{f_j^p, f_1^i, f_2^i, \ldots, f_k^i\}$ is a subset of $F$, and it is the union of a primary feature and the non-empty subset of the influencing feature set.

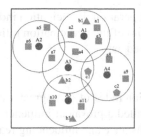

**Fig. 2.** An example of influencing cliques (Color figure online)

Size is used to identify the number of different influencing features that affect the co-location pattern. Since each influencing co-location pattern contains only one primary feature, the size of an influencing co-location pattern is its length minus 1, denoted as $size(c) = |c| - 1$.

**Definition 4 (Row Instance and Table Instance).** For an influencing co-location pattern $c$, if there is an influencing clique $IC$, we regard $IC$ as a row instance of $c$, if $IC$ includes all features in $c$, and no subset of $IC$ can contain all the features in $c$. The set including all the row instances of influencing co-location pattern $c$ is called table instance of $c$.

Figure 3 is the table instances of influencing co-location patterns in Fig. 2.

**(a)**

| Primary feature | Influencing feature |
| --- | --- |
| A | a |
| A1 | a1 |
| A1 | a2 |
| A1 | a3 |
| A1 | a4 |
| A2 | a5 |
| A2 | a6 |
| A2 | a7 |
| A3 | a4 |
| A3 | a7 |
| A4 | a8 |
| A4 | a9 |
| A5 | a10 |
| A5 | a11 |
| 1.00 | 1.00 |

| Primary feature | Influencing feature |
| --- | --- |
| A | b |
| A1 | b1 |
| A5 | b2 |
| A5 | b2 |
| A5 | b3 |
| 0.60 | 1.00 |

| Primary feature | Influencing feature |
| --- | --- |
| A | c |
| A3 | c1 |
| A4 | c1 |
| A4 | c2 |
| 0.40 | 1.00 |

**(b)**

| Primary feature | Influencing feature | |
| --- | --- | --- |
| A | a | b |
| A1 | a1 | b1 |
| A1 | a2 | b1 |
| A1 | a3 | b1 |
| A1 | a4 | b1 |
| A3 | a4 | b2 |
| A3 | a7 | b2 |
| A5 | a10 | b2 |
| A5 | a11 | b2 |
| A5 | a10 | b3 |
| A5 | a11 | b3 |
| 0.60 | 0.64 | 1.00 |

| Primary feature | Influencing feature | |
| --- | --- | --- |
| A | a | c |
| A4 | a8 | c1 |
| A4 | a9 | c1 |
| A4 | a8 | c2 |
| A4 | a9 | c2 |
| A3 | a4 | c1 |
| A3 | a7 | c1 |
| 0.40 | 0.36 | 1.00 |

| Primary feature | Influencing feature | |
| --- | --- | --- |
| A | b | c |
| A3 | b2 | c1 |
| 0.20 | 0.33 | 0.50 |

**(c)**

| Primary feature | Influencing feature | | |
| --- | --- | --- | --- |
| A | a | b2 | c |
| A3 | a4 | b2 | c1 |
| A3 | a7 | b2 | c1 |
| 0.20 | 0.18 | 0.33 | 0.50 |

**Fig. 3.** The table instances of influencing co-location patterns in Fig. 2, (a) 1-size influencing co-location patterns, (b) 2-size influencing co-location patterns, (c) 3-size influencing co-location patterns

**Definition 5 (Participation Ratio and Participation Index).** For a $k$-size influencing co-location pattern $c_k = \{f_j^p, f_1^i, f_2^i, \ldots, f_k^i\}$, The participation ratio $PR(c_k, f_i)$ of the influencing feature $f_i$ in $c_k$ is defined as the ratio of the number of non-repeated instances of $f_i$ in the table instance of $c_k$ to the total number of instances in $f_i$, it is expressed as:

$$PR(c_k, f_i) = \frac{|\pi_{f_i}(table\_instance(c_k))|}{|table\_instance(\{f_i\})|} \quad (6)$$

wherein $\pi$ is the projection operation. Participation index of influencing co-location pattern $c_k$ is defined the minimum of the participation ratios of all influencing features, which denoted as $I(c_k)$, expressed as:

$$PI(c_k) = min_{i=1}^{k}\{PR(c_k, f_i)\} \qquad (7)$$

When $PI(c_k)$, the participation index of $c_k$, is greater than or equal to the user given threshold $min\_prev$, it can be called a prevalent influencing co-location pattern, which represents the primary feature instance and influencing feature instances have prevalent association relationships in space.

**Definition 6 (Influencing Degree of Influencing Features (IFID)).** For a $k$-size influencing co-location pattern $c_k$, any $k$-1 size sub-pattern of $c_k$ is record as $c_{k-1} = \{f_j^p, f_1^i, f_2^i, \ldots, f_{k-1}^i\}$. The loss ratio of any influencing feature $f_i^i$ from $c_{k-1}$ to $c_k$, that is, the ratio of the number of lost instances of $f_i^i$ from $c_{k-1}$ to $c_k$ to the total number of instances in $f_i^i$, recorded as $LR$, expressed as:

$$LR\left(c_k, c_{k-1}, f_i^i\right) = \frac{|\pi_{f_i^i}(table\_instance(c_{k-1})| - |\pi_{f_i^i}(table\_instance(c_k))|}{|table\_instance(\{f_i^i\})|} \qquad (8)$$

The minimum of $LR$ from all influencing features in the co-location pattern from $c_{k-1}$ to $c_k$ is called the loss degree of pattern from $c_{k-1}$ to $c_k$. It is recorded as $LD(c_k, c_{k-1})$, expressed as:

$$LD(c_k, c_{k-1}) = min_{i=1}^{k-1}\{LR\left(c_k, c_{k-1}, f_i^i\right)\} \qquad (9)$$

The loss degree $LD(c_k, c_{k-1})$ indicates that the possibility that pattern $c_{k-1}$ is not affected by the influencing feature $f_k^i(f_k^i = c_k - c_{k-1})$ and appear by itself. With the increasing of $LI$, influencing feature $f_k^r$ has less effect to pattern $c_{k-1}$, that is, $f_k^r$ has less ability to influence the primary feature.

In order to more intuitively express the degree of influence of each influencing feature on the pattern, the concept of the influencing degree of the influencing feature is introduced. The influencing degree $IFID(c_k, f_i^r)$ is defined as:

$$IFID\left(c_k, f_i^r\right) = 1 - LI(c_k, c_{k-1}) \qquad (10)$$

The bigger the $IFID\left(c_k, f_i^i\right)$ is, the greater the influence of influencing feature $f_i^i$ on the primary feature in $c_k$.

**Definition 7 (Difference Degree of Influencing Feature (IFDD)).** In order to measure the difference of influencing degree between different influencing features, we propose the concept of influencing feature difference degree, that is, the difference between the influencing degree of the influencing feature and the minimum feature. The minimum feature influencing degree is recorded as $min\_IFID(c_k)$, and the influencing feature difference degree is recorded as $IFDD(c_k, f_i^i)$, expressed as:

$$min\_IFID(c_k) = min_{i=1}^{k}(IFID\left(c_k, f_i^i\right)) \qquad (11)$$

$$IFDD(c_k, f_i^i) = IFID\left(c_k, f_i^i\right) - min\_IFID(c_k) \qquad (12)$$

**Definition 8 (Influencing Co-location Pattern with Dominant Influencing Feature).**
For a $k$-size influencing co-location pattern $c_k$, given a minimum participation index
threshold *min_prev* and a minimum influencing feature difference degree threshold
*min_ifdd*. If the participation index $PI(c_k)$ of influencing co-location pattern is greater
than *min_prev*, and there are one or more influencing feature difference degree greater
than or equal to *min_ifdd*, then these influencing features can be called dominant features,
and the pattern $c_k$ is called an influencing co-location pattern with dominant influencing
feature, hereinafter referred to as the dominant influencing pattern. Dominant features
are marked with '*' in the pattern.

Figure 4 is an example of determining whether a pattern is a dominant influence
pattern:

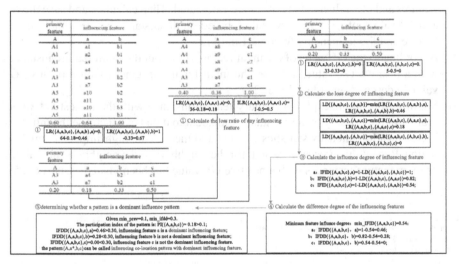

**Fig. 4.** The process to mine the dominant influencing pattern

## 3 Mining Algorithms

According to the above definition, in order to complete the research of this paper, we
need to classify the dataset first. According to the Definition 1 and Definition 2, we
divide the spatial dataset into primary feature set and influencing features set. Then we
propose an algorithm called AHIASCRM to obtain the high-impact spatial co-location
anomalous regions. Then in order to further explore whether there are one or more influencing features have a decisive influence on the production of the primary features, we
propose the ADICPM algorithm to mine the influencing co-location patterns with dominant influencing features in such regions. Finally, the prevalent influencing co-location
patterns and its dominant influencing features are obtained. The mining framework is
given in Fig. 5.

**Fig. 5.** The mining framework

First, we propose an algorithm AHIASCRM (Algorithm of high-impact anomalous spatial co-location region Mining) to obtain high-impact anomalous regions, the algorithm pseudocode is shown in the Fig. 6 and the steps are as follows:

(1) Initialization. For feature instances in space, the proximity relationship between each of the primary features and the proximity relationship between the primary features and the influencing features are calculated. Then calculate and save the *ICI* and $G*$ values of each primary feature based on the distance between the primary feature and the influencing feature. (Line 1)

(2) According to the $G*$ value of the main feature, filter out the main feature instances with a positive $G*$ value, and sort the main feature instances with a positive $G*$ value in descending order. (Lines 2–7)

(3) Starting from the unexplored primary feature instance with the largest $G*$ value, search for the primary feature instances adjacent to it that can increase the influencing co-location intensity statistics $G*$, and add them to the high-impact anomalous region. Repeat this process to search for the neighbors of the neighbors until all the primary feature instances with a positive $G*$ value has been visited. (Lines 8–25)

| | |
|---|---|
| **Input:** primary features set *PF* and influencing features set *IF*, distance threshold *d* | 5:     **end if** |
| | 6:   **end for** |
| **Output:** spatial data set *S* in anomalous regions | 7:   $desc\_sort\_queue$ = Descending_sort_with_$G*$ |
| **Variable:** | ($high\_level\_PF$) |
| *ICI*: Influencing co-location intensity of the primary feature instance | 8:   $n = 0$ |
| *Rn*: One of the anomalous regions | 9:   **while** $desc\_sort\_queue \neq \emptyset$ **do** |
| *n*: The serial number of anomalous regions, *n* starts from 0 | 10:    $pf$ = desc_sort_queue_pop() |
| $G*(Rn)$: $G*$ value of region *Rn* | 11:    add($Rn, pf$) |
| $G*(Rn \cup pf)$: $G*$ value of region Rn plus primary feature instance *pf* | 12:    push($cheacked\_queue, neighbors\_pf$) |
| | 13:    delet($desc\_sort\_queue, neighbors\_pf$) |
| $neighbors\_pf$: All primary feature instance neighbors of *pf* | 14:    **while** $cheack\_queue \neq \emptyset$ **do** |
| $high\_level\_PF$: Primary feature set with positive $G*$ value | 15:      $pf$ = cheacked_queue_pop() |
| $desc\_sort\_queue$: The queue of primary feature instances in descending order of $G*$ value | 16:      **if** $G*(Rn \cup pf) > G*(Rn)$ |
| | 17:        add($Rn, pf$) |
| $checked\_queue$: A queue for checking | 18:          push($cheacked\_queue, neighbors\_pf$) |
| *S*: spatial data set in anomalous regions | 19:          delet($desc\_sort\_queue, neighbors\_pf$) |
| **Method:** | 20:      **end if** |
| 1:   init_PF_ICI_and_$G*$($PF, RF, d$) | 21:    **end while** |
| 2:   **for** all $pf \in PF$ **do** | 22:    add($S, Rn$) |
| 3:     **if** $G*(pf) > 0$ | 23:    $n = n + 1$ |
| 4:       add($high\_level\_PF, pf$) | 24: **end while** |
| | 25: return($S$) |

**Fig. 6.** Pseudocode of AHIASCRM

After obtaining the high-impact anomalous region, we can intuitively understand and observe the status of the primary feature instance being affected by the influencing feature instance. In order to further explore which influencing features play a decisive role in the prevalent appearance of the primary features, this paper proposes a method to mine the influencing co-location with dominant influencing features in the spatial anomalous co-location region, referred to as dominant influencing co-location patterns or DICPs. The algorithm to mine DICPs is called ADICPM (Algorithm of Dominant Influencing Co-location Pattern Mining) The algorithm pseudocode is shown in the Fig. 7 and it can be described the following 3 steps:

(1) For the feature instances in the spatial anomalous region, calculate the neighbor relationship between the primary feature instance and the influencing feature instance and save it into the neighbor list, (Line 1)
(2) Generate the influencing co-location pattern table instance based on the neighbor relationship and filter it. Pattern that do not meet the minimum participation index threshold are pruned, and finally the prevalent influencing co-location pattern is obtained. (Lines 3 to 10)
(3) For prevalent influencing co-location patterns, calculate the influencing degree and difference degree of each influencing features, and judge whether each influencing feature is the dominant influencing feature, and whether the pattern is an influencing co-location pattern containing the dominant influencing feature. (Lines 11 to 32)

```
Input: primary features set PF and influencing features set IF.        11:      for all p ∈ P_{k-1}(c) do
spatial data set S in anomalous regions, distance threshold d,         12:          LD(c,p) = calculate_LD(c,p)
minimal pattern threshold min_prev, minimal influencing                13:      end for
features difference degree min_ifdd.                                   14:      for all f^i ∈ c do
Output: influencing co-location patterns containing dominant           15:          IFID(c,f^i) = calculate_IFID(c,f^i)
features                                                               16:          if min_IFID ≥ IFID(c,f^i)
Variable:                                                             17:              min_IFID = IFID(c,f^i)
NCP: neighbor set centered on the primary feature instance            18:          end if
C_k: set of k-size candidate influencing patterns                     19:      end for
P_k: set of k-size prevalent influencing patterns                     20:      for all f^i ∈ c do
IPCDF: set of influencing patterns containing dominant features       21:          IFDD(c,f^i) = IFID(c,f^i) −
DIF: set of dominant influencing features                                         min_IFID
Method:                                                               22:          if IFDD(c,f^i) ≥ min_ifdd
1: NCP = generate_pf_neighborhoods(S,PF,IF,d)                         23:              flag = 1,add( DIF(c),f^i)
3: k = 2, IDCP_set = ∅                                                24:          end if
3: P_k = select_prevalence_influernce_colocation (NCP)               25:      end for
4: while P_k ≠ ∅ do                                                   26:      if flag == 1
5:    k = k + 1;                                                      27:          add( IPCDF, {c,DIF(c)})
5:    C_k = generate_candidate_influernce_colocation(P_{k-1})         28:      end if
6:    while C_k ≠ ∅ do                                            29:          end if
7:        for all c ∈ C_k do                                        30:      end for
8:            PI(c) = calculate_PI(c,RI_set)                        31:  end while
9:            if PI(c) ≥ min_prev                                   32:end while
10:               add(P_k(c),c),flag = 0                            33:return(IPCDF)
```

**Fig. 7.** Pseudocode of ADICPM

# 4  Experimental Evaluation

The computer configuration used in the experiment is as follow: Intel®CoreTMi7-8700KCPU, 16 GB memory; Windows10 operating system; development language: Java.

The experiment was completed on real and synthetic data sets. The real data set is information of tumor patients and pollution source in a province and some surrounding areas. According to Definitions 1 and Definition 2, we regard tumor patients as the primary feature and pollution source as influencing feature including 26 types of primary features and 9 types of influencing features, a total of 13,562 spatial feature instances, of which 7,959 are primary feature instances and 5603 are influencing feature instances. The synthetic data set is generated using a spatial data generator based on [12].

The parameters in the synthetic data set are shown in Table 1.

**Table 1.** Synthetic data sets

| Data sets | Number of instances | Number of primary features | Number of influencing features |
|-----------|--------------------|-----------------------------|---------------------------------|
| Data1 | 10000 | 10 | 10 |
| Data2 | 20000 | 10 | 15 |
| Data3 | 40000 | 15 | 20 |
| Data4 | 100000 | 15 | 20 |

## 4.1  Performance Evaluation and Comparison of Mining Results

**Performance Evaluation and Comparison of Mining Results of AHIASCRM.** First, we evaluate the performance of the high-impact anomalous region mining algorithm on synthetic data sets of different magnitudes. In Fig. 8(a), we set the distance thresholds to 10, 20, 30, 40, and we can find that the running time increases with the number of instances, this is because in the process of mining high-impact anomalous regions, we need to traverse each instance in the primary feature instance set and operate on it, so that the data set size will directly affect the running time. At the same time, the running time increases first and then decreases as the distance threshold d increases. Because the anomalous region mined under different distance thresholds are different, the impact of thresholds on running time is also different.

In order to evaluate the influence of the distance threshold $d$ on the mining results, we respectively show the number of primary feature instances and the number of influencing features contained in the high-impact anomalous regions mined by Data3 and Data4 under different thresholds in Fig. 8(b) and Fig. 8(c). It can be found that the number of primary feature instances will decrease as $d$ increases, and the number of influencing feature instances will increase as $d$ increases. This is because when $d$ increases, the more influencing feature instances are considered to be adjacent to the primary feature instance, the average influencing co-location intensity of the entire region will increase, and the *ICI* value of the primary feature instance will also tend to average strength of influencing co-location intensity. Then the primary feature instances in the anomalous region will decrease, and the number of influencing feature instances will increase. If the value of $d$ is too small, some meaningful influencing feature instances will be ignored. If the value of $d$ is too large, some instances that have little influence on the primary feature instances will also be wrongly considered. Therefore, the value of $d$ needs to be determined according to the actual situation, otherwise the validity of the mining results will be affected.

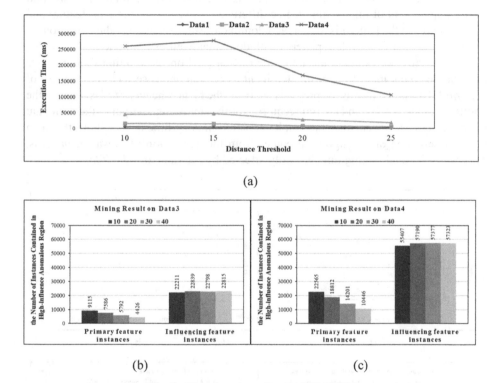

(a)

(b)                                    (c)

**Fig. 8.** Experiment results of AHIASCRM

**Performance Evaluation and Comparison of Mining Results of ADICPM.** After mining the high-impact anomalous region, we evaluate the performance of ADICPM algorithm on synthetic data sets of different magnitudes. In Fig. 9(a), let *min_prev* = 0.6, as the distance threshold $d$ increases, the running time of the algorithm increases, because the larger the value of $d$, the more instances are considered to be adjacent to each other, the more prevalent patterns would be generated by the algorithm. And the more connected operations will be done, the more time will be cost.

In Fig. 9(b), let $d = 11.5$. As the minimum participation threshold increases, the running time of the algorithm decreases, because the larger the *min_prev* value is, the more patterns will be judged as not prevalent patterns, and the algorithm will prune to reduce subsequent connection operations and the time consumption.

In order to evaluate the influence of the minimum difference degree threshold on the mining results and algorithm performance, we compare the differences on the data set Data3 with a moderate amount of data.

In Fig. 9(c), let $min\_prev = 0.6$ and $d = 11.5$, with the threshold of minimum difference degree $min\_ifdd$ increases, the total number of prevalent patterns remains unchanged, while the number of prevalent patterns containing dominant features decreases. This is because $min\_ifdd$ does not affect the generation of prevalent patterns, but only affects the judgment and storage of dominant features. The higher the $min\_ifdd$ is, the requirement of influencing degree difference between the dominant feature and other features in the pattern is higher, and dominant features that meet the conditions is fewer, the number of prevalent patterns with dominant features is smaller.

In Fig. 9(d), we evaluate the influence of $min\_ifdd$ on the running time, and we can find that the mining time of the dominant mode will decrease with the increase of $min\_ifdd$. This is because as the previous analysis, the higher the $min\_ifdd$ is, the less the dominant mode will be, and the time required for the storage and output of the dominant pattern is less. However, since the dominant pattern mining time is very short, the main time consumption of the algorithm comes from the generation of prevalent patterns. Therefore, the total mining time is hardly affected by the value of $min\_ifdd$.

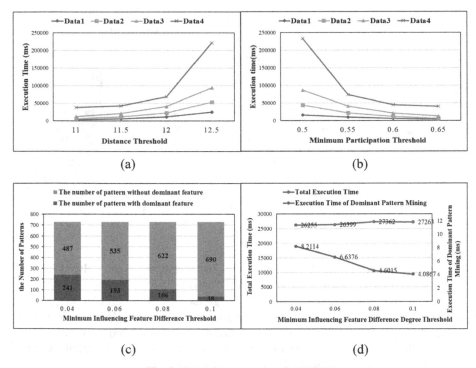

**Fig. 9.** Experiment results of ADICPM

**Comprehensive Performance Evaluation and Mining Results Analysis.** The previous two subsections compare the performance of AHIASCRM and ADICPM respectively. And in the subsequent case analysis, we need to use the two together. Therefore, this section will conduct a complete mining on the real data set coherently, and analyze the proportion of time spent in different steps to evaluate the comprehensive performance of the algorithm.

In Fig. 10(a), let $d = 2000$, $min\_prev = 0.25$, $min\_ifdd = 0.1$. The first step of the algorithm is to mine high-impact anomalous regions. Since the real data set contains only a small number of instances, this step takes less time, accounting for about 12% of the running time. The second step is to generate prevalent affecting co-location patterns. Due to a large number of row instances and table instances is generated in this step, it takes the most time, accounting for nearly 90% of total. The third step is to explore the dominant pattern. Since this step requires less operations, it takes very little time and the running time accounts for less than one ten thousandth. In general, the algorithm can obtain high-impact anomalous regions in a short time.

In Fig. 10(b), we can see that when $d = 2000$, the anomalous regions are reduced by nearly 20% compared with the conventional regions. According to the previous section, we can know that the fewer the number of instances is, the faster the mining time is. Therefore, the acquisition of high-impact anomalous regions not only improves the pertinence of our mining results, but also saves time for subsequent prevalent pattern mining.

Although the algorithm takes more time to generate prevalent patterns, compared with traditional algorithms, such as Join-less[13], we divide the feature set to only generate patterns that contain both primary features and influencing features. We will not generate patterns that include only primary features or influencing features, which undoubtedly saves a lot of time and overhead. Finally, the algorithm only needs to spend a very low cost to mine the dominant pattern, and it can get more instructive results.

In summary, our method is not only efficient, but also more targeted and instructive than traditional algorithms.

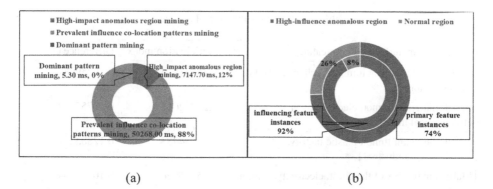

(a)                                    (b)

**Fig. 10.** Overall experiment results

## 4.2 Case Study

In the previous section, we evaluated the efficiency and effectiveness of the algorithm. In this section, we will analyze and explain the mining results on real data sets through case analysis to further verify the guidance and practicality of the algorithm.

First, we conduct the mining of high-impact anomalous regions, and the results obtained are shown in Fig. 11.

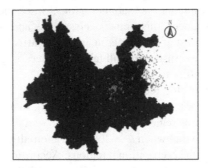

**Fig. 11.** High-impact anomalous regions on a real data set

Then we use the ADICPM algorithm to mine the above-mentioned high-impact anomalous regions, and obtain the influencing co-location pattern with dominant influencing features, so as to extract the industrial pollution sources that have the dominant influence on malignant tumors.

When the distance threshold $d = 2000$, the minimum participation index threshold $min\_prev = 0.25$, and the minimum difference degree threshold $min\_ifdd = 0.1$, the three-order dominant pattern obtained is shown in the following table (take multi-system tumors, abdominal malignancies, and biliary malignancies as examples), The ones with '*' are the dominant influencing features in the co-location pattern:

**Table 2.** Mining results on real data set

| Patterns that include dominant feature | Participation index | Difference degree |
|---|---|---|
| {Abdominal malignant tumor, oil chemical industry, printing factory*} | 0.25531915 | 0.19859687 |
| {Renal malignant tumor, electric power, petrochemical industry*} | 0.25862069 | 0.10757212 |
| {Renal malignant tumor, plastic factory, petrochemical industry*} | 0.31034483 | 0.12885274 |
| {Malignant tumors of the chest, electricity, printing houses*} | 0.25925926 | 0.10432244 |
| {Malignant tumors of the breast, coal industry, food processing plants*} | 0.26851852 | 0.11086198 |

It can be seen from Table 2 that on the basis of obtaining prevalent influencing co-location patterns, the algorithm further explores whether there are one or more pollution sources that play a decisive factor in the generation of patterns and the emergence of cancer patients.

For example, in the pattern {Renal malignant tumor, plastic factory, petrochemical industry*}, we can not only find that plastic factory and petrochemical industry enterprises often exist near patients with renal malignant tumor, but also know that compared with plastic factory, the pollution generated by petrochemical industry is more likely to be the primary factor causing the disease of patients with renal malignant tumor. Because as the dominant feature, the difference degree of petrochemical industry is greater than the given threshold. This shows that the effect of petrochemical industry on renal malignancy is obviously higher than that of plastic factory, which means that where there is petrochemical industry without plastic factory, there will still be kidney malignant tumor patients often appear, and where there is no petrochemical industry but only plastic factory, the incidence of renal malignancy is lower.

At present, in terms of medicine, the pathogenic factors of kidney malignancy are still to be studied, but according to The National Kidney Foundation (NKF) [14], we can know that it is likely to be related to air particle pollution. The various industrial processes of the petrochemical industry and the plastics factory will emit a lot of smoke and dust, resulting in the increase of PM2.5 in the air. This is consistent with our mining results. And since our results contain dominant features, it can further guide us to explore the causes of malignancy. For example, we may get some interesting results by analyzing the differences in the composition and emissions of the pollution produced by petrochemical and plastic factories.

Combined with the above analysis, we can find that the mining of high-impact anomalous regions eliminates the blindness of global data, and improves our pertinence in exploring the association between pollution sources and cancer patients. The proposal of the ADICPM algorithm makes our mining results more instructive, and can be applied well in pollution control, cancer etiology investigation and cancer prevention. When multiple pollution sources affect certain types of cancer patients at the same time, we can more accurately identify the leading pollution factors, and focus on them and strictly monitor them during pollution prevention and control.

For some cancers with ambiguous causes, we can analyze the pollutant emissions of leading pollution sources to obtain potential environmental influence factors and provide research directions for exploring the external causes of cancer.

# 5  Conclusions

Traditional spatial pattern mining determines the connection between a certain pollution source and a certain disease based on the proximity, while ignoring the common effect of multiple pollution sources on the same disease. This paper excavates high-impact regions where multiple pollution sources work together on certain patients based on the mining algorithm for anomalous co-location patterns. With an improved join-less algorithm, the dominant pollution source among multiple pollution sources affecting the same patient was discovered. It is more interesting than the traditional spatial pattern mining results, and can better reflect the objective laws of the real world.

In addition, this article also has some shortcomings. It still takes a lot of time to generate prevalent patterns. It can be improved in the future with new methods. The simulation of the distance attenuation effect is rough. If more experimental support can be obtained, it can be further improved.

# References

1. Bai, Y., Ni, Y., Zeng, Q.: A meta-analysis on association between PM2.5 exposure and lung cancer based on cohort studies. J. Public Health Prevent. Med. **31**(4), 5–8 (2020)
2. Wang, Z., et al.: Relationship between quality of drinking water and gastric cancer mortality from 11 counties in Fujian Province. Chin. J. Public Health **16**(2), 79–80 (1997)
3. Li, J., Adilmagambetov, A., Mohomed Jabbar, M.S., Zaïane, O.R., Osornio-Vargas, A., Wine, O.: On discovering co-location patterns in datasets: a case study of pollutants and child cancers. GeoInformatica **20**(4), 651–692 (2016). https://doi.org/10.1007/s10707-016-0254-1
4. Xiong, H, Shekhar, S, Huang, Y, Kumar, V, Ma, X, Yoo, J.S.: A framework for discovering co-location patterns in data sets with extended spatial objects. In: Proceeding of the 2004 SIAM International Conference on Data Mining (SDM 2004), Lake Buena Vista, pp.78–89 (2004)
5. Priya, G., Jaisankar, N., Venkatesan, M.: Mining co-location patterns from spatial data using rulebased approach. Int. J. Glob. Res. Comput. Sci. **2**(7), 58–61 (2011)
6. Manikandan, G., Srinivasan, S.: Mining of spatial co-location pattern implementation by FP growth. Indian J. Comput. Sci. Eng. (IJCSE) **3**(2), 344–348 (2012)
7. Manikandan, G., Srinivasan, S.: Mining spatially co-located objects from vehicle moving data. Eur. J. Sci. Res. **68**(3), 352–366 (2012)
8. Fang, Y., Wang, L., Wang, X., Zhou, L.: Mining co-location patterns with dominant features. In: Bouguettaya, A., et al. (eds.) WISE 2017. LNCS, vol. 10569, pp. 183–198. Springer, Cham (2017). https://doi.org/10.1007/978-3-319-68783-4_13
9. Cai J., Deng M., Guo Y., Xie, Y., Shekhar, S.: Discovering regions of anomalous spatial co-locations. Int. J. Geogr. Inf. Sci. (2020). https://doi.org/10.1080/13658816.2020.1830998
10. Hu, K., Yuan, H., Chen, D., Yi, Z.Y.: Study on mathematical model of urban pollution diffusion law. J. Shaoxing Coll. Arts Sci. Nat. Sci. **33**(04), 18–22 (2013)
11. Getis, A., Ord, J.K.: The analysis of spatial association by use of distance statistics. In: Perspectives on Spatial Data Analysis, pp. 127–145 (2010). https://doi.org/10.1007/978-3-642-01976-0_10
12. Huang, Y., Shekhar, S., Xiong, H.: Discovering colocation patterns from spatial data sets: a general approach. IEEE Trans. Knowl. Data Eng. **16**(12), 1472–1485 (2004)
13. Yoo, J.S., Shekhar, S.: A joinless approach for mining spatial colocation patterns. IEEE Trans. Knowl. Data Eng. **18**(10), 1323–1337 (2006)
14. Bowe, B., Xie, Y., Li, T., Yan, Y., Xian, H., Al-Aly, Z.: Particulate matter air pollution and the risk of incident CKD and progression to ESRD. J. Am. Soc. Nephrol. **29**(1), 218–230 (2018)

# Activity Organization Queries for Location-Aware Heterogeneous Information Network

C. P. Kankeu Fotsing[1,2]([✉]) [iD], Ya-Wen Teng[3], Sheng-Hao Chiang[3],
Yi-Shin Chen[1], and Bay-Yuan Hsu[4]

[1] Institute of Information Systems and Applications, National Tsing Hua University,
Hsinchu, Taiwan
yishin@cs.nthu.edu.tw
[2] Social Networks and Human-Centered Computing Program,
Taiwan International Graduate Program, Taipei, Taiwan
[3] Institute of Information Science, Academia Sinica, Taipei, Taiwan
{ywteng,jiang555}@iis.sinica.edu.tw
[4] Department of Computer Science and Information Engineering,
National Taipei University, Taipei, Taiwan
byhsu@mail.ntpu.edu.tw

**Abstract.** Activity organization query for Location-Based Social Networks is an important research problem, which aims at selecting a suitable group of people relevant to the query user and activity according to their social and spatial information. However, current activity organization queries mainly consider simplistic and direct relationships to measure the relevance. Although Heterogeneous Information Networks capture complex relationships by meta-structures, the relevance is seldom measured from both social and spatial aspects and does not take the distinctiveness of meta-structure into account. To fill this gap, we first propose a new relevance measurement, named *SIMER* to more accurately measure the connection strength. Then, we formulate a new query, named *MHS2Q*, which considers the social and spatial factors as well as the distinctiveness of meta-structures. Furthermore, we extend the MHS2Q to *Subsequent MHS2Q*, to consider a series of queries with varying spatial constraints. We design an efficient algorithm MS2MU to answer the (Subsequent) MHS2Q, which exploits a new index structure named *d-Table* to boost the computation for subsequent queries, and a pruning strategy, *MSR-pruning* to avoid unnecessary computation. Experiments on real LBSNs show that MS2MU is more effective to retrieve a social group that is both relevant and socially tight to the query.

**Keywords:** LBSN · Heterogeneous Information Networks ·
Meta-structure · Relevance measurement

Supported by MOST Taiwan.

C. S. Jensen et al. (Eds.): DASFAA 2021 Workshops, LNCS 12680, pp. 283–304, 2021.
https://doi.org/10.1007/978-3-030-73216-5_20

# 1   Introduction

Location Based Social Networks (LBSNs) platforms such as Yelp, Meituan and Instagram have witnessed a surge in popularity in the past and the ongoing decade. Abundant user information such as user friendships, check-ins, and reviews is kept on these platforms. Benefited from the information, many applications for marketing, recommendation, prediction, and so on are pervasive in our daily life [10,12,17,27]. Among these applications, the activity organization query is one of the most important and actively studied research problems, which aims at selecting a suitable group of people according to their social and spatial information for a wide spectrum of applications, e.g., organizing social activities (concert, trip), planning marketing events, completing certain tasks [8,11,20], and forming communities [8,14,19,22,27].

However, current activity organization queries mainly consider simplistic and direct relationships (i.e., liked, check-ins, and reviews) within the input network, which cannot fully capture the complex relationships between users and other social network entities. For instance, to hang out at a newly opened swimming pool, we may want to invite those who are likely to enjoy this swimming pool by some intuition, e.g., as illustrated in Fig. 1(a), a suitable attendee should visit a location that has a common attribute with the swimming pool but also is visited along with the swimming pool by another person. In this case, the acquired social group from previous works may not satisfy the need because previous works fail to considered more realistic, complex and semantically representative relationships as the one described on Fig. 1(a).

To better understand the users and capture the complex relationship from LBSNs, Heterogeneous Information Networks (HINs) [2,6,26] have been introduced to represent real-world relationships with various semantic meanings among different types of entities, e.g., person, restaurant, attribute. Consider the scenario that Gina is looking for someone to hang out at a newly opened swimming pool. To ensure that everyone enjoys, the invited attendees should have some connections with the swimming pool (e.g., they like workout) as well as is socially close to each other. In Gina's opinion, one may join if 1) he/she has a close relationship with Gina, and 2) some location he/she has been to shares a common attribute with this swimming pool and has been also visited together with this swimming pool by someone else (described as Fig. 1(a)). For example, in Fig. 1(b), Cindy has been to a gym which is similar to the swimming pool due to the same attributes, *healthy* and *indoor*, and Alice has been to both the gym and the swimming pool. So, Cindy is regarded as having a connection with this swimming pool.

Despite the abundant information on an HIN to capture various semantic meanings of entities and their relationships and the ability of meta-structures to describe the complex knowledge of relevance, we argue that there are still some issues not addressed in the previous works on measuring the relevance between an individual and an entity through complex semantic meaning. Specifically, two important aspects are not well considered and utilized in previous works for HIN, i.e., 1) the social and spatial factors, and 2) the distinctiveness of meta-structure.

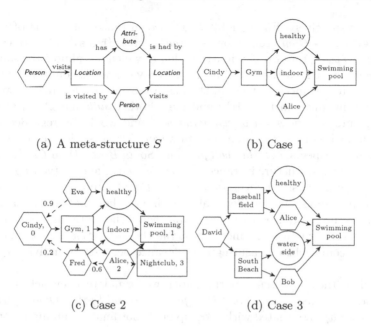

(a) A meta-structure $S$        (b) Case 1

(c) Case 2        (d) Case 3

**Fig. 1.** An illustrating example.

**Social and Spatial Factors.** Specifically, the social influence and the spatial distance are not incorporated into the measurement via meta-structures, which greatly affects people's perception [1,7,25]. In the real world, the neighborhood of people, both socially and geographically, usually encapsulates more valuable information, which suggests the connection is supposed to be established through or related to those that are socially and geographically close. For example, in Fig. 1(c), Cindy has a connection with the swimming pool if Alice has a powerful enough impact on her. On the other hand, in Fig. 1(d), the connection established by "South Beach", "waterside", and "Bob" may be meaningless to David since South Beach is not in David's living sphere. Hence, measuring the relevance between an individual and an entity from both structural, social, and geographical aspects helps well capture his/her perception of the entity.

**Distinctiveness of Meta-structure.** Moreover, the distinctiveness of meta-structure instances is not considered in previous works [2,6,26]. Formally, a meta-structure [6], denoted as $S = (A_S, R_S, a_s, a_t)$, is defined as a directed acyclic graph to describe the relationship between a source type $a_s \in A_S$ and a target type $a_t \in A_S$ through types in $A_S$ and relations in $R_S$, e.g., Fig. 1(a) illustrating the relationship between the types person and location. A meta-structure instance $M_S = (V_S, E_S, v_s, v_t)$ [6] of $S$ is a subgraph of an HIN, which represents that a source entity $v_s$ has the relationship $S$ with a target entity $v_t$, where each entity in $V_S$ and each relation in $E_S$ correspond to a unique type in $A_S$ and $R_S$, respectively. For example, consider the meta-structure $S$ in Fig. 1(a), where the source type $a_s$ and the target type $a_t$ are person and location, respectively. Figure 1(b) shows two instances of $S$, one of which is *Cindy* (person) $\rightarrow$ *Gym* (location) $\rightarrow$ *healthy* (attribute) and *Alice* (person) $\rightarrow$ *Swimming pool* (location) while the other of which has almost the same entities but a different attribute *indoor*. In the previous works, only the number of meta-structure instances matters in the measurement of relevance between two entities. For instance, in Figs. 1(b) and 1(d), both *Cindy* and *David* share two meta-structure instances with *Swimming pool* and would be regarded equally relevant in the previous measurements. However, *David* is relevant to *Swimming pool* from more aspects, i.e., *Baseball field* and *South Beach*, than *Cindy* is, which indicates the meta-structure instances shared by *David* and *Swimming pool* contain more comprehensive information. Hence, *David*'s and *Cindy*'s relevance to *Swimming pool* should not be identical even though they share the same number of meta-structure instances with *Swimming pool*. Therefore, in addition to the number of meta-structure instances, we argue the distinctiveness should also be taken into account when measuring the relevance between an individual and an entity.

To address the above issues, in this paper, we formulate a new activity organization query, referred to as *Meta Heterogeneous Social Spatial Query (MHS2Q)*. Given an activity (associated with geographical coordinates), an input user (e.g., organizer), a meta-structure $S$ (encapsulating the complex relationship of each attendee and the activity), and a parameter $k$, MHS2Q selects a group of $k$ attendees (i.e., users) such that they are socially close to the input user and their relevance to the activity with respect to $S$ is the highest. Since MHS2Q measures the relevance in terms of the spatial and social factors as well as the distinctiveness and extracts the top-$k$ users with the highest relevance, a suitable group of attendees is thus returned. The previous works that focus on either the structural relevance, social and/or spatial tightness are incapable of retrieving a proper user group for MHS2Q. To solve MHS2Q, we propose a measurement of relevance, named *Social-Spatial and Distinctiveness aware Meta-structure Relevance (SIMER)*, of a user to the query user and the set of query entities. Specifically, SIMER incorporates the social/spatial meta knowledge acquired from the influence propagation probability between users and the spatial distance and is aware of the distinctiveness of meta-structure instances by emphasizing the number of distinct entities rather than the number of instances. We also exploit a pruning strategy named *Minimum Social Ratio Pruning (MSR-pruning)* which takes advantage of the current results to avoid SIMER computation of the users who must not be reported as the top-$k$ ones due to lack of a high enough SIMER value.

However, the living sphere in measuring SIMER is usually uncertain due to the lack/availability of transportation mean in different areas. This entails that because of user's moving patterns, different geographical areas need different spatial distance requirement to select the most suitable group of users. Therefore, identifying the area of people's living sphere for the query is challenging for the query user. To address this issue, we further propose the use of subsequent queries with varying spatial distance requirements which consist

of a series of identical queries but different spatial distance requirements. The
optimal distance requirement is found when the subsequent queries results converge. We propose a table-based index structure named $d$-Table which records
intermediate results from previous queries to boost the processing of subsequent
queries and the computation of SIMER. Extensive experiments conducted on
real LBSNs show that MS2MU is more effective to retrieve a social group that
is both relevant and socially tight to the query.

The contributions of this paper are as follows.

1. We propose a new relevance measurement named *SIMER* to more accurately
   measure the connection strength between a users and the activity.
2. We formulate a new query, named *MHS2Q* to consider the spatial and social
   factors as well as the distinctiveness of meta-structure. We further propose
   a Subsequent MHS2Q to allow a series queries with varying spatial distance
   requirements.
3. We design an efficient algorithm, named *MS2MU* to answer the (Subsequent)
   MHS2Q with a new index structure named *d-Table* and a pruning strategy,
   *MSR-pruning* to boost the performance.
4. Experiments on real dataset show that our approaches significantly outperform the other baselines.

## 2   Related Work

Considering the social relationships among the entities returned (users), social
group queries in social networks have been addressed in existing literature considering different factors such as social distance and skills [4,5,15,20] for group
formation, later enhanced with the incorporation of the social influence [8]. Users'
social connection coupled with user interests [19] as well as geographical locations
[27] have been considered for activity planning. However, these works only consider simplistic and direct relationships in the input networks. Therefore, they
fail to take advantage of the rich semantics provided by complex relationships
in HIN for suitable activity and event organization.

Many previous researches have used Heterogeneous Information Network
(HIN) to compute the relevance between two entities. Liu et al. propose the
use of Meta-path in order to compute the relevance between two objects [9],
whereas, Shi et al. propose the concept of weighted meta-path for personalized
recommendation [17]. Yu et al. propose a probabilistic approach for path-based
relevance and the notion of cross-meta-path synergy for computing entities relevance [18]. To increase the accuracy of relevance by taking into consideration
complex relationships, Huang et al. introduce three meta-structure based i.e.,
StructCount, SCSE, BSCSE with additional data structure to efficiently compute the relevance between entities on HIN [6]. However, the previous works
do not consider the social relationships of the selected entities the social influence or the spatial dimension in the HIN which have been shown to improve
the accuracy of the relevance in activity organization [13] and recommendation
tasks respectively [12,16]. Additionally, the previous works do not consider the

distinctiveness of meta-structure instances, which reflects the comprehensiveness of the meta-structure instances.

In summary, the previous works on social activity organization and HIN are not applicable to our scenario which aims to select a group of users relevant to a set of input items while considering the social relationship between users, the social influence, geographical location in the HIN and the meta-structure distinctiveness necessary to form a suitable group of attendees.

## 3    Formulation

In order to take into consideration the social relationship among individuals, we propose an extension of an HIN named *Integrated Information Network (IIN)* by including the social information into the conventional HIN. Formally, an integrated information network is denoted as $G = (V, E, W)$ with schema $(A, R)$, where $V$ is a set of vertices representing the entities, $E$ is the set of edges representing the relations between entities, and $W$ is another set of edges which capture the social influence between entities specifically for the type **person**. In addition, $A$ is the type set of entities, and $R$ is the type set of relations for edges in $E$.

For each entity $v \in V$, we denote its type as $\psi(v) \in A$, and for each directed relation $(u, v) \in E$ between $u, v \in V$, its type is denoted as $\phi(u, v) \in R$. For entities $u, v \in V$ where $\psi(u)$ and $\psi(v)$ are both **person**, $p_{uv} \in [0, 1]$ in $W$ is also an directed edge representing the activation probability of $u$ on $v$ to model the social influence among them [7]. Note that, without loss of generality, each entity in $V$ is unique and there is no duplicate relation between the same pair of entities.

According to the information from an IIN, the relevance between an individual and an entity can be measured by their structural, social, and spatial connections [7, 24, 26].

By intuition, a meta-structure $S$ is given to illustrate the structural relation between two types. However, as mentioned earlier, simply counting the number of instances satisfying a meta-structure $S$ is insufficient to fully represent the relevance between an individual and an entity since the social influence, distinctiveness, the spatial distance, and entity similarity are not considered. Therefore, in the following, we first detail the proposed measurement to capture each factor mentioned above in Sect. 3.1. Then, based on these measurement for each factor, we propose a new measurement to integrate all the factors and formulate the research problem in Sect. 3.2.

### 3.1    Proposed Measurement for Each Factor

**Social Influence Factor.** As the social influence affects people's perception in the real world, the relevance between an individual $v_s$ and an entity $v_t$ established by or related to more influential individuals is more representative [7]. In other words, given a meta-structure $S$ and a source $v_s$, an entity $v_t$ is more relevant to $v_s$ if it is connected to $v_s$ through a meta-structure instance (i.e., $M_S$) of $S$ containing entities related to those who can influence $v_s$ more. Inspired by the concept of maximum influence path [21], which is a path possessing the maximum propagation probability from one to another, the probability of someone to activate $v_s$ is adopted to quantify the impact of his/her social influence on $v_s$. Therefore, the *social meta knowledge* capturing the extent of an entity socially related to $v_s$ is thus defined as follows.

**Definition 1. Social Meta Knowledge.** *Given a source individual (i.e., of type* **person***) $v_s$ and an entity $v \neq v_s$, let $\hat{v} \neq v_s$ denote the closest individual to $v$, i.e., $\hat{v}$ has the shortest distance to $v$ in IIN, where $\hat{v} = v$ if $\psi(v)$ is* **person***. Assume the maximum influence path from $\hat{v}$ to $v_s$ is $\mathcal{P}(\hat{v}, v_s) = \langle \hat{v} = v_1, v_2, \cdots, v_m = v_s \rangle$. The social meta knowledge of an entity $v$ to $v_s$, denoted as $\omega(v, v_s)$, is the propagation probability on $\mathcal{P}(\hat{v}, v_s)$, i.e., $\omega(v, v_s) = \prod_{i=1}^{m-1} p_{v_i v_{i+1}}$. Note that for $v = v_s$, the social meta knowledge is defined as 1 (the maximum value) without loss of generality.*

**Distinctiveness Factor.** On the other hand, as the diversity of relationships is concerned, the strength of relationship between an individual $v_s$ and an entity $v_t$ established by more distinct instances is stronger [23]. To evaluate the distinctiveness of instances, the number of distinct entities in these instances is crucial since the larger the number is, the more distinct these instances are due to the uniqueness of each entity in the IIN. Specifically, for a source $v_s$ and a target $v_t$, assume there are $n$ instances of the meta-structure $S$ and $M_S^i(v_s, v_t) = (V_S^i, E_S^i, v_s, v_t)$ denotes the $i^{th}$ instance of $S$. Accordingly, we have $|\bigcup_i V_S^i|$ entities in these $n$ instances while there supposedly are at most $n|A_S|$ entities for $n$ instances. Hence, the distinctiveness is the ratio of distinct entities to all entities, which is defined as follows.

**Definition 2. Distinctiveness.** *Given a meta-structure $S$, a source individual $v_s$, and a target entity $v_t$, let $\mathcal{M}_S(v_s, v_t)$ denote the set of meta-structure instances of $S$ with source $v_s$ and target $v_t$, where $M_S^i(v_s, v_t) = (V_S^i, E_S^i, v_s, v_t)$ is the $i^{th}$ instance in $\mathcal{M}_S(v_s, v_t)$. The distinctiveness of $\mathcal{M}_S(v_s, v_t)$ is*

$$\delta(\mathcal{M}_S(v_s, v_t)) = \frac{|\bigcup_i V_S^i|}{|\mathcal{M}_S(v_s, v_t)||A_S|}. \tag{1}$$

Intuitively, $\delta(\mathcal{M}_S(v_s, v_t)) \in [0, 1]$, where a larger value implies the relationship between $v_s$ and $v_t$ is thus stronger.

**Spatial Factor.** The spatial distance also plays an important role in measuring the relevance since the knowledge is usually regional, e.g., some scenic spots or fantastic restaurants are only known to locals. A connection established by or related to those entities located within people's living sphere is thus more reliable. In particular, given the radius of people's living sphere $d$, a meta-structure instance $M_S^i(v_s, v_t)$ counts for a spatial concern if and only if $v_s$ has a less distance than $d$ to all entities, i.e., $\forall v \in V_S^i$, $D_{spa}(v_s, v) \leq d$, where $D_{spa}(v_s, v)$ is the spatial distance between $v_s$ and $v$ [24,25]. Note that an entity not attached to a physical location, e.g., entities of type `attribute`, always satisfies the spatial criterion without loss of generality. Therefore, only the set of meta-structure instances meeting the spatial distance requirement, denoted as $\mathcal{M}_S(v_s, v_t, d)$, contributes to the relevance measurement of $v_s$ to $v_t$.

With the measurements for each factor in hand, now we are ready to propose an integrated measurement for all these factors and formulate the research problem, *MHS2Q*.

## 3.2    Integrated Measurement and Problem Formulation

As the knowledge of entities is fully expressed in an IIN, the relevance between an individual $v_s$ and an entity $v_t$ can be derived more comprehensively through the entity similarity. For example, iPhone XS and iPhone XS Max are quite similar from many aspects. The relevance between one and iPhone XS (or iPhone XS Max) thus can be the reference for that between him/her and the other iPhone. In order to extensively consider the entities similar to $v_t$, the set of extensive meta-structure instances $\mathcal{M}_S(v_s, *, d)$ are used instead of $\mathcal{M}_S(v_s, v_t, d)$. Therefore, the relevance between an individual $v_s$ and an entity $v_t$ with respect to a meta-structure $S$ is measured as the distinctiveness of $\mathcal{M}_S(v_s, *, d)$, the average social meta knowledge of entities in $\mathcal{M}_S(v_s, *, d)$, and the average similarity between target entities in $\mathcal{M}_S(v_s, *, d)$ and $v_t$.

**Definition 3.** *Influence-based and Meta-structure Distinctiveness-aware Relevance (IMDR). Given a meta-structure $S$, a source individual $v_s$, a target entity $v_t$, and the radius of people's living sphere $d$, the IMDR is*

$$IMDR(v_s, v_t, d \mid S) = \delta(\mathcal{M}_S(v_s, *, d)) \cdot \frac{\sum\limits_{v \in \bigcup_i V_S^i} \omega(v, v_s)}{|\bigcup_i V_S^i|} \cdot \frac{\sum_i Sim(v_t^i, v_t)}{|\mathcal{M}_S(v_s, *, d)|}, \quad (2)$$

*where $Sim(\cdot, \cdot)$ can be any similarity function.*

Please note $IMDR(v_s, v_t, d \mid S) \in [0, 1]$, where a larger value indicates that $v_s$ is more relevant to $v_t$ in terms of the social, spatial, and distinctiveness factors to $S$.

*Example 1.* Given the meta-structure $S$ in Fig. 1(a) and Jaccard function to measure the neighborhood similarity between two entities, consider the source Cindy and the target *Swimming pool* in Fig. 1(c), where the number in each entity represents the spatial distance to Cindy. Given the spatial distance requirement $d = 3$, since Cindy has two instances with *Swimming pool* and one instance with *Nightclub*, $|\mathcal{M}_S(\text{Cindy}, *, d)| = 3$. The Jaccard similarity between *Nightclub* and *Swimming pool* is $\frac{2}{3}$. The IMDR of Cindy to *Swimming pool* within $d = 3$ with respect to $S$ is thus

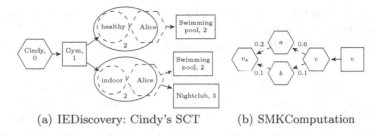

(a) IEDiscovery: Cindy's SCT        (b) SMKComputation

**Fig. 2.** Examples for MS2MU.

$$IMDR(\text{Cindy}, \text{Swimming pool}, 3 \mid S) = \frac{1 + 0.9 + 0.2 \cdot 2 + 0.12 \cdot 3}{3^2 \times 5} \times (1 + 1 + \frac{2}{3})$$

$$= 0.16.$$

∎

In addition to measuring the relevance between a source individual $v_s$ and the target entities, the connectivity between $v_s$ and the target user $u \in V$ is quite important as well. In order to acquire a social group who are familiar to the target user $u$, a more social distant source individual $v_s$ is less relevant. Formally, the Social-Spatial and Distinctiveness aware Meta-structure Relevance (SIMER) is defined as follows.

**Definition 4. *SIMER*.** *Given a target entity $v_t$, a target user $u$, a corresponding meta-structure $S$ illustrating the structural relationship between **person** and $v_t$'s type, and the spatial distance requirement $d$, the relevance of a source user $v_s$ ($v_s \neq u$) to $v_t$ and $u$ within $d$ with respect to $S$ is*

$$SIMER(v_s, v_t, u, d \mid S) = \frac{IMDR(v_s, v_t, d \mid S)}{D_{soc}(v_s, u)}, \tag{3}$$

*where $D_{soc}(v_s, u)$ is the social distance between $v_s$ and $u$.*

Therefore, $SIMER(v_s, v_t, u, d \mid S) \in [0, 1]$, where a larger SIMER value indicates $v_s$ is more suitable as an attendee.

*Example 2.* Given a target entity *Swimming pool*, a target user $u$, a meta-structure $S$ shown in Fig. 1(a), and the spatial distance requirement $d = 3$, consider the source Cindy in Fig. 1(c). Follow Example 1 and assume $D_{soc}(\text{Cindy}, u) = 2$.

$$SIMER(\text{Cindy}, \text{Swimming pool}, u, 3 \mid S) = \frac{1}{2} \times 0.16 = 0.08.$$    ∎

Given the above definition, now we can formulate the *Meta Heterogeneous Social-Spatial Query (MHS2Q)* problem which answers the top-$k$ relevant users to a specific user and a specific activity associated with geographical coordinates.

**Definition 5. *MHS2Q*.** *Given an IIN $G = (V, E, W)$ with schema $(A, R)$, a target entity $v_t \in V$, a target user $u \in V$, a meta-structure $S$, and a spatial distance requirement $d$, MHS2Q extracts the top-k users $v \in V$ having the highest $SIMER(v, v_t, u, d \mid S)$.*

MHS2Q is challenging since the relevance between any returned user and the query (including the query user and the activity) as well as the spatial distance to the activity of the returned users should be considered simultaneously. Actually, MHS2Q can be easily proved to be NP-Hard (with a reduction from the Hamiltonian path problem [3]).

# 4    Algorithm MS2MU for MHS2Q

In this section, we first design an efficient algorithm for MHS2Q named *Maximum Spatial and Social Meta-knowledge Relevant Users (MS2MU)* which retrieves the top-$k$ users with the highest SIMER values. To derive the value of SIMER of a specific source individual $v_s$ with respect to a specified meta-structure $S$, MS2MU calculates the value of IMDR towards each target entity in two steps. The first step, named Instance Entity Discovery (IEDiscovery), finds all distinct entities and meta-structure instances of $v_s$ within the spatial distance requirement with respect to $S$ while the second step, referred to as Social Meta knowledge Computation (SMKComputation), computes the social meta knowledge of those distinct entities to $v_s$. Then, the SIMER value as well as the IMDR values can be calculated according to Eqs. (3) and (2), respectively.

## 4.1    Instance Entity Discovery (IEDiscovery)

In this step, for a source user, IEDiscovery identifies all meta-structure instances and distinct entities among them. To achieve this goal, IEDiscovery traverses from the source user with respect to the given meta-structure and records the discovered meta-structure instances in a tree structure, where the entities in the same level corresponding to the meta-structure are regarded as a composite node and the source user is the root node. In order to fulfil the spatial distance requirement and save more space, we adapt the Compressed ETree (CT) [6], which merges the same (composite) nodes with the same ancestors in the same level together, and propose the Spatial Compressed ETree (SCT) to maintain the spatial information. The formal definition of a SCT is described as follows.

**Definition 6. *Spatial Compressed-ET (SCT)*.** *Given an IIN $G$, a meta-structure $S$, and a source individual $v_s$, the Spatial Compressed-ETree of $S$ from $v_s$, denoted as $SCT(v_s \mid S) = (V_{SCT}(v_s \mid S), E_{SCT}(v_s \mid S))$, is a tree, where each node $v \in V_{SCT}(v_s \mid S)$ in the $i^{th}$ level represents the composite of entities in the $i^{th}$ level of some instances. Note that all nodes with the same ancestors in the same level are merged together for saving space. Besides, for each node, a minimum spatial distance requirement is obtained to record the least spatial distance requirement for traversing from $v_s$ to this node. The least spatial distance* *requirement of a node v comes from the maximum of (1) the least spatial distance requirement of v's ancestors and (2) the maximum spatial distance from v's entities to $v_s$.*

*Example 3.* Figure 2(a) is the SCT of the meta-structure $S$ in Fig. 1(a) on the IIN in Fig. 1(c). In Fig. 1(c), there are three instances of $S$ and the entities of all instances in levels 0, 1, 2, and 3 include {*Cindy*}, {*Gym*}, {*healthy, indoor, Alice*}, and {*Swimming pool, Nightclub*}, respectively. The SCT in Fig. 2(a) has *Cindy* as the root with minimum spatial distance requirement 0 while *Gym* is placed in level 1 with minimum spatial distance requirement 1. Then, two nodes {*healthy, Alice*} and {*indoor, Alice*} are placed in level 2 both with minimum spatial distance requirement 2, where {*indoor, Alice*} is merged by two instances ended with *Swimming pool* and *Nightclub* since it and its ancestors, i.e., *Gym* and *Cindy*, are the same. Finally, level 3 contains the target entity of each instance, i.e., *Swimming pool, Swimming pool*, and *NightClub*, with minimum spatial distance requirements 2, 2, and 3, respectively. ∎

By definition, each path from the root node of SCT to a leaf node tracks a unique meta-structure instance. If the leaf node is associated with a minimum spatial distance requirement lower than or equal to the given spatial distance requirement $d$, this instance thus satisfies the spatial distance bound. Therefore, we can obtain the set of distinct entities in all the meta-structure instances satisfying the distance bound (i.e., $\bigcup_{(V_S^i, E_S^i) \in \mathcal{M}_S(v_s, *, d)} V_S^i$) by traversing the SCT using BFS or DFS algorithms. So far, by Equation (1), the distinctiveness is calculated.

## 4.2 Social Meta Knowledge Computation (SMKComputation)

According to Definition 1, the social meta knowledge is modeled as the propagation probability of the maximum influence path between two entities. Observed from the property that any subpath of a maximum influence path is also a maximum influence path, we have the social meta-knowledge of $v$ is a function of the social meta knowledge of $v$'s out-neighbor $v'$ that maximizes $p_{v,v'} \cdot \omega(v', v_s)$. Let $U$ denote the set of entities with type **person** in $V$. To compute the social meta-knowledge of each individual in $U$ to $v_s$, we dynamically compute the social meta-knowledge of each node in $U$ starting from $v_s$ by exploring the social edges between individuals using a BFS-based manner.

Specifically, initially, we have $\omega(v_s, v_s) = 1$ and $\omega(v, v_s) = 0, \forall v \in U$. At each iteration, the node $v' \in U$ with the highest social meta knowledge to $v_s$ is selected. The social meta knowledge to $v_s$ of each $v'$'s in-neighbor $v \in U$ becomes the product of the social meta knowledge of $v'$ to $v_s$ and the influence from $v$ to $v'$ in case the product is higher than the current social meta knowledge value i.e., $\omega(v, v_s) = p_{v,v'} \cdot \omega(v', v_s)$ if $p_{v,v'} \cdot \omega(v', v_s) > \omega(v, v_s)$ Each individual is selected exactly once and SMKCompution terminates when all individuals are selected. As all social meta knowledge of individuals in $U$ is discovered, SMKComputation updates the social meta knowledge of distinct entities not belonging to **person** by Definition 1.

*Example 4.* Consider Fig. 2(b) for example where the hexagon represents an entity with type **person**, the rectangle represents the entity whose type is not **person**, and the dashed edges represent the social relationships with social influence. SMKComputation first extracts all individuals in $U = \{v_s, a, b, c\}$, and has $\omega(v_s, v_s) = 1$ and the value of the social meta knowledge from each node (different from $v_s$) to $v_s$ is 0. At the first iteration, $v_s$ is selected and the values of social meta knowledge of all $v_s$'s in-neighbors i.e., $a$ and $b$, are computed as 0.2 and 0.1, respectively. At the second iteration, $a$ is selected because $\omega(a, v_s) = 0.2$ is the highest. Hence, $\omega(c, v_s)$ is set to $0.6 \cdot \omega(a, v_s) = 0.12$. The process continues until all individual have been selected. Finally, the social meta-knowledge of $v$ is set as $\omega(c, v_s) = 0.12$ due to $c$ is the closet individual to $v$.    ∎

As the average social meta knowledge is derived, the IMDR value is calculated by Eq. (2). Then, the SIMER value is also derived by Eq. (3). MS2MU thus reports the top-$k$ users with the highest SIMER as the answer. In the following, we prove that i) SMKComputation correctly computes the social meta knowledge from any $v_t \in V$ to $v_s \in V$.

### 4.3    Correctness of SMKComputation

**Theorem 1.** *Let $v_s \in V$, SMKComputation accurately computes the social metaknowledge from $v \in V$ to $v_s$.*

*Proof.* Let $v_s \in V$ and $O = V \setminus U$, where $U$ is the set of users. Since (1) SMKComputation selects each user node once, (2) each step the user node $\hat{u} = \arg\max_{u \in U} \omega(u, v_s)$ with the maximum SMK value computed so far is selected, and (3) each entity not belonging to **person** takes the SMK value of the neighboring user node with the maximum SMK value, we prove Theorem 1 by showing the induction that at each step of the SMKComputation, the SMK value of the user node $\hat{u}$, i.e., $\omega(\hat{u}, v_s)$, is accurate.

At iteration 1, $\omega(v_s, v_s) = 1$ which is maximal and therefore accurate. At iteration 2, the SMK values of the neighboring nodes to $v_s$ in the IIN are set to be 1 for $o \in O$ and to $p_{u,v_s}$ for each neighboring user $u$. Let $U_i$ be the set of user nodes that have not been selected yet and whose SMK values have been computed at step $i$. Let $\hat{u}_2 = \arg\max_{u \in U_2} \omega(u, v_s)$, and we have $\omega(\hat{u}_2, v_s) = \omega(v_s, v_s) \cdot p_{\hat{u}_2, v_s} = p_{\hat{u}_2, v_s}$. Let $N(v_s) \subseteq U$ be the set of user neighbors to $v_s$. $\omega(u, v_s), \forall u \in N(v_s)$ is always the product of a term bounded by $[0, 1]$ and $p_{\hat{u}, v_s}$ which implies that $\omega(u, v_s) \leq p_{\hat{u}, v_s}$. Therefore at step 2, $\omega(\hat{u}_2, v_s)$ is maximal and thus accurate.

Assume that at iteration $n$, $\hat{u}_n = \arg\max_{u \in U_n} \omega(u, v_s)$ is accurate. In the following we show that $\hat{u}_{n+1} = \arg\max_{u \in U_{n+1}} \omega(u, v_s)$ is also accurate at iteration $n + 1$.

Since $\hat{u}_{n+1} = \arg\max_{u \in U_{n+1}} \omega(u, v_s)$ at iteration $n + 1$, $p_{\hat{u}_{n+1}, t} \in [0, 1], \forall t \in N(\hat{u}_{n+1})$, and $\omega(v, v_s) \in [0, 1], \forall v \in V$, then $\omega(\hat{u}_{n+1}, v_s) \geq \omega(t, v_s) \cdot p_{\hat{u}_{n+1}, t}$,

$\forall t \in N(\hat{u}_{n+1}) \cup U_{n+1}$. Therefore, $\omega(\hat{u}_{n+1})$ is maximal and thus accurate at iteration $n+1$. Recall that SMKComputation pick a new node at each iteration and that SMKComputation stops after selecting each user node socially connected to $v_s$. Moreover, the SMK value of an entity in $O$ is the SMK value of the neighboring node with the highest SMK value and all nodes not socially connected to $v_s$ have an SMK value set to be 0. Therefore, the SMKComputation step accurately computes the SMK values of from each node to $v_s$. The theorem follows.

# 5  Subsequent MHS2Q

In the real world, since the radii of people's living spheres are various in different geographical areas, it is challenging for the query user to obtain the knowledge about the spatial distance requirement. In particular, the user moving pattern differs according to areas partly because of the influence of the lack/availability of transportation means in different areas. Therefore, the optimal spatial distance requirement will vary according to different areas and thus will be harder to determine. This can make the query user unsatisfied with the resulting top-$k$ users due to the lack of knowledge about user moving patterns in this area. To ensure a suitable top-$k$ list returned to the query user, we propose the Subsequent MHS2Q with varying spatial distance requirements until convergence of the resulting top-$k$ list.

A naïve approach to answer the Subsequent MHS2Q is to perform Algorithm MS2MU multiple times regardless of previous query results. However, the previous queries might generate intermediary results that can be used for subsequent queries in order to improve the efficiency. We notice that as the spatial distance requirement $d$ increases, some intermediary results from previous queries with smaller $d$ can be re-used to compute the new results of the subsequent queries. Inspired by this observation, we extend Algorithm MS2MU to incorporate a table-based index structure, referred to as the $d$-Table, which records the results of the exploration of a SCT regarding the minimum spatial distance requirement. By using the $d$-table, we further propose the MSR-pruning strategy for MS2MU to avoid the SIMER computation for those who must not be the top-$k$ users. Finally, the results of a series of queries are aggregated to form a top-$k$ list, where the number of times appearing in all top-$k$ lists is the first priority while the highest rank in the top-$k$ list is the second priority.

| 2 | Swimming pool, Swimming pool | $NULL$ |
|---|---|---|
| 3 | Nightclub | $NULL$ |

(a) Initial

| 2 | Swimming pool, Swimming pool | Swimming pool, healthy, Alice, Gym, Cindy, indoor |
|---|---|---|
| 3 | Nightclub | Nightclub |

(b) Updated

Fig. 3. An example of the $d$-table for Cindy's SCT.

## 5.1  $d$-Table

The $d$-Table is an index structure of an SCT of a source user with respect to a specified meta-structure. As the spatial factor is taken into account in calculating the IMDR, the computation will be boosted if the meta-structure instances satisfying the spatial distance requirement can be directly accessed. To address the need, the $d$-Table is thus designed to index all leaf nodes by their minimum spatial distance requirements. Furthermore, the $d$-table also records intermediate results, i.e., the distinct entities traversed in queries, so that the subsequent queries do not need to traverse the SCT again to obtain the same distinct entities. Hence, the $d$-Table is built during the IEDiscovery step and updated when MS2MU answers queries.

In particular, the $d$-Table contains three columns, including the minimum spatial distance requirement, the leaf nodes of the corresponding SCT, and the distinct entities that have been ever traversed. Initially, the $d$-Table only fills in the first two columns by indexing the leaf nodes of the SCT in an ascending order of their minimum spatial distance requirements while the third column contains $NULL$, indicating no leaf node associated with the current minimum spatial distance requirement is traversed. For example, the initial $d$-table of Cindy's SCT in Fig. 2(a) is shown in Fig. 3(a), where the two *Swimming pool* point to different leaf nodes in Fig. 2(a). When answering a query with the spatial distance requirement $d$ and computing the IMDR of some source individual $v_s$, MS2MU reads $v_s$'s $d$-Table for the rows with minimum spatial distance requirements less than or equal to $d$ in an ascending order. For a row with $NULL$ in the third column, indicating the leaf nodes in this row have not been traversed yet, MS2MU back traverses $v_s$'s SCT from each leaf node in this row, marks the nodes traversed in the SCT as visited, fills in the third column of the $d$-Table with the entities that do not appear in the third column of any row, and stops when a visited nodes in the SCT is met. If the third column of a row is not $NULL$, which means the leaf nodes of this row have been traversed, MS2MU can directly obtain the distinct entities in the third column for calculating the IMDR. Note that the third columns of all rows are disjoint since only distinct entities are recorded during traversal.

*Example 5.* Consider Cindy's SCT in Fig. 2(a) and the initial $d$-Table in Fig. 3(a). Given the spatial distance requirement $d = 3$, MS2MU first reads the first row associated with minimum spatial distance requirement 2. Since $2 < d$ and the third column is $NULL$, MS2MU back traverses the two leaf nodes *Swimming pool*, respectively. For the first *Swimming pool*, MS2MU records *Swimming pool, healthy, Alice, Gym*, and *Cindy* in the third column of the first row in order. Then, for the second *Swimming pool*, MS2MU only records *indoor* in the third column and stops traversal since the node *Gym* is visited. After that, since the second row associated with minimum spatial distance requirement $3 = d$ and the third column is $NULL$, MS2MU back traverses the leaf node *Nightclub*, records only *Nightclub* in the third column, and stops due to the visited node {*indoor, Alice*}. The updated $d$-Table is shown in Fig. 3(b). ∎

## 5.2    MSR-Pruning Strategy

For better efficiency, MS2MU exploits the minimum SIMER of the current top-$k$ list to avoid the SIMER computation for those that must not be the top-$k$ users. By Definition 4, the value of SIMER comes from the value of IMDR divided by the social distance $D_{soc}(v_s, u)$. We thus propose a pruning strategy named Minimum Social Ratio pruning (MSR-pruning) that takes the minimum SIMER, denoted as $s_{min}$, of the current top-$k$ list and the upper bound of SIMER of an individual $v_s$ into account for pruning. Specifically, if $v_s$'s upper bound of SIMER is less than $s_{min}$, indicating it is impossible for $v_s$ to be the top-$k$ users, $v_s$ can be pruned satisfactory. In order to infer the upper bound of SIMER, the upper bounds of IMDR should be first derived.

**Theorem 2.** *The IMDR value is upper bounded by the average similarity.*

*Proof.* Since all distinctiveness, average social meta knowledge, and average similarity are bounded in $[0, 1]$, the IMDR is upper bounded by the minimum value among them. If the average similarity is the minimum value among them, the IMDR is exactly upper bounded by the average similarity. Otherwise, there exists the minimum value among them less than the average similarity. In this case, IMDR is upper bounded by this minimum value, which is still upper bounded by the average similarity. The theorem is thus proved.

According to Theorem 2, the upper bound of an IMDR value can be derived by the average similarity of all leaf nodes and the corresponding query entity. By using the $d$-Table, MS2MU can easily obtain the leaf nodes satisfying the spatial distance requirement and compute the average similarity. The upper bound of SIMER is thus the average similarity divided by the social distance $D_{soc}(v_s, u)$. If the average similarity is less than $s_{min} \cdot D_{soc}(v_s, u)$, which is equivalent to that the upper bound of SIMER is less than $s_{min}$, MS2MU can prune $v_s$ satisfactory. The pseudo code of Algorithm MS2MU is in Algorithm 1.

---

**Algorithm 1.** MS2MU

---

**Require:** An IIN $G = (V, E, W)$ with $(A, R)$; a query user $u$; an query activity $v_t$; a meta-structure $S$; a spatial distance requirement $d$; a parameter $k$                    ▷ the output is a top-$k$ list
1: Build offline the SCT and $d$-Table for each individual with respect to $S$
2: $Q \leftarrow$ an empty priority queue with size $k$; $s_{min} \leftarrow 0$
3: **for** $v_s \in V$ and $v_s$ belongs to person **do**
4:     **if** avg. similarity $> s_{min} D_{soc}(v_s, u)$ **then**
5:         $s \leftarrow v_s$'s SIMER value by Equation (3)
6:         **if** $s > s_{min}$ **then**
7:             Add $v_s$ into $Q$ and update $s_{min}$ if necessary
8:         **end if**
9:     **end if**
10: **end for**
11: **return** $Q$

---

**Table 1.** The summary of datasets.

| Dataset | $|V|$ | $|E|$ | $|W|$ | $|A|$ | $|R|$ | # of type **person** |
|---------|-------|-------|-------|-------|-------|----------------------|
| Yelp | 33,025 | 970,818 | 196,134 | 8 | 4 | 7,990 |
| Weeplaces | 43,023 | 189,879 | 119,939 | 3 | 3 | 16,022 |

### 5.3   Results Aggregation

In order the answer the Subsequent MHS2Q, MS2MU needs to aggregate the results from a series of queries with different spatial distance requirements to ensure a suitable top-$k$ list. The aggregated top-$k$ users should be aware of the number of times appearing in all results as well as the highest rank. In other words, a user with greater frequency of appearance has a higher rank in the top-$k$ list. If two or more users have the same frequency, the rank in the top-$k$ list determined by their ever highest rank.

As the aggregated top-$k$ list converges, indicating the optimal spatial distance requirement is used for query, MS2MU reports the aggregated one as the answer. In the following, we prove that M2SMU obtains the optimal solution to MHS2Q.

### 5.4   Optimal Solution Guarantee

**Theorem 3.** *M2SMU returns the optimal solution for MHS2Q.*

*Proof.* MS2MU uses SMKComputation to compute IMDR and consequently SIMER. MS2MU discards all the users violating the social and spatial bound. Moreover, the IEDiscovery step finds all the instances of $S$ in $G$ and filters the found instances according to the spatial bound. Therefore, IMDR and SIMER are accurately computed. After the computation of SIMER for each new user, MS2MU inserts the new user in the top-$k$ list if the SIMER value of the new user is higher than the smallest SIMER value in the top-$k$ list. Furthermore, the returned top-$k$ list is connected. Therefore, MS2MU returns the optimal solution to MHS2Q. The theorem follows.

## 6   Experiments

In this section, we conduct experiments on two real location-based social networks Yelp and Weeplaces, whose summary is listed in Table 1. The performance of our algorithm MS2MU is evaluated in terms of the SIMER value, the social distance, and the running time. In particular, we implement three competitive approaches from the meta-structure, social, and spatial aspects, respectively. The details are described as follows. (1) BSCSE [6] reports the most relevant users to the query activity with respect to the given meta-structure as the result. (2) SDSSel [13] reports a social group interested in the query activity and guaranteeing the social tightness. (3) SPATIAL is a straightforward approach to consider

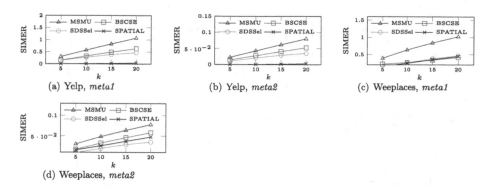

**Fig. 4.** The SIMER comparison as $k$ varies from 5 to 20 for $d = \infty$.

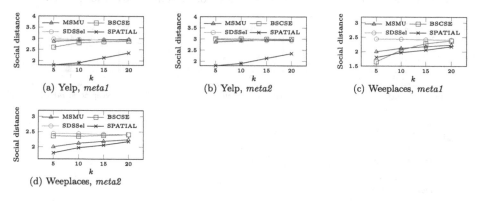

**Fig. 5.** The social distance comparison as $k$ varies from 5 to 20 for $d = \infty$.

the spatial factor, which reports the top-$k$ socially close users who are not only interested in the query activity but also satisfy the spatial constraint. For the experiments, we randomly select a pair of user and location for each query. We form 60 of such pairs on each dataset for each measurement. There are two corresponding meta-structures, *meta1* and *meta2*, both of which contains three levels. The spatial distance requirements include 25 km, 30 km, 35 km, and 40 km while the parameter $k$ is set as 5, 10, 15, and 20. All the programs are implemented in C++ and run on an Intel core E7-4850 2.20 GHz PC with 52 GB RAM using CentOS Linux release 7.3.1611.

## 6.1 Effectiveness

Figure 4 shows the SIMER value among MS2MU, BSCSE, SDSSel, and SPATIAL for $d = \infty$ when $k$ varies from 5 to 20 on Yelp and Weeplaces. We have two observations. First, for all $k$, MS2MU has the highest SIMER value, followed by BSCSE, and SDSSel is worse than BSCSE while SPATIAL is usually the worst. This is because MS2MU considers the relevance in terms of both social and spatial factors as well as the distinctiveness of meta-structures. Second, as $k$

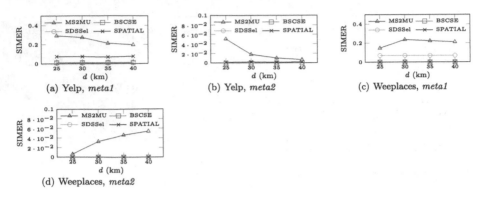

**Fig. 6.** The SIMER comparison as $d$ varies from 25 km to 40 km for $k = 20$.

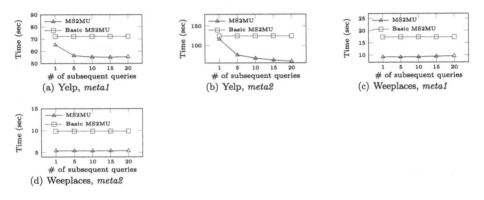

**Fig. 7.** The average response time comparison as the number of subsequent queries increases for $k = 20$.

increases, the values of SIMER also increase for all approaches as well. However, the increase velocity of MS2MU is faster than the competitive approaches, which shows that the a newly selected by MS2MU usually has a large SIMER while competitive approaches do not always a select user with a large SIMER.

Next, Fig. 5 demonstrates the social distance of the reported group to the query user among MS2MU, BSCSE, SDSSel, and SPATIAL for $d = \infty$ when $k$ varies from 5 to 20 on Yelp and Weeplaces. The results show that SPATIAL usually has the least social distance since the social distance to the query user is its main criterion for ranking. Apparently, for selecting socially close users, MS2MU can be comparable or even better than SDSSel, which aims at social tightness. On the other hand, BSCSE does not consider any social concept so that the social distance is unstable. Besides, there is usually an upward trend of social distance for most approaches as $k$ increases. This is because the number of users usually grows exponentially with the social distance so that it is more likely to select a socially distant user when $k$ increases.

On the other hand, Fig. 6 shows the SIMER value among MS2MU, BSCSE, SDSSel, and SPATIAL for $k = 20$ when $d$ varies from 25 km to 40 km on Yelp and Weeplaces. First of all, MS2MU is the best due to the comprehensive consideration, especially from the spatial aspect. The other approaches are comparable and cannot find a good solution. Since neither BSCSE nor SDSSel considers the spatial factor, the returned top-$k$ users are very likely to be irrelevant to the query when $d$ is constrained. For SPATIAL, even though the spatial factor is considered, the lack of comprehensive consideration of the other factors still makes SPATIAL fail to return a good solution. Moreover, the SIMER value is not sensitive to $d$ because the larger $d$ allows more meta-structure instances considered, which makes both higher and lower distinctiveness as well as average similarity possible. Note that the social distance as well as the top-$k$ users returned by BSCSE and SDSSel for different $d$ are the same as that for $d = \infty$. Besides, unlike varying $k$, the trend of the social distance for MS2MU and SPATIAL is not related to $d$ since a larger/less spatial distance requirement does not prefer selecting socially close or distant users. Hence, the social distance comparison for varying $d$ is thus omitted.

**Table 2.** The time comparison for $k = 20$ and $d = \infty$.

| (Sec) | Yelp, $meta1$ | Yelp, $meta2$ | Weeplaces, $meta1$ | Weeplaces, $meta2$ |
|---|---|---|---|---|
| MS2MU | 72.19 | 124.31 | 17.39 | 9.87 |
| BSCSE | 14.49 | 42.00 | 2.15 | 0.66 |
| SDSSel | 32.59 | 32.59 | 9.25 | 9.25 |
| SPATIAL | 6.53 | 6.53 | 10.61 | 10.61 |

## 6.2    Efficiency

In this subsection, we compare the query response time of all approaches. Since the response time is sensitive to neither $k$ nor $d$, Table 2 only shows the response time comparison for $k = 20$ when $d = \infty$. The results show that MS2MU takes a little longer time than the competitive approaches. This is because MS2MU carefully examines the social meta knowledge and distinctiveness of each candidate user while SDSSel and SPATIAL do not take the knowledge of meta-structures into account and BSCSE directly retrieves the relevant users to the query activity.

In addition, we also compare the average response time for Subsequent MHS2Q of our MS2MU algorithm with and without $d$-Table and MSR-pruning. Figure 7 shows the average response time as the number of subsequent queries increases, where the spatial distance requirement $d$ in the subsequent queries is increasing. Note that MS2MU without $d$-Table and MSR-pruning is denoted as Basic MS2MU in the figure. Obviously, with the help of $d$-Table and MSR-pruning, the response time is largely reduced. On Yelp, the spatial distance requirement $d$ in the first query is more than most spatial distance between entities so that it takes more time for the first query while the response time for subsequent queries becomes less. By contrast, the running time of each query is evenly distributed on Weeplaces since the spatial distance requirement $d$ for each query allows a number of additional entities to be traversed for SIMER computation.

In summary, compared to the baselines, MS2MU always reports a group of users with the highest SIMER and the users in the reported social group are socially close to the query user. Even though the response time of MS2MU is a little longer, MSMU can still answer the query in reasonable time. Moreover, by exploiting $d$-Table and MSR-pruning, MS2MU is efficient for Subsequent MHS2Q. Consequently, MS2MU is capable of answering the (Subsequent) MHS2Q effectively and efficiently.

## 7   Conclusion

In this paper, we proposed a new relevance measurement SIMER to more accurately measure the connection strength between a user and the activity. Then, we formulated a novel query MHS2Q to consider the spatial and social factors as well as the distinctiveness of meta-structure. We further proposed a Subsequent MHS2Q to allow a series queries with varying spatial distance requirements. To answer the (Subsequent) MHS2Q, we designed an efficient algorithm MS2MU with a new index structure $d$-Table and a pruning strategy MSR-pruning to boost the performance. Experiments on real dataset showed that our approaches significantly outperform the other baselines.

## References

1. Du, X., Liu, H., Jing, L.: Additive co-clustering with social influence for recommendation. In: Proceedings of the Eleventh ACM Conference on Recommender Systems, pp. 193–200. ACM (2017)
2. Fionda, V., Pirrò, G.: Meta structures in knowledge graphs. In: d'Amato, C., et al. (eds.) ISWC 2017. LNCS, vol. 10587, pp. 296–312. Springer, Cham (2017). https://doi.org/10.1007/978-3-319-68288-4_18
3. Gurevich, Y., Shelah, S.: Expected computation time for Hamiltonian path problem. SIAM J. Comput. **16**(3), 486–502 (1987)
4. Hsu, B.Y., Lan, Y.F., Shen, C.Y.: On automatic formation of effective therapy groups in social networks. IEEE Trans. Comput. Soc. Syst. **5**(3), 713–726 (2018)
5. Hsu, B.Y., Shen, C.Y., Chang, M.Y.: WMEgo: willingness maximization for ego network data extraction in online social networks. In: Proceedings of the 29th ACM International Conference on Information & Knowledge Management, pp. 515–524 (2020)
6. Huang, Z., Zheng, Y., Cheng, R., Sun, Y., Mamoulis, N., Li, X.: Meta structure: computing relevance in large heterogeneous information networks. In: Proceedings of the 22nd ACM SIGKDD International Conference on Knowledge Discovery and Data Mining, pp. 1595–1604. ACM (2016)

7. Hung, H.J., et al.: When social influence meets item inference. In: Proceedings of the 22nd ACM SIGKDD International Conference on Knowledge Discovery and Data Mining, pp. 915–924. ACM (2016)
8. Li, C.-T., Huang, M.-Y., Yan, R.: Team formation with influence maximization for influential event organization on social networks. World Wide Web **21**(4), 939–959 (2017). https://doi.org/10.1007/s11280-017-0492-7
9. Liu, X., Yu, Y., Guo, C., Sun, Y.: Meta-path-based ranking with pseudo relevance feedback on heterogeneous graph for citation recommendation. In: Proceedings of the 23rd ACM International Conference on Conference on Information and Knowledge Management, pp. 121–130. ACM (2014)
10. Nandanwar, S., Moroney, A., Murty, M.N.: Fusing diversity in recommendations in heterogeneous information networks. In: Proceedings of the Eleventh ACM International Conference on Web Search and Data Mining, pp. 414–422. ACM (2018)
11. Rahman, H., Roy, S.B., Thirumuruganathan, S., Amer-Yahia, S., Das, G.: Optimized group formation for solving collaborative tasks. VLDB J. **28**(1), 1–23 (2018). https://doi.org/10.1007/s00778-018-0516-7
12. Ren, Z., Liang, S., Li, P., Wang, S., de Rijke, M.: Social collaborative viewpoint regression with explainable recommendations. In: Proceedings of the tenth ACM International Conference on Web Search and Data Mining, pp. 485–494. ACM (2017)
13. Shen, C.Y., Fotsing, C.K., Yang, D.N., Chen, Y.S., Lee, W.C.: On organizing online soirees with live multi-streaming. In: Thirty-Second AAAI Conference on Artificial Intelligence (2018)
14. Shen, C.Y., Yang, D.N., Huang, L.H., Lee, W.C., Chen, M.S.: Socio-spatial group queries for impromptu activity planning. IEEE Trans. Knowl. Data Eng. **28**(1), 196–210 (2016)
15. Shen, C.Y., Yang, D.N., Lee, W.C., Chen, M.S.: Spatial-proximity optimization for rapid task group deployment. ACM Trans. Knowl. Discovery Data (TKDD) **10**(4), 1–36 (2016)
16. Shi, C., Li, Y., Zhang, J., Sun, Y., Philip, S.Y.: A survey of heterogeneous information network analysis. IEEE Trans. Knowl. Data Eng. **29**(1), 17–37 (2017)
17. Shi, C., Zhang, Z., Ji, Y., Wang, W., Philip, S.Y., Shi, Z.: SemRec: a personalized semantic recommendation method based on weighted heterogeneous information networks. World Wide Web **22**(1), 153–184 (2019)
18. Shi, Y., Chan, P.W., Zhuang, H., Gui, H., Han, J.: Prep: path-based relevance from a probabilistic perspective in heterogeneous information networks. In: Proceedings of the 23rd ACM SIGKDD International Conference on Knowledge Discovery and Data Mining, pp. 425–434. ACM (2017)
19. Shuai, H.H., Yang, D.N., Philip, S.Y., Chen, M.S.: A comprehensive study on willingness maximization for social activity planning with quality guarantee. IEEE Trans. Knowl. Data Eng. **28**(1), 2–16 (2016)
20. Ting, L.P.-Y., Li, C.-T., Chuang, K.-T.: Predictive team formation analysis via feature representation learning on social networks. In: Phung, D., Tseng, V.S., Webb, G.I., Ho, B., Ganji, M., Rashidi, L. (eds.) PAKDD 2018. LNCS (LNAI), vol. 10939, pp. 790–802. Springer, Cham (2018). https://doi.org/10.1007/978-3-319-93040-4_62
21. Wang, C., Chen, W., Wang, Y.: Scalable influence maximization for independent cascade model in large-scale social networks. Data Min. Knowl. Disc. **25**(3), 545–576 (2012). https://doi.org/10.1007/s10618-012-0262-1
22. Yang, D.N., Chen, Y.L., Lee, W.C., Chen, M.S.: On social-temporal group query with acquaintance constraint. Proc. VLDB Endow. **4**(6), 397–408 (2011)

23. Yang, Z., Fu, A.W.C., Liu, R.: Diversified top-k subgraph querying in a large graph. In: Proceedings of the 2016 International Conference on Management of Data, pp. 1167–1182. ACM (2016)
24. Ye, M., Yin, P., Lee, W.C., Lee, D.L.: Exploiting geographical influence for collaborative point-of-interest recommendation. In: Proceedings of the 34th International ACM SIGIR Conference on Research and Development in Information Retrieval, pp. 325–334. ACM (2011)
25. Yu, Y., Chen, X.: A survey of point-of-interest recommendation in location-based social networks. In: Workshops at the Twenty-Ninth AAAI Conference on Artificial Intelligence, vol. 130 (2015)
26. Zhao, H., Yao, Q., Li, J., Song, Y., Lee, D.L.: Meta-graph based recommendation fusion over heterogeneous information networks. In: Proceedings of the 23rd ACM SIGKDD International Conference on Knowledge Discovery and Data Mining, pp. 635–644. ACM (2017)
27. Zhu, Q., Hu, H., Xu, C., Xu, J., Lee, W.C.: Geo-social group queries with minimum acquaintance constraints. VLDB J. Int. J. Very Large Data Bases **26**(5), 709–727 (2017)

# Combining Oversampling with Recurrent Neural Networks for Intrusion Detection

Jenq-Haur Wang[1]([✉]) and Tri Wanda Septian[2]

[1] National Taipei University of Technology, Taipei, Taiwan
jhwang@csie.ntut.edu.tw
[2] Sriwijaya University, Palembang, Indonesia

**Abstract.** Previous studies on intrusion detection focus on analyzing features from existing datasets. With various types of fast-changing attacks, we need to adapt to new features for effective protection. Since the real network traffic is very imbalanced, it's essential to train appropriate classifiers that can deal with rare cases. In this paper, we propose to combine oversampling techniques with deep learning methods for intrusion detection in imbalanced network traffic. First, after preprocessing with data cleaning and normalization, we use feature importance weights generated from ensemble decision trees to select important features. Then, the Synthetic Minority Oversampling Technique (SMOTE) is used for creating synthetic samples from minority class. Finally, we use Recurrent Neural Networks (RNNs) including Long Short-Term Memory (LSTM) and Gated Recurrent Unit (GRU) for classification. In our experimental results, oversampling improves the performance of intrusion detection for both machine learning and deep learning methods. The best performance can be obtained for CIC-IDS2017 dataset using LSTM classifier with an F1-score of 98.9%, and for CSE-CIC-IDS2018 dataset using GRU with an F1-score of 98.8%. This shows the potential of our proposed approach in detecting new types of intrusion from imbalanced real network traffic.

**Keywords:** Class imbalance · Oversampling · Feature selection · Long short-term memory · Gated recurrent unit

## 1 Introduction

Nowadays, new variants of security threats in the cyber world are massively increasing on the Internet. It is the main focus for the system administrator to protect the network infrastructure from malicious behaviors such as new intrusions and attacks. Therefore, intrusion detection has become an important research area in network security. Intrusion detection systems aim to actively detect attacks and identify the critical illegal behaviors from network traffic. There are some challenges in effective classification for intrusion detection. First, most existing research on analyzing the characteristic of attack patterns use popular datasets such as KDD CUP'99, NSL-KDD, and ISCX2012, which need some improvement since they are out-of-date. With the development of Internet technology, there are increasing amount of new cyber-attacks. To deal with the issues of unreliable datasets that are out of date, we utilize new intrusion detection datasets

© Springer Nature Switzerland AG 2021
C. S. Jensen et al. (Eds.): DASFAA 2021 Workshops, LNCS 12680, pp. 305–320, 2021.
https://doi.org/10.1007/978-3-030-73216-5_21

including CIC-IDS2017 dataset [1], and CSE-CIC-IDS2018 dataset [2] which are developed by Sharafaldin et al. [3]. These new public datasets are based on real-time network traffic captured around the world, which are helpful for intrusion detection research. Second, due to the changing characteristics of new attacks in real data, we need to select features that can capture the most important characteristics. Louppe et al. [4] introduced the variable importance derived from tree-based methods and can be implemented as feature selection methods to improve classification accuracy. Third, we are faced with large-scale imbalanced datasets since only a small percentage of real network traffic are attacks or illegal traffic. To address this issue, we utilize Synthetic Minority Oversampling Technique (SMOTE) [5] to improve the prediction accuracy for the imbalanced dataset. Finally, for the classification algorithm, we compare classical learning methods, such as Random Forest [6], Decision Tree [7], and Naïve Bayes [8], with deep learning methods such as Long Short-Term Memory (LSTM) [9], and Gated Recurrent Unit (GRU) [10] in their classification performances. The contributions of this paper include:

1. We evaluate classification performance of recurrent neural networks (RNNs) for intrusion detection on two new publicly available datasets, which are captured from real work traffic in large scale.
2. We improve the performance of intrusion detection for the imbalanced dataset by using SMOTE oversampling technique as the feature selection method for both classical machine learning and deep learning methods.

The remainder of this paper are as follows. First, related work is reviewed in Sect. 2. Then, the proposed method is described in Sect. 3, and our experimental results are analyzed and discussed in Sect. 4. Finally, we give conclusions in Sect. 5.

## 2 Related Work

Intrusion detection has been an important research topic in information security. Many conventional machine learning methods have been used for intrusion detection. Albayati et al. [11] discussed the intelligent classifier suitable for automatic detection, such as Naïve Bayes, Random Forest, and decision tree algorithm. The best performance can be obtained for Random Forest classifiers with an accuracy of 99.89% when using all of the features from the NSL-KDD dataset. Almseidin et al. [12] evaluated the intrusion detection using machine learning methods: SVM, Random Forest, and decision tree algorithm. Random forest classifier registered the highest accuracy of 93.77%, with the smallest false positive rate for the KDD CUP'99 dataset. Khuphiran et al. [13] researched on detecting Distributed Denial of Services (DDoS), as the most common attack, using DARPA 2009 DDoS datasets, and implementing a traditional SVM and Deep Feed Forward (DFF) algorithm. Deep Feed Forward got the highest accuracy of 99.63% and F1-score is 0.996 while SVM got an accuracy rate of 81.23% and F1-score is 0.826.

Recently, deep learning methods especially recurrent neural networks, such as Long Short-Term Memory (LSTM), and Gated Recurrent Unit (GRU) have been implemented in intrusion detection research area. Althubiti et al. [14] applied LSTM algorithm for multi-classification using the *rmsprop* parameter in the CIDDS-001 dataset, which specializes in web attacks. The best accuracy of 84.83% can be obtained for LSTM, which is better than SVM and Naïve Bayes. Xu et al. [15] proposed a study in IDS with GRU, which uses softmax function for multiclass classification. The best accuracy of 99.42% using KDD CUP'99 and 99.31% using NSL-KDD can be obtained.

For an IDS to accurately detect unauthorized activities and malicious attacks in network traffic, different features might have different importance to distinguish between attacks and normal traffic. On the one hand, feature selection is needed since it is useful in analyzing complex data, and for removing features excessive or irrelevant. On the other hand, network traffic is extremely imbalanced since normal traffic accounts for most of the traffic, while intrusion or attacking traffic are very rare. For feature selection, Alazzam et al. [16] proposed a pigeon inspired optimizer for feature selection, and achieved good accuracy of 0.883 and 0.917 for NSL-KDD and UNSW-NB15 datasets respectively when reducing the feature size to 5. This shows the importance of feature selection in classification.

Regarding the class imbalance problem, some techniques have been proposed. Wu et al. [17] dealt with imbalanced health-related data with deep learning approaches using RNNs. Shuai et al. [18] devised a multi-source learning approach to extract common latent factors from different sources of imbalanced social media for mental disorders detection. To mitigate the problem of overfitting for the imbalanced class with random oversampling, the technique of SMOTE generates synthetic examples by k-nearest neighbor algorithm rather than simply replicating existing instances. Smiti and Soui [19] explored the idea of employing SMOTE and deep learning to predict bankruptcy. Seo and Kim [20] proposed to handle the class imbalance problem of KDD CUP'99 dataset by finding the best SMOTE ratios in different rare classes for intrusion detection.

Due to the growing types of new attacks, we focus on intrusion detection for the new datasets CIC-IDS2017 and CSE-CICIDS2018. Kurniabudi et al. [21] analyzed the features of CIC-IDS2017 dataset with information gain, and achieved the best accuracy of 99.86% for Random Forest. But they only used 20% of the full dataset, and cannot detect some types of traffic, for example, Infiltration attack. Kim et al. [22] compared the performance of intrusion detection on CSE-CICIDS2018 dataset using Convolutional Neural Networks (CNNs) and RNNs. They only focused on DoS category, and achieved the best accuracy of 91.5% and 65% for CNN and RNN, respectively.

In this paper, we apply deep learning methods to classify imbalanced network traffic for intrusion detection, and compare the performance with conventional machine learning methods using the two new datasets. Specifically, we compare variants of RNNs including LSTM and GRU. Then, we apply SMOTE technique to deal with class imbalance problem. To further improve the classification accuracy, we propose to use variable importance derived from tree-based methods [4] for feature selection, because it has fast calculation and suitable for large data size. We used the full datasets, and the best F1 score of 98.9% and 98.8% can be achieved for CIC-IDS2017 and CSE-CICIDS2018 datasets, respectively.

## 3  The Proposed Method

In our proposed method for intrusion detection using deep learning approach, there are three stages: data preprocessing, feature selection and oversampling technique, and classification. The proposed framework for intrusion detection is shown in Fig. 1.

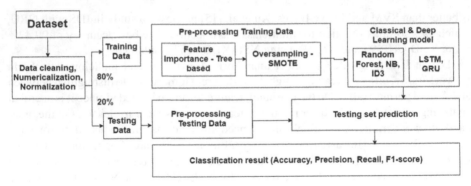

**Fig. 1.** The proposed framework for intrusion detection.

As shown in Fig. 1, in the preprocessing stage, missing and undefined values are fixed in data cleaning stage, and separate datasets are combined into a single one. In feature selection stage, we use totally randomized trees to find the important features in the dataset. Then, the SMOTE oversampling technique is used to deal with the class imbalance problem. Finally, we compare the classification performance of classical machine learning methods including Random Forest (RF), Iterative Dichotomiser 3 (ID3), and Naïve Bayes (NB) with recurrent neural networks, including LSTM and GRU on the large scale network traffic data. In the following subsections, we describe each stage in more details.

### 3.1 Preprocessing

There are several preprocessing tasks needed for the new datasets. First, we remove unnecessary information from the original dataset including the socket information of each data instance, such as source IP address "src_ip", destination IP address "dst_ip", Flow ID "flow_id," and "protocol." The reason to remove these is to provide unbiased detection. Second, we remove unreadable data which might include some noise in class labels such as: 'Web Attack Â\x96 Brute Force', 'Web Attack Â\x96 XSS', 'Web Attack Â\x96 SQL Injection', which can be replaced to distinct Unicode characters. Then, we also remove invalid numbers, such as Not a Number (NaN) and 'Infinity'. The missing values and other errors in the dataset are fixed, such as in "flow_bytes_per_s" and "flow_pkts_per_s" features. Regarding the data types, the dataset consists of categorical, strings, and numeric data types such as float64 and int64. The categorical data type in the label consists of benign, and all attack types. In the CSE-CICIDS2018 dataset, the data types of some features are not appropriate, which were changed from int64 to the float64 data type. Finally, for training purpose, the numeric attributes need to be normalized, since the difference of scale in numbers or values can degrade the performance of classification. For example, some of the features with large numeric values, e.g., 'flow_duration' can dominate small numeric values such as 'total_fwd_packets' and 'total_fwd_pckts'. Thus, we use min-max normalization to convert values into a normalized range.

## 3.2 Feature Selection

To select the most important features, we adopt the feature selection method proposed by Louppe et al. [4] to estimate feature importance using Mean Decrease Impurity (MDI) from randomized ensemble trees. Let $V = \{X_1, X_2, ....X_p\}$ denote categorical input variables, and Y means a categorical output, Shannon entropy is used as impurity measure on totally randomized trees as follows:

$$VarImp(X_m) = \sum_{k=0}^{p-1} \frac{1}{C_p^k} \frac{1}{p-k} \sum_{B \in P_k(V^{-m})} I(X_m; Y|B) \tag{1}$$

$$\sum_{m=1}^{p} VarImp(X_m) = I(X_1, X_2, ....X_p; Y) \tag{2}$$

Where $V^{-m}$ denotes the subset $V\backslash\{X_m\}$, $P_k$ ($V^{-m}$) denotes subsets of $V^{-m}$ of cardinality k, and $I(X_m;Y|B)$ is the conditional mutual information of $X_m$ and Y given the variables in B.

In this paper, X defines the input features in training data, and Y defines the output class of Benign and Attack. We adopt MDI for feature selection since it calculates each feature importance as the sum over the number of splits (across all trees) that include the features, proportionally to the number of samples it splits. In addition, ensembles of randomized trees are used to select the best subset of features for classification. This reduced feature set is then employed to implement an intrusion detection system.

## 3.3 Oversampling

In intrusion detection datasets, there is class imbalance problem, where the minority class of attack has much fewer instances than the benign class. The distribution of all classes in CIC-IDS2017 and CSE-CICIDS2018 is shown in Fig. 2.

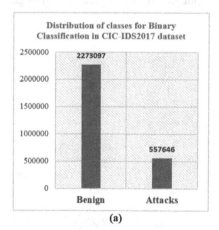
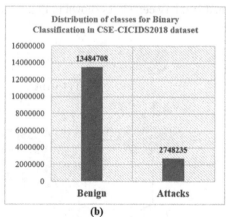

**Fig. 2.** Distribution of all classes in (a) CIC-IDS2017 dataset (b) CSE-CIC-IDS2018 dataset.

As shown in Fig. 2, the distribution of benign and attack classes show the class imbalance problem in both datasets. To tackle the problem, we adopt SMOTE [5] to improve our prediction of the minority class. The idea is to take each minority class sample, and add synthetic examples on the line segments which join the k-minority class nearest neighbors. This can be done as the following steps:

Step 1: Assume $x \in A$, where A is the set of the minority class. For each x of the k-nearest neighbors, it is obtained from the Euclidean distance calculation between x and samples from the set A.

Step 2: The number of samples N is chosen according to the sample proportion of imbalanced data. For instance, given $x_1, x_2 \dots, x_N$ ($N \leq k$) that are randomly selected by k-nearest neighbors, we can build a new set $A_1$.

Step 3: For each instance $x_k \in A_1$, where k is 1, 2,..., N, the formula is used to create a new instance $x_{new}$ as follows:

$$x_{new} = x + random(0, 1) * \|x - x_k\| \tag{3}$$

The amount of oversampling is influenced by the number of randomly selected samples from the k-nearest neighbors. It has been shown to perform better than simple under-sampling technique because this algorithm creates new instances of the minority class by using convex combinations of neighboring instances.

### 3.4 Classification

After preprocessing and oversampling the dataset, we use two types of RNNs, including LSTM and GRU, and compare with conventional machine learning classifiers such as Random Forest, ID3, and Naïve Bayes, for intrusion detection.

LSTM is a variation of RNNs to deal with the vanishing gradient problem in sequential data. The architecture of LSTM consists of input gate It, forget gate Ft, output gate Ot, and memory cell Ct, as shown in Fig. 3.

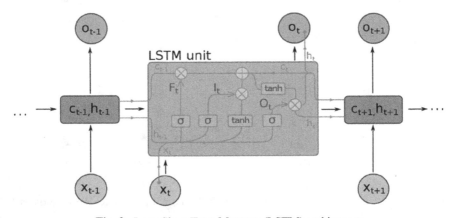

**Fig. 3.** Long Short Term Memory (LSTM) architecture.

The equations for the operations of LSTM architecture are given below:

$$F_t = \sigma(W_F x_t + U_F h_{h-1} + b_F) \tag{4}$$

$$I_t = \sigma(W_I x_t + U_I h_{h-1} + b_I) \tag{5}$$

$$O_t = \sigma(W_O x_t + U_O h_{h-1} + b_o) \tag{6}$$

$$C_t = F_t \odot c_{t-1} + I_t \odot tanh(W_c x_c + U_c h_{t-1} + b_c) \tag{7}$$

$$h_t = O_t \odot tanh(C_t) \tag{8}$$

$$O_t = f(W_o h_t + b_o) \tag{9}$$

where $\sigma$ denotes a sigmoid function, $x_t$ means an input vector at time $t$, $h_t$ denotes a hidden state vector at time $t$, $W$ denotes the hidden weight matrix from an input, $U$ means the hidden weight matrix from hidden layers, and $b$ means a bias term.

GRU is an LSTM without an output gate, in which the contents are fully written from its memory cell to the output at each time-step. Its internal structure is simpler and therefore considered faster to train as there are fewer computations needed to make updates to its hidden state. GRU has two types of gates: reset gate $r$, and update gate $z$. The reset gate determines the new input with the previous memory cell, and the update gate defines how much of the previous memory cell to keep.

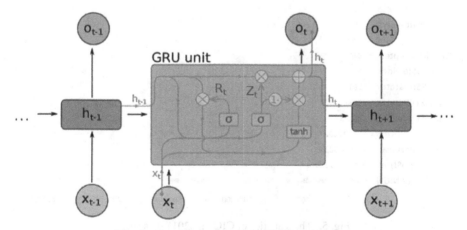

**Fig. 4.** Gated Recurrent Unit (GRU) architecture

Equations the operations of GRU architecture are given below:

$$Z_t = \sigma(W_z x_t + U_z h_{t-1} + b_z) \tag{10}$$

$$R_t = \sigma(W_R x_t + U_R h_{t-1} + b_R) \tag{11}$$

$$h_t = (1 - Z_t) \odot h_{t-1} + Z_t \odot tanh(W_h x_t + U_h(R_t \odot h_{t-1}) + b_h) \tag{12}$$

where $Z_t$ is the update gate, $R_t$ is the reset gate, and $h_t$ is the hidden state. $\odot$ is a multiplication element-wise, and $\sigma$ is the sigmoid activation function. W and U are denoted as learned weight matrices.

## 4    Experiments

In this paper, we use two new datasets CIC-IDS2017 and CSE-CIC-IDS2018, because they are up-to-date and offer broader attack types and protocols. We want to implement the intrusion detection system using real network traffic data with machine learning and deep learning methods. After the preprocessing stage, we obtained a total of 2,830,743 data instances containing 2,273,097 "benign" and 557,646 "attacks" in CIC-IDS2017. In CSE-CIC-IDS2018 there's a total of 16,232,943 data instances containing 13,484,708 "benign" and 2,748,235 "attacks". The detailed statistics of data distribution in different classes for the two datasets are shown in Figs. 5 and 6.

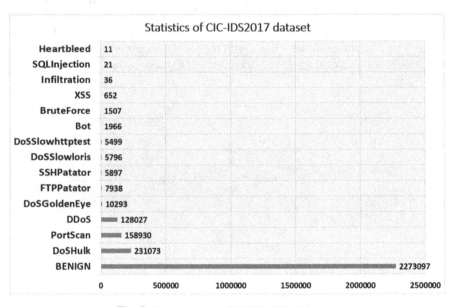

**Fig. 5.** The statistics of CIC-IDS2017 dataset.

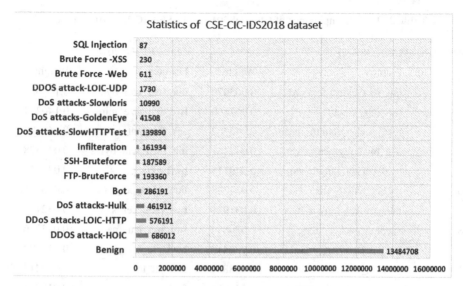

**Fig. 6.** The statistics of CSE-CIC-IDS2018 dataset.

In this paper, we divided the original 15 categories into two groups: 0 - benign, and 1 - attack. They are further separated into training and test sets as shown in Table 1.

**Table 1.** Training and test sets from CIC-IDS2017 and CSE-CIC-IDS2018 dataset.

| Dataset | Training set | Test set |
|---|---|---|
| CIC-IDS2017 | 2,264,694 | 566,149 |
| CSE-CIC-IDS2018 | 12,917,016 | 3,229,255 |

Then, we applied feature importance by MDI to select the top features as shown in Table 2.

After applying SMOTE for CIC-IDS2017 dataset, the number of minority instances increases from 445,820 to 1,818,774, and for CSE-CIC-IDS2018 it increases from 2,197,368 to 10,719,648.

In order to implement the LSTM and GRU models, we use the modules from the Keras Python library. Sequential model is a linear stack of layers to initializing the neural network. Dense is a regular layer of neurons in the neural network. A dropout layer is used for implementing regularization technique, which aims to reduce the complexity of the model to prevent overfitting. The architectures of LSTM and GRU both consist of three dimensional input array, one dropout layer, two dense layers, and the output layer which uses softmax function for classification.

The parameters of our model are as follows: Firstly, in the sequential model, and one layer of LSTM or GRU consists of 64 units, which are the dimensionality of the output space. The 3D input shape is the shape of our training set with the format [input

**Table 2.** Feature importance of CIC-IDS2017 and CSE-CIC-IDS2018 dataset.

| Number | CIC-IDS2017 | | CSE-CIC-IDS2018 | |
|--------|-------------|--------|-----------------|--------|
| | Features | Weight | Features | Weight |
| 1 | init_win_bytes_forward | 0.065970 | init_fwd_win_bytes | 0.158607 |
| 2 | psh_flag_count | 0.061660 | fwd_seg_size_min | 0.140294 |
| 3 | bwd_packet_length_mean | 0.046262 | ack_flag_cnt | 0.048100 |
| 4 | avg_bwd_segment_size | 0.042485 | init_bwd_win_bytes | 0.044378 |
| 5 | bwd_packet_length_std | 0.040715 | bwd_pkts_per_s | 0.037927 |
| 6 | packet_length_std | 0.034778 | flow_pkts_per_s | 0.035585 |
| 7 | bwd_packet_length_max | 0.031926 | fwd_pkts_per_s | 0.032158 |
| 8 | average_packet_size | 0.030894 | fwd_pkt_len_max | 0.023756 |
| 9 | bwd_packet_length_min | 0.030410 | bwd_pkt_len_max | 0.019874 |
| 10 | fwd_iat_max | 0.028837 | fwd_iat_tot | 0.019413 |
| 11 | min_seg_size_forward | 0.027920 | fwd_iat_mean | 0.018725 |
| 12 | flow_iat_max | 0.026415 | flow_iat_min | 0.018556 |
| 13 | packet_length_mean | 0.025863 | fwd_iat_max | 0.018508 |
| 14 | packet_length_variance | 0.022699 | flow_duration | 0.017454 |
| 15 | ack_flag_count | 0.022489 | flow_iat_mean | 0.016863 |

samples, time steps, features]. Secondly, we add a dropout layer with a dropout rate of 0.2, meaning that 20% of the layers will be dropped. Next, the dense layer specifies the output of 2 units (number of classes), and activated with softmax function which normalizes the output to a probability distribution over each output class.

Next, we compile our model using the Adaptive moment estimation (Adam), and sparse categorical cross-entropy loss function to obtain the output. Adam optimizer is implemented for maintaining a learning rate for updating each network weight separately, which can automatically decrease the gradient size steps towards minima based on the exponential moving average of gradients and squared gradients. Sparse categorical cross-entropy loss function is used for our classification since its efficiency and the use of integers as our class labels. Finally, a fitting function is used to fit the model on the data, and we ran the model for ten epochs, with the batch size of 1,000.

To evaluate the performance of intrusion detection, we use evaluation metrics including: Accuracy, Precision, Recall (sensitivity), and F1-score, as shown below.

$$Accuracy = \frac{TN + TP}{FP + TN + TN + FN} \tag{12}$$

$$Precision = \frac{TP}{FP + TP} \tag{13}$$

$$Recall = \frac{TP}{TP + FN} \tag{14}$$

$$F1 - Score = \frac{2*precision*recall}{precision + recall} \tag{15}$$

First, the evaluation results of the classification performance using the full dataset of CIC-IDS2017 are shown in Table 3.

**Table 3.** Evaluation results using full data (72 features) – CIC-IDS2017.

| Classifier | Precision | Recall | F1-score | Accuracy |
|---|---|---|---|---|
| RF | 92.7% | 92.7% | 92.7% | 92.7% |
| ID3 | 93.1% | 93.1% | **93.1%** | **93.1%** |
| NB | 78.2% | 68.6% | 71.5% | 68.6% |
| LSTM | 89.3% | 88.4% | 86.5% | **95.7%** |
| GRU | 93.3% | 93.2% | **92.8%** | 95.1% |

As shown in Table 3, for classical learning methods, ID3 gives better performance than Random Forest and Naïve Bayes with an accuracy of 93.1% and an F1-Score of 93.1%. For deep learning methods, better performance can be obtained for LSTM with an accuracy of 95.7%, and GRU with an F1-score of 86.5%.

Next, the evaluation results of CIC-IDS2017 dataset using SMOTE are shown in Table 4.

**Table 4.** Evaluation results using oversampling (72 features) – CIC-IDS2017.

| Classifier | Precision | Recall | F1-score | Accuracy |
|---|---|---|---|---|
| RF | 93.8% | 93.7% | 93.7% | **93.7%** |
| ID3 | 93.8% | 93.8% | **93.8%** | 93.7% |
| NB | 78.6% | 77.8% | 78.2% | 77.8% |
| LSTM | 96.6% | 96.2% | 96.3% | 96.2% |
| GRU | 96.9% | 96.5% | **96.6%** | **96.5%** |

As shown in Table 4, in classical learning, ID3 gives better performance, with an accuracy of 93.7% and an F1-Score of 93.8%. In deep learning, better performance can be obtained for GRU with an accuracy of 96.6% and an F1-score of 96.5%.

If we applied feature selection by MDI, the evaluation results using 20 selected features are shown in Table 5.

**Table 5.** Evaluation results using 20 selected features, SMOTE – CIC-IDS2017.

| Classifier | Precision | Recall | F1-score | Accuracy |
|------------|-----------|--------|----------|----------|
| RF | 95.2% | 94.2% | **94.4%** | **94.2%** |
| ID3 | 95.1% | 94.1% | 94.1% | 94.1% |
| NB | 88.8% | 82.5% | 83.4% | 82.5% |
| LSTM | 98.9% | 98.9% | **98.9%** | **98.9%** |
| GRU | 98.4% | 98.4% | 98.4% | 98.4% |

As shown in Table 5, in classical learning, Random Forest gives better performance, with an accuracy of 94.2% and an F1-score of 94.4%. In deep learning, LSTM gives better performance than GRU, with an accuracy of 98.9% and an F1-score of 98.4%.

If we further reduce the number of selected features, the evaluation results are shown in Table 6.

**Table 6.** Evaluation results using 10 selected features, SMOTE – CIC-IDS2017.

| Classifier | Precision | Recall | F1-score | Accuracy |
|------------|-----------|--------|----------|----------|
| RF | 95.1% | 94.1% | **94.1%** | **94.1%** |
| ID3 | 94.8% | 93.9% | 93.9% | 94.0% |
| NB | 88.5% | 80.9% | 81.9% | 80.9% |
| LSTM | 98.6% | 98.6% | **98.6%** | **98.6%** |
| GRU | 98.1% | 98.1% | 98.1% | 98.1% |

As shown in Table 6, in classical learning, Random Forest gives better result, with an accuracy of 94.1% and an F1-score of 94.1%. In deep learning, we found LSTM gives better performance than GRU with an accuracy of 98.6% and an F1-score of 98.6%.

From the performance comparison of results from Tables 5 and 6, we found in classical learning, Random Forest gives the best result with an accuracy of 94.4% and

**Table 7.** Evaluation results full data (72 features) – CSE-CIC-IDS2018.

| Classifier | Precision | Recall | F1-score | Accuracy |
|------------|-----------|--------|----------|----------|
| RF | 89.0% | 88.0% | 89.0% | 89.1% |
| ID3 | 93.3% | 93.2% | **93.2%** | **93.2%** |
| NB | 62.9% | 50.1% | 55.0% | 49.1% |
| LSTM | 81.1% | 81.0% | **89.9%** | **85.0%** |
| GRU | 87.1% | 86.8% | 83.7% | 84.7% |

an F1-score of 94.2%. In deep learning, LSTM gives the best result, with an accuracy of 98.9% from 20 selected features, and an F1 score of 98.9%. This shows the effectiveness of the proposed feature selection and deep learning methods. Next, we do the same for the CSE-CIC-IDS2018 dataset as shown in Table 7.

As shown in Table 7, in classical learning, the best performance can be obtained for ID3 with an accuracy of 93.2% and an F1-score of 93.2%. In deep learning, the best performance can be obtained for LSTM with an accuracy of 85.0%, and an F1-score of 89.9%. Then, evaluation results using SMOTE oversampling are shown in Table 8.

**Table 8.** Evaluation results of using oversampling (72 features) – CSE-CIC-IDS2018.

| Classifier | Precision | Recall | F1-score | Accuracy |
|------------|-----------|--------|----------|----------|
| RF         | 90.0%     | 88.9%  | 86.6%    | 88.9%    |
| ID3        | 93.9%     | 93.7%  | **93.8%**| **93.7%**|
| NB         | 85.7%     | 50.7%  | 55.0%    | 50.7%    |
| LSTM       | 91.1%     | 91.0%  | **89.9%**| **95.2%**|
| GRU        | 87.1%     | 86.8%  | 83.7%    | 94.7%    |

As shown in Table 8, in classical learning, ID3 gives better performance, with an accuracy of 93.7% and an F1-score of 93.8%. It's better than deep learning methods in F1-score, where LSTM gives better accuracy of 95.2%. Then, the evaluation results using 20 selected features are shown in Table 9.

**Table 9.** Evaluation results of using 20 selected features, SMOTE – CSE-CIC-IDS2018.

| Classifier | Precision | Recall | F1-score | Accuracy |
|------------|-----------|--------|----------|----------|
| RF         | 92.1%     | 92.1%  | 92.1%    | 92.1%    |
| ID3        | 94.6%     | 94.6%  | **94.7%**| **94.6%**|
| NB         | 82.9%     | 75.1%  | 75.1%    | 75.1%    |
| LSTM       | 98.0%     | 97.9%  | 97.9%    | 97.9%    |
| GRU        | 98.9%     | 98.8%  | **98.8%**| **98.8%**|

As shown in Table 9, in classical learning, ID3 gives better performance, with an accuracy of 94.6% and an F1-score of 94.7%. In deep learning, GRU gives better performance, with an accuracy of 98.8% and an F1-score of 98.8%.

Finally, the evaluation results using 10 selected features are shown in Table 10.

**Table 10.** Evaluation results using 10 selected features, SMOTE – CSE-CIC-IDS2018.

| Classifier | Precision | Recall | F1-score | Accuracy |
|---|---|---|---|---|
| RF | 91.5% | 91.5% | 91.5% | 91.4% |
| ID3 | 94.2% | 94.2% | **94.2%** | **94.2%** |
| NB | 82.5% | 74.5% | 74.5% | 74.5% |
| LSTM | 97.7% | 97.6% | 97.6% | 97.6% |
| GRU | 98.1% | 98.1% | **98.1%** | **98.1%** |

As shown in Table 10, in classical learning, ID3 gives better result with an accuracy of 94.2% and an F1-score of 94.2%. In deep learning, we found GRU gives better performance with an accuracy of 98.1% and an F1-score of 98.1%.

When comparing Tables 9 and 10, GRU shows the best F1-score and accuracy of 98.8%. In classical learning, ID3 gives the best performance with an accuracy of 94.6% and an F1-score of 94.7%.

In summary, when we compare the evaluation results for the two datasets, the best performance can be obtained using different methods: LSTM and RF for CIC-IDS2017 dataset, and GRU and ID3 for CSE-CICIDS2018 dataset. There's only slight difference between the best performance of LSTM and GRU. Also, we can see comparable performance when using only 10 selected features. This shows the effectiveness of combining the MDI feature selection method, and SMOTE oversampling method in recurrent neural networks.

## 5  Conclusions

In this paper, we aimed at intrusion detection using deep learning methods. In this context, the CIC-IDS2017 and CSE-CICIDS2018 datasets were used since they are up-to-date with wide attack diversity, and various network protocols (e.g., Mail services, SSH, FTP, HTTP, and HTTPS). First, by using a feature selection method, we can determine the most important features in both datasets. Then, it is combined with oversampling technique to deal with imbalanced data. The experimental results show that our results are better than existing works to classify and detect intrusions. In CIC-IDS2017 dataset, the best performance obtained for the proposed method is an accuracy of 98.9 and an F1-score of 98.9% by LSTM. Second, in CSE-CIC-IDS2018 dataset, the best performance can be obtained for GRU with an accuracy of 98.8% and an F1-score of 98.8%. Third, by using the top 10 selected features, the performance is better than using all features. This shows the effectiveness of our proposed method for using feature selection and oversampling for intrusion detection in large scale network traffic.

In future, we plan to use other datasets which include new variants of attacks like malware and backdoor activity in real network traffics. Besides, we want to compare with other feature selection methods. For under-sampling or over-sampling technique, to adjust the class distribution in the dataset, we can use the weight of distribution of minority class, to generate more synthetic data for the minority class. Finally, we plan to combine deep learning with other classification methods for improving the performance.

**Acknowledgements.** This work was partially supported by research grants from Ministry of Science and Technology, Taiwan, under the grant number MOST109-2221-E-027-090, and partially supported by the National Applied Research Laboratories, Taiwan under the grant number of NARL- ISIM-109-002.

# References

1. IDS 2017 | Datasets | Research | Canadian Institute for Cybersecurity | UNB. www.unb.ca, 2017. https://www.unb.ca/cic/datasets/ids-2017.html. Accessed 15 June 2019
2. A Realistic Cyber Defense Dataset (CSE-CIC-IDS2018) (2018). https://registry.opendata.aws/cse-cic-ids2018/. Accessed 15 June 2019
3. Sharafaldin, I., Lashkari, A.H., Ghorbani, A.A., Habibi Lashkari, A., Ghorbani, A.A.: Toward generating a new intrusion detection dataset and intrusion traffic characterization. In: Proceedings of 4th International Conference. Information System Security Privacy, pp. 108–116 (2018)
4. Louppe, G., Wehenkel, L., Sutera, A., Geurts, P.: Understanding variable importances in forests of randomized trees. In: Proceedings of the 26th International Conference on Neural Information Processing Systems - Volume 1 (NIPS 2013), pp. 431–439 (2013)
5. Chawla, K.W., Bowyer, L., Hall, O., Kegelmeyer, W.P.: SMOTE: synthetic minority over-sampling technique. J. Artif. Intell. Res. **16**, 321–357 (2002)
6. Breiman, L.: Random forests. Mach. Learn. **45**(1), 5–32 (2001)
7. Chen, J., Luo, D., Mu, F.: An improved ID3 decision tree algorithm. In: 2009 4th International Conference on Computer Science & Education, pp. 127–130 (2009)
8. Friedman, N., Geiger, D., Goldszmidt, M.: Bayesian network classifiers. Mach. Learn **29**, 131–163 (1997)
9. Hochreiter, S., Schmidhuber, J.: Long short-term memory. Neural Comput. **9**(8), 1735–1780 (1997)
10. Chung, J., Gülçehre, Ç., Cho, K., Bengio, Y.: Empirical evaluation of gated recurrent neural networks on sequence modeling. In: NIPS 2014 Workshop on Deep Learning (2014)
11. Albayati, M., Issac, B.: Analysis of intelligent classifiers and enhancing the detection accuracy for intrusion detection system. Int. J. Comput. Intell. Syst. 841–853 (2015)
12. Almseidin, M., Alzubi, M., Kovacs, S., Alkasassbeh, M.: Evaluation of machine learning algorithms for intrusion detection system. In: 2017 IEEE 15th International Symposium Intelligent System Informatics (SISY), pp. 277–282 (2017)
13. Khuphiran, P., Leelaprute, P., Uthayopas, P., Ichikawa, K., Watanakeesuntorn, W.: Performance comparison of machine learning models for DDoS attacks detection. In: 2018 22nd International Computer Science and Engineering Conference (ICSEC), pp. 1–4 (2018)
14. Althubiti, S.A., Jones, E.M., Roy, K.: LSTM for anomaly-based network intrusion detection. In: 2018 28th International Telecommunication Networks and Applications Conference (ITNAC), pp.1–3 (2018)
15. Xu, C., Shen, J., Du, X., Zhang, F.: An intrusion detection system using a deep neural network with gated recurrent units. IEEE Access **6**, 48697–48707 (2018)

16. Alazzam, H., Sharieh, A., Sabri, K.E.: A feature selection algorithm for intrusion detection system based on Pigeon Inspired Optimizer. Expert Syst. Appl. **148** (2020)
17. Wu, M.Y., Shen, C.-Y., Wang, E.T., Chen, A.L.P.: A deep architecture for depression detection using posting, behavior, and living environment data. J. Intell. Inf. Syst. **54**(2), 225–244 (2018). https://doi.org/10.1007/s10844-018-0533-4
18. Shuai, H.-H., et al.: A comprehensive study on social network mental disorders detection via online social media mining. IEEE Trans. Knowl. Data Eng. (TKDE) **30**(7), 1212–1225 (2018)
19. Smiti, S., Soui, M.: Bankruptcy prediction using deep learning approach based on borderline SMOTE. Inf. Syst. Front. **22**(5), 1067–1083 (2020). https://doi.org/10.1007/s10796-020-100 31-6
20. Seo, J.-H., Kim, Y.-H.: Machine-learning approach to optimize SMOTE ratio in class imbalance dataset for intrusion detection. Comput. Intell. Neurosci. 1–11 (2018)
21. Kurniabudi, D.S., Darmawijoyo, M.Y.B.I., Bamhdi, A.M., Budiarto, R.: CICIDS-2017 dataset feature analysis with information gain for anomaly detection. IEEE Access **8**, 132911–132921 (2020)
22. Kim, J., Kim, J., Kim, H., Shim, M., Choi, E.: CNN-based network intrusion detection against denial-of-service attacks. Electronics **9**(6) (2020)

# Multi-head Attention with Hint Mechanisms for Joint Extraction of Entity and Relation

Chih-Hsien Fang[1,2], Yi-Ling Chen[1(✉)], Mi-Yen Yeh[2], and Yan-Shuo Lin[1]

[1] National Taiwan University of Science and Technology, Taipei, Taiwan
{yiling,m10815052}@mail.ntust.edu.tw
[2] Institute of Information Science, Academia Sinica, Taipei, Taiwan
miyen@iis.sinica.edu.tw

**Abstract.** In this paper, we propose a joint extraction model of entity and relation from raw texts without relying on additional NLP features, parameter threshold tuning, or entity-relation templates as previous studies do. Our joint model combines the language modeling for entity recognition and multi-head attention for relation extraction. Furthermore, we exploit two hint mechanisms for the multi-head attention to boost the convergence speed and the F1 score of relation extraction. Extensive experiment results show that our proposed model significantly outperforms baselines by having higher F1 scores on various datasets. We also provide ablation tests to analyze the effectiveness of components in our model.

## 1 Introduction

In this paper, we aim to study the joint model to extract both entities and relations from raw texts, such that the F1 score of both tasks can be boosted mutually. Specifically, the task of name entity recognition (NER) [12] recognizes the boundary and the type of entities, and the relation extraction (RE) task [28] determines the relation categories over entity pairs. Figure 1 shows the extraction results on an example sentence, where the bounding boxes, the arrow links, and the colors identify different entities, relations between entities, and their corresponding types, respectively.

Traditional approaches treat NER as a predecessor to RE in a pipeline model [4] while recent studies have shown that joint extractions of entities and relations with end-to-end modeling can boost the quality of both tasks [17]. It is because the relations are constructed by entity pairs, the knowledge of relation types can increase the accuracy of entity extraction and vice versa. Owing to the advance of deep learning technologies, the neural network models have gained popularity for the joint learning task. Miwa and Bansal [17] presents a deep neural network model, of which the promising results on end-to-end relation extraction require additional NLP features such as POS and dependency trees. Later, Katiyar and Cardie [10] utilizes the pointer network to extract semantic relations between

C. S. Jensen et al. (Eds.): DASFAA 2021 Workshops, LNCS 12680, pp. 321–335, 2021.
https://doi.org/10.1007/978-3-030-73216-5_22

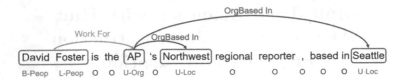

**Fig. 1.** Gold-standard entity and relation annotations for an example sentence from CoNLL04 dataset.

entity mentions without accessing dependency trees. However, the performance depends on a threshold parameter, and setting a proper threshold value can be quite challenging. Li et al. [14] uses a multi-turn question answering system to accomplish the joint extraction task. However, the proposed method requires some prior knowledge to build the templates about relations between entities and needs to enumerate all possible templates in advance, which may not be practical when used in real applications.

To solve the aforementioned issues, we propose to design a joint extraction model without the prior knowledge about additional NLP features, parameter thresholds, or entity-relation templates. First, we exploit the powerful language modeling of XLNet [25] to obtain word embeddings for higher accuracy of entity recognition in the joint task. For the relation extraction part, we exploit the multi-head encoder-decoder attention for deeper extraction of semantic relations.

To achieve even higher extraction F1 score, we design two hint mechanisms, the positional hint and the pair-wise hint. Due to the design of the encoder-decoder attention, at each time step, the query is always the last token in the sequence to the attention, and the key/value vector is the entire sequence. Since the index of query keeps growing as the time step increases, our proposed positional hint is to reverse the original positional encoding in the attention such that the positional information of the query can always be indexed first. In this way, the model clearly understands the first index of the positional encoding is the query we defined and is responsible for calculating the relation with all other tokens in the sequence. This deeper semantic information helps the model converge faster and have higher extraction F1 score.

In addition, the multi-head attention mechanism itself can capture the related information of the query to the entire sequence. Our proposed pair-wise hint further transforms such information in a clearer way by constructing triplets of two tokens (the query and the other one) and its corresponding relations. Such structure provides the model all existing relations and thus improves the recall rate of relation extraction.

Furthermore, our proposed method is able to extract multiple relations from one entity to others, and can decode all entities and relations within a sentence in one pass. The entity boundaries do not need to be specified first, and all possible relations can be decoded out rather than being limited to a predetermined entity pair as previous works do.

We have compared our method with baselines on various benchmark datasets. Extensive experiment results show that our model outperforms existing methods on all datasets. It is worth noting that the F1 score of our model for entity and relation extraction is 8.6% higher compared to the multi-turn QA model [14] on the ACE04 dataset. In the ablation tests, when replacing the language modeling with the one used in Li et al. [14] (i.e., BERT [7]), our model still achieves a higher F1 score for the joint extraction task compared to their model on CoNLL04, which demonstrates the advantages of two hint mechanisms.

The remainder of this paper is organized as follows. Section 2 details related works. We describe the proposed model in Sect. 3, the datasets, setting and experimental results in Sect. 4, and analysis in Sect. 5. Section 6 concludes our paper.

## 2   Related Works

**NER and RE.** Two basic NLP tasks, named entity recognition (NER) [12] and relation extraction (RE) [27,28], recently exploit neural networks to reach better performance. RNNs have been used for many sequential modeling and predictive tasks with time series, such as NER [19] and machine translation [1]. It has been found that variants such as adding a CRF-like objective on top of LSTMs produce more advanced results on several sequence prediction NLP tasks [5]. Moreover, traditional methods tend to deal with NER and RE tasks separately, assuming that the entity boundary has been given to perform the RE task. Several studies find that extracting entities and relations jointly can benefit both tasks [13] [3], and we follow this line of work in this study.

Note that many existing NER and RE methods rely on the availability of NLP tools (e.g., POS taggers and dependency tree parsers) [8] or manually-designed features, leading to additional complexity. Li et al. [14] proposes a new approach that combines NER and RE tasks with a QA system, but such a model design requires manually-formulated QA templates for different datasets. Creating these templates bring considerable costs, and if the template is not well-fitted to the dataset, it would hurt the model performance.

Miwa and Bansal [17] uses bi-directional LSTM as an encoder to learn the hidden word representations, and uses tree-structured LSTM to get relations between entities. Global optimization and normalization have been successfully applied on the RE using neural networks [26]. It maximizes the cumulative score of the gold-standard label sequence for one sentence as a unit. Katiyar and Cardie [10] proposes a new approach for RE that does not need dependency trees. Instead, they utilizes the pointer network [24] to extract semantic relations between entity mentions. In the inference time, all the labels with probability values above the threshold are outputted as results. Bekoulis et al. [2] uses adversarial training to improve robustness of neural network methods by adding small perturbations in training data, and such modifications substantially improve the performance. Sun et al. [22] proposes Minimum Risk Training to optimize global loss function to jointly train the NER and RE. The aforementioned approaches have various focuses and advantages, and they are included for comparison in Sect. 4.3. According to the experimental results, our model is able to obtain competitive or state-of-the-art scores on various datasets.

**Language Model.** The pre-trained Language Model (LM) has become very popular in recent years, and because of the breakthrough of BERT [7], the pre-trained LM reaches another peak. BERT is a pre-trained LM that can be fine-tuned on many NLP tasks to get better performance. The commonly used embedding method is Word2Vec [16] or GloVe [20], but a recent study [7] proves that BERT scores well in multiple NLP tasks. Another LM that surpasses BERT is XLNet [25]. Yang et al. [25] uses Permutation Language Modeling as the objective and builds Two-Stream Self-Attention as the core architecture. In order to have a fair comparison with the previous study [14], we evaluate our model with both of BERT and XLNet as LM in the ablation test.

**Multi-head Attention.** Using multi-head attention, the core concept of Transformer [23], can yield more interpretable models. While single-head attention works for many tasks, multi-head attention further helps capture more effective information related to the syntactic and semantic structure of sentences. In our model, this concept is leveraged to enrich the semantic information in the relation extraction task.

**Positional Encoding.** Positional encoding is a representation of the position of a word in a sentence by using sinusoidal function. This technique is used because there is no notion of word order in the proposed architecture (unlike common RNN or ConvNet architectures). Therefore, a position-dependent signal is added to each word-embedding to help the model incorporate the information about the order of words. Vaswani et al. [23] finds that using sinusoidal function to represent position has nearly identical results with using learned positional embeddings, and using sinusoidal version can extrapolate positional information to sentence length longer than the training phase.

## 3 Model

In this study, we propose a joint model that can extract all entities and the corresponding relations within the sentence at once. Our model contains a multi-layer bi-directional RNN that learns a representation for each token in the sequence from the output of LM. We treat named entity recognition as a sequence labeling task and relation extraction as a table filling task, respectively. Figure 1 and Table 1 show an example of annotating the entity tags and relation tags. At each time step $t$, the multi-head attention layer utilizes the information from the previous time steps to get the representation. After that, the current token representation points to all previous token representations to get the relation label. The left half of Fig. 2 represents the architecture for NER, and the other half is for RE. In this section, we introduce our model from the embeddings to training in a bottom-up way.

**Table 1.** Gold-standard relation annotation for the example sentence in Fig. 1. The entire relation table is first initialized with the zero vector, and then the table is filled with predicted results. The symbol ⊥ denotes "none relation" and "P" denotes padding tag. (If the token at time step $t$ is related to itself, the relation type is defined as "none relation", as the diagonal of this table. Note that the padding tags do not affect the calculation of loss, since the calculation only considers the lower left triangle of this table.)

| | David | Foster | is | the | AP | 's | Northwest | regional | reporter | , | based | in | Seattle |
|---|---|---|---|---|---|---|---|---|---|---|---|---|---|
| David | ⊥ | P | P | ... | | | | | | | | | |
| Foster | ⊥ | ⊥ | P | P | ... | | | | | | | | |
| is | ... | ⊥ | ⊥ | P | P | ... | | | | | | | |
| the | | ... | ⊥ | ⊥ | P | P | ... | | | | | | |
| AP | ⊥ | Work For → | ⊥ | ⊥ | ⊥ | P | P | ... | | | | | |
| 's | | | | ... | ⊥ | ⊥ | P | P | ... | | | | |
| Northwest | | | ... | ⊥ | OrgBased In → | ⊥ | ⊥ | P | P | ... | | | |
| regional | | | | | ... | ⊥ | ⊥ | ⊥ | P | P | ... | | |
| reporter | | | | | | | ... | ⊥ | ⊥ | P | P | ... | |
| , | | | | | | | | | ... | ⊥ | ⊥ | P | P |
| based | | | | | | | | | | ... | ⊥ | ⊥ | P | P |
| in | | | | | | | | | | | ... | ⊥ | ⊥ | P |
| Seattle | | | ... | ⊥ | OrgBased In → | ⊥ | ... | | | | | ⊥ | ⊥ |

## 3.1 Embeddings

One of the latest milestones in embeddings is the release of XLNet, which brakes eighteen records in various NLP tasks. The pre-trained LM representations can be fine-tuned with just one additional output layer. By using XLNet, we can save the time and computation resources that would have gone to training a model from scratch. XLNet is basically a trained "Transformer-XL" [6] stack. The Transformer-XL is a model that uses attention to boost the accuracy and consists of segment-level recurrence mechanism and relative positional encoding. Note that XLNet model has a large number of encoder layers, and we sum the output of all essential layers as the embedding vector.

The input of the model is a sequence of $n$ tokens (i.e., a sentence) $x = [x_1, \ldots, x_n]$. SentencePiece [11] is used here to effectively avoid the out-of-vocabulary (OOV) problem. We average the vectors of pieces of each word and convert it back to a vector to represent a word. We use $LM(\cdot)$ to denote the sum of all XLNet output layers and mean of word pieces of each word, and $v^{(LM)} \in \mathbb{R}^{n \times d_{LM}}$ is the word embeddings from $LM(\cdot)$ as below:

$$v^{(LM)} = LM(x). \tag{1}$$

## 3.2 Named Entity Recognition

We perform named entity recognition on multi-layer bi-directional LSTMs for sequence tagging since LSTMs are more capable of capturing long-term depen-

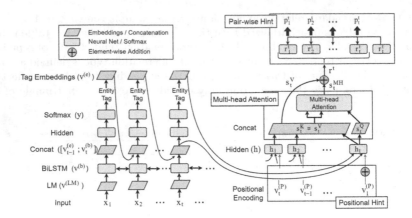

**Fig. 2.** The structure of our proposed end-to-end joint entity and relation extraction model, with sequence-wise multi-head attention and token-wise pointer network.

dencies between tokens. The vector $v^{(b)} \in \mathbb{R}^{n \times d_{LM}}$ denotes the output of BiL-STM as below:

$$v^{(b)} = BiLSTM(v^{(LM)}). \tag{2}$$

We treat entity detection as a sequence labeling task using BILOU (Begin, Inside, Last, Outside, and Unit) scheme similar to previous study [17]. Each entity tag represents the entity type and the position of a word in the entity. For example, in Fig. 1, we assign B-Peop and L-Peop to each word in David Foster to represent this phrase as a Peop (people) entity type, and we assign U-Org to represent the organization entity type of the word AP. For each token in the sequence, we formulate a softmax over all candidate entity tags to output the most likely entity tag as below:

$$y_t = softmax([v_{t-1}^{(e)}; v_t^{(b)}]W^E + b^E), \tag{3}$$

where $v_{t-1}^{(e)} \in \mathbb{R}^{d_{tag\_emb}}$ denotes the entity tag embeddings in the previous time step, the ";" symbol denotes concatenation, $v_t^{(b)}$ denotes the BiLSTM output in the time step $t$, $W^E \in \mathbb{R}^{(d_{tag\_emb}+d_{LM}) \times |E|}$ and $b^E \in \mathbb{R}^{|E|}$ are learnable parameters, $|E|$ denotes the entity size, and $y_t \in \mathbb{R}^{|E|}$ denotes the output of entity in the time step $t$. When $t = 0$, the $v_{t-1}^{(e)}$ is a zero vector. We decode the entity label from left to right in a greedy manner. To optimize the performance, we connect the entity embeddings in the previous time step and BiLSTM output in the current time step. Thus our outputs are not conditional independent from each other. Finally, we transform the output $y_{t-1}$ into the entity tag embeddings $v_{t-1}^{(e)}$.

## 3.3   Stacking Sequence

Here we concatenate the entity tag embeddings in the previous time step, the entity tag embeddings in the current time step, and the BiLSTM output in the current time step as the current information about the sequence into the relation layer as below:

$$h_t = [v_{t-1}^{(e)}; v_t^{(b)}; v_t^{(e)}]W^h + b^h,\qquad(4)$$

where $W^h \in \mathbb{R}^{(2d_{tag\_emb}+d_{LM}) \times d_{rel}}$ and $b^h \in \mathbb{R}^{d_{rel}}$ are learnable parameters, and $h_t \in \mathbb{R}^{d_{rel}}$ denotes the concatenated vector in lower dimension. We stack the time steps until current time step $t$ as the key $s_t^K$ and value $s_t^V$ (i.e., $s_t^K = s_t^V = [h_1; \ldots; h_t] \in \mathbb{R}^{t \times d_{rel}}$) into the multi-head attention. The query $s_t^Q \in \mathbb{R}^{1 \times d_{rel}}$ denotes the concatenated vector in the current time step into the multi-head attention.

### 3.4   Positional Hint (P-hint)

In [23], in order to subjoin some information about the positions of tokens in the sequence, they add positional encoding to input word embeddings. The positional encoding consists of sinusoidal function. Besides, using sinusoidal function is able to extrapolate the sequence lengths longer than the ones encountered during training. In the proposed model, we use positional encoding as a hint (called "Positional Hint") for relation extraction task, and we propose two strategies to utilize the hint. The first strategy (called "forward") is quite similar to the transformer positional encoding, where we add the positional encoding to the sequence in order. The second strategy (called "backward") is to reverse the positional encoding and add the reversed positional encoding to the sequence. In the designed model, we use the last token as the query to score all the previous tokens to attention function, so we reverse the positional encoding and give the first index of positional encoding to the last token, in order to maintain the consistency of the query position at each time step. In Sect. 5.2, comparison and analysis of these two strategies are provided.

### 3.5   Multi-head Attention

In our model, we adopt the multi-head attention on relation model. An attention function comprises a query and a set of key-value pairs. The output of attention function is computed as a weighted sum of the values, where the weight assigned to each value is computed by the compatibility function of the query with the corresponding key. We adopt encoder-decoder attention for the scaled dot-product as our attention function as below:

$$Attention(Q, K, V) = softmax(\frac{QK^T}{\sqrt{d_{rel}}})^T \otimes V,\qquad(5)$$

where $\otimes$ denotes the element-wise multiplication. A previous study [23] shows that multi-head attention is capable of processing information from different representation subspaces at different positions. We leverage this to enhance the context representation of the relation of entity pairs in every time step as below:

$$s_t^{MH} = MultiHead(s_t^Q, s_t^K, s_t^V)$$
$$= Concat(head_1, \ldots, head_m)W^O,\qquad(6)$$

$$head_j = Attention(s_t^Q W_j^Q, s_t^K W_j^K, s_t^V W_j^V),\tag{7}$$

where $W_j^Q, W_j^K, W_j^V \in \mathbb{R}^{d_{rel} \times d_v}$, $W^O \in \mathbb{R}^{md_v \times d_{rel}}$ denote the projection matrices and $m$ is the number of attention heads, and $md_v = d_{rel}$. We then apply a residual connection [9] on $s_t^V$ and $s_t^{MH}$ to obtain the vector $r^t$ as below:

$$r^t = s_t^V \oplus s_t^{MH},\tag{8}$$

where $r^t \in \mathbb{R}^{t \times d_{rel}}$ and $\oplus$ denotes the element-wise addition. To facilitate the residual connection, the dimensions of $s_t^V$ and $s_t^{MH}$ are the same.

### 3.6 Pair-Wise Hint (Pw-hint)

In our model, we formulate relation extraction as a table filling task. For each token, we want to find the relation between the current token and the token at the previous time step. In Table 1, the Peop entity type "David Foster" is related to the Org entity type "AP". For simplicity, we follow the previous study [18] where only the tail of entity (L tag or U tag) is related to each other, i.e., "Foster" is related to "AP" via the "Work For" relation. Because of this, our model is able to decode multiple relations through table filling.

Besides the positional hint, we further adopt additive attention as "Pair-wise Hint". Through this attention function, the pair-wise hint is to remind the model that the relation of token pair is prepared to be classified. In addition, it can also reduce the dimension to relation size via $u$. In every time step, the current token points to the previous tokens, in order to get the probability of relation of this entity pair as below:

$$\alpha_i^t = \tanh(r_i^t W^{P_1} + r_t^t W^{P_2})u,\tag{9}$$

$$p_i^t = softmax(\alpha_i^t),\tag{10}$$

where $i \in (1, \ldots, t)$ denotes time series until current time step $t$. $W^{P_1}$ and $W^{P_2} \in \mathbb{R}^{d_{rel} \times d_{pair}}$ are weight matrices of the model, and $u \in \mathbb{R}^{d_{pair} \times (2|R|+1)}$ is a learnable matrix to transform the hidden representations into attention scores. $r_i^t$ denotes the $i$-th token of the vector after residual connection, $r_t^t$ denotes the vector in current time step, and $\alpha_i^t$ denotes the score computed from the compatibility function with $r_i^t$ and $r_t^t$. Finally, $p_i^t \in \mathbb{R}^{2|R|+1}$ represents the result of relation extraction in the entity pair, i.e., $R$ relation types in bi-direction and "none relation". In the ablation tests (Sect. 5.1), there is a clear performance drop when Pw-hint is removed[1], indicating that Pw-hint is a key component to improve the information extraction of relations in the vector.

---

[1] When Pw-hint is removed, we directly multiply $r^t$ by a matrix to reduce its dimension to $\mathbb{R}^{t \times 2|R|+1}$ and then connect to softmax.

## 3.7   Training

The final objective for the joint task is to minimize $\mathcal{L}_{joint} = \mathcal{L}_{NER} + \lambda\mathcal{L}_{RE}$, where $\lambda$ is the hyperparameter controlling the proportion of relation loss to improve the model. Given a sentence $S$, we minimize the cross-entropy loss between this output distribution with the gold-standard distribution as below:

$$\mathcal{L}_{NER} = -\frac{1}{|S|} \sum_{i \in |S|} \log p(e_i|e_{<i}, S, \theta), \tag{11}$$

$$\mathcal{L}_{RE} = -\frac{1}{|S|} \sum_{i \in |S|} \log p(r_i|e_{\leq i}, S, \theta), \tag{12}$$

where $\theta$ represents the network parameters, $e$ denotes the prediction of entity tags, and $r$ denotes the prediction of relation tags. Please note that our model is designed to be capable of extracting all entities and the corresponding relations within the sentence at once. Our model design is also suitable for multi-label tasks, allowing multiple relations per entity.

## 4   Experiments

### 4.1   Data and Evaluation Metrics

We evaluate our proposed model and other baselines on four datasets, which are CoNLL04, ADE, ACE04, and ACE05. In order to conduct fair comparisons, we follow the preprocessing procedure used in the baseline methods. Specifically, for CoNLL04, we follow the previous study [18] to split the data into training and test corpora, and then divide 10% of the training corpus for development. For ADE, we perform 10-fold cross-validation as Bekoulis et al. [2] do. For ACE04 and ACE05, we perform 5-fold cross-validation as Miwa and Bansal [17] do on ACE04, and we also follow Miwa and Bansal [17] to split the ACE05 dataset into training, development and test sets.

For the evaluation metrics, we follow the previous studies [2,17] and report micro F1-scores, precision and recall on both entities and relations. An extracted entity is considered correct only if both the entity boundaries and the entity type are detected correctly. Similarly, a relation is considered correct only if we detect both the entity pair and the relation type correctly. In addition, we make average of precision, recall and F1 three times on CoNLL04 and ACE05.

### 4.2   Hyperparameters and Training Details

We use the AdamW optimizer [15] to set hyperparameters on the validation sets. For multi-head attention, we use 32 heads to get best accuracy. For XLNet, we average the pieces of each word and combine them into a vector. As for the version of XLNet, the XLNet$_{BASE}$ model is used as input of our contextual embeddings. For the efficiency and better embedding performance, we adopt

**Table 2.** Performance comparison of our proposed model and other state-of-the-art approaches on the test set. The dash ("–") indicates that this score is not reported in the original paper.

| Dataset | System | Entity | | | Entity + Relation | | |
|---------|--------|--------|--------|--------|--------|--------|--------|
| | | P | R | F1 | P | R | F1 |
| CoNLL04 | Our model | 89.0 | 88.3 | **88.6** | 73.8 | 67.7 | **70.7** |
| | Li et al. (2019) [14] | 89.0 | 86.6 | 87.8 | 69.2 | 68.2 | 68.9 |
| | Bekoulis et al. (2018a) [2] | – | – | 83.6 | – | – | 62.0 |
| | Zhang et al. (2017a) [26] | – | – | 85.6 | – | – | 67.8 |
| ADE | Our model | 87.6 | 86.0 | **86.8** | 78.1 | 74.8 | **76.4** |
| | Bekoulis et al. (2018a) [2] | – | – | 86.7 | – | – | 75.5 |
| ACE04 | Our model | 87.7 | 87.7 | **87.7** | 58.6 | 57.4 | **58.0** |
| | Li et al. (2019) [14] | 84.4 | 82.9 | 83.6 | 50.1 | 48.7 | 49.4 |
| | Bekoulis et al. (2018a) [2] | – | – | 81.6 | – | – | 47.5 |
| | Miwa and Bansak (2016) [17] | 80.8 | 82.9 | 81.8 | 48.7 | 48.1 | 48.4 |
| ACE05 | Our model | 88.9 | 89.3 | **89.1** | 65.0 | 57.6 | **61.0** |
| | Li et al. (2019) [14] | 84.7 | 84.9 | 84.8 | 64.8 | 56.2 | 60.2 |
| | Sun et al. (2018) [22] | 83.9 | 83.2 | 83.6 | 64.9 | 55.1 | 59.6 |
| | Miwa and Bansal (2016) [17] | 82.9 | 83.9 | 83.4 | 57.2 | 54.0 | 55.6 |

XLNet with feature-based approach, where fixed features are extracted from the pre-trained model. To generate the embedding vector, we sum all layers in XLNet model output. Moreover, in our task, the special tokens [CLS] and [SEP] are not needed. We also regularize our model using dropout [21] after the XLNet output, BiLSTM layer, and relation model.

### 4.3    Results

Table 2 compares the performance of our model with other state-of-the-art methods. Due to the complex nature of NLP tasks, it can be observed that the reported improvements among recent studies are usually limited (e.g., 1% from 2016 [17] to 2019 [14] on ACE04). Despite the difficulty, in relation extraction task, our model significantly outperforms (8.7% higher in F1 score) the adversarial training model [2] and exceeds (1.8%) the multi-turn QA model [14] on the CoNLL04 dataset. For the ADE dataset, our entity F1 score is slightly higher than that of Bekoulis et al. [2], but there is a greater improvement on combined F1 score performance (0.9% higher). As for ACE04, substantial improvements can be found in F1 score for relation extraction tasks (8.6%) when compared with the multi-turn QA model [14]. Finally, for ACE05, our entity F1 score is the highest, and the relation F1 score is also higher (0.8%) than the state-of-the-art

**Table 3.** Performance comparison of the model variants with different LM, P-hint, and Pw-hint on CoNLL04 and ACE05.

| Settings | CoNLL04 | | | | | | ACE05 | | | | | |
|---|---|---|---|---|---|---|---|---|---|---|---|---|
| | Entity | | | Entity + Relation | | | Entity | | | Entity + Relation | | |
| | P | R | F1 | P | R | F1 | P | R | F1 | P | R | F1 |
| BERT$_{base}$+P-hint(backward)+Pw-hint | 86.7 | 85.7 | 86.2 | 72.4 | 66.4 | 69.3 | 86.9 | 87.5 | 87.2 | 64.6 | 54.5 | 59.1 |
| BERT$_{large}$+P-hint(backward)+Pw-hint | 85.3 | 83.7 | 84.5 | 74.7 | 63.0 | 68.3 | 86.9 | 86.8 | 86.9 | 64.0 | 52.7 | 57.8 |
| XLNet$_{base}$+P-hint(backward)+Pw-hint | 89.0 | 88.3 | 88.6 | 73.8 | 67.7 | 70.7 | 88.9 | 89.3 | 89.1 | 65.0 | 57.6 | 61.0 |
| XLNet$_{base}$+P-hint(forward)+Pw-hint | 87.9 | 88.3 | 88.1 | 72.7 | 68.0 | 70.3 | 88.8 | 88.9 | 88.8 | 65.2 | 55.9 | 60.2 |
| XLNet$_{base}$+Pw-hint | 88.9 | 88.0 | 88.4 | 72.1 | 68.5 | 70.3 | 88.6 | 88.7 | 88.6 | 64.5 | 55.1 | 59.4 |
| XLNet$_{base}$+P-hint(backward) | 88.2 | 88.0 | 88.1 | 76.8 | 64.5 | 70.1 | 87.5 | 88.7 | 88.1 | 66.8 | 54.7 | 60.1 |
| XLNet$_{base}$ | 88.9 | 87.9 | 88.4 | 72.3 | 68.0 | 70.1 | 87.7 | 88.6 | 88.2 | 64.1 | 54.1 | 58.7 |
| XLNet$_{large}$+P-hint(backward)+Pw-hint | 88.3 | 88.4 | 88.3 | 76.8 | 65.5 | 70.6 | 89.0 | 89.4 | 89.2 | 69.3 | 54.1 | 60.8 |
| Li et al. (2019) [14] | 89.0 | 86.6 | 87.8 | 69.2 | 68.2 | 68.9 | 84.7 | 84.9 | 84.8 | 64.8 | 56.2 | 60.2 |

model [14] in relation extraction task. Our code will be publicly available for testing the reproducibility of results[2].

# 5    Analysis

## 5.1    Ablation Tests

In this subsection, we conduct a series of ablation tests on the test set of CoNLL04 and ACE05. We evaluate our model with and without the two hint mechanisms and replace the LM with the one used in previous studies, in order to see the contribution of these components. According to the experimental results in Table 3, using these two hints considerably improves the ability of NER and RE. Note that we focus on the F1 score of E+R hereafter without specifying. Specifically, on the dataset with difficult predictions such as ACE05 (which has more complex relation types), the improvement of the result is more significant (2.3% higher) with using two hints than not using these two components. On CoNLL04, the result only increases by 0.6%. It means that the two hint mechanisms bring more improvements for complex situations.

Moreover, we find that the score of backward P-hint is slightly higher (0.4%) than that of forward P-hint on CoNLL04, and it is also higher (0.8%) than that of forward P-hint on ACE05. It shows that the backward strategy of P-hint has certain impact on the results, confirming that it is effective to give query a fixed positional information while the query is the last token.

Furthermore, when using only Pw-hint on ACE05, the F1 score on relation is reduced by 1.6%. It means that the backward P-hint brings more suitable information to the relation task. For the relation decoding method, the last token of the sequence at each time step, i.e., the query, is given to the information of the first position, and it is effective to maintain the positional information of the query on the first position. When using only P-hint on ACE05, the F1

---

[2] Due to the anonymity requirement, the GitHub link will be provided after the anonymous review period ends.

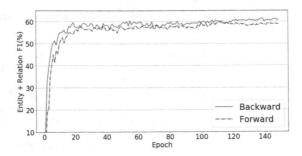

**Fig. 3.** Performance comparison of the P-hint strategies on ACE05 dev set. (The curves are plotted with LOWESS smoothing for better readability.)

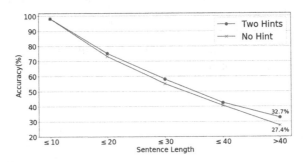

**Fig. 4.** Sentence-level accuracies with respect to sentence length.

score on relation is reduced by 0.9%, and the recall is reduced by 2.9%. It shows that Pw-hint helps improve recall of relation extraction, because this component confirms what kind of relation exists for each token pair. Finally, using two hint mechanisms together can make the results even better.

We also use different LMs (e.g., BERT$_{base}$, BERT$_{large}$ and XLNet$_{large}$) on the ablation tests and keep other hyper-parameters unchanged. The experimental results show that the F1 scores of base models slightly exceed the large models on two datasets and two different LMs, indicating that using large model may cause the overfitting problem. In addition, in order to have the same comparison standard with Li et al. [14], we also use BERT as the LM. The results show that our BERT version model slightly exceeds the multi-turn QA model [14] in F1 score (0.4%) on CoNLL04, and on ACE05 the relation F1 score is also competitive to their method.

## 5.2 P-Hint Strategies

In Sect. 3.4, we introduce two strategies for P-hint, i.e., "forward" and "backward". Figure 3 shows that the model using forward P-hint has a slower convergence speed and lower scores than the backward P-hint. This indicates that using backward P-hint brings higher F1 score and converges faster in the early training. Due to the model design, the query is always the last token of the sequence, and we must decode all the relations about the query at each time step. If we use backward P-hint, the hint will always gives the positional information of the first index to the query. It means that the backward P-hint can maintain the consistency of positional information about the query. On the other hand, forward P-hint always gives the last positional information to the query in every time step, and the positional information of query would not be consistent in every time step. The experimental results also support that it is important to maintain the consistency of the positional information of the query in our model.

## 5.3   Accuracy at Sentence Level

Generally speaking, longer sentences may have more entities and relations, so it is more difficult to extract all the entities and relations correctly. To show that two hint mechanisms indeed help improve the accuracy at the sentence level, we compare the corresponding accuracy when using P-hint and Pw-hint in the sentence level on the test set of ACE05. When evaluating the accuracy at the sentence level, a sentence is considered correct only if all the labels in the result table are correct. According to the results in Fig. 4, when the sentence length increases, the accuracy drops dramatically. The results further show that the sentence-level accuracy of using two hint mechanisms is higher than not using any hint. When the sentence length is greater than 40 words, the difference in accuracy between the two methods is more than 5%, indicating that our proposed model has more advantages in longer sentences.

## 6   Conclusions

In this paper, we propose a novel end-to-end joint NER and RE model, which uses the neural network based on multi-head attention, positional hint and pair-wise hint. Extensive experiments have shown the effectiveness of using these two hint mechanisms in the relation layer. Analysis and ablation tests further demonstrate that backward positional hint speeds up the convergence and pair-wise hint improves the recall of relation extraction. According to the experimental results, our final model achieves the state-of-the-art performances on various benchmark datasets.

## References

1. Bahdanau, D., Cho, K., Bengio, Y.: Neural machine translation by jointly learning to align and translate. arXiv preprint arXiv:1409.0473 (2014)
2. Bekoulis, G., Deleu, J., Demeester, T., Develder, C.: Adversarial training for multi-context joint entity and relation extraction. arXiv preprint arXiv:1808.06876 (2018)
3. Bekoulis, G., Deleu, J., Demeester, T., Develder, C.: Joint entity recognition and relation extraction as a multi-head selection problem. Expert Syst. Appl. **114**, 34–45 (2018)

4. Chan, Y.S., Roth, D.: Exploiting syntactico-semantic structures for relation extraction. In: Proceedings of the 49th Annual Meeting of the Association for Computational Linguistics: Human Language Technologies-Volume 1, pp. 551–560. Association for Computational Linguistics (2011)
5. Collobert, R., Weston, J., Bottou, L., Karlen, M., Kavukcuoglu, K., Kuksa, P.: Natural language processing (almost) from scratch. J. Mach. Learn. Res. **12**(Aug), 2493–2537 (2011)
6. Dai, Z., et al.: Transformer-XL: attentive language models beyond a fixed-length context. arXiv preprint arXiv:1901.02860 (2019)
7. Devlin, J., Chang, M.W., Lee, K., Toutanova, K.: BERT: pre-training of deep bidirectional transformers for language understanding. arXiv preprint arXiv:1810.04805 (2018)
8. Gupta, P., Schütze, H., Andrassy, B.: Table filling multi-task recurrent neural network for joint entity and relation extraction. In: Proceedings of COLING 2016, the 26th International Conference on Computational Linguistics: Technical Papers, pp. 2537–2547 (2016)
9. He, K., Zhang, X., Ren, S., Sun, J.: Deep residual learning for image recognition. In: Proceedings of the IEEE Conference on Computer Vision and Pattern Recognition, pp. 770–778 (2016)
10. Katiyar, A., Cardie, C.: Going out on a limb: joint extraction of entity mentions and relations without dependency trees. In: Proceedings of the 55th Annual Meeting of the Association for Computational Linguistics (Volume 1: Long Papers), pp. 917–928 (2017)
11. Kudo, T., Richardson, J.: SentencePiece: a simple and language independent subword tokenizer and detokenizer for neural text processing. arXiv preprint arXiv:1808.06226 (2018)
12. Kuru, O., Can, O.A., Yuret, D.: Charner: character-level named entity recognition. In: Proceedings of COLING 2016, the 26th International Conference on Computational Linguistics: Technical Papers, pp. 911–921 (2016)
13. Li, Q., Ji, H.: Incremental joint extraction of entity mentions and relations. In: Proceedings of the 52nd Annual Meeting of the Association for Computational Linguistics (Volume 1: Long Papers), vol. 1, pp. 402–412 (2014)
14. Li, X., et al.: Entity-relation extraction as multi-turn question answering. arXiv preprint arXiv:1905.05529 (2019)
15. Loshchilov, I., Hutter, F.: Fixing weight decay regularization in adam. arXiv preprint arXiv:1711.05101 (2017)
16. Mikolov, T., Sutskever, I., Chen, K., Corrado, G.S., Dean, J.: Distributed representations of words and phrases and their compositionality. In: Advances in Neural Information Processing Systems, pp. 3111–3119 (2013)
17. Miwa, M., Bansal, M.: End-to-end relation extraction using LSTMs on sequences and tree structures. arXiv preprint arXiv:1601.00770 (2016)
18. Miwa, M., Sasaki, Y.: Modeling joint entity and relation extraction with table representation. In: Proceedings of the 2014 Conference on Empirical Methods in Natural Language Processing (EMNLP), pp. 1858–1869 (2014)
19. Nadeau, D., Sekine, S.: A survey of named entity recognition and classification. Lingvisticae Investigationes **30**(1), 3–26 (2007)
20. Pennington, J., Socher, R., Manning, C.: Glove: global vectors for word representation. In: Proceedings of the 2014 Conference on Empirical Methods in Natural Language Processing (EMNLP), pp. 1532–1543 (2014)

21. Srivastava, N., Hinton, G., Krizhevsky, A., Sutskever, I., Salakhutdinov, R.: Dropout: a simple way to prevent neural networks from overfitting. J. Mach. Learn. Res. **15**(1), 1929–1958 (2014)
22. Sun, C., et al.: Extracting entities and relations with joint minimum risk training. In: Proceedings of the 2018 Conference on Empirical Methods in Natural Language Processing, pp. 2256–2265 (2018)
23. Vaswani, A., et al.: Attention is all you need. In: Advances in Neural Information Processing Systems, pp. 5998–6008 (2017)
24. Vinyals, O., Fortunato, M., Jaitly, N.: Pointer networks. In: Advances in Neural Information Processing Systems, pp. 2692–2700 (2015)
25. Yang, Z., Dai, Z., Yang, Y., Carbonell, J., Salakhutdinov, R., Le, Q.V.: XLNet: generalized autoregressive pretraining for language understanding. arXiv preprint arXiv:1906.08237 (2019)
26. Zhang, M., Zhang, Y., Fu, G.: End-to-end neural relation extraction with global optimization. In: Proceedings of the 2017 Conference on Empirical Methods in Natural Language Processing, pp. 1730–1740 (2017)
27. Zhang, Y., Qi, P., Manning, C.D.: Graph convolution over pruned dependency trees improves relation extraction. arXiv preprint arXiv:1809.10185 (2018)
28. Zhang, Y., Zhong, V., Chen, D., Angeli, G., Manning, C.D.: Position-aware attention and supervised data improve slot filling. In: Proceedings of the 2017 Conference on Empirical Methods in Natural Language Processing, pp. 35–45 (2017)

# Maximum $(L, K)$-Lasting Cores
# in Temporal Social Networks

Wei-Chun Hung and Chih-Ying Tseng[✉]

Department of Computer Science, National Tsing Hua University, Hsinchu, Taiwan

**Abstract.** Extracting dense structures in a social network is a fundamental task in graph mining and can find many real-world applications. The *temporal social network* augments the conventional social network with the temporal dimension, and extracting dense structures enables us to understand the period of time for which the dense structures exist. In this paper, we propose the new notion of $(L, K)$-*lasting core*, which is a densely connected subgraph lasting for a sufficiently long period of time in the temporal social network. We propose a polynomial-time algorithm to obtain the maximum $(L, K)$-lasting core with various processing strategies to boost the efficiency. We conduct extensive experiments on multiple datasets to validate the effectiveness and efficiency of the proposed approach. The experimental results show that our proposed approaches outperform the other baseline approaches in terms of solution quality and efficiency.

**Keywords:** Densest subgraph · Temporal social networks · Cores

## 1 Introduction

Extracting dense subgraphs has been actively studied in recent years. It is a basic and important work in graph analysis. There are many different definitions of dense subgroup, e.g., average degree [11,21,30], $k$-truss [22,40], clique [3,4,15], quasi-clique [1,27], $k$-core [12,24,32]. All of those different definitions refer to the same thing that vertices in the group are highly connected. Dense subgroup of different definitions have different ways to extract, such as linear programming [23,26], core decomposition [8,42], etc. It has a lot of application in practice, e.g., biological module discovery [20], story identification [2] and community detection [6,10].

Most of the previous work focus on the dense subgraph on single-layer graph which exists for just one time. However, they can not apply to multilayer graph. There were also many work finding dense subgraph on multilayer network. Some of them consider layers are different type of information [13,46]. Some of them consider that different layers represent different time [25,29] and we call them temporal networks. Each edge in temporal networks is associated with time. Densely connected vertices in a temporal network may correspond to a community. For example, in a collaboration network, dense subgroup may represent a

© Springer Nature Switzerland AG 2021
C. S. Jensen et al. (Eds.): DASFAA 2021 Workshops, LNCS 12680, pp. 336–352, 2021.
https://doi.org/10.1007/978-3-030-73216-5_23

research team or some researchers in the same domain publishing papers together continuously.

In this paper, we study the problem of long-lasting group in the temporal network. We not only consider the cohesiveness but also the time the group lasts. We adopt the definition of $k$-core, which is a group in which each member has at least $k$ other friends also in the group. We aim to find the largest group in the temporal network with given lasting-time constraint and connectivity constraint.

To achieve above goal, we present a model, called $(L, K)$-lasting core, based on the well-known concept of $k$-core. Our model can preserve a $k$-core lasting for $L$ time. For its application scenarios, we can leverage this model to find a team of researchers publishing paper every year in the same field, or a social group whose member interact with each other very frequently. We formally define the problem and propose effective algorithm to deal with it. We conduct extensive experiments to evaluate our model and study the impact of algorithm variations. In summary, the contributions of this paper are summarized as follows:

- We propose the notion of $(L, K)$-lasting core to integrate the social cohesiveness with the temporal dimension to identify important groups in temporal social networks.
- We develop effective algorithms and techniques to extract $(L, K)$-lasting cores.
- We conduct extensive experiments to evaluate the performance of our algorithms.

The rest of this paper is organized as follows. After reviewing related work on dense subgraph and temporal network in Sect. 2, we provide the notation and formulate our problem in Sect. 3. The details of the proposed algorithms are described in Sect. 4. We provide the experimental results in Sect. 5. We conclude this paper in Sect. 7.

## 2    Related Work

### 2.1    Dense Subgraph

Our work is related to the problem of extracting dense subgraphs, which has been actively studied for years. There are many measurements of dense subgraphs, including average degree [11], $k$-core [12], clique [4]. For example, Epasto et al. [11] proposed the method of maintaining a densest subgraph and quickly updating while an edge insertion or deletion. Sariyuce et al. [32] also studied on dynamic graph but what they maintained is the $k$-core decomposition. Bomze et al. [4] proposed several methods to deal with clique problem. In this paper, we propose the notion of $(L, K)$-lasting core, which extends the idea of $k$-core.

### 2.2    Multilayer Network

We study the problem on temporal networks, which is a special type of multilayer graph and its layers represent continuous time. There are many related work of multilayer graphs [13,25,29,46,47]. Zhang et al. [46] studied the problem on two layer graph which is a special case of multilayer graph. One layer is friendship of entities and the other is similarity between entities. Rozenshtein et al. [29] searched several groups of vertices with different time interval to maximize the sum of group density. Galimberti et al. [13] propose a method to find a subgroup that maximizes the minimum density of selected layers. Zhu et al. [47] aimed to find several multilayer cores that cover the largest number of vertices. Li et al. [25] deals with temporal network. They aimed to find dense subgraph with three constraints, $\theta$, $\tau$, and $k$ called $(\theta, \tau)$-persistent $k$-core. Note that $\tau$ is larger than $\theta$. The union graph of each $\theta$ length time interval in a $\tau$ length interval contains a k-core. We give an example of this work. In 1, we give $\theta = 2$, $\tau = 3$ and $k = 3$. For $\tau = 3$, we first choose $G_1$, $G_2$ and $G_3$ and then intersect two consistent snapshots in them for $\theta = 2$, e.g., $G_1$, $G_2$ and $G_2$, $G_3$. Then we find $k$-core of each intersection graph. The $k$-core of $G_1$, $G_2$ is $a, b, c, d$ and the one of $G_2$, $G_3$ is $a, b, c, d, e, f$. And the $(2, 3)$-persistent 3-core of this interval is $a, b, c, d$. The work we mentioned can't apply to our problem which we want to find a group lasting for a consistent time. We formulate the problem in the next section.

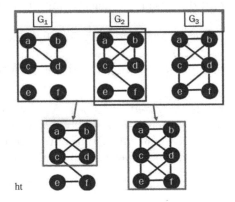

**Fig. 1.** An example of $(\theta, \tau)$-persistent $k$-core.

## 2.3   Community Search

Community search in social networks is an active research field [5,9,18,19,31, 41,43]. These works study various community search problems, including the enumeration of $k$-vertex connected components [41], extracting dense subgraphs, i.e., small-diameter $k$-plexes [9], identifying the maximum clique in sparse social networks [5], and proposing the UCF-Index to extract $(k, \eta)$-core in linear time for uncertain graphs [43]. In addition to finding dense communities in social networks, recent works also discuss finding sparse anti-communities in social networks [16,34,36,37], which has a wide spectrum of application scenarios.

## 2.4   Dense Subgraphs in Heterogeneous Social Networks

Extracting dense subgraphs in heterogeneous social networks have attracted research attentions [7,14,17,25,33,35,39,45,47]. These works propose new ideas and enable many new applications, such as enumerating the spatial cliques in the two-dimensional space for dense subgraph extraction [45]. Moreover, a set of socio-spatial group queries aim at identifying the socially-dense groups and the corresponding meeting points [14,38,44]. In addition to social and spatial relations, SDSQ is proposed for live multi-streaming scenarios in social networks that considers both the social tightness and preference of the users, as well as the diversity of multi-streaming channels [33].

## 3    Problem Definition

We are given a temporal network $G = (V, E)$, where $V$ is a set of vertices, and $E = \{(u, v, t)\}$, such that $u, v \in V$, timestamp $t \in \mathbb{N}$, indicating that edge $(u, v)$ exists at time $t$. Given $t \in \mathbb{N}$, $E_t = \{(u, v, t)\}$, which contains edges in timestamp $t$, we call each graph associated with certain time is a snapshot: Let $G_t = (V, E_t)$, $\forall t \in \mathbb{N}$. Given a subset $C \subseteq V$, edges induced by $C$ at timestamp $t$ is denoted $E_t(C) = \{(u, v, t)\}$ for $u, v \in C$. Then the degree of vertex $u \in C$ at time $t$ is denoted $d_t(u, C) = |\{u \in C | (u, v, t) \in E_t(C)\}|$.

**Definition 1** *(L-lasting time). L-lasting time means a time sequence which has $L$ continuous snapshot, e.g., $[0, 1, ..., L-1]$.*

**Definition 2** *((L, K)-lasting core). The $(L, K)$-lasting core of a temporal network $G = (V, E)$ is a non-empty set of vertices $C_{(L,K)} \subseteq V$, such that $\forall u \in C_{(L,K)}$ and $d_t(u, C(L, K)) \geq K$, $\forall t \in [t_0, t_1, ..., t_{L-1}]$, $K \in \mathbb{N}^+$.*

In other words, Definition 2 is saying that a $(L, K)$-lasting core is a $K$-core with $L$-lasting time. A maximum $(L, K)$-lasting core means it is a $(L, K)$-lasting core which has the most vertices. Then we formulate the first problem in this work which is to search a maximum $(L, K)$-lasting core.

*Problem 1* (Maximum $(L, K)$-lasting core). Given a temporal network $G = (V, E)$, two parameters $L$ and $K$, find the maximum $(L, K)$-lasting core of G.

We give an example of $(L, K)$-lasting core. In Fig. 2, we have a temporal network of 4 snapshots. Given $L = 2$ and $K = 3$, we can observe that a, b, c, d circled by red line is the maximum $(2, 3)$-lasting core in this temporal network. In the next section, we will propose basic algorithm for Problem 1 and how to speed up using advanced techniques.

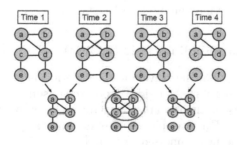

**Fig. 2.** An example of $(L, K)$-lasting core.

## 4    Algorithm Design

In this section, we introduce our proposed approaches and techniques to extract $(L, K)$-lasting core in details. First, we give a naive idea and a basic algorithm that is also the baseline in experiments. And then, we provide some simple but powerful techniques to speed up online search time. Finally, we explore the order of snapshots permutation to minimize the intersection cost.

### 4.1    Naive Algorithm

The naive algorithm of Problem 1 is detailed in Algorithm 1. First, we use the concept of sliding window and let $L$ be the length of the sliding window, i.e., it contains $L$ snapshot. The basic algorithm consists of two steps. The first step is to slide the window by changing starting time $t$. Let $C$ be $G_t$, the snapshot of time $t$. Then we intersect $C$ with the rest of $L - 1$ snapshot. The second step is to find $C_k$, the $K$-core of the intersection result $C$. If the size of $C_k$ is larger than the current maximum size of $(L, K)$-lasting core, $C_k$ become the current maximum $(L, K)$-lasting core.

---

**Algorithm 1.** Naive algorithm

---

**Input:** A temporal network $G = (V, E)$, $L$ and $K$
**Output:** The maximum $(L, K)$-lasting core $C_{(L,K)}$ of $G$
1:  $C_{(L,K)} \leftarrow \emptyset$;
2:  **forall** $t \in [0, 1,..., t_{max}\text{-}L]$ **do**
3:      $C \leftarrow G_t$;
4:      **forall** $i \in [t+1, t+2,..., t+L-1]$ **do**
5:          $C \leftarrow C \cap G_i$;
6:      $C_k \leftarrow$ find_kcore($C$);
7:      **if** $|C_k| > |C_{(L,K)}|$ **then**
8:          $C_{(L,K)} \leftarrow C_k$;
9:  **return** $C_{(L,K)}$;

---

Here, we omit the details of the $K$-core algorithm. It is to recursively remove the vertex with degree smaller than $K$ and check its neighbors' degrees until no vertex has degree smaller than $K$ in the subgroup $C$, i.e., $d_t(u, C) \geq K$. Finally, in line 9, we output the maximum $(L, K)$-lasting core.

### 4.2    Temporal Core Finding-Basic (TCFB)

In the previous subsection, the naive approach performs many redundant intersections. To tackle this issue, our idea is to reuse some parts of intersection results. Once we reuse them, we are able reduce the number intersections and improve the efficiency. For example, the current time sequence being processed is $[0, 1, 2, 3, 4]$ and the next time sequence is $[1, 2, 3, 4, 5]$. We can observe that the timestamp $[1, 2, 3, 4]$ is repeated. If we first do intersection of $[1, 2, 3, 4]$ which means $G_1, G_2, G_3$ and $G_4$ and have the result $C$, then the next action for current time sequence is to intersect $C$ and $G_0$ and for next time sequence is to intersect $C$ and $G_5$. Thus we can reduce one time of intersection of $G_1, G_2, G_3$ and $G_4$. The detailed description is outlined in Algorithm 2.

## 4.3   Min-Degree Pruning (MDP)

In this subsection, we focus on reducing vertices. Given a time sequence, each vertex has a degree on different snapshot. After the process of intersection and $k$-core, we get the result subgraph $C$, the degree of each vertex in $C$ will less than or equal to the vertex's smallest degree of all snapshots of the time sequence. In other words, given a time sequence $T$, $d(u, C) \leq min_t^{t \in T} d(u, G_t)$. Based on this, we can remove the vertex $u$ which $min_t^{t \in T} d(u, G_t) < K$ before the process. Equipped with this technique, we may traverse less vertices for each time sequence to reduce the cost. Here we just eliminate the vertex which does not satisfy the K constraint before the process, then we can further execute Algorithm 2.

---

**Algorithm 2.** Temporal Core Finding-Basic (TCFB)

---

**Input:** A temporal network $G = (V, E)$, $L$ and $K$
**Output:** The maximum $(L, K)$-lasting core $C_{(L,K)}$ of $G$
1:  $C_{(L,K)} \leftarrow \emptyset$;
2:  $C \leftarrow \emptyset$;
3:  **forall** $t \in [0, 1, ..., t_{max}-L]$ **do**
4:     $r \leftarrow t \bmod 2$
5:     **if** $r = 0$ **then**
6:        $C \leftarrow G_{t+1}$;
7:        **forall** $i \in [t+2, t+3, ..., t+L-1]$ **do**
8:           $C \leftarrow C \cap G_i$;
9:     $index \leftarrow i + (L - 1) * r$
10:    $C \leftarrow C \cap G_{index}$
11:    $C_k \leftarrow$ find_kcore($C$);
12:    **if** $|C_k| > |C_{(L,K)}|$ **then**
13:       $C_{(L,K)} \leftarrow C_k$;
14: **return** $C_{(L,K)}$;

---

## 4.4   Reordering for Intersection Minimization (RIM)

Now if we have a time sequence, we intersect snapshots chronologically in Algorithm 2. We observed that if we disrupt the order of original time sequence, the result of intersection graph remain the same, but the number of intersection times will be different. Here we can formulate a Subproblem. If we know the order having the least number of intersection times, then we have minimum cost. The subproblem is formulated as follows.

*Problem 2* (Minimum times of intersection). Given a time sequence of $L$ snapshot on $G$, we would like to find a permutation of given time sequence that minimize the sum of edge numbers of the intersection graph $[I_0, I_1, ..., I_{L-2}]$, such that $I_0 = G_0$, $I_1 = $ intersection$(I_0, G_1)$, $I_2 = $ intersection$(I_1, G_2)$,..., $I_{L-2} = $ intersection$(I_{L-2}, G_{L-1})$.

Clearly, if we can solve Subproblem 1, then we can solve Problem 1 more efficiently. A straightforward method to find optimal solution of Subproblem 1 is to enumerate all the permutation of given time sequence and find the best one. However, it is time-consuming to search in $L$ factorial number of permutation when $L$ grows. We propose a method, which choose snapshot greedily based on the previous snapshot. The standard of choosing next snapshot is by edge difference of two snapshots. Edge difference means the number of edges of previous snapshot that do not exist in the next snapshot. The detailed description is outlined in Algorithm 3. We compute edge difference offline. In line 4, edgediff$(G_i, G_j)$ means the number of edges of $G_i$ that do not exist in $G_j$. It should be noted that edgediff$(G_i, G_j)$ may be not same as edgediff$(G_j, G_i)$. After the computation, we get a map $Diff$ giving us information of edge difference of each snapshot pair. We then apply it on Algorithm 2 which is Algorithm 3. In Figs. 3, 3(a) is an example of snapshots. We can compute edge difference: $Diff[(1,2)] = 3$, $Diff[(1,3)] = 2$, $Diff[(1,4)] = 1$, $Diff[(2,3)] = 4$, $Diff[(2,4)] = 1$. If we are using greedy order, and we first choose snapshot 1, and next we choose snapshot 2 because $Diff[(1,2)] = 3$ is the largest. Then we choose snapshot 3. Now we see Fig. 3(b) to compute the cost: snapshot 1 has 4 edges; Intersection of 1, 2 has 1 edge; Intersection of 1, 2, 3 has 1 edge and Intersection of 1, 2, 3, 4 has 0 edge. Cost will be $4 + 1 + 1$. Finally, we have an advanced algorithm called Temporal Core Finding (TCF) which applies MDP and RIM on TCFB.

---

**Algorithm 3.** Edge Difference

---

**Input:** A temporal network $G = (V, E)$
**Output:** Edge difference of each pair of snapshots
  1: **forall** $i \in [0, 1,..., t_{max}]$ **do**
  2:    **forall** $j \in [0, 1,..., t_{max}]$ **do**
  3:      **if** $i \neq j$ **then**
  4:        $Diff[(i, j)] = $ edgediff$(G_i, G_j)$;
  5: **return** $Diff$;

---

---

**Algorithm 4.** RIM

---

**Input:** A temporal network $G = (V, E)$, $L$ and $K$
**Output:** The maximum $(L, K)$-lasting core $C_{(L,K)}$ of $G$;
1: $C_{(L,K)} \leftarrow \emptyset$;
2: **forall** $t \in [0, 1,..., t_{max}\text{-}L]$ **do**
3:    $C \leftarrow G_t$;
4:    $h \leftarrow t$;
5:    $S \leftarrow set(t{+}1, t{+}2,..., t{+}L\text{-}1)$;
6:    **forall** $i \in [1, 2,..., L\text{-}1]$ **do**
7:       $h \leftarrow argmax_{s \in S} Diff(h, s)$;
8:       $S \leftarrow S \setminus s$;
9:       $C \leftarrow C \cap G_h$;
10:    $C_k \leftarrow \text{find\_kcore}(C)$;
11:    **if** $|C_k| > |C_{(L,K)}|$ **then**
12:       $C_{(L,K)} \leftarrow C_k$;
13: **return** $C_{(L,K)}$;

---

## 4.5    Time Complexity and Optimality

Then we analyze the time complexity of naive algorithm that we mentioned in Sect. 4.1. We use a sliding window to traverse given time. Assume the length of given time is $N$, then we need to process $N - L + 1$ sliding window. For each sliding window, we have L snapshots and intersect graphs by iterating edges of graph of starting time, so the time complexity of this part is $O(EL)$. The next part is to find $k$-core of intersection graph. We recursively remove edges and vertices, thus the time complexity is $O(V{+}E)$. Then we can derive the time complexity of naive algorithm which is $O(NLE^2)$.

Next, we analyze the time complexity of TCF. The number of sliding window is still $N - L + 1$. We can see the part of reusing intersection of overlapped snapshots can decrease intersection times by about 2 times, which do not affect time complexity. The part of MDP remove vertices which can not be in the solution, but it may not remove any vertices in the worst case. The part of RIM change the intersection order of snapshots, but it has no effect on time complexity. Therefore, the time complexity of intersection part is still $O(EL)$. The part of finding $k$-core didn't change. Then we can derive the time complexity of TCF which is still $O(NLE^2)$. Though the time complexity of TCF doesn't change, the better performance can be see in later Sect. 5.

Here we discuss the optimality of basic algorithm and TCF. The basic algorithm slide a window to search in each interval, do intersection, and find $k$-core. Obviously, the basic algorithm can find optimal solution. Then, in TCF, the part of reusing intersection of overlapped snapshots just reduce intersection cost, it process the same thing. MDP part eliminates the vertex not in the solution which doesn't affect the result. RIM part change the order of snapshots, but the intersection graph of them is same as non-ordering one's. Thus we can observe that both naive algorithm and TCF can find optimal solution.

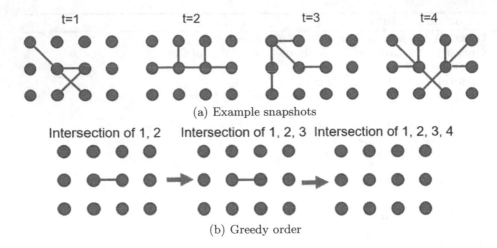

(a) Example snapshots

(b) Greedy order

**Fig. 3.** Example of greedy order and optimal order.

## 5   Experiments

In this section, we present experimental results and performance comparisons of our methods on different datasets.

**Datasets.** We use 3 real-world datasets recording interactions with temporal information. Each dataset is with a window size which defining how much continuous time each snapshot contains. If multiple interactions between two vertices appear in the same window, they just counted as one interaction. The characteristics of the datasets are reported in Table 1.

Last.fm records the co-listening history of its streaming platform. DBLP is the co-authorship network of the authors of scientific papers from DBLP computer science bibliography. Youtube [28] is a video-sharing web site that includes a social network and we pick two window size for this dataset. Synthetic datasets are all generated by Barabási–Albert preferential attachment model. Synthetic2000 is generated as 2000 vertices with 100 snapshots and the average degree of each snapshot ranges from 50 to 700. Synthetic5000 is generated as 5000 vertices and the average degree ranges from 80 to 900. Synthetic10W is generated as 100000 vertices and the average degree ranges from 60 to 300.

**Method.** We compare our approach (TCFB, TCF and two techniques) with brute force finding order of least intersection times (BF), greedy density method (Jethava) in [23] and maximal-span-cores algorithm (Galimberti) in [12]. Techniques are MDP and RIM. TCFB is in Sect. 4.2 and TCF is our best method applying MDP and RIM.

**Implementation.** All methods are implemented in C++. The experiments run on a machine equipped with Intel CPU at 3.7 GHz and 64 GM RAM.

**Table 1.** Datasets.

| Dataset | $|V|$ | $|E|$ | $|T|$ | Window size | Type |
|---|---|---|---|---|---|
| Last.fm | 992 | 4M | 77 | 21 days | Co-listening |
| DBLP | 1M | 11M | 78 | 1 year | Co-authorthip |
| Youtube_1 | 3M | 12M | 78 | 1 day | Friendship |
| Youtube_2 | 3M | 11M | 78 | 2 day | Friendship |
| synthetic2K | 2K | 39M | 100 | X | Synthetic |
| synthetic5K | 5K | 166M | 100 | X | Synthetic |
| synthetic10W | 10W | 636M | 50 | X | Synthetic |

## 5.1 Comparison of BF and TCF

First, we try to find optimal solution of Subproblem 1, finding the permutation of minimum number of times of intersection. We test all permutation of snapshots for each given time sequence. The running time is proportional to $L$ factorial. Then we conduct experiments with small $L$ on DBLP dataset to compare TCF and BF. In Fig. 4(a), $L$ remains the same, and when $K$ becomes larger, BF's running time decreases because the part of finding $k$-core removes most of the vertices that do not satisfy $K$ constraint and TCF's changes not much because most of the vertices are removed before processing and intersection time dominates the running time. In Fig. 4(b), we can see BF's running time grows with $L$ as expected. Figures 4(c) and (d) show that TCF has the smallest number of times of intersection in all conditions.

## 6 Methods with Different Techniques

We compare our methods applying different techniques in this subsection. In Fig. 5(a), the running time of intersection part decreases when $K$ grows, but the part of finding $k$-core dominates the running time, thus the total running time seems to remain the same. In Fig. 5(b), all the total running time of four methods decreases when $L$ grows because more vertices are removed before processing and more edges do not satisfy the $L$ constraint. Therefore, both running time of intersection part and finding $k$-core part decrease. Moreover, Figs. 5(c) indicates that TCFB+RIM's intersection times is smaller than TCFB's because reordering snapshots has effect on reducing the cost. They all remain the same because $L$ is fixed. Both TCFB+MDP's and TCF's intersection times decrease when $K$ grows because more vertices that not satisfy constraint $K$ are removed.

## 6.1 Synthetic Datasets

Then we conduct experiments on small and big synthetic datasets. For small synthetic datasets, in Figs. 6 and 7, we can observe that TCFB+MDP's and TCF's time and intersection times decrease when $K$ is larger than 60 in Figs. 6(a)

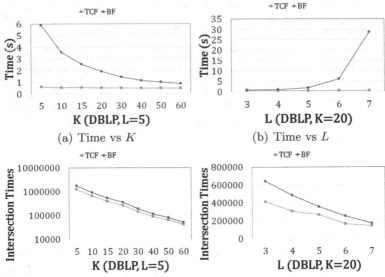

(a) Time vs $K$     (b) Time vs $L$

(c) Number of times of intersection (d) Number of times of intersection
vs $K$                              vs $L$

**Fig. 4.** Comparison of baseline and exhaustive - DBLP.

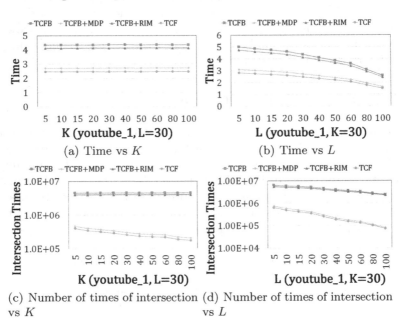

(a) Time vs $K$     (b) Time vs $L$

(c) Number of times of intersection (d) Number of times of intersection
vs $K$                              vs $L$

**Fig. 5.** Methods with different techniques - Youtube #1.

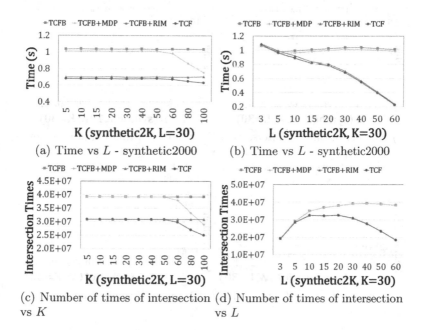

(a) Time vs $L$ - synthetic2000    (b) Time vs $L$ - synthetic2000

(c) Number of times of intersection (d) Number of times of intersection
vs $K$    vs $L$

**Fig. 6.** Methods with different techniques - synthetic datasets #1.

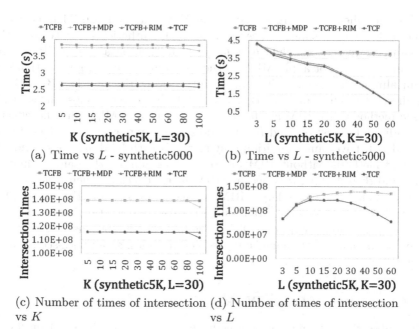

(a) Time vs $L$ - synthetic5000    (b) Time vs $L$ - synthetic5000

(c) Number of times of intersection (d) Number of times of intersection
vs $K$    vs $L$

**Fig. 7.** Methods with different techniques - synthetic datasets #2.

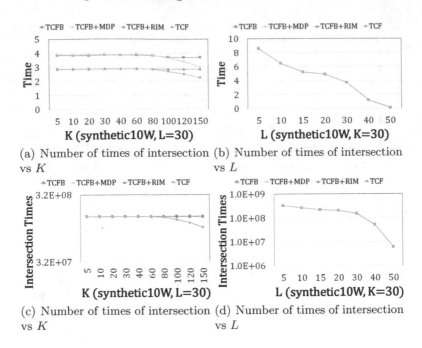

(a) Number of times of intersection (b) Number of times of intersection
vs $K$                                vs $L$

(c) Number of times of intersection (d) Number of times of intersection
vs $K$                                vs $L$

**Fig. 8.** Methods with different techniques - big synthetic datasets.

and (c) and when $K$ is larger than 80 in Figs. 7(a) and (c). That is because synthetic2K's vertex smallest degree is 60 and synthetic5K's is 80 and MDP only can remove vertices when $K$ is larger than each dataset's smallest degree. Other methods do not apply MDP so their running time and intersection times remain almost the same. In Figs. 6(b) and 7(b), TCFB and TCFB+MDP are lines almost the same, and TCFB+RIM and TCF are lines almost the same. We can see methods with RIM get large improvement and MDP have no effect on this condition because of fixed $K$ 30 is too small to remove any vertices. Figures 6(d) and 7(d) have same condition that methods with RIM are lines almost the same and methods without RIM are other lines almost the same. They first ascend and then decline, it is because RIM is less effective when $L$ is small. On the other hand, when $L$ is large, RIM reduces a lot of cost. For big synthetic dataset, in Figs. 8(a) and (c), MDP only decreases intersection times when $K$ is large enough. In Figs. 8(b) and (d), RIM is less effective because this dataset is too sparse to make great change on intersection times, so all lines seems almost the same.

## 6.2  Comparison of Other Work

In this subsection, we implement greedy density method (Jethava) in [23] and maximal-span-cores algorithm (Galimberti) in [12] and compare with our methods on Last.fm dataset. In Fig. 9(a), we can see Jethava takes more time than

Galimberti and TCF. For Galimberti and TCF, we can see more clear in Fig. 9(b) that Galimberti takes more time than TCF. Then we turn to judge the solution each method found. In Fig. 9(c), we can see Jethava's solution is always the biggest size and Galimberti's solution is same as TCF which is the optimal solution. Jethava find the group with largest density according to $L$, which do not consider $K$. And its solution is too big to find meaningful or interesting pattern. Although Galimberti finds optimal solution, TCF is more effective which has less running time. Therefore, our method TCF has better efficiency and effectiveness.

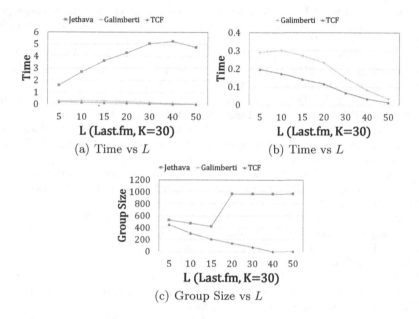

(a) Time vs $L$          (b) Time vs $L$

(c) Group Size vs $L$

**Fig. 9.** Comparison of other work.

Finally, we make a conclusion. MDP can reduce graph size when $K$ becomes larger thus reduce number of times of intersection and then running time. When $L$ becomes larger, RIM can intensively strengthen the performance. In other words, our methods can give good speedup in running time.

## 7   Conclusions

In this paper, we introduced a model called $(L, K)$-lasting core to detect the lasting group in temporal networks. We proposed efficient algorithms and applied advance techniques to solve this problem. Experiments in different datasets show us the efficiency and scalability. We can find interesting group by using our algorithms. In future work, we can extend our algorithms to temporal hypergraphs, which edges in the graph are arbitrary sets of nodes.

# References

1. Abello, J., Resende, M.G.C., Sudarsky, S.: Massive quasi-clique detection. In: Rajsbaum, S. (ed.) LATIN 2002. LNCS, vol. 2286, pp. 598–612. Springer, Heidelberg (2002). https://doi.org/10.1007/3-540-45995-2_51
2. Angel, A., Sarkas, N., Koudas, N., Srivastava, D.: Dense subgraph maintenance under streaming edge weight updates for real-time story identification. Proc. VLDB Endowment **5**(6), 574–585 (2012)
3. Balasundaram, B., Butenko, S., Hicks, I.V.: Clique relaxations in social network analysis: the maximum k-plex problem. Oper. Res. **59**(1), 133–142 (2011)
4. Bomze, I.M., Budinich, M., Pardalos, P.M., Pelillo, M.: The maximum clique problem. In: Handbook of Combinatorial Optimization, pp. 1–74. Springer (1999). https://doi.org/10.1007/978-1-4757-3023-4_1
5. Chang, L.: Efficient maximum clique computation over large sparse graphs. In: Proceedings of the 25th ACM SIGKDD International Conference on Knowledge Discovery & Data Mining, pp. 529–538. ACM (2019)
6. Chen, J., Saad, Y.: Dense subgraph extraction with application to community detection. IEEE Trans. Knowl. Data Eng. **24**(7), 1216–1230 (2010)
7. Chen, Y.-L., Yang, D.-N., Shen, C.-Y., Lee, W.-C., Chen, M.-S.: On efficient processing of group and subsequent queries for social activity planning. IEEE Trans. Knowl. Data Eng. **31**(12), 2364–2378 (2018)
8. Cheng, J., Ke, Y., Chu, S., Özsu, M.T.: Efficient core decomposition in massive networks. In: 2011 IEEE 27th International Conference on Data Engineering, pp. 51–62. IEEE (2011)
9. Conte, A., De Matteis, T., De Sensi, D., Grossi, R., Marino, A., Versari, L.: D2k: scalable community detection in massive networks via small-diameter k-plexes. In: Proceedings of the 24th ACM SIGKDD International Conference on Knowledge Discovery & Data Mining, pp. 1272–1281. ACM (2018)
10. Dourisboure, Y., Geraci, F., Pellegrini, M.: Extraction and classification of dense communities in the web. In: Proceedings of the 16th International Conference on World Wide Web, pp. 461–470. ACM (2007)
11. Epasto, A., Lattanzi, S., Sozio, M.: Efficient densest subgraph computation in evolving graphs. In: Proceedings of the 24th International Conference on World Wide Web, pp. 300–310. International World Wide Web Conferences Steering Committee (2015)
12. Galimberti, E., Barrat, A., Bonchi, F., Cattuto, C., Gullo, F.: Mining (maximal) span-cores from temporal networks. In: Proceedings of the 27th ACM International Conference on Information and Knowledge Management, pp. 107–116. ACM (2018)
13. Galimberti, E., Bonchi, F., Gullo, F.: Core decomposition and densest subgraph in multilayer networks. In: Proceedings of the 2017 ACM on Conference on Information and Knowledge Management, pp. 1807–1816. ACM (2017)
14. Ghosh, B., Ali, M.E., Choudhury, F.M., Apon, S.H., Sellis, T., Li, J.: The flexible socio spatial group queries. Proc. VLDB Endowment **12**(2), 99–111 (2018)
15. Himmel, A.-S., Molter, H., Niedermeier, R., Sorge, M.: Enumerating maximal cliques in temporal graphs. In: Proceedings of the 2016 IEEE/ACM International Conference on Advances in Social Networks Analysis and Mining, pp. 337–344. IEEE Press (2016)
16. Hsu, B.-Y., Lan, Y.-F., Shen, C.-Y.: On automatic formation of effective therapy groups in social networks. IEEE Trans. Comput. Soc. Syst. **5**(3), 713–726 (2018)

17. Hsu, B.-Y., Shen, C.-Y.: On extracting social-aware diversity-optimized groups in social networks. In: 2018 IEEE Global Communications Conference (GLOBE-COM), pp. 206–212. IEEE (2018)
18. Hsu, B.-Y., Shen, C.-Y.: Willingness maximization for ego network data extraction in online social networks. In: Proceedings of the 29th ACM International Conference on Information and Knowledge Management, pp. 515–524 (2020)
19. Hsu, B.-Y., Tu, C.-L., Chang, M.-Y., Shen, C.-Y.: Crawlsn: community-aware data acquisition with maximum willingness in online social networks. Data Mining Knowl. Disc. **34**(5), 1589–1620 (2020)
20. Hu, H., Yan, X., Huang, Y., Han, J., Zhou, X.J.: Mining coherent dense subgraphs across massive biological networks for functional discovery. Bioinformatics **21**(suppl-1), i213–i221 (2005)
21. Hu, S., Wu, X., Chan, T.: Maintaining densest subsets efficiently in evolving hypergraphs. In: Proceedings of the 2017 ACM on Conference on Information and Knowledge Management, pp. 929–938. ACM (2017)
22. Huang, X., Cheng, H., Qin, L., Tian, W., Yu, J.X.: Querying k-truss community in large and dynamic graphs. In: Proceedings of the 2014 ACM SIGMOD International Conference on Management of Data, pp. 1311–1322. ACM (2014)
23. Jethava, V., Beerenwinkel, N.: Finding dense subgraphs in relational graphs. In: Appice, A., Rodrigues, P.P., Santos Costa, V., Gama, J., Jorge, A., Soares, C. (eds.) ECML PKDD 2015. LNCS (LNAI), vol. 9285, pp. 641–654. Springer, Cham (2015). https://doi.org/10.1007/978-3-319-23525-7_39
24. Li, R.-H., Qin, L., Yu, J.X., Mao, R.: Influential community search in large networks. Proc. VLDB Endowment **8**(5), 509–520 (2015)
25. Li, R.-H., Su, J., Qin, L., Yu, J.X., Dai, Q.: Persistent community search in temporal networks. In: 2018 IEEE 34th International Conference on Data Engineering (ICDE), pp. 797–808. IEEE (2018)
26. Luenberger, D.G., Ye, Y., et al.: Linear and Nonlinear Programming, vol. 2. Springer (1984). https://doi.org/10.1007/978-3-319-18842-3
27. Pei, J., Jiang, D., Zhang, A.: On mining cross-graph quasi-cliques. In: Proceedings of the Eleventh ACM SIGKDD International Conference on Knowledge Discovery in Data Mining, pp. 228–238. ACM (2005)
28. Rossi, R.A., Ahmed, N.K.: The network data repository with interactive graph analytics and visualization. In: AAAI (2015)
29. Rozenshtein, P., Bonchi, F., Gionis, A., Sozio, M., Tatti, N.: Finding events in temporal networks: segmentation meets densest-subgraph discovery. In: 2018 IEEE International Conference on Data Mining (ICDM), pp. 397–406. IEEE (2018)
30. Rozenshtein, P., Tatti, N., Gionis, A.: Finding dynamic dense subgraphs. ACM Trans. Knowl. Disc. Data (TKDD) **11**(3), 27 (2017)
31. Sanei-Mehri, S-V., Das, A., Tirthapura, S.: Enumerating top-k quasi-cliques. In: 2018 IEEE International Conference on Big Data (Big Data), pp. 1107–1112. IEEE (2018)
32. Saríyüce, A.E., Gedik, B., Jacques-Silva, G., Wu, K.-L., Çatalyürek, Ü.V.: Streaming algorithms for k-core decomposition. Proc. VLDB Endowment **6**(6), 433–444 (2013)
33. Shen, C.-Y., Fotsing, C.K., Yang, D.-N., Chen, Y.-S., Lee, W.-C.: On organizing online soirees with live multi-streaming. In: Proceedings of the AAAI Conference on Artificial Intelligence, vol. 32 (2018)

34. Shen, C.-Y., Huang, L.-H., Yang, D.-N., Shuai, H.-H., Lee, W.-C., Chen, M.-S.: On finding socially tenuous groups for online social networks. In: Proceedings of the 23rd ACM SIGKDD International Conference on Knowledge Discovery and Data Mining, pp. 415–424. ACM (2017)
35. Shen, C.-Y., Shuai, H.-H., Hsu, K.-F., Chen, M.-S.: Task-optimized group search for social internet of things. In: EDBT, pp. 108–119 (2017)
36. Shen, C.-Y., et al.: Forming online support groups for internet and behavior related addictions. In: Proceedings of the 24th ACM International on Conference on Information and Knowledge Management, pp. 163–172 (2015)
37. Shen, C.-Y., et al.: On extracting socially tenuous groups for online social networks with k-triangles. IEEE Tran. Knowl. Data Engineering (2020)
38. Shen, C.-Y., Yang, D.-N., Huang, L.-H., Lee, W.-C., Chen, M.-S.: Socio-spatial group queries for impromptu activity planning. IEEE Trans. Knowl. Data Eng. **28**(1), 196–210 (2015)
39. Shen, C.-Y., Yang, D.-N., Lee, W.-C., Chen, M.-S.: Activity organization for friend-making optimization in online social networks. IEEE Trans. Knowl. Data Eng. (2020)
40. Wang, J., Cheng, J.: Truss decomposition in massive networks. Proc. VLDB Endowment **5**(9), 812–823 (2012)
41. Wen, D., Qin, L., Zhang, Y., Chang, L., Chen, L.: Enumerating k-vertex connected components in large graphs. In: 2019 IEEE 35th International Conference on Data Engineering (ICDE), pp. 52–63. IEEE (2019)
42. Wu, H., et al.: Core decomposition in large temporal graphs. In: 2015 IEEE International Conference on Big Data (Big Data), pp. 649–658. IEEE (2015)
43. Yang, B., Wen, D., Qin, L., Zhang, Y., Chang, L., Li, R.-H.: Index-based optimal algorithm for computing k-cores in large uncertain graphs. In: 2019 IEEE 35th International Conference on Data Engineering (ICDE), pp. 64–75. IEEE (2019)
44. Yang, D.-N., Shen, C.-Y., Lee, W.-C., Chen, M.-S.: On socio-spatial group query for location-based social networks. In: Proceedings of the 18th ACM SIGKDD International Conference on Knowledge Discovery and Data Mining, pp. 949–957 (2012)
45. Zhang, C., Zhang, Y., Zhang, W., Qin, L., Yang, J.: Efficient maximal spatial clique enumeration. In: 2019 IEEE 35th International Conference on Data Engineering (ICDE), pp. 878–889. IEEE (2019)
46. Zhang, F., Zhang, Y., Qin, L., Zhang, W., Lin, X.: When engagement meets similarity: efficient (k, r)-core computation on social networks. Proc. VLDB Endowment **10**(10), 998–1009 (2017)
47. Zhu, R., Zou, Z., Li, J.: Diversified coherent core search on multi-layer graphs. In: 2018 IEEE 34th International Conference on Data Engineering (ICDE), pp. 701–712. IEEE (2018)

# The 2021 International Workshop on Mobile Ubiquitous Systems and Technologies

# A Tablet-Based Game Tool for Cognition Training of Seniors with Mild Cognitive Impairment

Georgios Skikos and Christos Goumopoulos(✉)

Information and Communication Systems Engineering Department, University of the Aegean, Samos, Greece
goumop@aegean.gr

**Abstract.** The purpose of this study is to examine the acceptability and effectiveness of a cognitive training game application targeting elderly adults with Mild Cognitive Impairment. Such kind of impairment signifies one of the earliest stages of dementia and Alzheimer diseases. Ten serious games were designed and developed in the Android platform to train cognitive functions such as Attention, Visual Memory, Observation, Acoustic Memory, Language, Calculations, Orientation and Sensory Awareness. In particular this paper examines the feasibility of playing such games with the participation of a group of seniors (N = 6) in a pilot study. Usability assessment was also performed by collecting qualitative and quantitative data. Another dimension that investigated was the possibility of using this game tool as an alternative to traditional methods for evaluating cognitive functions. The results show that participants could learn quickly and understand the game mechanics. They also found the games easy to use and showed high enjoyment.

**Keywords:** Healthcare · Serious games · Elderly · Mild cognitive impairment · User experience

## 1 Introduction

As our society tends to age, new needs and problems emerge both at the individual and at the socio-economic level. Therefore, a significant part of the research is now focused on improving the quality of life for the elderly people [1]. In this light, one area that is of particular interest to scholars is the maintenance of a satisfactory level of cognitive function, which tends to decline progressively with age [2]. Cognitive impairment in the elderly is a major social issue as it is associated with reduced independence, well-being and an increased need for caregivers [3]. Researchers agree that normal aging is accompanied by functional impairment in many different areas of perception, reasoning and memory. However, despite the fact that various theories have been proposed to interpret the mechanisms responsible for this reduction, its exact causes have not yet been clarified.

© Springer Nature Switzerland AG 2021
C. S. Jensen et al. (Eds.): DASFAA 2021 Workshops, LNCS 12680, pp. 355–364, 2021.
https://doi.org/10.1007/978-3-030-73216-5_24

A number of studies have focused on the effectiveness of cognitive empowerment in healthy older adults, but the collected data are unclear. Cognitive intervention appears to be effective in both improving the cognitive function of healthy older adults [4] as well as limiting the cognitive impairment seen in the elderly with dementia [5]. However, in surveys conducted to improve memory through cognitive training, only half reported a statistically significant improvement with the rest not exceeding the control group [6]. These data highlight the need for more research in the field of cognitive training.

On the other hand, Mild Cognitive Impairment (MCI) is considered as the transition stage between the normal senility and Alzheimer's disease (AD). It is characterized by impaired cognitive functions to a greater level than expected for a specific age, without meeting the criteria, in order to diagnose dementia, putting the patient, however, in danger for future worsening in terms of AD [7].

Until now, handwritten methods have been mainly used to identify and determine the level of cognitive ability. The classical methods are time-consuming and require the presence of personalized and full-time supervision by experts. In addition, classical methods have the form of a test, which can affect alertness, effort, motivation and, of course, the end result. Technology advancements and the attempt to combine scientific fields to find solutions to such problems led to the introduction of computer technology to address them.

The evolution of technology now provides the possibility of cognitive interventions through computer programs, electronic games and mobile applications. These innovations seem to be gaining ground in the field of rehabilitation compared to traditional methods, as they are less costly, more flexible, and more and more people have access to these techniques [8].

In an attempt to use these results, serious games have been developed with the aim of strengthening the brain and cognitive abilities. Serious games include focused activities derived from specific scientific measures of cognitive functions. This new kind of games has been extremely popular around the world, especially in elderly adults as these games have beneficial effects [9].

## 2  Related Work

Previous studies showed that playing serious games could improve cognitive functions such as memory, orientation, attention and more. Valladares-Rodriguez et al. introduced Episodix [10], a serious game application, designed to detect MCI and AD. It was used to assess cognitive domains such as memory, attention and knowledge, with a set of six games based mainly on classical psychological tests. The application was tested by a group of 16 individuals comprising eight healthy users, three with MCI and five with AD.

Boletsis and McCallum developed Smartkuber, a serious game that uses Augmented Reality [11]. The main purpose of the game was cognitive screening aiming mainly at people with MCI. The game consists of many minigames for a variety of puzzles and aimed at different areas of cognitive behavior, including audiovisual logic, memory, attention, problem solving and logic. The application was evaluated with a sample of 13 individuals over 60 years old. The results showed a good overall correlation (0.81) between Smartkuber and Montreal Cognitive Assessment (MoCA) test.

Leduc-McNiven et al. presented a serious game called WarCAT [12]. The aim of the game was to detect changes in cognitive behavior in an elderly patient with MCI. They have also successfully used a machine learning model to detect deviations in cognitive behavior.

Chignell and Tong developed a serious game for cognitive status assessment in the elderly [13]. Evaluating the application resulted in statistically significant correlations between game performance and the results of MoCA cognitive test measurement. They also found that the serious game could be used as a first check diagnosis of delirium.

Tapbrain is a mobile and tablet application that combines 13 serious games, in order to stimulate brain activity and 4 serious games to induce physical activity [14]. The goal is to influence cognitive areas such as memory, attention, problem solving, decision making, and games to develop physical activity, developing five successive stages of patient activation.

Finally, Manera et al. examined the acceptance degree of the Kitchen and Cooking serious game by elderly patients with MCI [15]. The study sample consisted of 21 elderly people with and without cognitive impairment. At the end, by evaluating the results, it was found that patients with AD were slower regarding the completion time of scenarios and recorded lower scores in the memory scenarios than the MCI patients. Participants with AD had a higher apathy score than participants with MCI.

## 3   MCI Rehab Application

Our work presents and evaluates an application with ten serious games for improving cognitive health for elderly people with MCI. The MCI Rehab application can provide an informal measurement of the user's cognitive performance and an assessment of cognitive improvement by monitoring the total success game time.

### 3.1   Games Development

The MCI Rehab application includes ten serious games (Table 1). Each game targets one or more cognitive areas. The cognitive areas that were examined are: Attention, Visual Memory, Observation, Acoustic Memory, Language, Calculations, Observation, Orientation and Sensory Awareness.

All games can be played on a tablet using typical touchscreen interaction movements. For example, the user can select a desired object and drag it to the correct position. The application was developed using Android Studio as the main development environment.

In order to facilitate the users the design of the game screens have common characteristics (Fig. 1). For example, the button "EXIT" is always on the top right of the screen and allows users to return to the main menu whenever they like. The time the user has to complete the game, is always at the top left of the tablet screen in an effort to make it easier for the users, to become familiar with the functionality of the application. An introduction screen was created in each game containing instructions in order to help the users to understand faster the logic of the game. Because our users are elderly people with cognitive issues and in order to facilitate their engagement to the application, softer

**Table 1.** Description of the 10 games in MCI rehab application

| Game name | Task | Cognitive function |
|---|---|---|
| Puzzle | The user must put the correct pieces of a photo to the correct place | Attention |
| Chronological order | The user is asked to create a story in chronological order by ordering a number of image pieces | Spatial and temporal orientation |
| Recall | The user must memorize a series of digits displayed in the screen and then must select it from a group of numbers | Visual memory |
| Mathematics | The user is asked to solve arithmetic problems | Calculations |
| Observation | The user is asked to group the images appearing on the screen | Observation |
| Sounds | The user is asked to match the sound heard with the correct image | Acoustic memory |
| Maze | The user is asked to lead a mouse on the cheese through a maze | Visual motor and attention |
| Memory cards | The user is asked to flip the cards and match the tiles together in pairs | Short memory |
| Language | The user is asked to find antonyms and synonyms of words | Language |
| Logical sequence | The user is asked to find the correct choice that continues a given sequence | Perception |

colors and a larger font were used. Also, in each successful attempt the transition to the next level is done automatically to lower the interaction burden.

The application has been simplified as much as possible, as it was targeting people with cognitive impairments. As a design choice game screens should be as easy to use and enjoyable as possible. In this way, complexity should be avoided and easiness and clarity should be promoted, helping the user to be more focused on the game goal. Furthermore, the application allows the user to choose to play the games at different levels of difficulty depending on his/her capabilities. The choice of difficulty level is individual per user, affects all games, and is continuous in terms of time, as it is valid continuously until it changes from another option.

Three different basic difficulty levels have been defined: easy, medium and advanced. Each level adds a relative playing difficulty compared to the previous one. When the user successfully completes each level then a custom dialog box appears, encouraging the user to continue to the next level. At each level, the user has the possibility of three

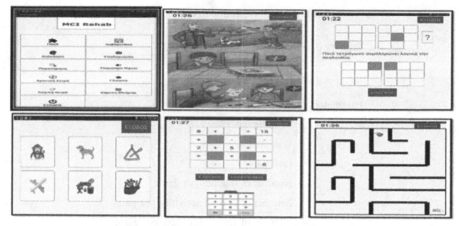

**Fig. 1.** Screen shots of the MCI Rehab game application (main menu, chronological order, logical sequence, sounds, mathematics and maze)

attempts; on the third unsuccessful attempt the game is completed and recorded as a failure, returning the user to the main menu.

In any successfully completed level, the transition to the next level is automatic. In this way we want to avoid complexity and promote usability, helping the user to be more focused on the gameplay.

### 3.2 Data Collection

Collection of user performance data is an important feature of the application allowing the analysis of users' cognitive state. For this reason the application records all user actions for each game separately. The parameters collected by the game are summarized in Table 2. Each game must be completed at a maximum allowable time. If this time elapses then the game is completed and identified as a failure otherwise the user has successfully completed the game. We are interested in completion times and points gained by the users as these metrics show their performance and adherence to using the application. A positive trend is that future game session results (e.g. after two/three/four weeks) should be better than past ones.

Consequently, conclusions could be drawn on which games the performance was improved and therefore which cognitive field has trained best. On the other hand, cognitive stagnation can be identified through long game times and unsuccessful attempts by analyzing performance data collected per game and per level.

**Table 2.** Data collection in each game

| Primary data | Player Anonymous Id |
|---|---|
| | Session Id |
| | Login date/Logout date |
| Parameters | Total time per game |
| | Total time per level |
| | Win games |
| | Lost games |
| | Number of touches per game |
| | Number of correct activities per game |
| | Number of wrong activities per game |

## 4 Evaluation

In total, the user group in the pilot study consisted of 6 participants (equal number of male and female), aged 66–79 years (Avg = 72.0 years, Stdev = 5.10) and was assembled as an opportunistic sample. According to Nielsen and Landauer [16], five users is a sufficient number to reveal 85% of the usability problems. The education level varied between secondary (66.6%) and higher (33.3%). Exclusion criteria for all potential participants were defined as i) age lower than 65 years; ii) the presence of irreversible hearing problems; iii) loss of vision; and iv) any other serious physical, mental and neurological illness (e.g., cardiovascular disease, stroke, etc.). An additional exclusion criterion was the presence of cognitive impairment. A score lower than 26 on the Montreal Cognitive Assessment (MoCA) test [17] is considered indicative of notable cognitive impairment. Finally, the Mini Mental State Examination (MMSE) test [18] was administered as additional tool for describing the cognitive level of the participants and not as an exclusion tool.

At the beginning of the process the participants signed a written consent form to participate in the research. The pilot phase regarding the usage and testing of the MCI Rehab application lasted 33 days (March-April 2019). Each participant completed 10 sessions of 30 min during the pilot study as a simulation of a cognitive rehabilitation process.

By employing standard experimental evaluation methods in terms of a semi structured interview and a usability questionnaire both qualitative and quantitative data were collected. As a suitable tool for assessing the perceived usability of the game tool the System Usability Scale (SUS) was selected [19]. The administration of the questionnaire was performed digitally through the application. Each participant (with the help of the researcher) scored on a 5 Likert scale ten statements regarding their experience using the application. A total of three evaluations were performed by each participant during the study. Fig. 2 shows the results of a total of 18 evaluations. An average SUS score for all participants was estimated as 89.4 in the scale of 100, suggesting a higher user acceptance [20].

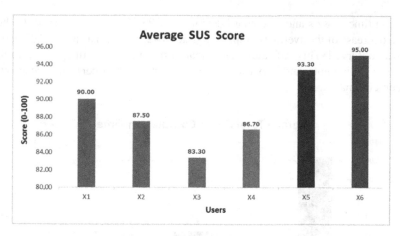

**Fig. 2.** Average SUS score per user

By analyzing the performance data collected by the application other interesting observations could be made. For example, for the user with id X1 even though a good overall cognitive assessment was originally recorded as attested by the MMSE and MoCA scores, by analyzing the results of specific cognitive domain tests separately, a lower performance was also noticed, especially in the domain of arithmetic calculations. While using the application the same user demonstrated similar poor cognitive performance in terms of successfully completing games that were related to arithmetic operations. As shown in Fig. 3 the user's playing time was quite high and had continuous failures to complete sessions of the Mathematics game. This result, demonstrates the potential effectiveness of games in detecting specific cognitive deficiencies. In more general, this is a promising indication that the use of such game applications can be utilized by experts as a tool to test specific cognitive abilities, as an alternative to traditional methods.

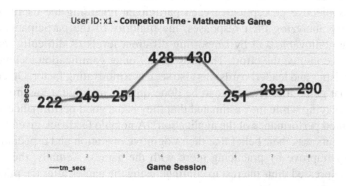

**Fig. 3.** Mathematics game completion times for participant with id X1

Figure 4 shows the average completion time improvement for all the participants in the beginning and in the end of the pilot study. During the first session an average completion time of 82 s and a standard deviation of 23.20 s were observed. In the last session a decrease of the average completion time to 42.50 s with a standard deviation of 4.26 s was observed. This indicates a significant improvement in the game completion times of the participants, and also a stability in the level of their performance as outliers have been eliminated.

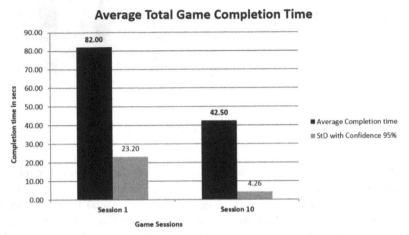

**Fig. 4.** Performance improvement in terms of successful game completion time due to practice between the first and last sessions

Overall, these measurements may be indicative that the game screens and tasks were well designed for the elderly people and that they were also sufficiently demanding to engage them to put additional effort for achieving improved performance throughout continuous training.

Finally, all participants attended at the end of the study an interview session to acquire qualitative data. There are no right or wrong answers. The feedback acquired reflects the subjective perception of each participant regarding the usefulness of the game tool. By analyzing their responses, the majority of the participants agree that the games are enjoyable and by supporting different levels of difficulty makes them challenging to continue the effort. The presence of other gamification techniques, such as points, statistics and leaderboards was also seen as a motivating factor. Other feedback given was that game tool learning can be done quickly and that a low mental load is required for playing while they estimated that they had a good performance (confirmed by the measured performance of the application). A notable feedback given by the more elder participants was their belief that their cognitive operation and especially memory will probably improve by practicing more with the games. Actually, the majority of participants expressed their interest to continue using the games after the pilot even just for the fun.

# 5 Conclusions

Given the growing need to implement more effective cognitive interventions in MCI patients or healthy groups of the population, research should focus on new technological developments, such as mobile applications, which combine cognitive training with fun, enjoyment, adaptation and easy access. This study confirms previous studies that properly designed serious games can be used as cognitive training programs for the elderly. A limitation of the present study is that the sample is small to draw safe conclusions. A larger, non-opportunistic sample would provide more consistent measurements regarding the feasibility of the approach.

Regarding the MCI rehabilitation dimension of the game tool, an extensive evaluation is underway in the context of a research project. The methodology includes a control group and an intervention group applying a randomized controlled trial. For all users, measurements of cognitive functions will be recorded before using the game tool, while after the intervention, the same measurements will be recorded only for the intervention team. The intervention team will use the game tool 1–2 times a week until each participant completes 24 user sessions.

**Acknowledgements.** This research has been co-financed by the European Regional Development Fund of the European Union and Greek national funds through the Operational Program Competitiveness, Entrepreneurship and Innovation, under the call ERA-NETS 2018 (ID:T8EPA2–00011, grant MIS:5041669). The authors would like to thank the volunteers that took part in the evaluation study.

# References

1. Jin, K., Simpkins, J.W., Ji, X., Leis, M., Stambler, I.: The critical need to promote research of aging and aging-related diseases to improve health and longevity of the elderly population. Aging Dis. **6**(1), 1 (2015)
2. Craik, F.I., Bialystok, E.: Cognition through the lifespan: mechanisms of change. Trends Cogn. Sci. **10**(3), 131–138 (2006)
3. Logsdon, R.G., Gibbons, L.E., McCurry, S.M., Teri, L.: Assessing quality of life in older adults with cognitive impairment. Psychosom. Med. **64**(3), 510–519 (2002)
4. Smith, G.E., et al.: A cognitive training program based on principles of brain plasticity: results from the Improvement in memory with plasticity-based adaptive cognitive training (IMPACT) study. J. Am. Geriatr. Soc. **57**(4), 594–603 (2009)
5. Valenzuela, M., Sachdev, P.: Can cognitive exercise prevent the onset of dementia? Systematic review of randomized clinical trials with longitudinal follow-up. Am. J. Geriatr. Psychiatry **17**(3), 179–187 (2009)
6. Reijnders, J., van Heugten, C., van Boxtel, M.: Cognitive interventions in healthy older adults and people with mild cognitive impairment: a systematic review. Ageing Res. Rev. **12**(1), 263–275 (2013)
7. Kelley, B.J., Petersen, R.C.: Alzheimer's disease and mild cognitive impairment. Neurol. Clin. **25**(3), 577–609 (2007)
8. Valladares-Rodríguez, S., Pérez-Rodríguez, R., Anido-Rifón, L., Fernández-Iglesias, M·(2016) Trends on the application of serious games to neuropsychological evaluation: a scoping review. J. Biomed. Inform. **64**, 296–319 (2016). https://doi.org/10.1016/j.jbi.2016.10.019. Epub Nov 1 PMID: 27815228

9. Ben-Sadoun, G., Manera, V., Alvarez, J., Sacco, G., Robert, P.: Recommendations for the design of serious games in neurodegenerative diseases. Front. Aging Neurosci. **10**, 13 (2018)

10. Valladares-Rodriguez, S., Fernández-Iglesias, M., Anido-Rifón, L., Facal, D., Pérez-Rodríguez, R.: Episodix: a serious game to detect cognitive impairment in senior adults. A psychometric study. PeerJ **6**(9), E5478 (2018)

11. Boletsis, C., McCallum, S.: Smartkuber: a serious game for cognitive health screening of elderly players. Games Health J. **5**(4), 241–251 (2016)

12. Leduc-McNiven, K., White, B., Zheng, H., McLeod, R.D., Friesen, M.R.: Serious games to assess mild cognitive impairment: 'The game is the assessment'. Res. Rev. Insights **2**(1) (2018)

13. Tong, T., Chignell, M., Tierney, M.C., Lee, J.: A serious game for clinical assessment of cognitive status: validation study. JMIR Serious Games **4**(1), e7 (2016)

14. Kang, K., Choi, E.J., Lee, Y.S.: Proposal of a Serious Game to Help Prevent Dementia. In: Bottino, R., Jeuring, J., Veltkamp, R. (eds.) Games and Learning Alliance. GALA 2016. LNCS, vol. 10056, pp. 415–424. Springer, Cham (2016)

15. Manera, V., et al.: 'Kitchen and cooking', a serious game for mild cognitive impairment and Alzheimer's disease: a pilot study. Front. Aging Neurosci. **7**, 24 (2015)

16. Nielsen, J., Landauer, T.K.: A mathematical model of the finding of usability problems. In: Proceedings of the INTERACT 1993 and CHI 1993 Conference on Human Factors in Computing Systems, pp. 206–213, May 1993

17. Nasreddine, Z.S., et al.: The montreal cognitive assessment, MoCA: a brief screening tool for mild cognitive impairment. J. Am. Geriatr. Soc. **53**(4), 695–699 (2005)

18. Folstein, M.F., Folstein, S.E., McHugh, P.R.: "Mini-mental state": a practical method for grading the cognitive state of patients for the clinician. J. Psychiatr. Res. **12**(3), 189–198 (1975)

19. Brooke, J.: SUS: a quick and dirty'usability. Usability evaluation in industry, p. 189 (1996)

20. Bangor, A., Kortum, P.T., Miller, J.T.: An empirical evaluation of the system usability scale. Int. J. Hum.-Comput. Interact. **24**(6), 574–594 (2008)

# AntiPhiMBS-Auth: A New Anti-phishing Model to Mitigate Phishing Attacks in Mobile Banking System at Authentication Level

Tej Narayan Thakur(iD) and Noriaki Yoshiura(✉)(iD)

Department of Information and Computer Sciences,
Saitama University, Saitama 338-8570, Japan
yoshiura@fmx.ics.saitama-u.ac.jp

**Abstract.** In the era of digital banking, the advent of the latest technologies, utilization of social media, and mobile technologies became prime parts of our digital lives. Unfortunately, phishers exploit digital channels to collect login credentials from users and impersonate them to log on to the victim systems to accomplish phishing attacks. This paper proposes a novel anti-phishing model for Mobile Banking System at the authentication level (AntiPhiMBS-Auth) that averts phishing attacks in the mobile banking system. This model employs a novel concept of a unique id for authentication and application id that is known to users, banking app, and mobile banking system only. Phishers and phishing apps do not know the unique id or the application id, and consequently, this model mitigates the phishing attack in the mobile banking system. This paper utilized a process meta language (PROMELA) to specify system descriptions and security properties and built a verification model of **AntiPhiMBS-Auth.** The verification model of **AntiPhiMBS-Auth** is successfully verified using a simple PROMELA interpreter (SPIN). The SPIN verification results prove that the proposed **AntiPhiMBS-Auth** is error-free, and financial institutions can implement the verified model for mitigating the phishing attacks in the mobile banking system at the authentication level.

**Keywords:** Mobile banking system · Authentication · Anti-phishing model · Verification · Model checking

## 1 Introduction

Phishing is a growing social engineering asynchronous attack in which phishers exploit digital channels such as mobile banking, Internet banking, ATM (Automated Teller Machine), social media platform (such as Facebook, Line, Viber, WhatsApp, etc.) for the attacks. Phishers craft an email, SMS (Short Messaging Service), or voicemail and wait for the victims to log onto their phishing site or phishing application. Subsequently, phishers collect the stolen credentials and impersonate them to log onto the digital channels. Attackers have a list of validated banking credentials for manual account takeover on the banking site and banking applications. Shape Security's 2018 credential

© Springer Nature Switzerland AG 2021
C. S. Jensen et al. (Eds.): DASFAA 2021 Workshops, LNCS 12680, pp. 365–380, 2021.
https://doi.org/10.1007/978-3-030-73216-5_25

spill report [28] shows 2.3 billion credentials breaches in 2017, and online retail loses about $6 billion per year while the US consumer banking industry faces over $50 million per day in potential losses from the attacks.

In the era of digital banking, the advent of the latest technologies, the utilization of social media, and mobile technologies have become prime parts of our digital lives. People use mobile applications, email, SMS, web, and social media every day, but unfortunately, they are all abused for phishing attacks. Phishers employ all of these platforms to exploit information, harm society, and globally dispute attacker activities, thus leading to financial loss and cybercrime. According to the findings of Phishlab's 2019 phishing trends and intelligence report [29], phishing attack volume raised to 40.9% in 2018, and 83.9% of attacks targeted credentials for financial, email, cloud, payment, and SaaS services. The financial institutions, being on top as the single most targeted industry, accounted for 28.9% of all phishing websites in 2018, and the most credential theft achieved using phishing-based links accounted for 88% [29]. According to F5 Labs' 2020 phishing and fraud report [30], phishing incidents rose by a staggering 220% compared to the yearly average during the height of global pandemic fears.

The phishing attack is the main challenge all around the world, and it has become one of the main burdens for the full-fledged implementation of mobile banking in financial institutions. Researchers have been working for the mitigation of such attacks universally. Some of the researchers have focused on machine learning models. Machine learning models [1–6] and artificial neural networks [7] can mitigate the known phishing attacks up to some extent. The advantages of these models are that they can be suitable for the known phishing websites but may not work for newly conceived phishing attacks. Besides, algorithms [8, 9] can mitigate only the specified pattern of phishing attacks. Drury and Meyer [10] proposed an email account separation for the detection of phishing emails but the proposed method can detect phishing emails in an organization where the mail server is configured securely for operation. Miller, Miller, Zhang, and Terwilliger [13] presented a three-pillared strategy for the prevention of phishing attacks using one-time passwords, multi-level desktop barrier applications, and behavior modification which can be advantageous for the organization where employees do not misconduct the information technology policy and guidelines of the organization.

Some of the researchers focused on multifactor authentication for the mitigation of phishing attacks. The advantage of using the counter challenge authentication method in [17] is that the phisher cannot provide the challenge enforced by the web application and thus, mitigates the phishing attacks. The use of various multifactor authentication methods in [18–27] can help in mitigating phishing attacks in online banking systems.

Generally, banking users install banking applications on their mobile to get banking services using mobile. The users might install a phishing app on their mobile misguidedly and enter the login credentials for the mobile banking system. Moreover, the users might follow the links of a phishing email or phishing SMS and enter the login credentials unknowingly about the phishing. In this way, phishers collect the login credentials from banking users and exploit them for phishing attacks in the mobile banking system. Some of the above research adopted machine learning models for the mitigation of phishing attacks using web pages and some employed multifactor authentication methods for the prevention of phishing attacks within the Internet banking systems. However, these approaches are inefficient for mitigating phishing attacks within the mobile banking system. According to the above studies, we found that the existing approaches are insufficient to account for the phishing attacks in the mobile banking system at the authentication level. A phishing attack can be pandemic in the future if proper actions are not taken in time by the financial institutions. This paper presents a new anti-phishing model for mobile banking system at the authentication level (AntiPhiMBS-Auth) to beat this gap, and the objective of this research is to build a new anti-phishing model, to the best of our knowledge, is the first attempt to mitigate phishing attacks in the mobile banking system at the authentication level.

Financial institutions can implement this model globally to mitigate phishing attacks in the mobile banking industry. Phishers might succeed in collecting the login credentials using phishing apps, phishing email, phishing SMS, or social media platform but could not complete the authentication process of the mobile banking system if this model is implemented in the financial institutions correctly. Henceforth, phishers could not succeed in executing the transactions in the mobile banking system using stolen login credentials, and financial institutions can save millions of dollars globally. The model will play a significant role in increasing mobile banking transactions and will contribute to the transformation towards a cashless society in the era of digital banking. The paper is further structured as follows: Sect. 2 describes the related studies, Sect. 3 presents the new anti-phishing model for the mobile banking system at the authentication level, Sect. 4 presents the results and discussion, and Sect. 5 describes conclusions and future work.

## 2   Background

Phishing has become one of the foremost attacks nowadays and many researchers have been proposing different solutions to mitigate the ongoing phishing attacks in cyberspace. Authors of [1–7] employed different machine learning models and artificial neural networks to classify websites as legitimate or phishing using the knowledge base of the phishing attacks. Tchakounte, Molengar, and Ngossaha [1] employed a formal description logic to prepare the knowledge base of phishing attacks and designed an ontology-oriented approach to add semantics in the knowledge base of phishing attacks. Subasi and Kremic [2] presented an intelligent phishing website detection framework where different machine learning models such as AdaBoost and Multiboost are employed to classify websites as legitimate or phishing websites. Ozker and Sahingoz [3] proposed a machine learning model and implemented a content-based phishing detection system that analyzes the text and additional properties of the web page and tries to understand whether there is a fraudulent web page or not on the websites. Priya, Selvakumar, and Velusamy [4] proposed a radial basis function (RBF) network with enhanced hyper parameters for classifying and predicting the phishing websites. K-modes clustering algorithm along with the proposed dissimilarity evaluation is used in [4] to select the RBF centers and spread constant of the network for better learning. Odeh, Alarbi, Keshta, and Abdelfattah [5] presented an intelligent model for detecting phishing websites on the Internet that applies multilayer perceptron to the system which classifies the inputted URL and applies the single attribute evaluator to eliminate irrelevant attributes to detect the phishing attacks.

Hossain, Sarma, and Chakma [6] analyzed different machine learning techniques that can be implemented over a dataset of features regarding websites and their corresponding details to detect a possible phishing website. Su [7] adopted long short-term memory (LSTM) and optimized the training method of the model in combination with the characteristics of recurrent neural networks (RNN) for the detection of phishing websites. Authors of [8, 9] used algorithms to classify the incoming URL (Uniform Resource Locator) into a phishing site or non-phishing site. Abiodun, Sodiya, and Kareem [8] employed an algorithm design to extract link characteristics from loading URLs to determine their legitimacy. Sharathkumar, Shetty, Prakyath, and Supriya [9] proposed a system that extracts features of the inputted URL and classifies them as a phishing site or a non-phished site using the random forest algorithm. Drury and Meyer [10] proposed email account separation as a possible approach to detect phishing emails by analyzing the collection process of email addresses. Awan [11] discussed different types of phishing attacks and various defenses against the attacks. Alabdan [12] presented a review of the approaches used during the phishing attacks and analyzed the characteristics of the existing classic, modern, and cutting-edge phishing attack techniques. Miller, Miller, Zhang, and Terwilliger [13] presented a three-pillared strategy for the prevention of phishing attacks in which the strategy is based on one-time passwords, multi-level desktop barrier applications, and behavior modification. Ustundag Soykan and Bagriyanik [14] implemented deterministic and randomized attack scenarios to demonstrate the success of the attack using a state-of-the-art simulator on the IEEE (Institute of Electrical and Electronics Engineers) European low voltage feeder test system in which authors identified threats, conducted impact analysis and estimated the likelihood of the attacks for various attacker types and motivations. Natadimadja, Abdurohman, and Nuha [15] used Hadoop (A Java-based open-source platform under apache to support applications that run on big data) and MapReduce (A programming model aimed at processing large datasets) for the detection of phishing websites. Chaudhry, Chaudhry, and Rittenhouse [16] explained various methods used in phishing attacks and pointed out the prevention from such attacks using a combination of client-side tools and server-side protection. Shaik [17] presented a counter challenge authentication method that uses a counter challenge from a user to a web application asking to provide certain information from one or more user details recorded at the time of registration.

Different authentication methods are suggested in [18–22] for additional security during the authentication in the banking systems. Aravindh, Ambeth Kumar, Harish, and Siddartth [18] used the method of pass matrix which allows the user to select an image from a set of pre-defined images in combination with the password for authentication in banking systems. Sukanya and Saravanan [19] showed a safe graphical confirmation framework named pass grid to be used for a password by the user during authentication in the banking systems. Modibbo and Aliyu [20] emphasized multifactor biometric authentication systems to transform into a cashless society and to decrease the electronic payment system fraud in the Nigerian financial service industry. Tam, Chau, Mai, Phuong, Tran, and Hanh [21] pointed out various types of cybercrimes in the banking industry of Vietnam and made preventive recommendations for commercial banks, policymakers, stakeholders, and customers for the proper mitigation of cybercrimes in the banking system. Lakshmi Prasanna, and Ramesh [22] proposed a proficient and handy client confirmation plot utilizing individual gadgets that use distinctive cryptographic natives such as encryption, computerized signature, and hashing for authentication in the Internet banking system.

Authors of [23–27] emphasized the multifactor authentication system in mobile banking systems. Aldwairi, Masri, Hassan, and ElBarachi [23] implemented a three-stage authentication system for mobile applications in which the first stage is the user-name with device serial number, the second stage is the selection of the correct square from a large grid of independent squares and the final stage is the selection of particular images in the same order as picked them during the registration. Srinivasa Rao, Deepashree, Pawaskar, Divya, and Drakshayini [24] used geolocation in addition to the existing two-factor authentication scheme using the user ID, password, and OTP for additional security in mobile banking transactions. Miiri, Kimwele, and Kennedy [25] utilized keystroke dynamics and location for authentication in the mobile banking system. Likewise, Song, Lee, Jang, Lee, and Kim [26] proposed a face recognition authentication scheme in which distance between the point of eyes, nose, and mouth from captured user's face is compared with the stored facial features and Macek, Adamovic, Milosavljevic, Jovanovic, Gnjatovic, and Trenkic [27] proposed a cryptographically secured iris biometrics for authentication in the mobile banking system.

Our paper proposes a new anti-phishing model for the mobile banking system that prevents phishing attacks in mobile banking systems at the authentication level. Banks and financial institutions can implement this model for the mitigation of enduring phishing attacks in Electronic banking (E-Banking) globally.

## 3 Proposed Anti-phishing Model for Mobile Banking System at Authentication Level

The proposed model is an **anti-phi**shing model for Mobile Banking System at the **auth**entication level (**AntiPhiMBS-Auth**). AntiPhiMBS-Auth aims to mitigate phishing attacks in the mobile banking system at the authentication level for two categories of banking users. The first category is the banking users who download the phishing app misguidedly and start using the phishing app inadvertently in place of a genuine bank app. The second category is the banking users who may receive phishing emails or phishing SMS or social media platform messages and may click the link of the phishing login interface. Phishers design the phishing app and phishing login interface looking similar to that of the banks. Banking users do not understand the phishing mechanisms and may input login credentials in the phishing app or the phishing login interface. AntiPhiMBS-Auth safeguards banking users from the phisher's use of stolen login credentials for authentication in the mobile banking system as in Fig. 1.

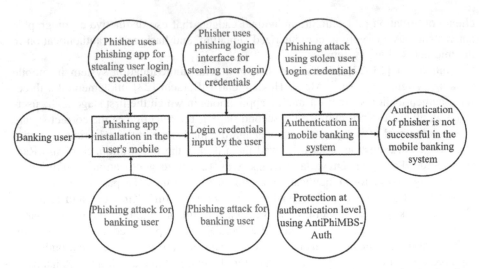

**Fig. 1.** Phishing attack protection in the mobile banking system at the authentication level

This paper proposes the architecture of the anti-phishing model AntiPhiMBS-Auth. It also verifies that AntiPhiMBS-Auth satisfies the system specification and security properties and is error-free to implement in financial institutions.

### 3.1   Architecture of Anti-phishing Model AntiPhiMBS-Auth

The architecture of the anti-phishing model AntiPhiMBS-Auth consists of the model for defending against phishing attacks in the mobile banking system at the authentication level. Participating entities in the model are mobile user, bank, bank application, mobile banking system, and phishing application. A mobile user opens an account in a bank and receives login parameters for authentication in the mobile banking system. The bank develops a bank application that communicates with the user and the mobile banking system server. The bank shares required parameters with the bank application and the

**Table 1.** Notations used in AntiPhiMBS-Auth

| Notation | Description | Notation | Description |
|---|---|---|---|
| U | Mobile User | mobNo | Mobile Number |
| B | Bank | app | Application |
| MBS | Mobile Banking System | appId | Application Id |
| uId | User id | BA | Bank App |
| lgnPwd | Login Password | PA | Phishing App |
| unqIdAuth | Unique Id for Authentication | | |

mobile banking system. The phisher might develop a phishing application for impersonating the attacks in the mobile banking system. Notations of entities and various parameters used by AntiPhiMBS-Auth are in Table 1.

### Entities and Initial Conditions for Working of AntiPhiMBS-Auth

- A mobile user (U) opens an account in the Bank (B).
- Bank provides uId, lgnPwd, and unqIdAuth to the user for authentication in the Mobile Banking System (MBS).
- Bank shares uId, lgnPwd, unqIdAuth, appId, and mobNo of each user to MBS.
- Each of the banking applications is identified by an appId, and the bank administers the database of all appId.
- Only bank and MBS know the relationship between uId, and unqIdAuth.
- Only bank, bank app, and MBS know the relationship between appId and uId.
- Users do not reveal the information provided by the bank to others.

### Model for Defending Against Phishing Attacks in the Mobile Banking System at the Authentication Level

This model proposes a unique id for authentication (unqIdAuth) and an application id (appId) system in addition to the traditional login credentials to strengthen the security for authentication in the mobile banking system. The participating entities (User, bank, bank app, and mobile banking system) of the model must follow the necessary steps for the successful operation of AntiPhiMBS-Auth to mitigate the phishing attacks in the

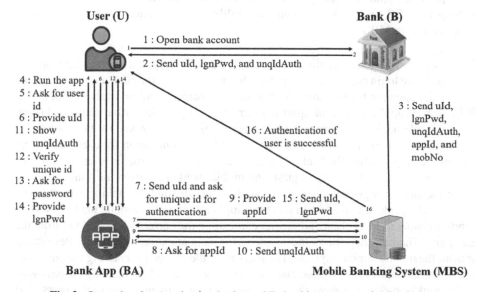

**Fig. 2.** Scenario of authentication in the mobile banking system using bank app

mobile banking system at the authentication level. The scenario of authentication in the mobile banking system using a bank app is in Fig. 2.

**Scenario of Authentication in the Mobile Banking System Using Bank App**
The following steps are necessary for authentication in MBS using a bank app.

- Step 1. A mobile user (U) opens a bank account in the Bank (B).
- Step 2. Bank sends user id (uId),login password (lgnPwd), and a unique id for authentication (unqIdAuth) to the user for authentication in the mobile banking system (MBS).
- Step 3. Bank sends user id (uId), login password (lgnPwd), unique id for authentication (unqIdAuth), application id (appId), and mobile number (mobNo) of each user to the mobile banking system (MBS).
- Step 4. A mobile user runs the installed mobile app.
- Step 5. Bank app (BA) asks for a user id from the user.
- Step 6. The user provides a uId to BA.
- Step 7. BA sends uId to MBS and requests for a unique id for authentication for uId.
- Step 8. MBS asks for an appId from BA.
- Step 9. BA provides an appId to MBS.
- Step 10. MBS searches the unique id for authentication based on the user id and sends a unique id for authentication to BA.
- Step 11. BA shows unqIdAuth to the user.
- Step 12. The user verifies the unqIdAuth.
- Step 13. BA asks for a login password from the user.
- Step 14. The user provides a lgnPwd to BA.
- Step 15. BA sends uId and lgnPwd to MBS for authentication.
- Step 16. MBS verifies the uId, lgnPwd, unqIdAuth, and authentication of the user is successful in MBS.

In the scenario of authentication in the mobile banking system using the bank app, in addition to the login credentials (user id and login password), the bank manages a unique id for authentication for each user. Furthermore, the bank also administers an application id for its bank application. Bank application knows its application id that is already known to the mobile banking system too. The mobile banking system knows the relationship among user id, login password, and a unique id for authentication for each user of the mobile banking system. The bank app asks the user to input the user id after the user runs the application. The bank app requests the mobile banking system for a unique id for authentication after getting the user id from the user. The mobile banking system wants to verify the bank app and asks to input the application id. The mobile banking system sends a unique id for authentication after confirming the correct application id from the bank app. The bank app shows the unique id for authentication to the user after getting it from the mobile banking system. The users have the choice for entering the password within the authentication process. The users have the choice of not entering the password if the bank app does not show the right unique id for authentication to the users. The user verifies the unique id and inputs the password as long as the unique id is correct. Then, the banking app requests the mobile banking system for authentication with login credentials. The mobile banking system authenticates the user for the operation of the mobile banking system if the login credentials are correct.

**Scenario of Authentication in the Mobile Banking system Using Phishing App**
Even though banking users might be knowing the correct procedure for downloading the bank app, they might receive phishing emails or phishing SMS, or phishing social media platform messages from the phishers. Users might download and start the phishing app unintentionally. Sometimes, users might click phishing links and are redirected to the phishing login screen to steal the login credentials. Thus, phishers collect the login credentials from the users either by a phishing app or by a phishing login screen. The scenario of authentication in the mobile banking system using a phishing app is in Fig. 3.

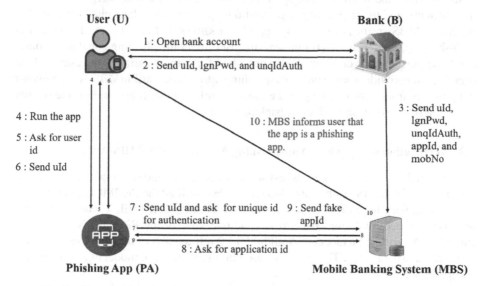

**Fig. 3.** Scenario of authentication in the mobile banking system using phishing app

The following steps may be executed in the scenario of authentication using a phishing app.

- Step 1. A mobile user (U) opens a bank account in the Bank (B).
- Step 2. Bank sends user id (uId), login password (lgnPwd), and a unique id for authentication (unqIdAuth) to the user for authentication in the mobile banking system (MBS).
- Step 3. Bank sends user id (uId), login password (lgnPwd), unique id for authentication (unqIdAuth), application id (appId), and mobile number (mobNo) of each user to the mobile banking system (MBS).
- Step 4. User may receive phishing emails or phishing SMS or phishing social networking messages with links to login into the phisher's phishing app. User may run the phishing app.
- Step 5. Phishing app asks for a user id from the user.

- Step 6. The user may provide uId to the phishing app.
- Step 7. Phishing app sends uId to the MBS and requests for a unique id for authentication for that user.
- Step 8. MBS asks for an app id to the phishing app.
- Step 9. Phishing app provides fake app id to the MBS.
- Step 10. MBS cannot find the phisher's app id in the database and informs the user that the app is a phishing app.

Banking users might use phishing apps or phishing login interface unknowingly. Phishing apps ask for a user id to fool the user by imitating the procedure of the banking app. The user may provide a user id to the phishing app. Phishing apps send valid user id to the MBS and request a unique id for authentication for that user. MBS asks for the application id to the phishing app. Phishing apps provide a fake phishing app id to the MBS. MBS verifies the phishing app id and knows that the unique id requester is a phishing app. After that, MBS informs the user about the phishing app. On the other hand, the phishing app does not show a unique id for authentication to the user, and the user does not provide a password in the phishing app. Thus, AntiPhiMBS-Auth prevents the phishers from authenticating in the mobile banking system and forbids them from doing the transactions in the mobile banking system.

## 3.2 Verification of Proposed Anti-phishing Model AntiPhiMBS-Auth

We specified the system properties and security properties and developed a verification model of AntiPhiMBS-Auth using PROMELA. We verified the PROMELA verification model employing the SPIN. This paper does not show the PROMELA codes (239 lines) because of space limitations but explains the overview of the PROMELA codes. The PROMELA verification model of AntiPhiMBS-Auth consists of the processes, message channels, and data types. The processes used in the verification model of AntiPhiMBS-Auth are in Table 2.

**Table 2.** Processes used in AntiPhiMBS-Auth

| Process name | Description |
| --- | --- |
| mobileUser | The process represents the end user of the mobile banking system |
| bank | The process represents the bank where the user opens the bank account, and it is responsible for the administration of the mobile banking system for all the banking users |
| mobileBankingSystem | The process represents the mobile banking system that offers banking services to the banking users using mobile |
| bankApp | The process represents the bank's genuine application which communicates with the user and the mobile banking system |
| phishingApp | The process represents the phisher's app which imitates the bank application and tries to fool the users to collect login credentials for performing the phishing attacks |

The processes in the PROMELA model communicate using message channels. Message channels are used to model the exchange of data between the processes. The message channels used for communication in the AntiPhiMBS-Auth model are in Table 3.

**Table 3.** Channels used in AntiPhiMBS-Auth

| Channel name | Description |
|---|---|
| mobileUser_bank | It is used to send messages from mobileUser to bank |
| bank_mobileUser | It is used to send messages from bank to mobileUser |
| bank_mobileBankingSystem | It is used to send messages from bank to mobileBankingSystem |
| mobileUser_bankApp | It is used to send messages from mobileUser to bankApp |
| bankApp_mobileUser | It is used to send messages from bankApp to mobileUser |
| bankApp_mobileBankingSystem | It is used to send messages from bankApp to mobileBankingSystem |
| mobileBankingSystem_bankApp | It is used to send messages from mobileBankingSystem to bankApp |
| mobileBankingSystem_mobileUser | It is used to send messages from mobileBankingSystem to mobileUser |
| mobileUser_phishingApp | It is used to send messages from mobileUser to phishingApp |
| phishingApp_mobileUser | It is used to send messages from phishingApp to mobileUser |
| phishingApp_mobileBankingSystem | It is used to send messages from phishingApp to mobileBankingSystem |
| mobileBankingSystem_phishingApp | It is used to send messages from mobileBankingSystem to phishingApp |

We defined the processes and message channels in the PROMELA code of AntiPhiMBS-Auth. All the processes of AntiPhiMBS-Auth communicate with each other using the above-defined message channels for the operation of AntiPhiMBS-Auth.

We also specified the following security properties using linear temporal logic (LTL) in the verification model of AntiPhiMBS-Auth.

[]((((usrId==bankUsrId)&&(lgnPwd==bankLgnPwd)&&(usrUnqIdAuth== bankUnqIdAuth))-><>(authenticationSuccess==true)).

Authentication of the user in the mobile banking system is successful only if (i) the user id provided by the user and received by MBS from the bank is equal, (ii) the login password provided by the user and received by MBS from the bank is equal, and (iii) the unique id for authentication provided by the user and received by MBS from the bank is equal.

# 4  Results and Discussion

This paper verifies the safety properties and LTL properties of the proposed model AntiPhiMBS-Auth. We accomplished experiments using SPIN Version 6.4.9 running on a computer with the following specifications: Intel® Core(TM) i5–6500 CPU@3.20 GHz, RAM 16 GB and windows10 64bit. We set advanced parameters in the SPIN environment for optimal results during the verification. We set physical memory available as 4096 (in Mbytes), maximum search depths (steps) as 1000000, estimated state space size as 1000, search mode as depth-first search (partial order reduction), and storage mode as bitstate/ supertrace for the verification. Besides, extra compile-time directives were set to DVECTORSZ as 9216 to avoid the memory error during the experiments. After that, we ran SPIN to verify the safety properties of AntiPhiMBS-Auth for up to 100 users. SPIN checked the state space for invalid end states and assertion violations during the verification of safety properties. The SPIN verification results for safety properties are in Table 4.

**Table 4.** Verification results for safety properties

| No. of users | Time (Seconds) | Memory (Mbytes) | Transitions | States stored | Depth | Safety properties verification status |
|---|---|---|---|---|---|---|
| 10 | 0.05 | 39.026 | 29786 | 1429 | 812 | Verified |
| 20 | 0.16 | 39.026 | 56906 | 2709 | 1472 | Verified |
| 30 | 0.34 | 39.026 | 84026 | 3989 | 2132 | Verified |
| 40 | 0.58 | 39.026 | 111146 | 5269 | 2792 | Verified |
| 50 | 0.88 | 39.026 | 138266 | 6549 | 3452 | Verified |
| 60 | 1.25 | 39.026 | 165386 | 7829 | 4112 | Verified |
| 70 | 1.69 | 39.026 | 192506 | 9109 | 4772 | Verified |
| 80 | 2.23 | 39.026 | 219626 | 10389 | 5432 | Verified |
| 90 | 2.72 | 39.026 | 246746 | 11669 | 6092 | Verified |
| 100 | 3.33 | 39.026 | 273866 | 12949 | 6752 | Verified |

Table 4 shows the results obtained from SPIN depicting the elapsed time, total memory usage, number of states transitioned, states stored, depth reached, and verification status for safety properties for various users. The SPIN verification results indicate that there is an unceasing rise in the verification time with the increase in the number of users, and the required memory remained constant for all the users during the verification of AntiPhiMBS-Auth. Moreover, an increasing trend is seen for the states stored, depth reached, and transitions for various users during the experiment. Also, the SPIN verification did not detect any deadlock or any errors during the runs of the AntiPhiMBS-Auth model.

After that, we executed SPIN in the same computing environment to verify the LTL properties for up to 100 users. SPIN checked the statespace for never claim and assertion violations in the run of LTL properties. The SPIN verification result for LTL properties is in Table 5.

**Table 5.** Verification results for LTL properties

| No. of users | Time (Seconds) | Memory (Mbytes) | Transitions | States stored | Depth | LTL properties verification status |
|---|---|---|---|---|---|---|
| 10 | 0.05 | 39.026 | 29786 | 1429 | 1487 | Verified |
| 20 | 0.16 | 39.026 | 56906 | 2709 | 2687 | Verified |
| 30 | 0.34 | 39.026 | 84026 | 3989 | 3887 | Verified |
| 40 | 0.58 | 39.026 | 111146 | 5269 | 5087 | Verified |
| 50 | 0.88 | 39.026 | 138266 | 6549 | 6287 | Verified |
| 60 | 1.26 | 39.026 | 165386 | 7829 | 7487 | Verified |
| 70 | 1.68 | 39.026 | 192506 | 9109 | 8687 | Verified |
| 80 | 2.18 | 39.026 | 219626 | 10389 | 9887 | Verified |
| 90 | 2.75 | 39.026 | 246746 | 11669 | 11087 | Verified |
| 100 | 3.32 | 39.026 | 273866 | 12949 | 12287 | Verified |

Table 5 depicts the results obtained from SPIN showing the elapsed time, total memory usage, states transitioned, states stored, and verification status for LTL properties for various users. The memory required for the verification of LTL properties for all the users is the same. The SPIN verification results show that there is a perpetual rise in the verification time with the increase in the number of users during the verification of LTL properties. Furthermore, there is also a continuous rise in the number of transitions, states stored, elapsed time, and depth with the increase in the number of users in the experiment. The SPIN verified the LTL properties of the AntiPhiMBS-Auth model successfully.

Table 4 shows the results after SPIN checked for the existence of deadlocks and assertion violations by generating the execution paths during the verification of the AntiPhiMBS-Auth model. Similarly, Table 5 shows the results after SPIN checked for temporal properties we expect the system behavior of the AntiPhiMBS-Auth model to conform during the system lifetime. The results of these experiments show that there is no error in the design of AntiPhiMBS-Auth. No counterexample was generated by SPIN during the experiments. Hence, the verified AntiPhiMBS-Auth is applicable for the development and implementation of the anti-phishing system within the banks and financial institutions globally to mitigate the continued phishing attacks in the mobile banking industry.

## 5  Conclusion and Future Work

The most conventional sort of phishing attack within the mobile banking industry is in the appearance of an authentication attack. Phishers employ a phishing app or phishing login interface to compile login credentials from the users and exploit the stolen credentials for authentication within the mobile banking industry. Even though credential thefts are soaring day by day, any anti-phishing model for the mitigation of such attacks has not been developed so far for the mobile banking industry. Therefore, this paper developed a new anti-phishing model for Mobile Banking System at the authentication level (AntiPhiMBS-Auth) to mitigate the phishing attacks in the mobile banking industry. A phisher might send phishing emails/SMS/social media messages to the banking users and redirect them to download the phishing app or input login credentials in the phishing login interface. Banking users might install, run, and input user id in the phishing app or the phishing login interface inadvertently. However, AntiPhiMBS-Auth applies a unique id for the authentication system, and users have the choice to not input the password without verifying the unique id for authentication. Besides, AntiPhiMBS-Auth employs an application id for the bank app so that the mobile banking system can differentiate the genuine bank app from the phishing app. Hence, AntiPhiMBS-Auth prevents phishing attacks in the mobile banking system as the phisher cannot disclose the unique id to the users, and the phishing app cannot evince its identity by rendering a genuine application id to the mobile banking system.

We observed from our experimental SPIN results of the AntiPhiMBS-Auth Promela program that the AntiPhiMBS-Auth does not encompass any deadlocks or errors within the model. Moreover, SPIN verified all the safety properties and LTL properties within the PROMELA model of AntiPhiMBS-Auth. Hence, financial institutions can implement this verified AntiPhiMBS-Auth model to mitigate the unending phishing attacks within the mobile banking industry and increase the mobile banking transactions to transform into a cashless society in this era of digital banking.

In future research, we will propose a new anti-phishing model to mitigate fraudulent transactions in the mobile banking system at the transaction level. Moreover, we will further extend the AntiPhiMBS-Auth model in mitigating phishing attacks in other digital systems utilizing login credentials for authentication.

## References

1. Tchakounte, F., Molengar, D., Ngossaha, J.M.: A description logic ontology for email phishing. Int. J. Inf. Secur. Sci. **9**(1), 44–63 (2020)
2. Subasi, A., Kremic, E.: Comparison of adaboost with multiboosting for phishing website detection. Procedia Comput. Sci. **168**, 272–278 (2020). https://doi.org/10.1016/j.procs.2020.02.251
3. Ozker, U., Sahingoz, O.K.: Content based phishing detection with machine learning. In: 2020 International Conference on Electrical Engineering (ICEE), Istanbul, Turkey, pp. 1–6. IEEE (2020). https://doi.org/10.1109/ICEE49691.2020.9249892
4. Priya, S., Selvakumar, S., Velusamy, R.L.: Detection of phishing attacks using radial basis function network trained for categorical attributes. In: 2020 11th International Conference on Computing, Communication and Networking Technologies (ICCCNT), Kharagpur, India, pp. 1–6. IEEE (2020). https://doi.org/10.1109/ICCCNT49239.2020.9225549

5. Odeh, A., Alarbi, A., Keshta, I., Abdelfattah, E.: Efficient prediction of phishing websites using multilayer perceptron (MLP). J. Theoret. Appl. Inf. Technol. **98**(16), 3353–3363 (2020)
6. Hossain, S., Sarma, D., Chakma, R.J.: Machine learning-based phishing attack detection. Int. J. Adv. Comput. Sci. Appl. **11**(9), 378–388 (2020)
7. Su, Y.: Research on website phishing detection based on LSTM RNN. In: 2020 IEEE 4th Information Technology, Networking, Electronic and Automation Control Conference (ITNEC), Chongqing, China, pp. 284–288. IEEE (2020). https://doi.org/10.1109/ITNEC48623.2020. 9084799
8. Abiodun, O., Sodiya, A.S., Kareem, S.O.: Linkcalculator – an efficient link-based phishing detection tool. Acta Informatica Malaysia **4**(2), 37–44 (2020). https://doi.org/10.26480/aim. 02.2020.37.44
9. Sharathkumar, T., Shetty, P.R., Prakyath, D., Supriya, A.V.: Phishing site detection using machine learning. Int. J. Res. Eng. Sci. Manag. **3**(6), 240–243 (2020)
10. Drury, V., Meyer, U.: No phishing with the wrong bait: reducing the phishing risk by address separation. In: 2020 IEEE European Symposium on Security and Privacy Workshops (EuroS&PW), Genoa, Italy, pp. 646–652. IEEE (2020). https://doi.org/10.1109/Eur oSPW51379.2020.00093
11. Awan, M.A.: Phishing attacks in network security. LC Int. J. STEM (Sci. Technol. Eng. Math) **1**(1), 29–33 (2020)
12. Alabdan, R.: Phishing attacks survey: types, vectors, and technical approaches. Future Internet **12**(10), 1–39 (2020). https://doi.org/10.3390/fi12100168
13. Miller, B., Miller, K., Zhang, X., Terwilliger, M.G.: Prevention of phishing attacks: a three-pillared approach. Issues Inf. Syst. **21**(2), 1–8 (2020)
14. Ustundag Soykan, E., Bagriyanik, M.: The effect of smishing attack on security of demand response programs. Energies **13**(17), 1–7 (2020). https://doi.org/10.3390/en13174542
15. Natadimadja, M.R., Abdurohman, M., Nuha, H.H.: A survey on phishing website detection using hadoop. Jurnal Informatika Universitas Pamulang **5**(3), 237–246 (2020). https://doi. org/10.32493/informatika.v5i3.6672
16. Chaudhry, J.A., Chaudhry, S.A., Rittenhouse, R.G.: Phishing attacks and defenses. Int. J. Secur. Its Appl. **10**(1), 247–256 (2016). https://doi.org/10.14257/ijsia.2016.10.1.23
17. Shaik, C.: Counter challenge authentication method: a defeating solution to phishing attacks. Int. J. Comput. Sci. Eng. Appl. **10**(1), 1–8 (2020). https://doi.org/10.5121/ijcsea.2020.10101
18. Aravindh, B., Ambeth Kumar, V.D., Harish, G., Siddartth, V.: A novel graphical authentication system for secure banking systems. In: 2017 IEEE International Conference on Smart Technologies and Management for Computing, Communication, Controls, Energy and Materials (ICSTM), Chennai, India, pp. 177–183. IEEE (2017). https://doi.org/10.1109/ICSTM. 2017.8089147
19. Sukanya, S., Saravanan, M.: Image based password authentication system for banks. In: 2017 International Conference on Information Communication and Embedded Systems (ICICES), Chennai, India, pp. 1–8. IEEE (2017). https://doi.org/10.1109/ICICES.2017.8070764
20. Modibbo, A., Aliyu, Y.: Cashless society, financial inclusion and information security in Nigeria: the case for adoption of multifactor biometric authentication. Int. J. Innov. Sci. Res. Technol. **4**(11), 872–880 (2019)
21. Tam, L.T., Chau, N.M., Mai, P.N., Phuong, N.H., Tran, V.K.H., Hanh, P.H.: Cybercrimes in the banking sector: case study of Vietnam. Int. J. Soc. Sci. Econ. Invention **6**(5), 272–277 (2020). https://doi.org/10.23958/ijssei/vol06-i05/207
22. Lakshmi Prasanna, A.V., Ramesh, A.: Secure Internet banking authentication. J. Eng. Serv. **11**(2), 152–161 (2020)
23. Aldwairi, M., Masri, R., Hassan, H., ElBarachi, M.: A novel multi-stage authentication system for mobile applications. Int. J. Comput. Sci. Inf. Secur. **14**(7), 389–396 (2016)

24. Srinivasa Rao, A.H., Deepashree, C.S., Pawaskar, D., Divya, K., Drakshayini, L.: GeoMob - a geo location based browser for secured mobile banking. Int. J. Res. Eng. Sci. Manag. **2**(5), 515–519 (2019)
25. Miiri, E.M., Kimwele, M., Kennedy, O.: Using keystroke dynamics and location verification method for mobile banking authentication. J. Inf. Eng. Appl. **8**(6), 26–36 (2018)
26. Song, J., Lee, Y.S., Jang, W., Lee, H., Kim, T.: Face recognition authentication scheme for mobile banking system. Int. J. Internet Broadcast. Commun. **8**(2), 38–42 (2016). https://doi.org/10.7236/IJIBC.2016.8.2.38
27. Macek, N., Adamovic, S., Milosavljevic, M., Jovanovic, M., Gnjatovic, M., Trenkic, B.: Mobile banking authentication based on cryptographically secured iris biometrics. Acta Polytechnica Hungarica **16**(1), 45–62 (2019)
28. Credential spill report. https://info.shapesecurity.com/rs/935-ZAM-778/images/Shape_Credential_Spill_Report_2018.pdf. Accessed 20 Nov 2020
29. 2019 Phishing trends and intelligence report. https://info.phishlabs.com/2019-pti-report-evolving-threat. Accessed 20 Nov 2020
30. 2020 phishing and fraud report. https://www.f5.com/content/dam/f5-labs-v2/article/articles/threats/22--2020-oct-dec/20201110_2020_phishing_report/F5Labs-2020-Phishing-and-Fraud-Report.pdf. Accessed 20 Nov 2020

# BU-Trace: A Permissionless Mobile System for Privacy-Preserving Intelligent Contact Tracing

Zhe Peng, Jinbin Huang, Haixin Wang, Shihao Wang, Xiaowen Chu,
Xinzhi Zhang, Li Chen, Xin Huang, Xiaoyi Fu, Yike Guo, and Jianliang Xu[✉]

Hong Kong Baptist University, Kowloon Tong, Hong Kong
{pengzhe,jbhuang,hxwang,shwang,
chxw,lichen,xinhuang,xujl}@comp.hkbu.edu.hk
{xzzhang2,xiaoyifu,yikeguo}@hkbu.edu.hk

**Abstract.** The coronavirus disease 2019 (COVID-19) pandemic has caused an unprecedented health crisis for the global. Digital contact tracing, as a transmission intervention measure, has shown its effectiveness on pandemic control. Despite intensive research on digital contact tracing, existing solutions can hardly meet users' requirements on privacy and convenience. In this paper, we propose BU-Trace, a novel permissionless mobile system for privacy-preserving intelligent contact tracing based on QR code and NFC technologies. First, a user study is conducted to investigate and quantify the user acceptance of a mobile contact tracing system. Second, a decentralized system is proposed to enable contact tracing while protecting user privacy. Third, an intelligent behavior detection algorithm is designed to ease the use of our system. We implement BU-Trace and conduct extensive experiments in several real-world scenarios. The experimental results show that BU-Trace achieves a privacy-preserving and intelligent mobile system for contact tracing without requesting location or other privacy-related permissions.

**Keywords:** Privacy-preserving · Permissionless · Intelligent · Contact tracing

## 1 Introduction

The eruption of the COVID-19 pandemic has drastically reconstructed the normality across the globe. Tracing and quarantine of close contacts is an important and effective non-pharmaceutical intervention (NPI) for reducing the transmission of COVID-19 [14]. A recent study shows that the COVID-19 pandemic can be stopped if 60% of the close contacts can be immediately identified, in particular combined with other preventive measures such as social distancing [13]. However, traditional manual contact tracing is inefficient because it is limited by a person's ability to recall all the close contacts since infection and the time it takes to reach these contacts. Thus, in a world coexisting with the infectious coronavirus, an effective and secure digital contact tracing system is much

© Springer Nature Switzerland AG 2021
C. S. Jensen et al. (Eds.): DASFAA 2021 Workshops, LNCS 12680, pp. 381–397, 2021.
https://doi.org/10.1007/978-3-030-73216-5_26

desired. With such a system, identified close contacts could be provided with early quarantine, diagnosis, and treatment to track and curb the spread of the virus.

There are several challenges in building an effective digital contact tracing system. First, user privacy protection, especially during contact data collection and processing, is known to have a significant impact on the uptake of such a system [18,27]. Thus, assuring user privacy should be the very first requirement for digital contact tracing. Most existing contact tracing systems [1,7,10,17,20] are designed with a centralized model, where personal data is uploaded to a central server for contact matching. However, in such systems, user privacy could be compromised and system security is not guaranteed. Based on the collected personal data (e.g., identities, locations, etc.), the server might be able to infer knowledge pertaining to users' interests. Moreover, the server is a valuable target for malicious attackers, which may result in serious data breach and leakage.

Second, to fully eliminate concerns about user privacy, contact tracing should be conducted without collecting user location data during normal operations. Even better, the contact tracing mobile app should be *permissionless*, i.e., neither accessing any restricted data nor performing any restricted action that requires users' location permissions. However, most prior solutions either need to access location data such as GPS and e-payment transaction records [1,17,20] or leverage Bluetooth for decentralized contract tracing [3–6,8]. As Bluetooth can be used to gather information about the location of a user, the use of Bluetooth must explicitly request the location permission on both Android and iOS platforms, which could affect users' willingness to install the mobile app and participate in contact tracing.

Third, making the contact tracing system *intelligent* is crucial to enhance user experience. For example, when a user leaves a venue, a check-out reminder would be automatically displayed on the screen to remind the recording of the leaving event. Moreover, to preserve data privacy and reduce privacy worries, the intelligent algorithm should be localized and realized merely based on permissionless data obtained from the local mobile phone. In other words, advanced sensor data such as step counts cannot be utilized because access to them requires privacy-related permissions, namely the *Activity Recognition* permission on Android and the *Health Data* or *Motion Data* permission on iOS.

**Contributions.** To meet these challenges, we propose BU-Trace, a permissionless mobile contact tracing system based on QR code and Near-Field Communication (NFC) technologies, which simultaneously offers user privacy protection and system intelligence. Concretely, we make the following major contributions in this paper.

- We conduct a user study to investigate and quantify the user acceptance of a mobile contact tracing system. Specifically, anonymous participants are provided with qualitative virtual focus group interviews and quantitative surveys. Based on the tailored investigation, we find two most desired properties of the system, *i.e.*, privacy and convenience. With these user study results, we are inspired and motivated to develop a privacy-preserving intelligent contact tracing system.

- We propose a decentralized system to enable contact tracing while protecting user privacy. Compared with other systems, BU-Trace leverages QR code and NFC technologies to record users' venue check-in events, which requires no location access permission at all. Moreover, the system enables participants to confidentially conduct contact matching on local mobile phones based on historical venue check-in records, *i.e.*, no local data is ever exposed during the record matching phase.

- We design an intelligent behavior detection algorithm to ease the use of the system. Our algorithm is (i) *automatic*, automatically recognizing the user's movement behavior to display a check-out reminder for recording the leaving event; and (ii) *localized*, leveraging the mobile phone's internal accelerator data for local behavior detection.

- We demonstrate the practicability of BU-Trace by fully implementing a system using readily-available infrastructural primitives. We also evaluate the effectiveness of our proposed intelligent behavior detection algorithm within two practical scenarios and hundreds of collected data records. The experimental results show that BU-Trace achieves a privacy-preserving and intelligent mobile system for contact tracing without requesting location or other privacy-related permissions.

## 2    Related Work

### 2.1    Contact Tracing

Existing approaches for contact tracing utilize various technologies, such as GPS, GSM, Bluetooth, and QR code, to decide a person's absolute and relative location with others. The current contact tracing systems can be classified into *centralized* and *decentralized* models based on their architectures. In centralized systems, users are required to upload their data to a central server, which is usually supported by the government authority [1,7,10,17,20]. The server will maintain the user data and perform data matching. However, user privacy cannot be guaranteed since the user data will be collected and uploaded to the central server. Oppositely, in decentralized systems, all data will be collected and stored in users' local devices. Existing decentralized systems mainly utilize the Bluetooth technology to determine the relative distances among users [3–6,8]. However, all these approaches need to request location access permission from the mobile operation system. Concerning user privacy, users may not be willing to install and use these mobile apps, which makes the contact tracing ineffective. For example, TraceTogether [10], the Bluetooth-based contact tracing app from Singapore, has only 1.4M active users (25% of population) after more than five months of release [9]. For a more comprehensive comparison, we summarize and contrast the features of various contact tracing systems in Table 1. Different from existing systems, our proposed BU-Trace in this paper leverages QR code and NFC technologies, which are permissionless. Moreover, the system simultaneously enables user privacy protection and system intelligence.

**Table 1.** Comparison with global contact tracing apps.

| Country/Region | Approach | Tracing technology | Privacy protection | No location permission needed | System intelligence |
|---|---|---|---|---|---|
| Mainland China (Health Code System) | Centralized | QR Code, GSM, e-Payment Transactions | ✗ | ✗ | ✗ |
| South Korea (Contact Tracing System) | Centralized | GPS | ✗ | ✗ | ✗ |
| India (Aarogya Setu) | Centralized | GPS, Bluetooth | ✗ | ✗ | ✗ |
| Singapore (SafeEntry) | Centralized | QR Code | ✗ | ✓ | ✗ |
| Singapore (TraceTogether) | Centralized | Bluetooth | ✗ | ✗ | ✗ |
| Australia (CovidSafe) Canada (COVID Shield) German (Corona-Warn) Switzerland (SwissCovid) Google and Apple (ENS) | Decentralized | Bluetooth | ✓ | ✗ | ✗ |
| **BU-Trace** | **Decentralized** | **QR Code, NFC** | ✓ | ✓ | ✓ |

## 2.2   Sensor Data Analysis

Many new types of sensors have been equipped with modern mobile phones, which can produce abundant sensing data about environment and human. In order to effectively analyze the sensor data to obtain valuable information, different methods have been proposed [15,21,26]. Specifically, in many human-centric intelligent applications, it is very significant to accurately recognize various user behaviors and motions from the sensor data. Among existing techniques, machine learning is proven to be a promising solution for processing time sequence data [24]. Many learning-based methods, such as recurrent neural networks (RNN) and long short term memory (LSTM), have been developed and applied in real-world scenarios successfully [16,19]. Santos et al. proposed a method to model human motions from inertial sensor data [23]. Shoaib et al. implemented a smartphone-based data fusion technique for detecting various physical activities [25]. Bedogni et al. came up with a solution to detect users' motion types in realtime by using sensor data collected from smartphones [11]. In addition, some studies have attempted to leverage sensor data to detect users' transportation modes. Fang et al. proposed a new method to classify different transportation modes by analyzing sensor data from multiple sources [12]. In order to effectively recognize various human activities, most existing approaches need to collect abundant sensor data and conduct computation on a powerful server. In contrast, considering user privacy, our approach will conduct sensor data analysis on local mobile phones.

# 3    User Study

With the voluntary principle, in order for BU-Trace to be widely adopted in communities to break the virus transmission chain, the developed approach should be able to effectively satisfy users' demands and concerns. Therefore, we start by understanding users' perception and acceptance of a mobile contact tracing system.

## 3.1    Method

A user study involving quantitative survey and qualitative focus group interviews were carried out in September 2020. We recruited 20 participants (8 male, 12 female, aged 18–55 years) via the purposive sampling method. Invitation letters for joining the study were sent to the Deans and Department Heads within a public university in Hong Kong, asking the Deans and Department Head to nominate three representatives from the Faculty. The participants included 8 university students and 12 staffs from five different Schools and Faculties, covering a wide range of disciplines (including science, social science, arts, and humanities).

The user study was conducted online. The participants received a link directed to the quantitative questionnaire. After finishing the questionnaire, they were asked to join a virtual focus group hosted on Slack. The participants were allocated into four different discussion channels (each channel included four to five participants) and could share their perceptions and concerns on a mobile contact tracing system, and discuss with other participants in an asynchronous manner.

The user study mainly focused on examining the factors that may make people adopt or resist the mobile contact tracing application. A total of 18 filter questions were settled, which meant different questions were displayed to the participants based on their answer of the current question. For all the quantitative questions, five-point Likert scales were used where 1 meant the most disagreement, and 5 meant the most agreement.

## 3.2    Results

To comprehensively investigate the user acceptance of a mobile contact tracing system, participants were divided into two groups based on whether they hold a positive or negative view towards the system. Concretely, 13 out of 20 participants (called *positive participants*) indicated that they would like to install and use such a mobile app for getting virus risk alerts, while others (called *negative participants*) indicated that they would not.

For positive participants, they were first requested to input the most important factors that make them use such a mobile app. Among their answers, the factors frequently mentioned include getting alerts, receiving latest news about the pandemic, and self-protection. In addition, participants were asked to input factors that might make them stop using the app. Privacy concern and inconvenience were two mentioned factors. Furthermore, quantitative questions were provided to evaluate the reasons of stopping using the app. As shown in Table 2, both privacy and convenience are crucial factors (mean values are larger than 3). Based on the survey for positive participants, we find that interviewees hope to receive timely infection risk alerts from the app, but they have certain privacy and inconvenience concerns.

**Table 2.** For positive participants (n = 13), the reasons of stopping using the app (1 = the most disagree; 5 = the most agree).

|  | Mean | SD | 1 | 2 | 3 | 4 | 5 |
|---|---|---|---|---|---|---|---|
| If privacy leakage is found | **4.23** | 1.09 | 0 | 1 | 3 | 1 | 8 |
| If the app is not convenient to use | **3.85** | 0.80 | 0 | 0 | 5 | 5 | 3 |

**Table 3.** For negative participants (n = 7) subjects, the reasons of changing the idea of the app usage (1 = the most disagree; 5 = the most agree).

|  | Mean | SD | 1 | 2 | 3 | 4 | 5 |
|---|---|---|---|---|---|---|---|
| If privacy is completely protected | **3.14** | 0.38 | 0 | 0 | 6 | 1 | 0 |
| If the app is convenient to use | **3.29** | 0.76 | 0 | 1 | 3 | 3 | 0 |

For negative participants, they were first requested to grade two scenarios where they might change the idea of using such an app. As shown in Table 3, privacy concern and inconvenience related conditions both received higher grades. After that, participants were invited to input some suggestions to increase the adoption rate of the app. Participants express very strong and salient opinions on the privacy issue and data usage. Some incentive mechanisms were also proposed, such as free drink or food coupons, free face masks, and free COVID-19 testing once getting a virus risk alert. Based on this survey, we find that privacy concern and inconvenience are also important concerns for negative interviewees.

Finally, for all participants, we provided some quantitative questions about options of permitting the app to check in and check out a venue, e.g., the canteen. For the check-in scenario, users can 1) scan a QR code every time when entering a venue (this option does not require the permission to access the user's location) or 2) use Bluetooth for auto check-in (this option requires authorizing the location permission, though the app does not collect the user's location data). As shown in Table 4, users preferred to scan a QR code every time instead of using Bluetooth for auto check-in. For the check-out scenario, besides the QR code and Bluetooth options, another option was provided, i.e., using intelligent technologies for auto check-out (this option does not require the permission to access the user's location). As shown in Table 4, participants preferred to choose

**Table 4.** Evaluation on options of permitting the app to check in and check out a venue, for all the participants (n = 20) (1 = the most disagree; 5 = the most agree).

|  |  | Mean | SD | 1 | 2 | 3 | 4 | 5 |
|---|---|---|---|---|---|---|---|---|
| Check In | Scan the QR code every time (no location access) | **3.55** | 1.15 | 1 | 3 | 4 | 8 | 4 |
|  | Bluetooth for auto check-in (need location access) | **2.95** | 1.23 | 2 | 6 | 6 | 3 | 3 |
| Check Out | Scan the QR code every time (no location access) | **3.15** | 1.23 | 2 | 4 | 6 | 5 | 3 |
|  | Bluetooth for auto check-in (need location access) | **3.05** | 1.32 | 3 | 4 | 5 | 5 | 3 |
|  | AI for auto check-out (no location access) | **3.40** | 1.39 | 3 | 2 | 4 | 6 | 5 |

intelligent technologies to estimate the duration of the stay in a particular venue and realise the auto check-out.

Therefore, based on the tailored investigation, we find two most desired properties of a contact tracing system, i.e., privacy and convenience. With these user study results, we are inspired and motivated to develop an effective contact tracing system.

## 4    System Overview

BU-Trace is a permissionless mobile system for privacy-preserving intelligent contact tracing. The system will send an alert message to users through a mobile app if they and an infected person have visited the same place within a time period that gives rise to risks of exposure. Figure 1 shows the BU-Trace system architecture. Our system mainly consists of the following actors: (i) *patient*, (ii) *client*, and (iii) *authority*. A patient is a virus-infected person, while a client refers to an unconfirmed person. An authority represents a government sector or an organization, which can provide a close proximity certification for further screening. In the following, we briefly describe three basic modules of the system.

*Permissionless Location Data Collection.* BU-Trace utilizes QR code and NFC technologies to record users' venue check-in information. When users scan the system's QR code or NFC tag before entering a venue, the venue ID is collected. Then, the collected venue ID, as well as the time of the visit, are saved on users' mobile phones. Different from existing systems, our app does not request the location permission from the mobile platforms but only a camera usage notification (no location access) will be displayed on the screen during the first attempt, which safeguards user privacy from the system level.

*Privacy-Preserving Contact Tracing.* The contact tracing module is designed as a decentralized approach to protect user privacy. In this module, confirmed patients need to upload their venue visit records within the past 14 days to the authority. Specifically, to enhance the security, the hashed value of a venue ID instead of the plain text is transmitted to the authority. The authority will further broadcast the encrypted venue records uploaded by confirmed patients to other clients. Upon receiving the encrypted venue records, all clients apply the same hash function to their own venue records and conduct a cross-check on local mobile phones. If a match is found, the app will display an alert notification and the client could report his/her case to the authority for further follow-up. In the whole process, both data storage and data computation are conducted in a decentralized manner to protect user privacy and make the system more scalable.

*Intelligent Behavior Detection.* We design an automatic check-out function based on an intelligent behavior detection module, which improves not only user experience but also time accuracy of check-out records. To avoid the location permission request and alleviate users' privacy concerns, the intelligent behavior detection module utilizes only inertial sensor data from the mobile phone's accelerometer. We design a simple yet effective sliding window-based detection algorithm to detect the behavior transition for auto check-out reminders. Overall, the data analysis procedure is strictly restrained in local mobile phones based on sensor data that does not request users' location or any other privacy-related permissions. We give more details about the intelligent behavior detection method in the next section.

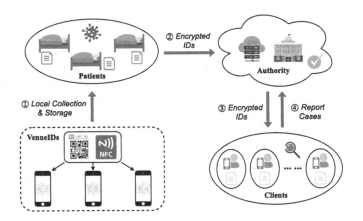

**Fig. 1.** The system architecture of BU-Trace

# 5    Intelligent Behavior Detection

For effective contact tracing, both venue check-in and check-out events should be precisely recorded. BU-Trace enables users to record the check-in event via scanning the QR code or tapping the NFC tag when they enter a venue. However, our pilot experiments showed that users could easily forget to record the check-out event when they left the venue. Therefore, we design an intelligent behavior detection method to facilitate the recording of check-out events.

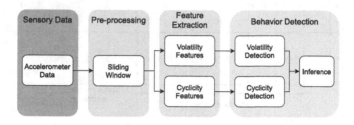

**Fig. 2.** The framework of accelerometer-based intelligent behavior detection

## 5.1 Accelerometer-Based Behavior Detection Framework

In this section, we introduce our accelerometer-based intelligent behavior detection framework for auto check-out reminders in BU-Trace, as illustrated in Fig. 2.

Our hypothesis is that there is usually a behavior transition when a user is leaving a venue (e.g., taxi, train, restaurant, theatre, etc.). BU-Trace monitors the accelerometer data in three orthotropic directions (i.e., X, Y, and Z axes, denoted as $\mathbf{A_x}$, $\mathbf{A_y}$, and $\mathbf{A_z}$, respectively) and intelligently infers the check-out event by detecting such a transition. To reduce the power consumption of BU-Trace, we design a simple yet effective sliding window-based algorithm to detect the behavior transition. As shown in Fig. 3, the original sensor data is separated into different windows (denoted by dashed line) with a window width $l$. For each window, taking the taxi ride as an example, it will be classified into two categories with a behavior detection algorithm, i.e., on-taxi window and off-taxi window. As such, the inferred check-out time will be located in the first recognized off-taxi window (highlighted by red dashed rectangle).

## 5.2 Behavior Detection Algorithm

Algorithm 1 describes the procedure of the behavior detection algorithm. The designed algorithm mainly consists of two behavior detection methods (i.e., volatility detection and cyclicity detection) in terms of (i) *walking volatility* and (ii) *walking cyclicity*, respectively. Then, the final behavior detection result is jointly decided by the two methods.

**Volatility Detection.** In the volatility detection method, we aim to identify the change of behavior patterns caused by the check-out action. In many practical scenarios, the check-out action happens in a short period and is followed by continuous walking for a period of time (e.g., getting off a taxi, or leaving a

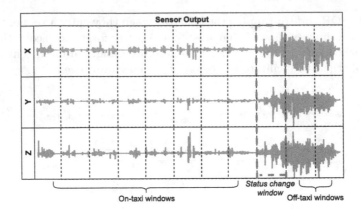

**Fig. 3.** Sensor data pre-processing based on sliding window (Color figure online)

restaurant). These actions will cause a distinct change of sensor readings in the time domain. With this intuition, we propose a simplified polynomial logistic regression algorithm to classify a time window. First, we use powered average to intensify the features (line **8** in Algorithm 1). For a specific window, the powered average $\bar{\mathbf{P}}$ of accelerometer readings in the three axes is calculated as

$$\bar{\mathbf{P}} = \frac{1}{l} \sum_{i=1}^{l} [(A_x^i)^k + (A_y^i)^k + (A_z^i)^k], \tag{1}$$

where $l$ is the window width, $k$ is an even integer hyperparameter, $A_x^i$ (resp. $A_y^i$, $A_z^i$) is the $i$th accelerometer reading in the window along the X (resp. Y, Z) axis. Specifically, $k$ is constrained to be even in order to generate positive values. Through extensive experiments, we find that $k = 4$ is a good choice in practice. Then, we identify the check-out window if $\bar{\mathbf{P}} > \mathbf{L_a}$ (line **12** in Algorithm 1), where $\mathbf{L_a}$ is a threshold learnt from our collected training data.

**Cyclicity Detection.** In the cyclicity detection method, we aim to recognize the unique and inherent behavior pattern of human walking. Based on our observation, the accelerometer readings of human walking usually present cyclicity. Thus, leveraging this property, we can recognize the walking windows for the check-out event through accurately capturing the periodic crests. With this intuition, we first strengthen the cyclicity features through the $k$th power of the original accelerometer data. Then, in order to reduce the influence of unintentional shakes, we further use wavelet transform [22] to filter the false crests in the frequency domain (line **10** in Algorithm 1). Specifically, for the accelerometer readings in each window $A_{win}$, we process the $k$th power sequence $A_{win}^k$ through

$$\mathbf{W_A} = \frac{1}{\sqrt{|a|}} \sum_i A_{win}^k(i) \overline{\Psi \left( \frac{t-b}{a} \right)}, \tag{2}$$

where $a = a_0^m$ ($a_0 > 1$, $m \in Z$) is the scale parameter, $b = n \cdot a_0^m \cdot b_0$ ($b_0 > 0$, $n \in Z$) is the translation parameter, $\overline{\Psi(t)}$ is the analyzing wavelet with a conjugate

---

**Algorithm 1:** Sliding Window-based Joint Detection Algorithm

---

**Input:** Accerelometer readings $A$ containing $A_x$, $A_y$, $A_z$;
Sliding window width $l$;
Powered average threshold $L_a$;
Crest amount threshold $L_p$;
Continuous check-out window amount threshold $L_w$;
**Output:** A boolean value $T$ indicating whether the user has checked out

1   **while** *there is input $A$* **do**
2     Window count $C_w = 0$;
3     Empty window $A_{win} = \{\}$;
4     **while** $|A_{win}| < l$ **do**
5       |   $A_{win}$.append($A$);
6     **end**
7     **for** $A_i$ in $A_{win}$ **do**
8       |   Calculate Powered Average $\bar{P} \mathrel{+}= \frac{(A_x^i)^k + (A_y^i)^k + (A_z^i)^k}{|A_{win}|}$;
9     **end**
10    $W_A \leftarrow$ WaveletTransform($A_{win}^k$);
11    $P_c \leftarrow$ FindCrests($W_A$);
12    **if** $\bar{P} > L_a$ && $P_c > L_p$ **then**
13      |   $C_w \mathrel{+}= 1$ ;
14    **else**
15      |   $C_w = 0$ ;
16    **if** $C_w > L_w$ **then**
17      |   return True;
18 **end**

---

operation. Moreover, based on the output $\mathbf{W_A}$, we detect the wave crests and count the amount $\mathbf{P_c}$ with the constraints of peak value $p_v$ and peak interval $p_i$ (line **11** in Algorithm 1). Finally, a threshold $\mathbf{L_p}$ for the amount of crests, which is learnt from training data, is used to classify the window. If $\mathbf{P_c} > \mathbf{L_p}$, this window will be categorized as the walking window (line **12** in Algorithm 1).

**Joint Detection.** To maximize the detection accuracy of check-out events, we combine the volatility detection and cyclicity detection methods. For each window of accelerometer readings, it will be finally inferred as a check-out window, only if both methods indicate the positive. Thus, in the sliding window framework, a boolean sequence $\mathcal{S}$ will be derived for each accelerometer data record, for example, $\mathcal{S} = \{\ldots, in, in, in, out, out, out, \ldots\}$. Our joint detection algorithm then uses a variable $\mathbf{C_w}$ to record the number of continuous check-out windows in the boolean sequence $\mathcal{S}$ (line **13** in Algorithm 1). Finally, once $\mathbf{C_w} > \mathbf{L_w}$ (line **16** in Algorithm 1), where $\mathbf{L_w}$ is an indicator threshold to infer the check-out event, a check-out reminder will be triggered in the mobile app.

Using the training data collected in the real-world, we employ grid searches to automatically find the optimal values of the hyper-parameters and thresholds used in the behavior detection algorithm.

# 6    Implementation and Evaluation

## 6.1    Experimental Setup

We implement the BU-Trace mobile app with Java and the back-end system with PHP. Four models of smartphones are used for evaluation, including Samsung A715F, Samsung A2070, Xiaomi 10 Lite 5G, and OPPO Reno4 Pro 5G. We conduct experiments in two representative real-world scenarios, i.e., taxi and canteen. Specifically, 156 and 110 accelerometer data records are collected in the taxi scenario and the canteen scenario, respectively. The data record length ranges from 1 min 48 s to 39 min 40 s. The sampling frequency of the accelerometer sensor is set 50 Hz.

To evaluate our proposed intelligent behavior detection algorithm, we use the collected datasets to measure its effectiveness with regard to different parameters, compare the performance with other methods, and test the power computation of our mobile app. Specifically, three Long Short-Term Memory (LSTM) [16] based methods are also implemented to compare the detection accuracy. In addition, we present the real-world deployment of our system.

## 6.2    Performance Evaluation

**Effectiveness Evaluation.** In this experiment, we evaluate the effectiveness of the intelligent behavior detection module on the whole taxi dataset and canteen dataset, respectively. We evaluate the effectiveness from two aspects, i.e., detection accuracy (ACC) and false positive rate (FPR). The ACC is the proportion of correct inferences for both check-out events and non-check-out events among the total number of records in the dataset. The FPR is calculated as the ratio between the number of check-out events wrongly detected in the in-venue windows (false positives) and the total number of records in the dataset.

Three algorithmic parameters, including window width $l$, # continuous windows $\mathbf{L_w}$, and powered average threshold $\mathbf{L_a}$, are evaluated on their influences on the detection performance. In the experiment, we first use grid searches to find the optimal values of these three parameters as their initial settings. Then, we measure the effectiveness by varying the setting of each parameter separately. Additionally, we compare our proposed joint detection method with only volatility detection (VD) and only cyclicity detection (CD).

Figure 4 (a) shows the ACC and FPR results of the three methods with regard to window width $l$ under the taxi scenario. In general, under various settings of $l$, our method can always have a higher ACC and a lower FPR, demonstrating the effectiveness of combining the volatility detection and cyclicity detection. If $l$ is set at $2s$ (the optimal value), our joint detection method can achieve the best performance with 85.26% ACC and 9.62% FPR. As a contrast, both the other two methods have worse performance with 78.21% ACC & 16.67% FPR and 15.38% ACC & 0.00% FPR, respectively. From the results, we can also find the accuracy of the joint method will be decreased with a longer window width.

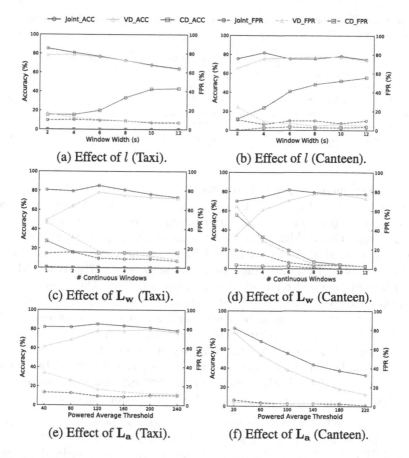

**Fig. 4.** Effectiveness evaluation of intelligent behavior detection algorithms.

This is mainly because the volatility feature becomes indiscriminative with a longer period in the taxi scenario.

Figure 4 (c) plots the ACC and FPR results with regard to # continuous windows $L_w$. As can be seen, the best performance is achieved when three continuous check-out windows are used to detect check-out events. Due to the inherent volatility of taxi riding, more false positive detection cases emerge with fewer windows, resulting in a lower ACC and a higher FPR. Moreover, benefited from the cyclicity detection, more efficient detection could be achieved with $2s \times 3$ *windows* $= 6s$, as compared to the case of only using volatility detection with $4s \times 3$ *windows* $= 12s$.

Figure 4 (e) shows the ACC and FPR results with regard to the powered average threshold $L_a$ used in the volatility detection. When the threshold is set at 120, the highest accuracy and a relatively low FPR can be achieved. The experiment results also show that our joint detection method could perform better than the volatility detection through jointly considering the cyclicity of human walking patterns.

**Table 5.** Overall performance comparison between different methods on both taxi and canteen dataset.

|         |                         | Status Change | Current Status | Current Status (Balanced) | Ours |
|---------|-------------------------|---------------|----------------|---------------------------|--------|
| Taxi    | Classification Accuracy  | Overfitting   | Overfitting    | 57.17%                    | –      |
|         | Classification Loss      | 0.0609        | 0.1559         | 0.1623                    | –      |
|         | Check-out Accuracy       | 0.00%         | 0.00%          | 40.61%                    | **90.91%** |
| Canteen | Classification Accuracy  | Overfitting   | Overfitting    | 87.99%                    | –      |
|         | Classification Loss      | 0.0364        | 0.0579         | 0.0722                    | –      |
|         | Check-out Accuracy       | 0.00%         | 0.00%          | 71.43%                    | **81.14%** |

Next, we evaluate the performance for the canteen scenario. The experiment results are shown in Fig. 4 (b), (d), and (f). The optimal values for window width $l$, # continuous windows $\mathbf{L_w}$, and powered average threshold $\mathbf{L_a}$ are 4$s$, 6, and 20, respectively. With these parameters, our method can achieve the best performance (i.e., 81.82% ACC and 6.36% FPR) on the whole canteen dataset. In contrast, the optimal volatility detection only achieves 77.27% ACC and 4.55% FPR with 8$s$ window width and eight continuous windows. These results suggest that in the canteen scenario, check-out events could be detected more efficiently with our joint method. This is because users' movement in canteens tends to be gentle and the behavior pattern change during the check-out is distinct for detection. Thus, benefiting from the cyclicity detection, the detection time of our method can be reduced compared with the volatility detection. In summary, the experiment results in both scenarios show that our joint method can always achieve higher effectiveness compared with the other non-joint methods.

**Comparison with LSTM.** To further evaluate our proposed algorithm, we compare the detection performance with three methods based on LSTM [16]. For the LSTM based methods, we randomly extract two thirds of the whole dataset as the training set, and the rest as the testing set. After training the parameters on the training set, we measure the algorithm performance on the testing set with the generated optimal parameter values. Each accelerometer data record is divided by a sliding window with 5$s$ width and further fed into the model. Three methods are tested, including detecting the status change window (SC), detecting the current status of the window (CS), and detecting the current status of the window with balanced data (CSB). In the SC method, windows are categorized into two classes, i.e., the status change window and the status unchange window. The CS method aims at recognizing two types of windows in terms of the current status, including the in-venue window and the out-venue window. For the CSB method, considering the unbalanced sample quantity, we manually synthesize data records by concatenating more out-venue windows and then classify windows as the CS method.

As shown in Table 5, our proposed approach outperforms the LSTM-based methods in both scenarios. The joint detection method achieves a higher check-out event detection accuracy on the testing dataset, because many discriminative features are successfully extracted and recognized. In contrast, the LSTM-based methods cannot learn effective features from the unbalanced dataset or the size-

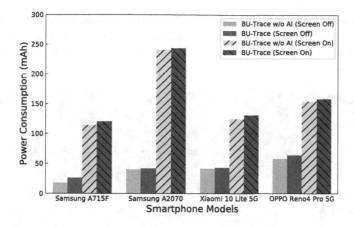

**Fig. 5.** Power consumption of BU-Trace

limited dataset, resulting in overfitting or lower accuracy. Specifically, for the SC and CS methods, the quantity of status change windows and out-venue windows in the dataset are both very limited. In our experiments, the overfitting is observed, where the two methods cannot converge and only output the dominant class (i.e., the status unchange window and the in-venue window, respectively). Thus, for the two methods, the check-out accuracy is only 0.00% for both of our datasets (without synthesized data records).

**Power Consumption.** In this experiment, we test the power consumption of our system on four models of mobile phones. Four testing cases are considered, i.e., (i) BU-Trace without intelligent detection (screen off), (ii) complete BU-Trace (screen off), (iii) BU-Trace without intelligent detection (screen on), and (iv) complete BU-Trace (screen on). For each case, we conduct the test for one hour to obtain more stable results. Figure 5 shows the results. When the phones are in screen-off status, the intelligent auto check-out module consumes 4.32 mAh– of extra power in one hour on average. In other words, the average extra power consumed by the intelligence module is 10.96% of the consumed power by working phones without this module. When the phones are in screen-on status, the power consumed by screen display dominates. The average power consumed by the intelligence module is 3.08% of that by working phones without this module. All these results indicate that our BU-Trace system does not consume much power during user behaviour monitoring.

**System Deployment.** Our developed BU-Trace system has been deployed on the university campus since September 2020 [2]. The mobile app is available for download from both Google Play (for Android users) and Apple's App Store (for iOS users). Students and staff can install and use the mobile app without enabling the location permission. Figure 6(a) shows the main interface of the BU-Trace app. Figure 6(b) shows some photos of the tailored QR code and NFC tag at a venue entrance. To effectively realize contact tracing, over 200 QR codes and NFC tags have been deployed at selected venues, such as canteens, libraries, labs, and meeting rooms. The system has been successful in tracing the contacts of an infected student and protecting the safety of the campus community.

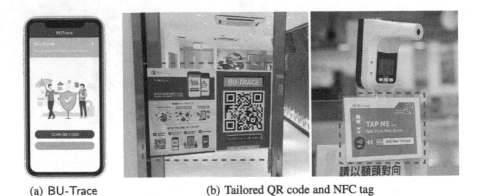

(a) BU-Trace                    (b) Tailored QR code and NFC tag

**Fig. 6.** BU-Trace deployment

## 7    Conclusion

In this paper, we propose BU-Trace, a permissionless mobile system for contact tracing based on QR code and NFC technologies. Compared with previous works, BU-Trace offers user privacy protection and system intelligence without requesting location or other privacy-related permissions. To realize this system, a user study is first conducted to investigate and quantify the user acceptance of a mobile contact tracing system. Then, a decentralized system is proposed to enable contact tracing while protecting user privacy. Finally, an intelligent behavior detection algorithm is developed to ease the use of our system. We implement BU-Trace and conduct extensive experiments in practical scenarios. The evaluation results demonstrate the effectiveness of our proposed permissionless mobile system.

**Acknowledgement.** This research is supported by a strategic development grant from Hong Kong Baptist University.

## References

1. Aarogya Setu. https://www.mygov.in/aarogya-setu-app/
2. BU-Trace. https://butrace.hkbu.edu.hk/
3. Corona-Warn. https://www.bundesregierung.de/breg-de/themen/corona-warn-app/corona-warn-app-englisch
4. COVID Shield. https://www.nhsinform.scot/illnesses-and-conditions/infections-and-poisoning/coronavirus-covid-19/coronavirus-covid-19-shielding

5. COVIDSafe. https://www.health.gov.au/resources/apps-and-tools/covidsafe-app
6. Exposure Notifications. https://www.google.com/covid19/exposurenotifications/
7. SafeEntry. https://www.safeentry.gov.sg
8. SwissCovid. https://www.bag.admin.ch/bag/en/home/krankheiten/ausbrueche-epidemien-pandemien/aktuelle-ausbrueche-epidemien/novel-cov/swisscovid-app-und-contact-tracing.html
9. TraceTogether. https://www.channelnewsasia.com/news/singapore/covid-19-singapore-low-community-prevalence-testing-13083194
10. Bay, J., Kek, J., Tan, A., Hau, C.S., Yongquan, L., et al.: BlueTrace: a privacy-preserving protocol for community-driven contact tracing across borders. Technical report, Government Technology Agency-Singapore (2020)
11. Bedogni, L., Di Felice, M., Bononi, L.: By train or by car? Detecting the user's motion type through smartphone sensors data. In: IEEE Wireless Days (2012)
12. Fang, S.H., et al.: Transportation modes classification using sensors on smartphones. Sensors 16(8), 1324 (2016)
13. Ferretti, L., Wymant, C., Kendall, M., Zhao, L., et al.: Quantifying SARS-CoV-2 transmission suggests epidemic control with digital contact tracing. Science 368(6491) (2020)
14. Flaxman, S., Mishra, S., Gandy, A., Unwin, H.J.T., et al.: Estimating the effects of non-pharmaceutical interventions on Covid-19 in Europe. Nature 584(7820), 257–261 (2020)
15. Gonzalez, J.A., Cheah, L.A., et al.: Direct speech reconstruction from articulatory sensor data by machine learning. IEEE/ACM Trans. ASLP 25(12), 2362–2374 (2017)
16. Hochreiter, S., et al.: Long short-term memory. Neural Comput. 9(8), 1735–1780 (1997)
17. Jeremy, H.: Contact tracing apps struggle to be both effective and private. IEEE Spectrum (2020)
18. Li, H.P., Hu, H., Xu, J.: Nearby friend alert: location anonymity in mobile geosocial networks. IEEE Pervasive Comput. 12(4), 62–70 (2012)
19. Medsker, L.R., Jain, L.: Recurrent neural networks. Des. Appl. 5 (2001)
20. Mozur, P., et al.: In coronavirus fight, china gives citizens a color code, with red flags (2020)
21. Peng, Z., Gao, S., Xiao, B., Wei, G., Guo, S., Yang, Y.: Indoor floor plan construction through sensing data collected from smartphones. IEEE IoTJ 5(6), 4351–4364 (2018)
22. Rein, S., Reisslein, M.: Low-memory wavelet transforms for wireless sensor networks: a tutorial. IEEE Commun. Surv. Tutor. 13(2), 291–307 (2010)
23. Santos, O.C.: Artificial intelligence in psychomotor learning: modeling human motion from inertial sensor data. IJAIT 28(04), 1940006 (2019)
24. Shi, X., Yeung, D.Y.: Machine learning for spatiotemporal sequence forecasting: a survey. arXiv preprint arXiv:1808.06865 (2018)
25. Shoaib, M., Bosch, S., Incel, O.D., Scholten, H., Havinga, P.J.: Fusion of smartphone motion sensors for physical activity recognition. Sensors 14(6), 10146–10176 (2014)
26. Yao, Y., et al.: An efficient learning-based approach to multi-objective route planning in a smart city. In: Proceedings of IEEE ICC (2017)
27. Zeinalipour-Yazti, D., Claramunt, C.: Covid-19 mobile contact tracing apps (MCTA): a digital vaccine or a privacy demolition? In: Proceedings of IEEE MDM (2020)

# A Novel Road Segment Representation Method for Travel Time Estimation

Wei Liu[1], Jiayu He[2], Haiming Wang[1,2], Huaijie Zhu[1,2(✉)], and Jian Yin[1,2]

[1] School of Computer Science and Engineering, Sun Yat-Sen University,
Guangzhou, China
{liuw259,zhuhuaijie,issjyin}@mail.sysu.edu.cn,
{hejy47,wanghm39}@mail2.sysu.edu.cn
[2] Laboratory of Big Data Analysis and Processing, Guangzhou 510006, China

**Abstract.** Road segment representation is important for evaluating travel time, route recovery and traffic anomaly detection. Recent works mainly consider topology information of road network based on graph neural network, while dynamic character of topology relationship is usually ignored. Especially, the relationship between road segments is evolving with time elapsing. To obtain road segment representation based on dynamic spatial information, we propose a model named temporal and spatial deep graph infomax network (ST-DGI). It not only captures road topology relationship, but also denotes road segment representation under different time intervals. Meanwhile, the global traffic status/flow will also affect local road segments' traffic situation. Our model would learn the mutual relationship between them, with maximizing mutual information between road segment (local) representation and traffic status/flow (global) representation. Furthermore, it would make road segment representation more distinguishable by this kind of unsupervised learning, and be helpful for downstream application. Extensive experiments are conducted on two important traffic datasets. Compared with the state-of-the-arts models, the experiment results demonstrate the superior effectiveness of our model.

**Keywords:** Road segment representation · Travel time estimation

## 1 Introduction

Travel time estimation of a path is an important task in recent years, which is helpful for route planning, ride-sharing, navigation and so on [10,21,23]. Nowadays, almost all the travel service applications have this function, including Google Map, Baidu Map, Uber and Didi. Based on accuracy travel time estimating service, user could obtain road's status, plan personalized trip and avoid wasting time on congested roads. Meanwhile, many researchers have devoted themselves on study of high quality travel time estimation [9,13,22]. However, as travel time estimation is a complex task and many factors need to consider, it is still a challenge to provide accuracy estimation.

© Springer Nature Switzerland AG 2021
C. S. Jensen et al. (Eds.): DASFAA 2021 Workshops, LNCS 12680, pp. 398–413, 2021.
https://doi.org/10.1007/978-3-030-73216-5_27

To achieve accuracy travel time estimation, an effective road segment representation is necessary. Recent works mostly utilized graph-based neural network to learning representation of road segment, for instance graph auto-encoder [9], DeepWalk [13], which would make road segment representation be similar to neighborhoods. There are three drawbacks in these methods:

1) These methods lead to adjacent road segments undistinguishable and will not be beneficial for downstream tasks. Such as in Fig. 1, the adjacent road segments $i$, $j$'s status would be similar. While, in fact, road segment $j$ would be more special, since it is apt to block up with the flows from neighbor road segments.

2) They only focuses on local feature in graph, and could not learn mutual influence between global traffic condition and each road segment status. Global traffic condition has influence on individual road segment, and some critical road segments' status may also have a great impact on global traffic condition. As in Fig. 1, the traffic flows which come from office areas $A$, $B$ to residential area $C$ would affect related road segments, making road segment $j$ congested. Meanwhile, road segment $j$'s congestion will also influence other road segments, not just the adjacent road segments. Sometimes, parts of road segments representation at some time intervals are absent, while global (or high-level) traffic condition is rather easily obtained. If we could infer road segment presentation from global traffic condition, it would be beneficial for overcoming data sparsity.

3) They couldn't consider road segment's dynamic status. Under different time interval, road segments will have unique status, which includes not only their dynamic status, but also the special relationship with corresponding global traffic condition. For instance, road segment $j$ would be unblocked in the morning, and the adjacent road segments are also unimpeded. But in the afternoon, $j$ might be congested, and affect the adjacent road segments. Therefore, it requires a model could consider local-global relationship and temporal factors in road network, simultaneously.

Since compared with large volumes of road segments, trajectories are too sparse to denote the distinguishable features of each road segment. Especially, when considering temporal factors, the sparsity problem is much more severe. Meanwhile, traffic system is integrated and complex, the mutual relations between global and local is difficult to model. To solve problems above, we propose a model based on deep graph infomax, which would maximize mutual information between road segment (local) representation and road network (global) representation. It is beneficial for denoising unrelated information and make road segment representation be consistent with entire road network's condition, which could make road segment representation unique and capture similar structures in the whole network. To denote road network's condition, we not only make use of representations from whole graph itself, but also take advantage of geographical traffic status and flows condition. Besides, since road segments' status is dynamic, we would character road segment's representation under different time interval.

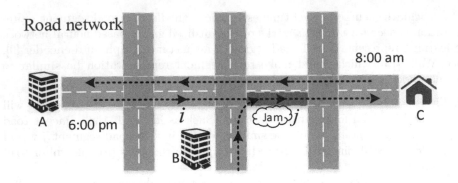

Fig. 1. An example.

In summary, our contribution could be concluded as below:

- A spatial and temporal deep graph infomax network is proposed to learn road segment representation, which could reflect local and global feature of road network simultaneously, especially capture similar structures from whole network view.
- Traffic condition and flows status are both introduced in our model, which make road representation could perceive global condition, section-level global condition with CNN. And dynamic road segment representation is designed to character the temporal traffic condition.
- Extensive experiments are conducted on three public traffic datasets. The results on downstream task (travel time estimation) consistently outperforms the state-of-the-art methods, which proves our road segment representation model is effective and performs more excellent in the sparse traffic setting.

## 2    Related Work

Traffic forecasting is a popular research topic in recent years. Firstly, we would introduce several main tasks in traffic forecasting and the related works. Especially, we focus the related works in travel time estimation task. Then we would analyze the difference between our work and the most related works.

Since plenty of data are from sensors and GPS, more and more researchers focus on the traffic forecasting to provide convenient service for individuals and traffic management [15,24,26]. In traffic forecasting, there are mainly three kinds of tasks:

**1) Travel Time Estimation.** The corresponding methods are categorized to two kinds: route-based methods and neighbor- based methods. Route-based methods would map trajectory onto road segments, estimate each segment's travel time and aggregate these time as final estimation. [23] treated travel time estimation as a regression problem and proposed wide-deep-recurrent network

to capture spatial feature of each road segment and sequential feature for estimating the travel time. Beside of GPS traces and road network, [5] utilized
smartphone inertial data for customized travel time estimation (CTTE). [20]
utilized convolutional neural network to extract spatial feature of road segment
and employs recurrent neural network to learn the sequential feature of trajectories. [13] utilized a deep generative model to learn travel time distribution,
considering spatial feature of road segments with DeepWalk in road network and
temporal feature with the real-time traffic. Different with route-based method,
neighbor-based methods utilize neighboring trips with a nearby origin and destination to estimate travel time. [21] found similar trip with the target trip and
utilized the travel time of those similar trips to estimate the travel time of target
trip.

**2) Travel Speed Estimation.** Although travel speed estimation is usually
related with travel time estimation, they are still different tasks in traffic forecasting. Because travel time estimation contains more factors, such as traffic
lights and making left/right turns, than travel speed estimation. In this kind of
task, graph neural network and recurrent neural network are generally utilized
for extracting topology feature and sequential feature of road network. [28] utilized graph convolutional network to extract spatial topological structure, and
gated recurrent unit to capture dynamic variation of road segments' speed distribution. It fuses these spatio-temporal features together to predict traffic speed, it
is also similar to [4,14,24,27] which introduced diffusion convolutional recurrent
neural network, a deep learning framework for traffic forecasting that incorporates both spatial and temporal dependency in the traffic flow. [3] used history
road status (speed) to predict next time's road speed by considering multi-hop
adjacent matrix and LSTM. [9] utilized graph convolutional weight completion
to learn each road's speed distribution. [11] proposed graph convolutional generative autoencoder to fully address the real-time traffic speed estimation problem.

**3) Traffic Flow Prediction.** In addition to travel speed, traffic flow is another
important sign of traffic condition. To represent the high-level feature of traffic flow, convolutional neural network are usually used. [17,25,26] transformed
traffic flow data into a tensor, and utilized a convolutional neural network and
residual neural network to extract spatio-temporal feature for urban traffic prediction. [16] employed a sequence-to-sequence architecture to make urban traffic
predictions step by step for both of traffic flow and speed prediction. Besides,
there are also some other tasks. [15] predicted the readings of a geo-sensor over
several future hours by considering multiple sensors' readings, meteorological
data, and spatial data. [22] developed a Peer and Temporal-Aware Representation Learning based framework (PTARL) for driving behavior analysis with
GPS trajectory data.

In traffic forecasting, road segment representation effect greatly on prediction
performance. This paper would focus on high-quality representation of road segment. The most related works with ours is GTT [13], GCWC [9], ST-MetaNet
[16]. The differences with them are mainly two points as below: 1) Our model first
propose road segment representation based on local and global traffic conditions

simultaneously. Since road network is a complex system, each road segment is not only linked with local adjacent neighborhoods, but also global traffic dynamics. GCWC, ST-MetaNet only considers local features, ignoring global traffic condition. Though GTT considers real-time traffic condition, it learns static road segmentation and global traffic condition independently, fusing them linearly with concatenation operation, which loses sight of the mutual impacts between them. 2) Our model considers the dynamic relationship between road network. In GTT and GCWC, the relationship between road segments is just the topology relationship in road network and changeless. However, in fact it would evolve with time elapsing. Different with ST-MetaNet learning sequential temporal features with gated recurrent unit which needs much denser data sets and more computation costs, we pay attention on periodic and non-uniform temporal features. In summary, our model would not only consider local and global features in traffic simultaneously and mutually, but also model the dynamic status of each road segment. Based on these factors, our model could get a more comprehensive and adaptive road segment representation, which is beneficial for travel time estimation.

# 3    Proposed Model

**Problem Definition:** Given a road network $G(V, E)$, historical trajectory dataset $\mathcal{H} = \{T_{(k)}\}_{k=1}^{K}$, our objective is to learn the representation $\mathbf{h}_i^{(t)}$ for each road segment $r_i$ under different time interval $t$.

We hope the road segment representation $\mathbf{r}_i^{(t)}$ not only could capture the topology relationship between road segment $r_i$ and joint roads under different time interval, but also could denote potential relationship between $r_i$ and global traffic condition. Road segment representation could be used for travel time estimation [13], traffic speed prediction [16] (seen in experiments), route planning and so on.

## 3.1    Model Overview

Our model would be introduced by four parts. Firstly, we introduce a basic graph constructed by road network, and propose a static road segment representation method based on maximization mutual information (Static Version). Secondly, the temporal factors are considered and fused into the model to obtain dynamic road segment representation (Temporal Version). Thirdly, the traffic status and flows from global view are modeled and mixed into our model (Dynamic Version). Finally, it is the optimization method for our model.

## 3.2    Static Road Segment Representation

Road segment has complex topology and spatial relations with other road segments. To capture the relations, graph convolutional network (GCN) is usually adopted. Especially, graph is constructed by adjacent topology relations, spatial information could not be captured by GCN. Since GCN excessively emphasizes proximity information between neighbors, the in-dependence of road segment is easily ignored. In this part, we first introduce a basic version of road segment representation. Then to consider spatial information simultaneously, a grid-based CNN is utilized to represent section which road segment belongs to. Based on the two steps, an optimized version for maximizing mutual information will be introduced.

**Feature Initialization.** The road network could be represented by a topology *graph*, in which *node* denotes road segment, *edge* between nodes means the link relationship between road segments. Accordingly, we assume a generic graph-based unsupervised machine learning setup: we are provided with a set of road segment features, $\mathbf{R} = \{\mathbf{r}_1, \mathbf{r}_2, ..., \mathbf{r}_N\}$, where $N$ is the number of road segments in the graph and $\mathbf{r}_i \in \mathbb{R}^F$ represents the features of road segment $r_i$, where $F$ is the dimension of road segment basic feature. In our datasets, there are totally 14 kinds of road types[1], 7 kinds of lane number (from 1 to 7), 2 kinds of way direction (one-way or bidirectional), so the $F$ would be 23. Directly, we could use one-hot encoding to represent each road segment, while it would miss the detail characteristics of each road segment, such as number of lanes, speed limit, road shape and so on. Also it could consider neighbor Point-of-Interests as road segment's meta information [20], which could be extended in future. Shown in Fig. 2(a), we would use multi-hot encoding to initialize each road segment feature which denotes road segment's special attribute, and reduce dimension with fully-connect layer, similar to the amortization technique [13].

**Feature Propogation Based on GCN.** Based on road network, we could obtain relational information between these road segments in the form of an adjacency matrix, $\mathbf{A} \in \mathbb{R}^{N \times N}$. In all our experiments we will assume the graphs to be unweighted, i.e. $A_{ij} = 1$ if there exists a connection $r_i \rightarrow r_j$ in the road network and $A_{ij} = 0$ otherwise. Here, the adjacency matrix could be obtained by the road network. However, it is static and could not reflect the road real-time relationship under special time interval. In fact, road relationship will be dynamic with time elapsing. In Sect. 3.3, we would introduce a temporal adjacency matrix $A(t)$ based on historical trajectories.

To learn road segment representation, we would build an encoder, $\mathcal{E} : \mathbb{R}^{N \times F} \times \mathbb{R}^{N \times N} \rightarrow \mathbb{R}^{N \times F'}$, such that $\mathcal{E}(\mathbf{R}, \mathbf{A}) = \mathbf{H} = \{\mathbf{h}_1, \mathbf{h}_2, ..., \mathbf{h}_N\}$ represents high-level representation $\mathbf{h}_i \in \mathbb{R}^{F'}$ for road segment $r_i$ . These representations may then be retrieved and used for travel time estimation task, speed prediction and so on. The definition of function $\mathcal{E}$ could be seen as following:

$$\mathcal{E}(\mathbf{R}, \mathbf{A}) = \sigma(\hat{\mathbf{D}}^{-\frac{1}{2}} \hat{\mathbf{A}} \hat{\mathbf{D}}^{-\frac{1}{2}} \mathbf{R} \mathbf{W}) \tag{1}$$

---

[1] i.e., living street, motorway, motorway link, primary, primary link, residential, secondary, secondary link, service, tertiary, tertiary link, trunk, trunk link, unclassifie.

where $\mathbf{\hat{A}} = \mathbf{A} + \mathbf{I}_N$ is the adjacent matrix with added self-connections, $\mathbf{I}_N$ is the identity matrix, $\mathbf{\hat{D}}$ is the degree matrix, $\mathbf{\hat{D}} = \sum_j \mathbf{\hat{A}}_{ij}$. $\sigma$ is a nonlinear activation function. $\mathbf{W} \in \mathbb{R}^{F \times F'}$ is a learnable linear transformation applied to every node.

Here we will focus on graph convolutional encoders – a flexible class of node embedding architectures, which generate road segment representations by repeated aggregation over local road neighborhoods [6]. A key consequence is that the produced road embeddings, $\mathbf{h}_i$, summarize a local patch of the graph centered around road segment $r_i$ rather than just the road segment itself. In what follows, we will often refer to $\mathbf{h}_i$ as local representation to emphasize this point.

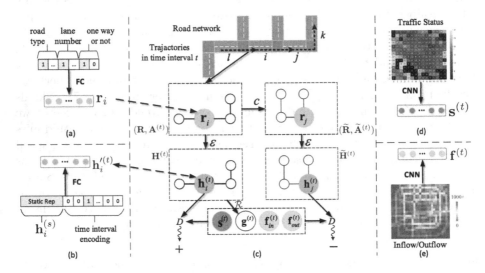

**Fig. 2.** Framework of our model. Road network (and trajectories) could be transformed to node representation. Based on GCN, each node obtains a local representation. A corrupted graph is also generated. Based on global feature, a discriminator is utilized to distinguish the real or fake local feature from different graph.

**Spatial Region Feature Construction.** Since graph can only capture topology structures, the spatial information is easily lost. We utilize grid-index of road segments to construct spatial relations between them. Each grid is denoted by a multi-hot embedding, $\mathbf{p} \in \mathbb{R}^N$. If a road segment crosses the grid, the corresponding element of embedding is set 1, and 0 otherwise. Multiplying with road segment representation matrix $\mathbf{R}$, grid representation is achieved $\mathbf{q} = \mathbf{pR}$, which would subsume representations of inner road segments.

Since road segment may cross several grids, we also consider adjacent regions to include the spatial related road segments with target road segment as much as possible. Figure 3 shows region-level embedding for target road segment.

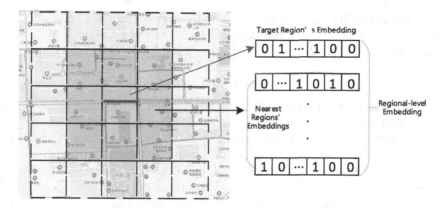

**Fig. 3.** Capturing region-level representation

**Feature Optimization with Mutual Information.** Since road network is an entire systems, a road segment is not only related to adjacent neighborhoods, but also the whole traffic system. The models based on only GCN are insufficient for road segment representation. Therefore, we need to obtain road segment (i.e., local) representations that capture the global information of the entire road network. As the general graph auto encoder (GAE) [2] could not realize this object, which directly optimizes the Euclidean Distance (or discrepancy) between input and output, we adopt maximizing local-global mutual information [8,19] between road segment (local) representation and entire road network (global) representation, which could make road segment representation not only unique but also containing global feature. To achieve this objective, there are 4 questions to solve:

1) How to represent the feature of road segment?
2) How to represent the feature of entire road network information?
3) How to compute the local-global mutual information?
4) How to maximize the local-global mutual information?

For **question 1)**, we would follow the previous method to represent road segment as $h_i$. Next, for **question 2)**, in order to obtain the global-level road network features, $\mathbf{g}$, we leverage a readout function, $\mathcal{R} : \mathbb{R}^{N \times F} \to \mathbb{R}^F$, and use it to gather the obtained local (road segment) representations into a global-level representation; i.e., $\mathbf{g} = \mathcal{R}(\mathcal{E}(\mathbf{R}, \mathbf{A}))$. A simple but efficient choice of $\mathcal{R}$ is average function, $\mathcal{R}(\mathbf{H}) = \sigma(\frac{1}{N} \sum_{i=1}^{N} h_i)$, where $\sigma$ is the activation function.

$$I(H; G) = H(H) - H(H|G), \tag{2}$$

$$= \int_{\mathcal{H} \times \mathcal{G}} log \frac{d\mathbb{P}_{HG}}{d\mathbb{P}_H \otimes \mathbb{P}_G} d\mathbb{P}_{HG}, \tag{3}$$

$$= D_{KL}(\mathbb{P}_{HG} \| \mathbb{P}_H \otimes \mathbb{P}_G) \tag{4}$$

For **question 3)**, the computation of local-global mutual information is shown in Eq. 4, where $\mathbb{P}_{HG}$ is the joint distribution of two variables, $\mathbb{P}_H \otimes \mathbb{P}_G$ is the product of margins, $D_{KL}$ is the KL divergency between two distributions.

$$I(H;G) \leq \mathbb{E}_{\mathbb{P}_{HG}}[\mathcal{D}_\theta(\mathbf{h},\mathbf{g})] - log(\mathbb{E}_{\mathbb{P}_H} \otimes [e^{\mathcal{D}_\theta(\mathbf{h},\mathbf{g})}]) \tag{5}$$

For **question 4)**, to maximize the local-global mutual information, the mutual information (ML) in KL divergency form could be transformed to Donsker-Varadhan (DV) representation as dual representations [1] in Eq. 5. A discriminator is employed to make DV representation to approximate MI, $\mathcal{D} : \mathbb{R}^F \times \mathbb{R}^F \to \mathbb{R}$, such that $\mathcal{D}(\mathbf{h},\mathbf{g})$ represents the probability scores assigned to this local-global pair. The training of discriminator and maximization MI would be processed simultaneously. Here contrastive method is adopt [7,12,18], which is to train the discriminator $\mathcal{D}$ to score contrastively between local representations (positive examples) that contain features of the whole and those undesirable local representations (negative examples). The discriminator is defined as following.

$$\mathcal{D}(\mathbf{h},\mathbf{g}) = \sigma(\mathbf{h}^T \mathbf{W}_2 \mathbf{g}) \tag{6}$$

where $\mathbf{W}_2 \in \mathbb{R}^{F' \times F'}$ is a learnable linear transformation applied to every node. Therefore, based on approximately monotonic relationship between Jensen-Shannon divergence and mutual information, a noise-contrastive type objective [8] is formularized with a standard binary cross-entropy (BCE) loss between the samples from the joint (positive examples) and the product of marginals (negative examples) for maximizing mutual information, as following equation:

$$\mathcal{L} = \frac{1}{N+M}(\sum_{i=1}^{N} \mathbb{E}_{(\mathbf{R},\mathbf{A})})[log\mathcal{D}(\mathbf{h}_i,\mathbf{g})]+ \\ \sum_{j=1}^{M} \mathbb{E}_{(\tilde{\mathbf{R}},\tilde{\mathbf{A}})}[log(1-\mathcal{D}(\tilde{\mathbf{h}}_j,\mathbf{g}))]) \tag{7}$$

This approach effectively maximizes mutual information between $\mathbf{h}_i$ and $\mathbf{g}$, based on the Jensen-Shannon divergence between the joint and the product of marginals.

Negative samples for $\mathcal{D}$ are provided by pairing the global $\mathbf{g}$ from $(\mathbf{R},\mathbf{A})$ with local representation $\tilde{h}_j$ of an alternative graph, $(\tilde{\mathbf{R}},\tilde{\mathbf{A}})$, where $\tilde{\mathbf{R}}$ and $\tilde{\mathbf{A}}$ are corruption versions of original data, respectively. For the road network graph, an explicit (stochastic) corruption function, $\mathcal{C} : \mathbb{R}^{N \times F} \times \mathbb{R}^{N \times N} \to \mathbb{R}^{M \times F} \times \mathbb{R}^{M \times M}$ ($M$ is the node number of corruption graph) is required to obtain a negative example from the original graph, i.e., $(\tilde{\mathbf{R}},\tilde{\mathbf{A}}) = \mathcal{C}(R,A)$. Corruption function $\mathcal{C}$ could be feature-based by row-wise shuffling feature matrices or adjacent-matrix-based by changing part of adjacent matrix elements, which would be discussed in experiments. The choice of the negative sampling procedure will govern the specific kinds of structural information that is desirable to be captured as a byproduct of this maximization.

As all of the derived local representations are driven to preserve mutual information with the global road network representation, this allows for discovering and preserving dependency on the local-level – for example, distant roads with related structural roles. For road network and traffic condition, our aim is for the road segment to establish link to related road segment across the road network, rather than enforcing the global representation to contain all of these correlations.

## 3.3   Temporal Adjacent Matrix

As mentioned before, road segment's status is not static. At different time intervals, road segment's status will be different. The same to relationship between adjacent road segments. To model the dynamic road segment representation, we propose to construct temporal adjacent matrix $A^{(t)}$, which is based on trajectories $T^{(t)} = \{tr_1, tr_2, ...tr_{|T_t|}\}$ in corresponding time interval $t$. For instance, $t = [t_s, t_e)$, where $t_s$ means the start of the time interval, $t_e$ means the end of the time interval. Based on the temporal adjacent matrix $A^{(t)}$ and encoder $\mathcal{E}$, dynamic road segment representation $\mathbf{h}_i^{(t)}$ could be obtained.

However, if each road segment has an independent representation at each time interval, it would cause overfit and need large storage memory. Moreover, not all road segments have trajectories at each time interval, a.k.a. data sparsity problem. Here, we utilize *amortization* technique, which makes road segment $r_i$'s static representation $\mathbf{h}_i^{(s)}$ and one-hot encoding of time interval $\mathbf{e}_l$ mapping into a low-dimensional representation by fully-connected layer to denote dynamic road segment representation $\mathbf{h}_i^{\prime(t)}$, shown in Eq. 8. To learn the parameters in the fully-connected layer, with $\mathbf{h}_i^{(t)}$ obtained by temporal adjacent matrix and the encoder $\mathcal{E}$ denoted in Eq. 9, we utilize $L_2$ to minimize the error between $\mathbf{h}_i^{\prime(t)}$ and $\mathbf{h}^{(t)}$, which would optimize parameters in the fully-connected layer. Based on the fully-connected layer, we could get road segment's dynamic representation.

$$\mathbf{h}_i^{\prime(t)} = \sigma(w(\mathbf{h}_i^{(s)} \oplus t) + b) \tag{8}$$

$$\mathbf{H}^{(t)} = \mathcal{E}(\mathbf{R}, \mathbf{A}^{(t)}) \tag{9}$$

$$\mathcal{L} = min||\mathbf{h}_i^{(t)}, \mathbf{h}_i^{\prime(t)}||_2^2 \tag{10}$$

## 3.4   Global Traffic Condition and Flows

To optimize the mutual information between local and global features in traffic graph, we utilize GCN to extract each road segment representation as the local feature. And the global feature is denoted by the average of all the road segment representation. As in Scct. 3.3, the local feature could be dynamic. Directly, we could also update the global feature with the temporal local feature. However, the global feature is insufficient by aggregating local features, which ignores important traffic information, especially traffic real-time status $s^{(t)}$ and flows $f^{(t)}$ under time interval $t$. Here, we utilize two vectors $\mathbf{s}^{(t)} \in \mathbb{R}^F$, $\mathbf{f}_{in/out}^{(t)} \in \mathbb{R}^F$ to denote them, respectively. Therefore, we could update the global feature $\mathbf{g}^{(t)}$ by unifying these traffic feature with $\mathbf{g}'^{(t)}$. There would be several methods to mix these features, the detail would be discussed in the experiment part.

$$\mathbf{s}^{(t)} = CNN(\mathbf{S}^{(t)}) \tag{11}$$

To calculate traffic real-time condition, we could split geographical space into grids. And we calculate the average speed of trajectories in each grid under time interval $t$ to denote the traffic condition $\mathbf{S}^{(t)} \in \mathbb{R}^{P \times Q}$, where $P$, $Q$ are the girds number in rows and columns, respectively. Then convolutional neural network would be utilized to extract the high-level traffic condition as shown in Fig. 2(d). To represent traffic flow, we could count each grids's flow-in and flow-out times, which would construct matrix $\mathbf{F}_{in}^{(t)}, \mathbf{F}_{out}^{(t)} \in \mathbb{R}^{P \times Q}$ to denote the traffic flow separately, as depicted in Fig. 2(e). The process of extracting high-level feature of traffic flow is similar to traffic real-time condition's. To avoid utilizing the future data in prediction, we could not directly utilize the current traffic data. Here, we adopt traffic condition $\mathbf{s}^{(t-1)}$ under time interval $t-1$ as the global traffic condition. In future, we could utilize LSTM (Long Short Term Memory) to predict current traffic condition $\mathbf{s}^{(t)}$ as global traffic condition by the history traffic condition $\{\mathbf{s}^{(t-k)}, ...\mathbf{s}^{(t-1)}\}$.

**Optimization.** Assuming the single-graph setup (i.e., $(\mathbf{R}, \mathbf{A}^{(t)})$) provided as input), we will now summarize the steps of the model optimization procedure:

1. Sample a negative example by using the corruption function: $(\tilde{\mathbf{R}}, \tilde{\mathbf{A}}^{(t)}) \sim \mathcal{C}(\mathbf{R}, \mathbf{A}^{(t)})$.
2. Obtain local representations, $\mathbf{h}_i^{(t)}$ for the input graph by passing it through the encoder: $\mathbf{H}^{(t)} = \mathcal{E}(\mathbf{R}, \mathbf{A}^{(t)}) = \{\mathbf{h}_1^{(t)}, \mathbf{h}_2^{(t)}, ..., \mathbf{h}_N^{(t)}\}$.
3. Obtain local representations, $\mathbf{h}_j^{(t)}$ for the negative example by passing it through the encoder: $\tilde{\mathbf{H}}^{(t)} = \mathcal{E}(\tilde{\mathbf{R}}, \tilde{\mathbf{A}}^{(t)}) = \{\tilde{\mathbf{h}}_1^{(t)}, \tilde{\mathbf{h}}_2^{(t)}, ..., \tilde{\mathbf{h}}_M^{(t)}\}$.
4. Summarize the input graph by passing its local representation through the readout function: $\mathbf{g}^{(t)} = \mathcal{R}(\mathbf{H}^{(t)})$, and form $\mathbf{g}'^{(t)}$ by unifying global traffic condition with $\mathbf{g}^{(t)}$.
5. Update parameters of $\mathcal{E}$, $\mathcal{R}$ and $\mathcal{D}$ by applying gradient descent to maximize Eq. 7.

This algorithm is fully depicted in Fig. 2(c).

# 4 Experiments

In this section, we will conduct extensive experiments to evaluate the effectiveness of our model on travel time estimation task. Firstly, we described three large scale datasets. Secondly, the evaluation metrics are introduced. Thirdly, the compared state-of-the-art models would be presented. Next, the results compared with baselines would be analyzed. Then, we study the effectiveness of hyper-parameters and each part in our model.

**Task Description:** Travel Time Estimation, based on each road segment's representation $\mathbf{h}_i^{(t)}$ in trajectory $tr$, we could initialize road segment representation of travel time estimation model, e.g., DeepGTT. After training DeepGTT, we could infer each road segment's travel speed $v_i^{(t)}$ from road segment's representation $\mathbf{h}_i^{(t)}$. Considering each road segment's distance $l_i$ as a weight, the average travel speed is obtained as following:

$$v_r = \sum_{i=1}^{n} w_i v_i, w_i = \frac{l_i}{\sum_{k=1}^{n} l_k} \tag{12}$$

then the travel time could be achieved as below:

$$t = \frac{\sum_i^n l_i}{v_r} \tag{13}$$

**Datasets.** We conduct experiments on two public datasets from Didi[2] and one dataset from Harbin [13].

- Chengdu Dataset: It consists of 8,048,835 trajectories (2.07 billion GPS records) of 1,240,496 Didi Express cars in Oct 2018 in Chengdu, China. The shortest trajectory contains only 31 GPS records (0.56 km), the longest trajectory contains 18,479 GPS records (96.85 km), the average of GPS records (distance) in trajectory is 257 (4.56 km).
- Xi'an Dataset: It consists of 4,607,981 trajectories (1.4 billion GPS records) of 728,862 Didi Express cars in Oct 2018 in Xi'an, China. The shortest trajectory contains only 31 GPS records (0.27 km), the longest trajectory contains 16,326 GPS records (166.12 km), the average of GPS records (distance) in trajectory is 316 (4.96 km).
- Harbin Dataset: It consists of 517,857 trajectories ( the sampling time interval between two consecutive points is around 30 s) of 13,000 taxis cars during 5 days in Harbin, China. The shortest trajectory contains only 15 GPS records (1.2 km), the longest trajectory contains 125 GPS records (60.0 km), the average of GPS records (distance) in trajectory is 43 (11.4 km).

In all datasets each trajectory is associated with the timestamp and driverID. For the first two dataset, we set trajectories in the first 18 days as the training set, trajectories in last 7 days as the testing set and the rest of trajectories as the validation set. For Harbin dataset, we set trajectories in the first 3 days as the training set, trajectories in last day as the testing set and the rest of trajectories as the validation set. We adopt Adam optimization algorithm to train

---

[2] http://bit.ly/366rlXf.

the parameters. The learning rate of Adam is 0.001 and the batch size during training is 128. Our model is implemented with PyTorch 1.0. We train/evaluate our model on the server with one NVIDIA RTX2080 GPU and 32 CPU (2620v4) cores.

**Baselines.** To prove the effectiveness of our model on travel time estimation task, we first compare our model with several baseline methods, including:

- DeepTTE: It treats road segments as tokens and compress a sequence of tokens (a route) to predict travel time by using RNN [20].
- ST-MetaNet: It employs a sequence-to-sequence architecture to make prediction step by step, in which it contains a recurrent graph attention network to capture diverse spatial correlations and temporal correlations [16].
- GCWC: It proposed a graph convolutional weight completion to fill each road's time-varying speed distribution, which would be helpful for travel time representation [9].
- DeepGTT: It utilizes a deep generative model to learn the travel time distribution for any route by conditioning on the real-time traffic [13].

**Evaluation Metrics.** To estimate the performance of different models on the prediction task, we adopt RMSE and MAE,

$$RMSE(t,\hat{t}) = \sqrt{\frac{1}{|t|}||t-\hat{t}||_2^2}, MAE(t,\hat{t}) = \frac{1}{|t|}||t-\hat{t}||_1 \qquad (14)$$

where $t$, $\hat{t}$ denote ground truth and estimated value, respectively.

**Performance.** Compared with the state-of-the-art models, our model outperforms other models at least 5.0%, for two reasons: 1) Our model makes the road representation consider local spatial relationship and global traffic status, simultaneously, rather than considering only local spatial relationship as in DeepTTE, ST-MetaNet and DeepGTT. 2) Our model learns road segment's temporal representation under different time interval, which would denote the dynamic spatial relationship under a certain time interval, which is also ignored by other models (Table 1).

**Hyperparameters and Ablation.** To demonstrate the performance of our model, we tested our model under different hyperparamters on Harbin dataset, including the dimension of features $F' \in \{40, 80, 120, 160, 200\}$, the time interval $|t| \in \{10, 15, 20, 30, 60\}$ min, as shown in Fig. 4.

*Dimension:* From the values of RMSE and MAE of our model on Harbin dataset, we could find as dimension increase, the model's error become less and accuracy become increasing. Especially, when dimension equals 200, the RMSE/MAE

**Table 1.** Performance results on travel time estimation

|            | Chengdu | | Xi'an | | Harbin | |
|------------|---------|---------|---------|---------|---------|---------|
|            | RMSE | MAE | RMSE | MAE | RMSE | MAE |
| DeepTTE    | 356.74 | 248.04 | 344.69 | 250.15 | 330.65 | 239.67 |
| ST-MetaNet | 274.90 | 182.89 | 266.72 | 193.78 | 254.96 | 184.97 |
| GCWC       | 280.72 | 195.43 | 270.15 | 200.12 | 264.92 | 193.93 |
| DeepGTT    | 236.44 | 165.36 | 228.46 | 166.77 | 219.10 | 159.11 |
| ST-DGI     | **222.92** | **154.45** | **216.33** | **156.64** | **208.43** | **150.53** |
| ST-DGI/S   | 229.27 | 160.56 | 222.82 | 162.15 | 213.44 | 154.61 |
| ST-DGI/T   | 231.34 | 161.59 | 224.53 | 163.85 | 215.37 | 158.02 |
| ST-DGI/G   | 228.73 | 159.32 | 221.12 | 161.93 | 212.28 | 153.70 |

both is least, our model achieves best performance. Therefore, in the experiments we set the dimension of features in our model as 200.

*Time Interval:* We also test the value of time interval's effect to our model. Compared with dimension, we could find time interval's change has bigger influence to model's performance. When time interval equals 20 min, our model has the best performance. When time interval become smaller, the performance decreases. It may be caused by the data sparsity problem, in which there are too few trajectories to learn temporal road segment representation.

To prove the effectiveness of each part in our model, we estimate our model with its variants:

- ST-DGI/S: It is a variant of our model without the statistical information of each road segments. Here, we adopt stochastic initialization of road segment representation.
- ST-DGI/T: It is a variant of our model without considering temporal factor. Here, we adopt the same representation for a road segment at different time intervals.
- ST-DGI/G: It is a variant of our model without considering global traffic status and traffic flows. It just utilizes summarized representation of nodes in road network graph.

From Table 1, we could find each component is beneficial for our model. Because without considering any part of our model, the RMSE/MAE of the variants both become larger. Meanwhile, we could find the variant without temporal factor, our model's performance decreases most. It denotes the temporal factor is very important for road segment representation and travel time estimation. Besides, as without considering global traffic status and the statistical information of road segment, it proves the two factors also play an important role in travel time estimation.

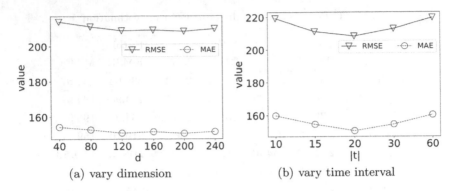

(a) vary dimension                    (b) vary time interval

**Fig. 4.** Effect of two hyper parameters: dimension and time interval.

## 5   Conclusion

Both road's statistical information and global traffic factors and road segment's dynamic status with time changing. A mutual information loss function is utilized to learn road segment representation, which could make the road segment representation characterize global traffic status at special time interval. The experiments's performance on travel time estimation demonstrates the effectiveness of our road representation method, compared with the state-of-the-arts. In future, we would extend the application of our work on other traffic prediction tasks, such as traffic speed prediction, route planning and so on.

**Acknowledgment.** This work is supported by the National Natural Sci- ence Foundation of China (61902438, 61902439, U1811264, U19112031), Natu- ral Science Foundation of Guangdong Province under Grant (2019A1515011704, 2019A1515011159), National Science Foundation for Post-Doctoral Scientists of China under Grant (2018M643307, 2019M663237), and Young Teacher Training Project of Sun Yat-sen University under Grant (19lgpy214,19lgpy223).

## References

1. Belghazi, M.I., et al.: Mutual information neural estimation. In: ICML, p. 530–539 (2018)
2. Berg, R.V.D., Kipf, T.N., Welling, M.: Graph convolutional matrix completion. CoRR abs 1706.02263 (2017)
3. Cui, Z., Henrickson, K., Ke, R., Wang, Y.: Traffic graph convolutional recurrent neural network: a deep learning framework for network-scale traffic learning and forecasting. IEEE TITS **21**, 4883–4894 (2019)
4. Defferrard, M., Bresson, X., Vandergheynst, P.: Convolutional neural networks on graphs with fast localized spectral filtering. In: NeurIPS, pp. 3844–3852 (2016)
5. Gao, R., et al.: Aggressive driving saves more time? Multi-task learning for customized travel time estimation. In: IJCAI, pp. 1689–1696 (2019)
6. Gilmer, J., Schoenholz, S.S., Riley, P.F., Vinyals, O., Dahla, G.E.: Neural message passing for quantum chemistry. arXiv:1704.01212 (2017)

7. Grover, A., Leskovec, J.: node2vec: scalable feature learning for networks. In: KDD, p. 855–864 (2016)
8. Hjelm, R.D., et al.: Learning deep representations by mutual information estimation and maximizationg. In: ICLR (2019)
9. Hu, J., Guo, C., Yang, B., Jensen, C.S.: Stochastic weight completion for road networks using graph convolutional networks. In: ICDE, pp. 1274–1285 (2019)
10. Ide, T., Sugiyama, M.: Trajectory regression on road networks. In: AAAI (2011)
11. Jian, J., Yu, Q., Gu, J.: Real-time traffic speed estimation with graph convolutional generative autoencoder. IEEE TITS **20**, 3940–3951 (2019)
12. Kipf, T.N., Welling, M.: Variational graph auto-encoders. CoRR abs/1611.07308 (2016)
13. Li, X., Cong, G., Sun, A., Cheng, Y.: Learning travel time distributions with deep generative model. In: WWW, pp. 1017–1027 (2019)
14. Li, Y., Yu, R., Shahabi, C., Liu, Y.: Diffusion convolutional recurrent neural network: data-driven traffic forecastings. In: ICLR (Poster) (2018)
15. Liang, Y., Ke, S., Zhang, J., Yi, X., Zheng, Y.: GeoMAN: multi-level attention networks for geo-sensory time series prediction. In: IJCAI, pp. 3428–3434 (2018)
16. Pan, Z., Liang, Y., Wang, W., Yu, Y., Zheng, Y., Zhang, J.: Urban traffic prediction from spatio-temporal data using deep meta learning. In: KDD, pp. 1720–1730 (2019)
17. Pan, Z., Wang, Z., Wang, W., Yu, Y., Zhang, J., Zheng, Y.: Matrix factorization for spatio-temporal neural networks with applications to urban flow prediction. In: CIKM, pp. 2683–2691 (2019)
18. Perozzi, B., Al-Rfou, R., Skiena, S.: DeepWalk: online learning of social representations. In: KDD, pp. 701–710 (2014)
19. Velickovic, P., Fedus, W., Hamilton, W.L., Lio, P., Bengio, Y., Hjelm, R.D.: Deep graph infomax. In: ICLR (Poster) (2019)
20. Wang, D., Zhang, J., Cao, W., Li, J., Zheng, Y.: When will you arrive? Estimating travel time based on deep neural networks. In: AAAI, pp. 2500–2507 (2018)
21. Wang, H., Tang, X., Kuo, Y.H., Kifer, D., Li, Z.: A simple baseline for travel time estimation using large-scale trip data. ACM Trans. Intell. Syst. Technol. **10**(2), 19:1-19:22 (2019)
22. Wang, P., Fu, Y., Zhang, J., Wang, P., Zheng, Y., Aggarwal, C.C.: You are how you drive: peer and temporal-aware representation learning for driving behavior analysis. In: KDD, pp. 2457–2466 (2018)
23. Wang, Z., Fu, K., Ye, J.: Learning to estimate the travel time. In: KDD, pp. 858–866 (2018)
24. Yu, B., Yin, H., Zhu, Z.: Spatio-temporal graph convolutional networks: a deep learning framework for traffic forecasting. In: IJCAI, pp. 3634–3640 (2018)
25. Zhang, J., Zheng, Y., Qi, D., Li, R., Yi, X., Li, T.: Predicting citywide crowd flows using deep spatio-temporal residual networks. Artif. Intell. **259**, 147–166 (2018)
26. Zhang, J., Zheng, Y., Qi, D.: Deep spatio-temporal residual networks for citywide crowd flows prediction. In: AAAI, pp. 1655–1661 (2017)
27. Zhang, Z., Li, M., Lin, X., Wang, Y., He, F.: Multistep speed prediction on traffic networks: a deep learning approach considering spatio-temporal dependencies. Transp. Res. Part C: Emerg. Technol. **105**, 297–322 (2019)
28. Zhao, L., et al.: T-GCN: a temporal graph convolutional network for traffic prediction. IEEE TITS **21**, 3848–3858 (2019)

# Privacy Protection for Medical Image Management Based on Blockchain

Yifei Li[1], Yiwen Wang[1], Ji Wan[2], Youzhi Ren[1], and Yafei Li[1(✉)]

[1] School of Information Engineering, Zhengzhou University, Zhengzhou, China
[2] School of Computer Science and Technology, Beihang University, Beijing, China
wanji@buaa.edu.cn

**Abstract.** With the rapid development of medical research and the advance of information technology, Electronic Health Records (EHR) has attracted considerable attention in recent years due to its characteristics of easy storage, convenient access, and good shareability. The medical image is one of the most frequently used data format in the EHR data, which is closely relevant to patient personal data and involves many highly sensitive information such as patient names, ID numbers, diagnostic information and telephone numbers. A recent survey reveals that about 24.3 million medical images have been leaked from 50 countries all over the world. Moreover, these medical images can be easily modified or lost during the transmission, which seriously hinders the EHR data sharing. Blockchain is an emerging technology which integrates reliable storage, high security and non-tamperability. In this paper, we propose a privacy protection model that integrates data desensitization and multiple signatures based on blockchain to protect the patient's medical image data. We evaluate the performance of our proposed method through extensive experiments, the results show that our proposed method achieves desirable performance.

**Keywords:** Privacy protection · Blockchain · Medical image · Data management

## 1 Introduction

With the rapid development of medical research and the advance of information technology, Electronic Health Record (EHR) [12], has attracted considerable attention in recent years due to the features of easy storage, convenient access, and good shareability. The EHR data generally contains laboratory sheet, examination result, and medical image plays an essential role for cliniciansand researcher in their daily life, which are of great significance to help cliniciansmake efficient and effective diagnosis and treatment plans, and to promote the prevention of diseases by researchers.

---

Y. Li and Y. Wang—Equal Contribution.

C. S. Jensen et al. (Eds.): DASFAA 2021 Workshops, LNCS 12680, pp. 414–428, 2021.
https://doi.org/10.1007/978-3-030-73216-5_28

Since the EHR data is closely relevant to the patient privacy and involves many highly sensitive information, many national laws clearly stipulate that the usage of EHR data should be offered strict attention to patient privacy protection and information security. Nevertheless, the patient privacy is often leaked during the sharing process of the EHR data. A recent survey reveals that about 24.3 million medical images have been leaked from 50 countries all over the world. Moreover, these medical images can be easily modified or lost during the transmission, which seriously hinders the EHR data sharing. Hence, how to effectively balance the sharing capability and the privacy for the EHR data is the critical task to be addressed. The medical image is one of the most frequently used data format in the EHR data, and contains many privacy information such as names, ID numbers, diagnostic information, and telephone numbers. For the reason of simplification, in this paper we mainly focus on the privacy protection for the medical image data management.

Recently, a novel technology, Blockchain [11], receives strategic attention from various countries, which integrates the characteristics of reliable storage, high security and non-tampering. Notably, the feature of immutable timestamp can protect the data integrity and ensure traceability of the source and the usage for medical images. In addition, the cryptography and signature technologies can realize the data privacy protection and solve the trust certification among users. However, the usage of blockchain for the medical image sharing still has some shortcomings, e.g., the storage speed of the chain is slow and the medical images require huge storage space, resulting in poor processing performance.

In this paper, we propose an efficient privacy protection model based on the blockchain for the medical image data management, which integrates the technologies of data desensitization and multiple signature. Firstly, since the medical images typically contain the patient's personal information, thus we parse and desensitize the medical image data to separate the privacy information from the medical images. Secondly, we propose an efficient data storage strategy based on the blockchain sharding technology [17] where the patient privacy data is stored on different nodes, each single node only saves part of the encrypted patient privacy information instead of the complete medical images, addressing the problem of huge storage cost. In addition, we adopt the Inter Planetary File System (IPFS) technology to store the medical images with non-sensitive information and the hash value of the return value and text records directly on the blockchain together. Finally, we have evaluated the efficiency and effectiveness of our proposed solution through extensive experiments.

## 2 Related Works

In this paper, we propose an efficient solution based on blockchain to protect the security of medical image data and strengthen the patient privacy. We next introduce several relevant works in this section.

## 2.1  Electronic Health Record

Electronic health record (EHR) is a record about health status and health behavior of individuals throughout the life cycle and is stored and managed electronically, which can improve the efficiency of medical services and promotes the sharing capability of health information.

Zyskind [18] proposed a decentralized information management system, through MIT's OPAL platform to complete the encrypted storage of data, and the system can also use the underlying architecture of Bitcoin to achieve access control of case information. Lazarovich et al. [8] proposed a privacy storage project of the opportunity blockchain, introducing an open source third-party database to store information, and using AES (advanced Encryption Standard) symmetric encryption algorithm to enhance the security of the project. Bhuiyan et al. [2] proposed a cross-institutional security sharing model of medical files using blockchain technology, dividing multiple hospitals into two groups with different permissions. If the hospital has dishonest behavior, it will be demoted to the lower-privileged group. Xia et al. [14] proposed a medical blockchain, and the system mainly uses the PBFT technology consensus algorithm, which also uses asymmetric encryption and public key infrastructure (Public key infrastructure, PKI) and other cryptographic technologies. In addition, the system also designed storage methods for structured and unstructured data in blockchain and external databases. Cai et al. [4] proposed to use the dual chains to solve the problem of privacy leakage. The transaction chain is responsible for the transaction and settlement of commercial resources, and the user chain only saves account information and does not involve related transactions. Gay et al. [7] proposed the idea of using multiple private keys and multi-person authorization to solve the problem of user privacy protection from the two aspects of decentralization and privacy protection of electronic health records. However, the storage and sharing methods of these electronic health files often require a large amount of storage space when processing medical pictures and images, and the processing performance is reduced, which is difficult to adapt to actual application requirements.

## 2.2  Medical Privacy Protection

With the development of the Internet and cloud computing, more and more medical images are transmitted and stored through the Internet. However, the medical images have a lot of sensitive information, it is easy to cause the personal privacy of patients to be stolen or leaked. Many researchers have carried out Related research work on medical image privacy protection [15]. The traditional image encryption algorithms consists of the pixel scrambling [10], transform domain [13], homophobic [3], chaos-based system [5,6] approaches.

Based on pixel scrambling, the pixel original position is scrambled to destroy the correlation of the image to achieve encryption. The pixel scrambling does not change the size of original value and the statistical information of the original pixel is still retained. The biggest feature of homomorphic encryption is that the ciphertext domain can be directly operated, that is, the process of decrypting, calculating and re-encrypting the image is no longer needed, which simplifies the image processing process.

Maniccam [10] proposed a scrambling encryption algorithm based on SCAN, which destroyed the correlation of the image according to the scrambling rules of SCAN to achieve encryption. Yang [16] uses image scrambling technology to achieve encryption, and the secret key is stored in form of image, which improves the security of image transmission. Liu [9] realizes the effect of scrambling pixels through exclusive-or and modulo calculation, thereby encrypting the image. Bhatnagar [1] uses discrete transformation to obtain sparse matrix, and then uses the sparse matrix to implement image encryption. However, these encryption methods are mainly for traditional medical images, and there are few related researches on medical images.

## 3    System Framework

The privacy protection integrated with data desensitization and multi-signature for medical image data includes the upload and query processing. As the processing of upload involves data desensitization, IPFS star file system and blockchain node up chain operations, multiple systems are required to cooperate with each other. For the query processing, multi-signature and access control are involved to guarantee system data accurate and secure.

### 3.1    Update Processing

The upload processing for medical images is the key function of the system. Initially, we execute the upload operation of the patient's medical images, the system will execute strict desensitization for these images. To reduce the huge transmit workload, we use the IPFS system to store the desensitized medical images instead of the whole images to the blockchain, and the hash values of medical images are saved in the blockchain to achieve the permanent storage. The storage steps for medical images are as follows: The digital certification authentication center has the rights of key initialization and certificate management, which allocates different keys for different patients and institutions. When a patient requests adding new medical images in the system. To prevent the image being tampered, the patient's private key is used to sign the medical images. After receiving the request to add medical image data, the system first verifies the signature according to the public key to verify the requester. If the signature verification is finished, the system generates a receipt that is sent to the IPFS server of this institution. Afterwords, once the IPFS server receives the receipt, it uses the public key of the medical institution to verify the message. If the verification is correctness, it notifies the blockchain to participate in the consensus to record the characteristic value of medical images.

## 3.2   Query Processing

In this system, the patient's medical image is stored in the address that corresponds to the patient's public key. This address is used as the only index in the backend to identify a certain patient. The steps for querying and modifying the medical image data are as follows: When accessing a patient's medical image, the patient's private key and the doctor's private key are used to send a signed data access request to the IPFS server. Next, the IPFS server needs to verify the signed message. If the verification is success, the IPFS server should return the desensitized medical image. After that, the IPFS server will send the signed message to the blockchain. Then, if the medical images only needs to be accessed by doctor or patient, the above two steps are enough. However, if medical image needs to be modified, multiple-signatures are required through the private keys of doctor and patient. If the verification is success, the IPFS server will modify the medical image of the patient and save the message patient ID, doctor ID, modification time, hash value of the medical image on blockchain.

## 4   Image Desensitization

Digital Imaging and Communications in Medicine (DICOM) not only refers to digital medical imaging and communication in medicine but also an international standard for medical images and related information. In this section, we first introduce the specification of the DICOM standard medical image, then explains the procedures of desensitizing the information carried.

### 4.1   Content Analysis

A standard medical image file consists of the header information and the dataset body. The header information is composed of the introduction, prefix, and file header element. The dataset body comprises multimedia information and multiple data elements (such as images and videos).

The storage information part of both the header element and the dataset element has the same data structure which mainly contains the following four-folds: tag, type, length, value field, these four parts are described in Fig. 1.

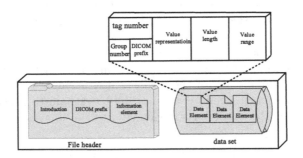

**Fig. 1.** The structure of DICOM file

## 4.2  Desensitization

In this section, we propose the method to parse the DICOM file, which consists of two steps, namely, the file resolution and file processing. The file resolution comprises three main parts: header resolution, content reading, and sensitive information. While the procedures of file processing include desensitization processing and saving file.

The file header resolution of DICOM standard medical image is bytecode data stream processing, because the first 128 bytes of the front file header does not contain valid data. Therefore the processing the byte part is skipped and the data value of 128–131 bytes is used to determine whether corresponds to the hexadecimal of the four letters "DICM" or not. If they are equal, the file is definitely DICOM standard medical image file and enter the step 2. Otherwise, the file is not a DICOM standard medical image file.

**Step 1:** (Data reading) The data elements in the file are stored in key-value format. The elements are read in turn and save them in the corresponding array collection. Then, go to Step 3.

**Step 2:** (Data filtering) Data elements are filtered out according to the DICOM standards, for example, group number '0010' is usually the patient's basic information, group number '0008' is usually the patient's condition information. The value of this part of the information is read in a group number index and the information is cached for the fourth step of desensitization.

**Step 3:** (Desensitization) Each byte is desensitized ac-cording to DICOM standards, such as masking sensitive labels (e.g. names), erasing hospital information and patient ID numbers, truncating dates and locations.

**Step 4:** (File Saving) A special tool is required while accessing standard DICOM files, which is not conducive to access to and sharing of medical images. We can choose whether to save the files as JPEG format or both DICOM and JPEG format when we work with files.

The method proposed in this section protects the patient's medical image information by using data processing and data desensitization. In the meanwhile, after the detailed analysis of medical image files, we can make the conclusion that, the anonymization of data can be better realized by two steps that are the file resolution and file processing, leading to a good share data on blockchain system.

## 5  Blockchain-Based Privacy Protection

The private data in the medical image needs to be protected and encrypted. This is particularly important when it refers to sensitive information, such as AIDS, hepatitis B, cancer, face-lifting, and psychosis. The research institutions should take relevant measures to make patients trust the institution fully, and build a privacy and confidentiality system with a complete data security protection mechanism. For the existing system, the personal health data of patients is managed by different hospitals or companies, and different medical institutions often

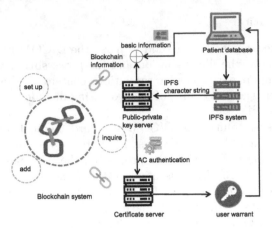

**Fig. 2.** The structure of DICOM file

use the information systems built by different companies, which makes it difficult to interact. Meanwhile, medical institutions are unwilling to share medical information for the sake of protecting their copyright. In this section, we propose an efficient blockchain-based solution with the features of privacy protection and reliable sharing. Putting IPFS and blockchain together improves the efficiency greatly, where IPFS stores massive amounts of data while blockchain saves IPFS addresses. At the same time, the privacy protection of the medical image data is realized through signature technology and access control mechanism (Fig. 2).

## 5.1  Digital Certificate Authentication

Certification Authority (CA) certification is a significant part of the system, which refers to the certification user's authority. This module is devoted to realize the function of key initialization and certificate management. It is applied to make sure the user's identity in the network. After CA authenticates the user's public key, a digital certificate will be generated, the status of which in the network is the same as that of the ID card in real life.

**Table 1.** Data access permissions

| Type | Authority |
|------|-----------|
| 1 | Personal information |
| 2 | Medical record |
| 3 | Medical images and videos |
| 4 | Desensitized medical images |
| 5 | Feature value of medical image (hash value) |

The authentication and management of user's public and private keys is the main function of CA authentication in the system, which can only be used after authentication. It includes functions such as generation and logout. The reason why the management of user's public and private keys is vital is that, the signature of user's public and private keys needed in the privacy protection and reliable sharing system of medical image data based on blockchain. Generally, each institution has its own distinct data access permission, the public and private key pairs in the CA include the following permissions which are shown in Table 1.

In order to realize the detailed division of the authority of different users, the authority control manages the authority for different users and organizations separately. Note that each patient is capable of accessing the IPFS server with the Key-1 secret key to obtain all permission of personal image data, and get the characteristic value through the blockchain consensus node server to verify whether the image data has been modified. The corresponding personal information, medical files, medical images and videos can be obtained by the research of medical institution by using Key-2. The insurance company can get the patient data through Key-3 after desensitization so as to audit the insurance claims. Due to the desensitization data without sensitive information, the risk of privacy leakage is small. By checking the hash value of image data, we can verify whether the user's data have been tampered. In case of doctor-patient disputes, the government can obtain the eigenvalue information of image data on the blockchain consensus node through Key-4, thus realizing the supervision on whether medical institutions tamper with the data of patient's image data. The research institutions can get medical images, videos, and corresponding written medical files through Key-5 to carry out scientific research. The permissions of each type of key are shown in Table 2.

**Table 2.** Permissions of different secret keys

| Type | Authority | Owner | Key |
|------|-----------|-------|-----|
| 1 | 1, 2, 3, 4, 5 | Patients | key-1 |
| 2 | 1, 2, 3 | Hospital | key-2 |
| 3 | 1, 4, 5 | Government | key-3 |
| 4 | 5 | Company | key-4 |
| 5 | 2, 3 | Institution | key-5 |

## 5.2  Blockchain and IPFS System Storage

The IPFS is a distributed file system that includes a mapping function. It can divide files into several same size blocks with the same size, and then calculate the combination of each block to build a file retrieval table, which can realize the purpose of storing file blocks in different server clusters. In order to achieve permanent, decentralized storage, and sharing of files, the medical image data can be identified by generating independent hash values from their content. Only one file with the same content can exist in the system, which saves storage space and strengthens the protection of user privacy.

The system used in this paper consists of three layers. The data layer includes the IPFS cloud storage cluster and the blockchain consensus node which is off-chain IPFS distributed cloud storage clusters. It is a blockchain consensus node cluster based on the alliance chain. The business layer mainly realizes the reliable access and desensitization storage functions of medical image data, and reserves smart contract interfaces, including data desensitization modules, data layer access interfaces, and smart contract modules. It can realize efficient desensitization processing of sensitive patient information, and provide access interface between user layer and data layer. The user layer consists of the medical data management system and the user mobile phone access system, which can realize the medical data query on the mobile phone, the registration of staff and patients in various medical institutions. The system architecture is as shown in Fig. 3.

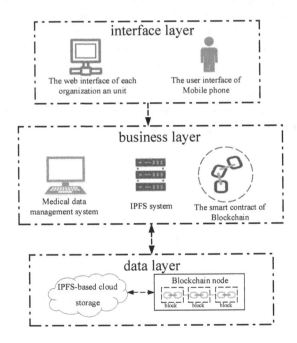

**Fig. 3.** System architecture based on IPFS

The steps to add medical image data are as follows:

**Step 1:** A patient $P$ initiates a request to add the medical image data to the institution $Dep$.

$$R_{Add} = AddMsg \| Sig_{R_{pk}}(H(AddMsg)) \tag{1}$$

$AddMsg = (ID_R, ID_P, ID_{Dep}, M_E, R_{pk}, t)$ contains the requester $ID_R$, patient $ID_P$, medical institution $ID_{Dep}$, medical image $M_E$, the public key of the requester $R_{pk}$, and the request time $t$. In order to prevent the requested content from being tampered, the private key $R_{sk}$ of the requester will be used for signature.

**Step 2:** After receiving a request to add a medical image, the medical institution $Dep$ first verifies the signature according to the public key $R_{pk}$ to check the identity of the data sender. If the signature verification is successful, the receipt information is generated and sent to the IPFS server of the institution, where the receipt information is: $T = (ID_R, ID_P, ID_{Dep}, M_E, R_{pk}, Dep_{pk}, t_2)$. Next, the private key of the institution can be used to encrypt the information and send it to the IPFS server. If the signature verification fails, the request will be rejected.

**Step 3:** After the IPFS server receives the bill information sent by the medical institution $Dep$, it uses the public key of the medical institution $Dep_{pk}$ to verify the message. If the verification is passed, the bill information is determined to be valid, and the IPFS server will send a notification $M = (ID_R, ID_P, ID_{Dep}, IPFS_{pk}, Hash_E, t_3)$ to the blockchain consensus node. Among them, $IPFS_{pk}$ is the public key of the IPFS server node. The blockchain node only needs to save the hash value of the file while does not need to save the complete file containing text, images, or other information.

## 5.3   Data Access Based on Multi-signature

The patient's medical image data is stored in a distributed manner through the IPFS system. In order to further protect the privacy of the patient and prevent the doctor from unilaterally using the patient's data information, we have designed a multi-signature algorithm (MSA), which makes it necessary to use multiple signature of doctors and patients to access the medical image data. Assuming that there are $n$ signers and verifying that they have received at least $m$ signatures, the signature is set to be valid, so it is especially suitable for scenarios where multiple people vote together. In the privacy protection and reliable sharing system, in order to protect the privacy of patients, at least one doctor and one patient are required to authorize at the same time to access the patient's medical image data. Therefore, the multi-signature algorithm can satisfy our application scenes well. Compared with threshold signatures, multiple signatures also have high security and are easier to design and implement. Next, we will explain the medical data access algorithm based on multiple signatures used in this paper.

Assuming that $M$ is a message to be signed. Firstly, $M$ is grouped into a bit string as $M = \{m_1, m_2, \cdots, m_n\}$, where $m_i$ represents the group that needs to be encrypted separately. The encryption system generates a key (GK) to encrypt each block. The encrypted bit group set constitutes a ciphertext signature (Sig), and finally the patient uses the private key (PK) to perform signature verification to obtain the plaintext. Generally, the MSA has three entities: key generation center (KGC), signer, and validator, which are mainly composed of the following three operations, It can be expressed as $MSA = (GK, Sig, Ver)$.

- Generate the key (GK). The CA center realizes key initialization, including the doctor's public key $PK_d = (n_d, e_d)$ and the doctor's private key $SK_d = (n_d, d_d)$ in the first layer of encryption. Then, the patient's public key $PK_p = (n_p, e_p)$ and the private key $SK_p = (n_p, d_p)$ in the second layer of encryption. Among them, $n_d$ and $e_d$ are the random numbers generated when the RSA algorithm generates the doctor's public key, $d_d$ is the unique corresponding number calculated by $(n_d, e_d)$. $n_p$ and $e_p$ are the random numbers generated when the RSA algorithm generates the patient's public key, $d_p$ is the unique corresponding number calculated by $(n_p, e_p)$.
- Generate signature (Sig). The doctor's public key $PK_d = (n_d, e_d)$ is used to encrypt and generate $Sig = M' = \{m'_1, m'_2, \cdots, m'_n\}$, where $m'_i$ represents the first encryption of each gruop. Then, we use the patient's public key $PK_p = (n_p, e_p)$ to encrypt and generate $Sig' = M'' = \{m''_1, m''_2, \cdots, m''_n\}$, where $m''_i$ represents the result of the second encryption of each group.
- Signature verification (Ver). After entering the patient's private key $SK_p = (n_p, d_p)$, and getting the public key of patient $P$ from the CA center, the signature result $M'$ is obtained. Then, the doctor's private key $SK_d = (n_d, d_d)$ is entered, the doctor's public key from the CA center can be used to verify the message $M'$. Finally, the final signature result $M'$ is obtained.

The basic process of the MSA encryption algorithm in our method is as follows:

**Step 1:** We first find two large numbers $p$ and $q$ randomly and verify their primality. After the verification, we can get the random number $nd = p*q$ in the doctor's key, and the random number $n_p$ and Euler function $f_{n_p}$ in the patient key. Then, we divide the message $M$ into $\{m_1, m_2, m_3, ..., m_n\}$.

**Step 2:** In this step, we generate the doctor's public key $PK_d = (n_d, e_d)$, where $e_d$ is an integer randomly selected in $(1, f_{nd})$ and needs to satisfy $gcd(e_d, f_{nd}) = 1$, where $gcd$ represents the greatest common divisor, the same is true for generating the patient's public key $PK_p = (n_p, e_p)$.

**Step 3:** In this step, we generate the doctor's private key $SK_d$ and match the unique $d_d$ according to the selected $e_d$, and then generate the patient's private key $SK_p$.

**Step 4:** We use the doctor's public key $PK_d$ to encrypt and generate $Sig = M'$ and use the patient's public key $PK_p$ to encrypt and generate $Sig' = M''$;

**Step 5:** If we enter the patient's private key $SK_p$, we can get the public key of patient $P$ from the CA center and get the signature result $M'$. If we enter the doctor's private key $SK_d$, we can get the public key of doctor $D$ from the CA center and then verify the message $M'$ to obtain the final signature result $M$.

# 6    Performance Evaluation

In this section, we have verified the performance of our proposed method that integrates data desensitization and multiple signatures. Firstly, we introduce the experimental settings including the actual simulation program and the construction of the experimental environment. Then, we test our method on real-life DICOM image dataset with size of $5k$ in terms of processing speed, access delay, and the desensitization quality.

## 6.1    Experimental Settings

In this paper, we build an IPFS server, a certificate authority center, and the smallest Byzantine system composed of four nodes (composed of $3F + 1$ surviving nodes, F is the number of malicious nodes). The hardware environment is as shown in Table 3.

**Table 3.** System settings

| Items | Environment |
| --- | --- |
| Blockchain Node | Aliyun, Xeon E5-2682 v4, 2.5 GHz 2G DDR4 |
| OS | Ubuntu 7.5.0-3ubuntu1 18.04 |
| Blockchain System | Fabric 1.4 |
| Language | python 3.8.5, nodejs v8.10.0, go1.9.5 |
| IPFS | go-ipfs v0.4.15 |

## 6.2    Experimental Result

In this section, we report the evaluation results from the perspective of the system's image processing speed, the access delay, and the storage cost.

**Effect of Processing Speed.** As shown in Fig. 4, the average processing time is becoming longer as the resolution increases. For the image with a resolution of 640*480, the average processing time is about 380 ms, which means that the system has a high ability to support concurrency. When the average number of images input by the system per second is less than 50, the time for the system to process a picture is relatively stable. When the average number of images per second is greater than 100, the average time for the system to process each image increases significantly, indicating that the system can simultaneously provide image processing capabilities for approximately 100 clients.

**Effect of Access Delay.** Since the medical images are stored in the IPFS system, there is no need for consensus among nodes. While the text medical record information is stored on the blockchain, which needs to reach consensus

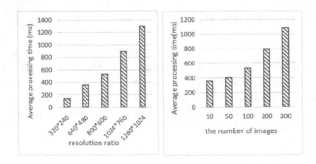

**Fig. 4.** Effect of image processing speed

among multiple nodes. The experimental results in Fig. 5 show that the IPFS system has high performance to obtain images, the average access delay for accessing text information and images increases with the increase of tasks. And the average access delay for medical images is significantly lower than the average access delay for obtaining text medical records.

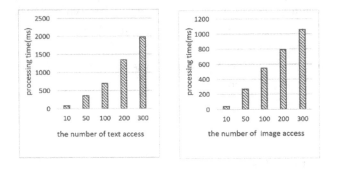

**Fig. 5.** Effect of access delay

**Effect of Storage Cost.** In Fig. 6, it can be seen that the algorithm proposed in this paper only needs to store the Hash value for the medical image data file on the blockchain. Therefore, compared with directly storing the medical image data on the blockchain, the processing time is reduced by 24.2%. When four blockchain nodes are used to store data, 77.5% of storage space is saved. The IPFS system can greatly improve storage space utilization and reduce system processing time.

**Effect of Throughput.** As shown in Fig. 7, with the increase in the number of transactions per second, the throughput of both increases linearly with the increase in the number of requests. However, the peak value of CA digital certification processing requests is 600 times per second, while the peak value of IPFS

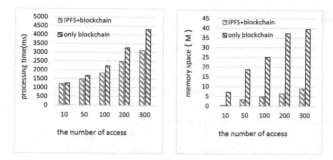

**Fig. 6.** Effect of storage cost

can handle about 1100 requests/second. We can also find that the concurrency of the system is mainly limited to the signature verification part. At the same time, the concurrency of image data in the blockchain system is much lower than that of the IPFS system.

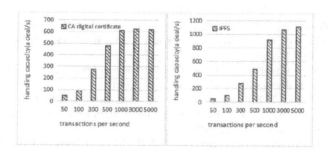

**Fig. 7.** Effect of throughput

## 7   Conclusion

In this paper, we propose a privacy protection scheme for medical image data that integrates data desensitization and multi-signature. First, we designed a solution to parse the file format for the standard medical image. We also propose a data storage method based on blockchain and IPFS, the IPFS system is used for storing medical images and hash value of medical record are stored on the block chain, which improves the data storage efficiency and strengthens patient privacy protection. Finally, we tested the performance of our proposed method through extensive experiments where the results show that the proposed method in this paper achieves desirable performance.

# References

1. Bhatnagar, G., Wu, Q.M.J., Raman, B.: Discrete fractional wavelet transform and its application to multiple encryption. Inf. Sci. **223**, 297–316 (2013)
2. Bhuiyan, M.Z.A., Zaman, A., Wang, T., Wang, G., Tao, H., Hassan, M.M.: Blockchain and big data to transform the healthcare. In: Proceedings of the International Conference on Data Processing and Applications, ICDPA 2018, Guangdong, China, 12–14 May 2018, pp. 62–68 (2018)
3. Brakerski, Z., Vaikuntanathan, V.: Efficient fully homomorphic encryption from (standard) LWE. SIAM J. Comput. **43**(2), 831–871 (2014)
4. Cai, G.G.: Channel selection and coordination in dual-channel supply chains. J. Retail. **86**(1), 22–36 (2010)
5. Chai, X., Fu, X., Gan, Z., Lu, Y., Chen, Y.: A color image cryptosystem based on dynamic DNA encryption and chaos. Signal Process. **155**, 44–62 (2019)
6. Chen, Q., Wu, T.T., Fang, M.: Detecting local community structures in complex networks based on local degree central nodes. Phys. A: Stat. Mech. Appl. **392**(3), 529–537 (2013)
7. Gay, J., Odessky, A.: Multi-person gestural authentication and authorization system and method of operation thereof (2017)
8. Lazarovich, A.: Invisible Ink: blockchain for data privacy. Ph.D. thesis, Massachusetts Institute of Technology (2015)
9. Liu, D., Zhang, W., Yu, H., Zhu, Z.: An image encryption scheme using self-adaptive selective permutation and inter-intra-block feedback diffusion. Signal Process. **151**, 130–143 (2018)
10. Maniccam, S.S., Bourbakis, N.G.: Lossless image compression and encryption using SCAN. Pattern Recognit. **34**(6), 1229–1245 (2001)
11. Nakamoto, S., et al.: Bitcoin: a peer-to-peer electronic cash system (2008)
12. Shahnaz, A., Qamar, U., Khalid, A.: Using blockchain for electronic health records. IEEE Access **7**, 147782–147795 (2019)
13. Sudharsanan, S.: Shared key encryption of JPEG color images. IEEE Trans. Consumer Electron. **51**(4), 1204–1211 (2005)
14. Xia, Q., Sifah, E.B., Smahi, A., Amofa, S., Zhang, X.: BBDS: blockchain-based data sharing for electronic medical records in cloud environments. Information **8**(2), 44 (2017)
15. Xie, X., Niu, J., Liu, X., Chen, Z., Tang, S.: A survey on domain knowledge powered deep learning for medical image analysis. CoRR abs/2004.12150 (2020)
16. Yang, Y.L., Cai, N., Ni, G.Q.: Digital image scrambling technology based on the symmetry of Arnold transform. J. Beijing Inst. Technol. **15**(2), 216–220 (2006)
17. Zamani, M., Movahedi, M., Raykova, M.: Rapidchain: scaling blockchain via full sharding. In: Proceedings of the 2018 ACM SIGSAC Conference on Computer and Communications Security, pp. 931–948 (2018)
18. Zyskind, G., Nathan, O., Pentland, A.: Decentralizing privacy: using blockchain to protect personal data. In: 2015 IEEE Symposium on Security and Privacy Workshops, SPW 2015, San Jose, CA, USA, 21–22 May 2015, pp. 180–184 (2015)

# Approximate Nearest Neighbor Search Using Query-Directed Dense Graph

Hongya Wang[1,2(✉)], Zeng Zhao[1], Kaixiang Yang[1], Hui Song[1],
and Yingyuan Xiao[3]

[1] Donghua University, Shanghai, China
hywang@dhu.edu.cn
[2] Shanghai Key Laboratory of Computer Software Evaluating and Testing,
Shanghai, China
[3] Tianjin University of Technology, Tianjin, China

**Abstract.** High-dimensional approximate nearest neighbor search (ANNS) has drawn much attention over decades due to its importance in machine learning and massive data processing. Recently, the graph-based ANNS become more and more popular thanks to the outstanding search performance. While various graph-based methods use different graph construction strategies, the widely-accepted principle is to make the graph as sparse as possible to reduce the search cost. In this paper, we observed that the sparse graph incurs significant cost in the high recall regime (close or equal to 100%). To this end, we propose to judiciously control the minimum angle between neighbors of each point to create more dense graphs. To reduce the search cost, we perform K-means clustering for the neighbors of each point using cosine similarity and only evaluate neighbors whose centroids are close to the query in angular similarity, i.e., query-directed search. PQ-like method is adopted to optimize the space and time performance in evaluating the similarity of centroids and the query. Extensive experiments over a collection of real-life datasets are conducted and empirical results show that up to 2.2x speedup is achieved in the high recall regime.

**Keywords:** Nearest neighbor search · Graph-based method · Query-directed search

## 1 Introduction

Nearest neighbor search (NNS) has been a hot topic over decades, which plays an important role in many applications such as data mining, machine learning and massive data processing. For high-dimensional NNS, due to the difficulty of finding exact results [8,9], most people turn to the approximate version of NNS, named Approximate Nearest Neighbor Search (ANNS). Recently, graph based methods have gained much attention in answering ANNS. Given a finite point set $S$ in $\mathbb{R}^D$, a graph is a structure composed of a set of nodes (representing

© Springer Nature Switzerland AG 2021
C. S. Jensen et al. (Eds.): DASFAA 2021 Workshops, LNCS 12680, pp. 429–444, 2021.
https://doi.org/10.1007/978-3-030-73216-5_29

a point in the dataset) and edges. If there is a neighbor relationship between two nodes, an edge is added between the two nodes. If each node links $K$ edges, the graph is a $K$NN graph. The way to construct the graph affects greatly the search efficiency and precision, so many researchers are committed to improving the performance using different heuristics in the construction of the graph.

The common wisdom in constructing an $K$NN graph is to reduce the average out-degree as much as possible because the search cost is determined by the number of hops during walking the graph times the average out-degree. By graph theory, average out-degree and the connectivity of graphs are conflicting design goals. Hence, low average out-degree will make the graph too sparse and thus increase the difficulty of finding high quality $k$NN.

In this paper, we argue that one could obtain low search cost and high answer quality at one shot with affordable extra memory. Our first observation is that the state-of-the-art algorithms such as HNSW and NSG cannot achieve this goal even given enough extra memory, which will be discussed in details in Sect. 2. To tackle this problem, we propose to (1) control the minimum angle between neighbors of each point judiciously to create dense $K$NN graphs and thus improve the connectivity, and (2) use the query to guide the evaluation of neighbors of the base point, which significantly reduce the search cost. Figure 1 illustrates our idea using a simple example. In Fig. 1(a), the search algorithm examines all neighbors of $o_1$ and chooses the nearest one to $q$ as the next base point. In contrast, $o_2$ has more neighbors than $o_1$ because the minimum angle between them is smaller. Suppose we are aware of the direction of $q$ then we can only compare $o_3$ and $o_4$ with $q$, which reduces the number of distance evaluation dramatically. Note that, for almost all search graph construction, the memory used to store the graph depends on the maximum out-degree (MOD) instead the average out-degree for implementation efficiency. Thus, our method do not increase the memory cost to store the graph itself as will be discussed in Sect. 2.

Knowing the direction of $q$ requires extra information associated with the graph. Our proposal is to partition neighbors of each point into clusters using standard clustering algorithms such as K-means. One modification is that we use *cosine similarity*, instead of the Euclidean distance, as the similarity measure. By comparing the cosine similarity between centroids and $q$ we can avoid access-ing distant neighbors and reduce the overall cost. The memory cost is very high (several times as much as the dataset) if we store the original centoids directly. Fortunately, slightly imprecise direction information is acceptable, which enables us to compress centroids using the product quantization method [18]. For exam-ple, an original centroid of dimension 128 needs 512 bytes to store and the compressed code occupies only 8 to 16 bytes.

To sum up, the main contributions of this paper are:

- We propose a novel query-directed dense graph (QDG) indexing method by controlling the minimum angle between neighbors and using clustering cen-troids to guide efficient search procedure. Please note that the design prin-ciples of QDG is orthogonal to specific graph construction algorithms, and thus are applicable for almost all graph-based methods.

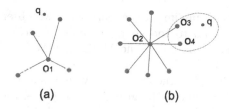

**Fig. 1.** An example to illustrate the main idea

- To improve space efficiency, we use modified product quantization (PQ) method to optimize the algorithm performance and reduce the index size.
- Extensive experiments show that QDG outperforms HNSW [24] and NSG [13], the two state-of-the-art graph algorithms in efficiency over a collection of real datasets. Particularly, up to 2.2x speedup is achieved at high recall regime.

This paper is organized as follows. Section 2 motivates our proposal. The details of QDG is presented in Sect. 3. Experimental results and analysis is given in Sect. 4. Related work is discussed in Sect. 5. Section 6 concludes this article.

## 2 Motivation

In recent graph-based methods, due to the high computational cost of building an exact $K$NN graph, many researchers turn to build an approximated $K$NN graph. Many experimental results such as Efanna [12] proved that the approximate $K$NN graph still performs well.

For almost all graph-based methods, the ANN search procedure is based on the same principle as follows. For a query $q$, start at an initial vertex chosen arbitrarily or using some sophisticated selection rule. Moves along an edge to the adjacent vertex with minimum distance to $q$. Repeat this step until the current element $v$ is closer to $q$ than all its neighbors, and then report $v$ as the NN of $q$. Figure 4(a) illustrates the searching procedure for $q$ in a sample graph starting from $p$.

In order to reduce the searching time on the graph, constructing an approximate $K$NN graph usually tends to reduce the out-degree of the graph. Out-degree refers to the number of neighbors connected to each node on the graph. For example, HNSW adopts the RNG's edge selection strategy to select the neighbors. It can reduce the out-degree to a constant $C_D + o(1)$, which is only related to the dimension $D$ [24]. However, this edge selection strategy is too strict to provide sufficient edges. NSG adopts the MRNG's edge selection strategy, which is based on a directed graph. It can better ensure that each node on the graph has sufficient neighbors than RNG, and the angle between any two edges sharing the same node is at least 60° [13].

Through a number of preliminary experiments we observed that such edge selection policies may lead to too sparse graphs, especially for datasets such as Trevi and Nuswide. Table 1 lists the MOD and AOD for HNSW, NSG and the proposed method QDG, where AOD denotes the average out-degree over all points in the graph. By graph theory we know that the connectivity of graph is closed related to its AOD and if the graph is too sparse, the traversal length of a query will increase, which in turn decreases the efficiency [25].

**Table 1.** Comparison of the out-degree of graph in three methods. HNSW contains multiple graphs and we only report the AOD and MOD of its bottom-layer graph (HNSW0) here.

| Dataset | $HNSW_0$ | | NSG | | QDG | |
|---------|-----|-----|-----|-----|-----|-----|
| | MOD | AOD | MOD | AOD | MOD | AOD |
| Audio | 70 | 13 | 70 | 17 | 70 | 65 |
| Sun | 70 | 13 | 70 | 24 | 70 | 66 |
| Cifar | 70 | 17 | 70 | 29 | 70 | 69 |
| Nuswide | 70 | 4 | 70 | 8 | 70 | 16 |
| Trevi | 70 | 5 | 70 | 8 | 70 | 46 |

However, simply increasing MOD does not make out-degree greater because the edge selection policies such as RNG and MRNG set the lower bound of the minimum angle between any two edges sharing the same node. Table 2 lists the recall and search time at high recall regime for four datasets using NSG[1], which suggests (1) increasing MOD does not change the average out-degree much, and (2) adding more memory to the index cannot trade space to speed by using the existing index structure alone. Please note the index size is determined by MOD, instead of AOD, for almost all existing graph-based algorithms for implementation efficiency.

**Table 2.** Comparison of recall and cost on different datasets by increasing MOD.

| MOD | Audio | | Sun | | Nuswide | | Trevi | |
|-----|--------|------|--------|------|--------|------|--------|------|
| | Recall | Cost | Recall | Cost | Recall | Cost | Recall | Cost |
| 70 | 0.999 | 0.4403 | 0.9997 | 0.07573 | 0.7685 | 0.01964 | 0.9920 | 0.03730 |
| 100 | 0.999 | 0.4403 | 0.9997 | 0.07774 | 0.7685 | 0.02037 | 0.9935 | 0.03850 |
| 160 | 0.999 | 0.4403 | 0.9997 | 0.07820 | 0.7700 | 0.02079 | 0.9942 | 0.03962 |
| 220 | 0.999 | 0.4403 | 0.9997 | 0.07827 | 0.7702 | 0.02085 | 0.9952 | 0.03999 |
| 500 | 0.999 | 0.4403 | 0.9997 | 0.07827 | 0.7700 | 0.02086 | 0.9952 | 0.04002 |

---

[1] HNSW exhibits similar trends.

These two observations motivate us to increase the out-degree to ensure that there are sufficient neighbors for each point, that is, improving the connectivity. The side-effect of dense graph is the increasing computational cost because all neighbors of each point along the search path will have to be examined. To solve this problem, we give higher priority to the neighbors closer to the query, which will be discussed in next section.

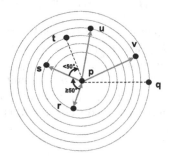

**Fig. 2.** The neighbor selection strategy at point $p$. All candidate neighbors are sorted by distance to $p$. Starting from the nearest point $r$, the neighbors whose angle is larger than the specified degree (e.g. 50°) will be reserved. The dotted line represents the abandoned neighbor, the directed arrows point to the reserved neighbor.

# 3    Query-Directed Dense Graph Algorithm

## 3.1    Graph Construction

QDG consists of three stages in search graph construction.

The first stage is to construct an approximate $K$NN graph. We use the same method as NSG in this stage [13]. After constructing the approximate $K$NN graph, the approximate center of the dataset will be calculated, which is called the Navigating Node. When we choose neighbor candidate sets for a point $p$, it will be treated as a query, and the greedy-search algorithm will be performed starting from the navigating node on the approximate $K$NN graph. During the search, the candidate set will be sorted by the distance to $p$ and used for neighbor selection in the second stage.

Instead of using MRNG's edge selection adopted by NSG, we adjust the number of neighbors for each point by controlling the minimum angle between its neighbors. The edge selection strategy in the second stage is shown in Fig. 2. Assume that the minimum angle is 50°. First, the point $r$ closest to $p$ is selected and put it into the result set. When selecting the remaining edges, it will be selected from the candidate set according to the distance ranking with respect to point $p$. If the angle between itself and the existing ones is greater than 50°, it will be kept (like $s$ in Fig. 2) and discarded otherwise (like $t$ Fig. 2). The choice of minimum angle directly affects the average out-degree of the graph and is left for user to determine according to the dataset property.

The third stage is illustrated in Fig. 3. Each point on the graph has a set of neighbors, then we use K-means algorithm to cluster neighbors that are close to each other in angular similarity. Since the standard K-means algorithm only support the Euclidean distance, we make the following pre-processing. As we all know, the Euclidean distance between point $A$ and point $B$ in high-dimensional space is calculated as follows:

$$\|A - B\|^2 = (A - B)^T(A - B) = \|A\|^2 + \|B\|^2 - 2\,A^T B.$$

If $A$ and $B$ are normalized to unit vectors, i.e., $\|A\|^2 = \|B\|^2 = 1$, then $\|A - B\|^2$ is equal to $2(1 - cos(A, B))$, which means there is a monotonic relationship between the Euclidean distance and cosine similarity. As shown in Fig. 3, we first transform all candidate neighbors of point $p$ into unit vectors w.r.t. $p$, then we use the K-means algorithm to cluster all unit vectors by the Euclidean distance (cosine similarity). The number of cluster centers $\mathcal{K}$ is specified by users and in Fig. 3 $\mathcal{K} = 4$.

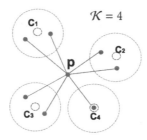

**Fig. 3.** The blue points are neighbors of point $p$ on the graph, and the number of cluster centroids is 4. (Color figure online)

## 3.2   $k$NN Search on QDG

Most graph-based search algorithms use the greedy-search algorithm to identify $k$NN of a query. The only difference between the general search method and QDG is that we focus on reducing the out-degree at search stage, instead of the index construction stage. Figure 4(a) and Fig. 4(b) depict examples of the general greedy-search algorithm and QDG's search strategy, respectively. As shown in Fig. 4(a), the general greedy-search algorithm initializes the dynamic candidate set as the starting point $p$ and its neighbors first. In the candidate set, the point closest to the query point is selected as the new starting point for the next iteration and visited points will be marked. The candidate set is of fixed size, which is often greater than $k$, and points in the candidate set are sorted according to the distance from the query point. This method can quickly reach the neighborhood of the query point. When all the points in the candidate set are examined, the iteration ends, and the algorithm returns the first $k$ points in candidate set as $k$NN.

The search procedure of QDG differs from the general one mainly in the neighbor selection policy. Particularly, we specify the number of clusters $k'$ to be checked during the search. As shown in Fig. 4(b), the number of clusters $\mathcal{K}$ is 3. Point 1, point 2 and point 3, 4 are in three different clusters, respectively. When starting from $p$, we calculate the cosine similarity between three cluster centroids and $q$. If we specify $k' = 1$, then we only need to check point 3 and point 4, which reduces the search cost significantly. $k'$ and $\mathcal{K}$ are two tuning knobs determined by users.

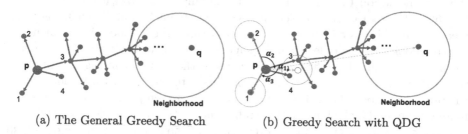

(a) The General Greedy Search          (b) Greedy Search with QDG

**Fig. 4.** (a) An example of the general search procedure. (b) The greedy-search algorithm of QDG. Point $p$ is the entry node, point $q$ is the query node, and the dark yellow circle represents the $k$NN neighborhood of $q$. The red dashed circle represents the cluster centroid of $p$ and $\alpha_1, \alpha_2, \alpha_3$ represent the angles between the cluster centroids and $q$, respectively. (Color figure online)

## 3.3   Space and Time Performance Optimization

Suppose that the dataset consists of $n$ points and the number of clusters is $\mathcal{K}$, additional space for storing $n * \mathcal{K}$ vectors is required, which is unacceptable. Suppose, for any point, the number of cluster centroids of its neighbors is the close to the number of its neighbors, it will become meaningless to do clustering. To solve this problem, we only cluster the points with more than $L$ neighbors, where $L$ is set to 10 by default in this paper.

For points of which the number of their neighbors are greater than $L$, we use PQ to compress the centroid vectors. Specifically, the cluster centroids are used to train a codebook $C$ and all the original centroid vectors are stored in compressed code form, which will greatly reduce the index storage cost. Figure 5 depicts a simple codebook trained using PQ, where the number of subvectors $m = 4$ and the number of centroids $k^* = 4$. Using this codebook, a vector of 16*4=64 bytes could be compressed into a one-byte code[2]. Please refer to Sect. 5.3 and [18] for more details about PQ.

Besides evaluating the Euclidian distance between candidate points and the query, the most time-consuming part of this algorithm is to calculate the cosine similarity between each cluster centroid and the query vector. To reduce such computation cost, we adopt a pre-calculation method similar to PQ.

At online search stage, suppose a cluster centroid $p$ of code 00010011 to be evaluated, the formula for calculating the cosine similarity between $q$ and $p$ is

---

[2] The dimension of the vector is 16 and it takes four bytes to store a float number.

as follows, where the dimension of $p$ and $q$ is 16 as illustrated in Fig. 5. The re-constructed vector of $p$ from its code is the elements with yellow background color.

$$\cos(\boldsymbol{p}, \boldsymbol{q}) = \frac{\boldsymbol{p} \cdot \boldsymbol{q}}{\|\boldsymbol{p}\| \cdot \|\boldsymbol{q}\|}$$

Since the length of $\boldsymbol{q}$ does not affect the ranking of the cosine similarity of different cluster centers, we do not computer $\|\boldsymbol{q}\|$. $\|\boldsymbol{p}\|$ can be obtained through the pre-calculation table constructed at indexing stage, which is illustrated in Fig. 6(a) (the root of sum of elements with yellow background color). Each element in this table is computed as the square sum of corresponding elements in the codebook. For example, the first element 0.50 in row one in Fig. 6(a) is equal to the square sum of the first four elements in the first row in Fig. 5. Similarly, the inner product pre-calculation table is illustrated in Fig. 6(b). By looking up the pre-calculation tables, the cosine similarity between $p$ and $q$ can be approximately computed by looking up these tables as $\cos(\boldsymbol{p}, \boldsymbol{q}) = \frac{11.03}{3.30 \times 3.38} = 0.98$ since $\boldsymbol{p} \cdot \boldsymbol{q} = 0.59 + 3.90 + 3.81 + 2.73 = 11.03$, $\|\boldsymbol{p}\| = \sqrt{0.50 + 4.01 + 3.79 + 2.63} = 3.30$ and $\|\boldsymbol{q}\| = 3.38$. The ranking of cosine similarity of all cluster centroids can be obtained with these approximations.

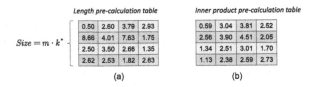

**Fig. 5.** Query vector and codebook. (best viewed in color)

| Length pre-calculation table | | | | Inner product pre-calculation table | | | |
|---|---|---|---|---|---|---|---|
| 0.50 | 2.60 | 3.79 | 2.93 | 0.59 | 3.04 | 3.81 | 2.52 |
| 8.66 | 4.01 | 7.63 | 1.75 | 2.56 | 3.90 | 4.51 | 2.05 |
| 2.50 | 3.50 | 2.66 | 1.35 | 1.34 | 2.51 | 3.01 | 1.70 |
| 2.62 | 2.53 | 1.82 | 2.63 | 1.13 | 2.38 | 2.59 | 2.73 |

$Size = m \cdot k^*$

(a)                              (b)

**Fig. 6.** (a) Length pre-calculation table and (b) inner product pre-calculation table. (best viewed in color)

## 4   Experiments

In this section, we conduct a detailed analysis using publicly available datasets to show the efficiency of QDG. The design principles of QDG are orthogonal to specific graph-based search methods. In this paper, we only report the results using NSG. We first describe the datasets and the parameters used, and then we present the results and analysis.

## 4.1   Datasets and Experiment Setting

Our experiment uses five datasets, Audio, Sun, Cifar, Nuswide and Trevi. All the datasets we used can be found on Github[3]. The detailed information on the datasets is listed in Table 3. A set of 200 queries are randomly chosen from each dataset and then removed from the original dataset. We carried out comprehensive experiments with different $k$ and the results exhibit similar trends. Due to space limitation, we only report the results for top-100 queries.

The MOD are all set to 70 for all three methods and other important parameter settings are listed in Table 3 as well. As we can see from Table 1, QDG graph are far more dense than HNSW and NSG because we decrease the minimum angle between neighbors of points. The number of clusters $\mathcal{K}$ in graph construction and the number of cluster centroid examined during the NN search $k'$ are tuned to be the optimal. For cluster centroid compression, each vector is partitioned into $m = 8$ subvectors and $k^*$ is set to 256. This incurs 8 bytes per cluster centriod extra memory for indexing. Please note the original index space cost for all three methods are determined by the MOD, which are all equal to 70 * 4 = 280 bytes.

**Table 3.** Statistics of datasets and parameter settings.

| Dataset | Dimension | No. of points | No. of queries | Minimum Angle | $\mathcal{K}$ | $k'$ | $m$ | $k^*$ |
|---------|-----------|---------------|----------------|---------------|---------------|------|-----|-------|
| Audio   | 192       | 53,387        | 200            | 50°           | 9             | 7    | 8   | 256   |
| Sun     | 512       | 79,106        | 200            | 50°           | 9             | 7    | 8   | 256   |
| Cifar   | 512       | 50,000        | 200            | 40°           | 8             | 6    | 8   | 256   |
| Nuswide | 500       | 268,643       | 200            | 45°           | 9             | 7    | 8   | 256   |
| Trevi   | 4096      | 99,900        | 200            | 50°           | 8             | 7    | 8   | 256   |

## 4.2   Evaluation Measures

In order to measure the performance of different algorithms, we use the *average recall* as a criterion for evaluating accuracy. Given a query point, all the algorithms are expected to return $k$ points. We need to compare how many of these $k$ points are in the true $k$ nearest neighbors. Suppose the returned set of $k$ points for a query is $R'$ and the true $k$ nearest neighbors set of the query is $R$, the recall is defined as:

$$\text{recall} = \frac{|R' \cap R|}{|R|}$$

The *average recall* is the average over all the query points.

Another performance measure is the *average cost*. At online search stage, the number of Euclidean distance calculation with the query will be counted[4].

---

[3] https://github.com/DBWangGroupUNSW/nns_benchmark.

[4] For QDG, the number of evaluation of cluster centroids and the query is also counted.

Suppose the number is $c$ and the total number of points in the dataset is $n$. Then the cost is defined as:

$$\text{cost} = \frac{c}{n}$$

The *average cost* is the average over all the query points. Usually, the smaller the average cost is, the shorter the search time will be.

### 4.3  Baseline Algorithms

The algorithms we choose to compare are the two state-of-the-art, i.e., NSG and HNSW. They are implemented in C++. We do not compare the non-graph methods because they have been shown less efficient by many researchers [13,20]. Since it is desirable to obtain high-precision results in real scenarios, we focus on the performance of all algorithms in the high-precision region.

There are many algorithms that do not support multi-threading at searching stage, so we use single thread setting to compare when searching. Most of these methods support multi-threading at indexing stage. To save time, we use eight threads when building the index.

HNSW is based on a hierarchical graph structure, which was proposed in [24]. In [22,23,27] authors have proposed a proximity graph $k$-ANNS algorithm called Navigable Small World (NSW). HNSW is an improved version of NSW and has a huge improvement in performance. HNSW has multiple implementation versions, such as Faiss, hnswlib[5]. We use hnswlib since it performs better than Faiss implementation.

NSG is a method based on $K$NN graph, in which the neighbor set of each point on this graph is pruned by the MRNG method. This method was first proposed in [13]. At search stage, each query point starts searching from the same navigating node. NSG can approximate MRNG very well and try to ensure a monotonic search path in the search procedure. Besides, NSG shows superior performance in the E-commercial search scenario of Taobao (Alibaba Group) and has been integrated into their search engine.

All methods, including QDG, are written in C++ and compiled by g++ 5.4 with "O3" option. The experiments on all datasets are carried out on a computer with i5-8300H CPU and 40 GB memory. Please note our design principles are also applicable for other graph-based search methods besides NSG.

### 4.4  Results and Analysis

**Recall Vs. Cost.** The recall-cost curves of three algorithms on different datasets are shown in Fig. 7. From these figures we can see:

1. The cost of HNSW is constantly inferior to NSG and our method. This agrees with the result reported in [13]. Since QDG can be viewed as an enhanced version of NSG, it also performs better than HNSW on all five datasets.

---

[5] https://github.com/nmslib/hnswlib.

2. For Nuswide dataset, QDG beats NSG all the time. This can be explained by the fact that the AOD of QDG is two times as much as that of NSG (Table 1). Too sparse graph leads to weak connectivity, which results in long search path and high cost. In contrast, the dense graph of QDG provides much stronger connectivity and thus lower cost. Particularly, NSG examined 5276 points on average whereas QDG vistited 4047 points (centroids included) at recall 76.7, which translates to 30% performance gain.

3. For the remaining four datasets, QDG performs almost the same as or slightly worse than NSG in the relatively low recall region. The reason is that the connectivity of NSG already could provide fine accuracy at low cost and QDG are far more dense, which incurs slightly higher cost even with the help of query-directed pruning. However, the trend changes after a critical point in the high recall regime. Particularly, the recalls at the transition point are around 99.65%, 99.9%, 99.95% and 98% for Audio, Sun, Cifar and Trevi, respectively. After the critical point, the cost of NSG increase dramatically whereas QDG enjoys more smooth incline. For example, QDG achieves 2.7x, 1.7x and 1.34x speedup over NSG at recall of 100% for Audio, Sun and Cifar, respectively. For Trevi, the cost of NSG is 1.53 times as much as that of QDG at recall of 98.95%. The main reason is that QDG is dense enough to provide high recall while NSG has to search a way longer path to achieve the same recall.

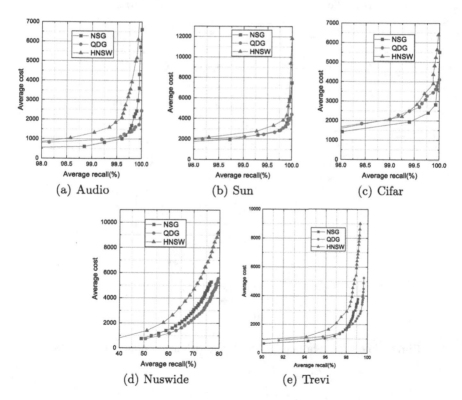

**Fig. 7.** The recall-cost curves of three algorithms on different datasets.

**Recall Vs. Time.** The time-recall curves of three algorithms on different datasets are shown in Fig. 8. Similar trends are observed as in Fig. 7 since the wall-clock search time are proportional to the cost. Particularly, QDG constantly outperforms NSG with around 10% performance gain on Nuswide and achieves 2.2x, 1.51x and 1.08x speedup over NSG at recall of 100% for Audio, Sun and Cifar, respectively. For Trevi, the cost of NSG is 1.29 times as much as that of QDG at recall of 98.95%. The speedup is slightly smaller than that in the case of Recall vs. Cost because it takes time to build the pre-calculation tables. More importantly, the accuracy of NSG saturates once reaching 99% whereas QDG achieve higher recall that the other algorithms cannot provide.

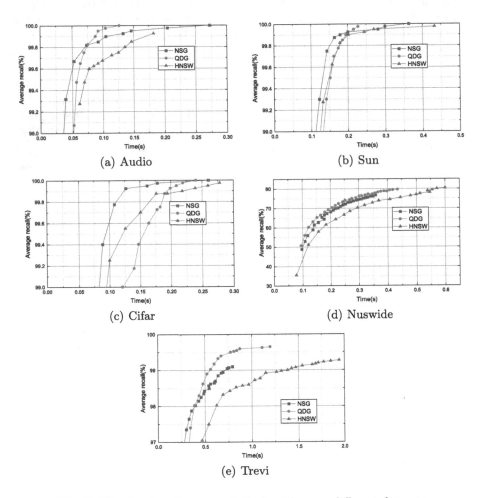

(a) Audio

(b) Sun

(c) Cifar

(d) Nuswide

(e) Trevi

**Fig. 8.** The time-recall curves of all algorithms on different datasets.

# 5   Related Work

Approximate nearest neighbor search (ANNS) has been a hot topic over decades, it provides fundamental support for many applications of data mining, databases and information retrieval [2,11,29,31]. There is a large amount of significant literature on algorithms for approximate nearest neighbor search, which are mainly divided into the following categories: tree-structure based approaches, hashing-based approaches, quantization-based approaches, and graph-based approaches.

## 5.1   Tree-Structure Based Approaches

Hierarchical structures (tree) based methods offer a natural way to continuously partition a dataset into discrete regions at multiple scales, such as KD-tree [6,7], R-tree [10], SR-tree [19]. These methods perform very well when the dimensionality of the data is relatively low. However, it has been proved to be inefficient when the dimensionality of data is high. It has been shown in [30] that when the dimensionality exceeds about 10, existing indexing data structures based on space partitioning are slower than the brute-force, linear-scan approach. Many new hierarchical-structure-based methods [12,26] are presented to address this limitation.

## 5.2   Hashing-Based Approaches

For high-dimensional approximate search, the well-known indexing method is locality sensitive hashing (LSH) [15]. The main idea is to use a family of locality-sensitive hash functions to hash nearby data points into the same bucket. After the query point goes through the same hash functions, it will get the corresponding bucket number, and only compare the distance between the point in the bucket and the query point. In the end, the $k$ approximate nearest neighbor results that are closest to the query point will be returned. In recent two decades, many LSH-based variants have been proposed, such as QALSH [17], Multi-Probe LSH [21], BayesLSH [28]. However, there is no guarantee that all the neighbor vectors will fall into the nearby buckets. In order to achieve a high recall (the number of true neighbors within the returned points set divides by the number of required neighbors), a large number of hash buckets need to be checked.

## 5.3   Quantization-Based Approaches

The most common of quantization-based methods is product quantization (PQ) [18]. It seeks to perform a similar dimension reduction to hashing algorithms, but in a way that better retains information about the relative distances between points in the original vector space. Formally, a quantizer is a function q mapping a $D$-dimensional vector $x \in \mathbb{R}^D$ to a vector $q(x) \in C = \{c_i; i \in \mathcal{I}\}$, where the index set $\mathcal{I}$ is from now on assumed to be finite: $\mathcal{I} = 0 \ldots k-1$. The reproduction values $c_i$ are called centroids. The set $\mathcal{V}_i$ of vectors mapped to given index i is referred to as a cell, and defined as

$$\mathcal{V}_i \triangleq \{x \in \mathbb{R}^D : q(x) = c_i\}$$

The k cells of a quantizer form a partition of $\mathbb{R}^D$. So all the vectors lying in the same cell $\mathcal{V}_i$ are reconstructed by the same centroid $c_i$. Due to the huge number of samples required and the complexity of learning the quantizer, PQ uses m distinct quantizers to quantize the subvectors separately. An input vector will be divided into m distinct subvectors $u_j$, $1 \leq j \leq m$. The dimension of each subvector is $D^* = D/m$. An input vector x is mapped as follows:

$$\underbrace{x_1, \ldots, x_{D^*}}_{u_1(x)}, \cdots, \underbrace{x_{D-D^*+1}, \ldots, x_D}_{u_m(x)} \rightarrow q_1(u_1(x)), \ldots, q_m(u_m(x))$$

where $q_j$ is a low-complexity quantizer associated with the $j^{th}$ subvector. And the codebook is defined as the Cartesian product,

$$\mathcal{C} = \mathcal{C}_1 \times \ldots \times \mathcal{C}_m$$

and a centroid of this set is the concatenation of centroids of the m subquantizers. All subquantizers have the same finite number $k^*$ of reproduction values, the total number of centroids is $k = (k^*)^m$.

PQ offers three attractive properties: (1) PQ compresses an input vector into a short code (e.g., 64-bits), which enables it to handle typically one billion data points in memory; (2) the approximate distance between a raw vector and a compressed PQ code is computed efficiently (the so-called asymmetric distance computation (ADC) and the symmetric distance computation (SDC)), which is a good estimation of the original Euclidean distance; and (3) the data structure and coding algorithm are simple, which allow it to hybridize with other indexing structures. Because these methods avoid distance calculations on the original data vectors, it will cause a loss of certain calculation accuracy. When the recall rate is close to 1.0, the required length of the candidate list is close to the size of the dataset. Many quantization-based methods try to reduce quantization errors to improve calculation accuracy, such as SQ, Optimal Product Quantization (OPQ) [14], Tree Quantization (TQ) [3].

## 5.4  Graph-Based Approaches

Recently, graph-based methods have drawn considerable attention, such as NSG [13], HNSW [24], Efanna [12], and FANNG [16]. Graph-based methods construct a $K$NN graph offline, which can be regard as a big network graph in high-dimensional space [4,5]. However, the construction complexity of the exact $K$NN graph will increase exponentially. Hence, many researchers turn to building an approximated $K$NN graph.

Many graph-based methods perform well in search time, such as Efanna [12], KGraph [1], HNSW and NSG. They all use different neighbor selection methods to reduce the average out-degree. As we have shown in this paper, too sparse graph may jeopardize the performance at the high recall region.

# 6    Conclusion

In this paper, we proposed a new approximate nearest neighbor search method called QDG. This method is constructed based on the approximate $KNN$ graph, and neighbors are selected according to the minimum angle between the neighbors of each point. To guide the search path using the query point, we cluster the neighbors of all points with cosine similarity in advance and only compare the clusters close to the query point in angular similarity at NN search stage. Extensive experiments indicates that our method perform better than the two state-of-the-art, NSG and HNSW, especially in the high recall regime.

**Acknowledgments.** The work reported in this paper is partially supported by NSFC under grant number (No: 61370205), NSF of Xinjiang Key Laboratory under grant number (No:2019D04024) and Tianjin "Project + Team" Key Training Project under Grant No. XC202022.

# References

1. KGraph. https://github.com/aaalgo/kgraph
2. Arora, A., Sinha, S., Kumar, P., Bhattacharya, A.: Hd-index: Pushing the scalability-accuracy boundary for approximate knn search in high-dimensional spaces. arXiv preprint arXiv:1804.06829 (2018)
3. Babenko, A., Lempitsky, V.: Tree quantization for large-scale similarity search and classification. In: Proceedings of the IEEE Conference on Computer Vision and Pattern Recognition, pp. 4240–4248 (2015)
4. Baranchuk, D., Babenko, A.: Towards similarity graphs constructed by deep reinforcement learning. CoRR abs/1911.12122 (2019)
5. Baranchuk, D., Persiyanov, D., Sinitsin, A., Babenko, A.: Learning to route in similarity graphs. ICML **97**, 475–484 (2019)
6. Beis, J.S., Lowe, D.G.: Shape indexing using approximate nearest-neighbour search in high-dimensional spaces. In: Proceedings of IEEE Computer Society Conference on Computer Vision and Pattern Recognition, pp. 1000–1006. IEEE (1997)
7. Bentley, J.L.: Multidimensional binary search trees used for associative searching. Commun. ACM **18**(9), 509–517 (1975)
8. Beyer, K., Goldstein, J., Ramakrishnan, R., Shaft, U.: When is "nearest neighbor" meaningful? In: Beeri, C., Buneman, P. (eds.) ICDT 1999. LNCS, vol. 1540, pp. 217–235. Springer, Heidelberg (1999). https://doi.org/10.1007/3-540-49257-7_15
9. Böhm, C., Berchtold, S., Keim, D.A.: Searching in high-dimensional spaces: index structures for improving the performance of multimedia databases. ACM Comput. Surv. (CSUR) **33**(3), 322–373 (2001)
10. Boston, M., et al.: A dynamic index structure for spatial searching. In: Proceedings of the ACM-SIGMOD, pp. 547–557 (1984)
11. Chen, L., Özsu, M.T., Oria, V.: Robust and fast similarity search for moving object trajectories. In: Proceedings of the 2005 ACM SIGMOD International Conference on Management of Data, pp. 491–502 (2005)
12. Fu, C., Cai, D.: Efanna: An extremely fast approximate nearest neighbor search algorithm based on knn graph. arXiv preprint arXiv:1609.07228 (2016)
13. Fu, C., Xiang, C., Wang, C., Cai, D.: Fast approximate nearest neighbor search with the navigating spreading-out graph. arXiv preprint arXiv:1707.00143 (2017)

14. Ge, T., He, K., Ke, Q., Sun, J.: Optimized product quantization for approximate nearest neighbor search. In: Proceedings of the IEEE Conference on Computer Vision and Pattern Recognition, pp. 2946–2953 (2013)
15. Gionis, A., Indyk, P., Motwani, R., et al.: Similarity search in high dimensions via hashing. Vldb **99**, 518–529 (1999)
16. Harwood, B., Drummond, T.: Fanng: fast approximate nearest neighbour graphs. In: Proceedings of the IEEE Conference on Computer Vision and Pattern Recognition, pp. 5713–5722 (2016)
17. Huang, Q., Feng, J., Zhang, Y., Fang, Q., Ng, W.: Query-aware locality-sensitive hashing for approximate nearest neighbor search. Proc. VLDB Endow. **9**(1), 1–12 (2015)
18. Jegou, H., Douze, M., Schmid, C.: Product quantization for nearest neighbor search. IEEE Trans. Pattern Anal. Mach. Intell. **33**(1), 117–128 (2010)
19. Katayama, N., Satoh, S.: The SR-tree: an index structure for high-dimensional nearest neighbor queries. ACM Sigmod Rec. **26**(2), 369–380 (1997)
20. Li, W., Zhang, Y., Sun, Y., Wang, W., Zhang, W., Lin, X.: Approximate nearest neighbor search on high dimensional data - experiments, analyses, and improvement (v1.0). CoRR abs/1610.02455 (2016)
21. Lv, Q., Josephson, W., Wang, Z., Charikar, M., Li, K.: Multi-probe LSH: efficient indexing for high-dimensional similarity search. In: Proceedings of the 33rd International Conference on Very Large Data Bases, pp. 950–961 (2007)
22. Malkov, Y., Ponomarenko, A., Logvinov, A., Krylov, V.: Scalable distributed algorithm for approximate nearest neighbor search problem in high dimensional general metric spaces. In: Navarro, G., Pestov, V. (eds.) SISAP 2012. LNCS, vol. 7404, pp. 132–147. Springer, Heidelberg (2012). https://doi.org/10.1007/978-3-642-32153-5_10
23. Malkov, Y., Ponomarenko, A., Logvinov, A., Krylov, V.: Approximate nearest neighbor algorithm based on navigable small world graphs. Inf. Syst. **45**, 61–68 (2014)
24. Malkov, Y.A., Yashunin, D.A.: Efficient and robust approximate nearest neighbor search using hierarchical navigable small world graphs. IEEE Trans. Pattern Anal. Mach. Intell. (2018)
25. Newman, M.: Networks: An Introduction. Oxford University Press (2010)
26. Nister, D., Stewenius, H.: Scalable recognition with a vocabulary tree. In: 2006 IEEE Computer Society Conference on Computer Vision and Pattern Recognition (CVPR 2006), vol. 2, pp. 2161–2168. IEEE (2006)
27. Ponomarenko, A., Malkov, Y., Logvinov, A., Krylov, V.: Approximate nearest neighbor search small world approach. In: International Conference on Information and Communication Technologies & Applications, vol. 17 (2011)
28. Satuluri, V., Parthasarathy, S.: Bayesian locality sensitive hashing for fast similarity search. arXiv preprint arXiv:1110.1328 (2011)
29. Teodoro, G., Valle, E., Mariano, N., Torres, R., Meira, W., Saltz, J.H.: Approximate similarity search for online multimedia services on distributed CPU-GPU platforms. VLDB J. **23**(3), 427–448 (2014)
30. Weber, R., Schek, H.J., Blott, S.: A quantitative analysis and performance study for similarity-search methods in high-dimensional spaces. VLDB **98**, 194–205 (1998)
31. Zheng, Y., Guo, Q., Tung, A.K., Wu, S.: Lazylsh: approximate nearest neighbor search for multiple distance functions with a single index. In: Proceedings of the 2016 International Conference on Management of Data, pp. 2023–2037 (2016)

# Author Index

Printed in the United States
by Baker & Taylor Publisher Services

Printed in the United States
by Baker & Taylor Publisher Services